P9-ECY-797

EVOLUTION
Process and Product

Third Edition

Edward O. Dodson
Emeritus, University of Ottawa

Peter Dodson
University of Pennsylvania

Prindle, Weber & Schmidt
Boston

PWS PUBLISHERS

Prindle, Weber & Schmidt • ♣ • Duxbury Press • ♠ • PWS Engineering • ♠
Statler Office Building • 20 Park Plaza • Boston, Massachusetts 02116

© Copyright 1985 by PWS Publishers.
© Copyright 1976, 1960 by Wadsworth, Inc.

All rights reserved. No part of this book may be reproduced or transmitted in any form or by any means, electronic or mechanical, including photocopying, recording, or any information storage and retrieval system, without permission, in writing, from the publisher.

PWS Publishers is a division of Wadsworth, Inc.

Evolution: Process and Product, third edition was prepared for publication by the following people:

Sponsoring Editor: Jean-François Vilain
Production and Art Editor: Robine Storm van Leeuwen
Cover Designer: Trisha Hanlon
Interior Designer: Glenna Collett

Typesetting by Grafacon, Inc.; text printed and bound by Halliday Lithograph; covers printed by John P. Pow Company.

Library of Congress Cataloging in Publication Data

Dodson, Edward O. (Edward Ottway)
   Evolution: process and product.

   Bibliography: p.
   Includes index.
   1. Evolution. I. Dodson, Peter. II. Title.
QH366.2.D6 1985   575   84-26476
ISBN 0-87150-826-5

46,314

ISBN 0-87150-826-5

Printed in the United States of America
89 88 87 86 85 — 1 2 3 4 5 6 7 8 9 10

PREFACE

Evolutionary biology, a rapidly developing science, continues to hold a central position in the biological sciences. It is essential for the understanding of botany, zoology, ecology, physical anthropology, and many aspects of agriculture. It is equally important in geology; indeed, it has some importance for most of the sciences. Our goal in this third edition of *Evolution: Process and Product* is to introduce undergraduate students to the principles that determine how plant and animal populations adapt—or fail to adapt—to their environment and to help beginning students achieve a basic understanding of evolution at the end of an average one-term course.

We have reorganized the book to conform to the preferences of many users and, incidentally, to the subtitle of the book. Part 1, *A Definition of Evolution*, summarizes the major lines of evidence for evolution, thus defining our subject in depth. Part 2, *The Origin of Variation*, discusses mutation, both point and chromosomal, as the source of evolutionary variability. Part 3, *The Origin of Species and of Higher Categories*, goes more deeply into evolutionary processes. Part 4, *Phylogeny: Evolution above the Species Level*, discusses the products of evolution. Thus, the processes of evolution (Parts 2 and 3) and the products of evolution (Part 4) now correspond to the order of the subtitle. Part 5, *Evolutionary Perspectives*, discusses what future developments we or our descendants might expect to see, both in evolutionary theory and in the process of evolution itself.

We have thoroughly revised and updated all parts of the book. Part 1 has many substantive changes, of which the most striking is the expansion of the chapter on the development of evolutionary thought, which traces the subject from its beginnings in antiquity. Part 2 has been expanded to include such recent developments as "jumping genes," punctuated equilibria, and molecular drive and to introduce the modern evolutionary synthesis. We have extensively rethought, expanded, and rewritten Part 3, which now goes into greater depth on the processes of evolution, such as natural selection, population genetics, isolating mechanisms, polyploidy, biogeography, and chance.

In Part 4 we describe the product of evolution as seen in the fossil record and in the comparative biology of the world of life today. An important aspect of this, and one of high interest to most of us, is the evolution of the order Primates, including that egocentric species, *Homo sapiens*. The chapter that deals with *H. sapiens* has been thoroughly rewritten and updated; and a chapter on generalizations derived from the evolutionary record has been added.

The concluding chapter brings together many threads from earlier chapters and weaves them into what we hope is a meaningful whole. It discusses some potential aspects of future evolution; and, for the first time, it includes a

CAMROSE LUTHERAN COLLEGE
LIBRARY

discussion of the current creationist movement, which we hope our readers will find both informative and objective.

Keeping in mind that this is an introductory course, we have assumed a minimal mathematical background; an elementary algebra course is sufficient for an understanding of the mathematics presented in the text. We have also made a conscious effort to draw examples from both botany and zoology so as to present a more complete and balanced picture of the subject.

The number of illustrations has been greatly increased. There are many excellent new photographs and drawings, and most of the drawings from earlier editions have been redrawn. We hope that the illustrations will now contribute more than ever to the comprehensibility of the text. In addition, we have provided students with the following aids: all new terms defined in context and appearing in the glossary at the back of the book; concise summaries, offering a brief review of the material covered in the chapters; and a selected list of references at the end of each chapter, providing a starting point for further study.

We are indebted to those users who wrote to alert us to specific sections of the book that needed correction or strengthening, and we are grateful to each of them. In particular, we want to thank the following professors who reviewed the manuscript: George A. Clark, University of Connecticut; Charles B. Curtin, Creighton University; John E. Hafernik, Jr., San Francisco State University; Frances C. James, Florida State University; Lazarus W. Macior, University of Akron; David A. West, Virginia Polytechnic Institute and State University; Norris H. Williams, The Florida State Museum, University of Florida. All of them read large parts of the manuscript and some of them read the entire manuscript in various stages of its preparation. Their suggestions have been of great value to us, and some of the excellent features of the book originated with them.

We thank Hermann W. Pfefferkorn, now of the University of Heidelberg, and Alan Mann, of the University of Pennsylvania, for important discussions concerning the content of the chapters on higher plants and on primates; and we also thank Joel A. Hammond of Biosis in Philadelphia, whose acute knowledge of and interest in evolutionary biology have been a valuable resource.

We are grateful to those scientists and institutions who have provided photographs. Special mention is due M. Georges Bentchavtchavadze for many excellent photographs. For their artistic contributions we thank D.B. Weishampel, P. Fortey, and B. Dalzell. Finally, we thank PWS Publishers, and particularly Jean-François Vilain and Robine Storm van Leeuwen, for the high level of interest they have shown in producing an attractive new edition of this book.

Edward O. Dodson

Peter Dodson

CONTENTS

Part 2
The Origin of Variation
125

Part 3
The Origin of Species and of Higher Categories
181

Part 4
Phylogeny: Evolution above the Species Level
343

1

A Definition of Evolution

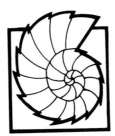

Evolution is a much misunderstood subject. Popular ideas about evolution often bear only a tenuous and misleading relationship to the subject as it is understood by evolutionary biologists. A brief definition would be insufficient to correct these misconceptions. In Part 1, comprising Chapters 1 through 6, we have therefore tried to define the subject in depth by discussing the career of Charles Darwin, by sampling evidence (from many fields) that has convinced most biologists that the great variety of living species with which the world now teems is the product of a historical process (an evolution), and by sketching the history of modern evolutionary thought. Thus, Part 1 lays the foundation for understanding the rest of the book.

CHAPTER

1

Evolution by Natural Selection: Darwin's Concept

Evolution may be defined as "descent with modification" (Darwin), with closely related species resembling one another because of their common inheritance and differing from one another because of the hereditary differences they have accumulated since the separation of their ancestors. Stated in another way, evolution is the process by which related populations diverge from one another, thereby giving rise to new species (or higher groups), and by which the adaptations of these new species to diverse conditions are subsequently refined.

These definitions, as well as others commonly given, are general. They do not imply a particular line of descent for any organism. The descent of humans from "monkeys" is not a point of definition for evolution. Nor do competent students of evolution regard humans as derived from any other organism now living. Instead they regard both humans and the great apes as coming from a common ancestor, an unknown primate. Humans do not even play as useful a role in the study of evolution as we might wish because we are not available for laboratory experimentation to the extent that other animals are and because primitive humans were rarely fossilized, though apparently more frequently than other primitive primates were fossilized. However, the student of evolution belongs to an egocentric species, so a chapter on human evolution is included in this book.

3

DARWINIAN PREMISES

Evolution, as conceived by Darwin, can be briefly summarized like this: all plants and animals reproduce in excess of the numbers that can actually survive, yet adult populations remain relatively constant. Hence, a struggle for survival must occur. Now the members of any species vary one from another. Some of the variations may be neutral, but others help or hinder the organism in its struggle for survival. As a consequence, the "survival of the fittest" (Spencer's phrase) variants will occur, and the less fit will be eliminated by either their physical or biotic environments. (Darwin thought of predation and mortal combat as the instruments of this natural selection, but today we know these to be only two among many factors in differential reproductive success.) Thus species will be gradually modified in the direction of the most advantageous variants.

Prodigality of Nature

The prodigality of nature with respect to reproduction is well known. A single codfish produces 10,000,000 eggs in a season; an oyster may pass as many as 114,000,000 eggs at a single spawning; *Ascaris lumbricoides* var. *suum,* a common parasite of hogs, has been observed to pass as many as 700,000 eggs in a single twenty-four-hour period under laboratory conditions; and the ocean sunfish *Mola mola* may spawn as many as 300,000,000 eggs in a season! That such immense numbers of individuals should survive and themselves reproduce in similar numbers is simply unimaginable. For example, a thorough study of a small sector of the Pacific coast just north of San Francisco revealed about one hundred starfish (mainly *Pisaster ocraceus,* but a few other species were included). If we assume that half of these were females and that each produced one million eggs (a modest estimate), the population in the next year would be about 50,000,000. This population would include about 25,000,000 females, all of which would again produce about one million eggs each. It is obvious that if the ordinary rate of reproduction were to continue for even a few generations with 100 percent survival of all offspring, the starfish would soon fill the seas and be pushed out across the lands by sheer pressure of reproduction. Indeed, at this rate of reproduction, it would take only fifteen generations for the number of starfish to exceed the estimated number of electrons in the visible universe (10^{79})!

We have intentionally chosen these animals from among the more prolific members of the animal kingdom. Essentially the same situation applies, however, to even the slowest-breeding animals. Frogs, although generally regarded as quite prolific, produce, at the most, 20,000 eggs annually (the bullfrog *Rana catesbeiana*). Most species of frogs produce fewer than 1,000 eggs annually, and a few (robber frogs, family Leptodactylidae) may lay as few as six eggs annually. Perhaps the slowest-breeding organism is the elephant. Darwin calculated the results of a minimal rate of reproduction for this animal. Assuming a life span of about 100 years, with active breeding life from about 30 to about 90 years of age, a single female will probably bear no fewer than 6 young. If all of these young survived and continued to reproduce at the same rate, then after only 750 years, the descendants of a single pair would number about 19,000,000.

Thus, regardless of the rate of reproduction of a species, its numbers would soon become impossibly large if all survived and reproduced, simply because the rate of increase is geometric. Because most do not survive, we must assume that there is a struggle for existence, with the majority of the participants losing. This struggle may take the form of overcoming adverse environmental conditions such as cold or drought, escaping predators, or obtaining an adequate share of a limited food supply for which there are many competitors. Darwin thought of the struggle as being most intense among members of the same species that must compete for identical resources.

The sizes of adult populations are not constant to the degree that Darwin thought; populations of wild species may vary tremendously from year to year. However, they never approach the maximal size calculated from the reproductive rate; clearly, a severe struggle for existence must account for the difference.

Organic Variation

The fact of variation among living things is so obvious that it need be proved to no one. Even the proverbial peas in a pod vary from one another visibly. With the exception of identical twins, any two individuals of a given species generally show easily recognizable differences, which can usually be measured. Not infrequently, a whole population may show a definite pattern of variation that differentiates it from the rest of its species. Such a population may be called a subspecies. Darwin regarded such subspecies as "incipient species," that is, new species in the process of formation. Many of these natural variations may be completely neutral and confer on their bearers neither advantage nor disadvantage in the struggle for existence. Others, however, may influence the chances of survival of their bearers. For example, a variation that tends to reduce water loss will favor a desert plant, one that increases speed will aid an ungulate in escaping its predators, and one that increases the sensitivity of sense organs will aid a predator in detecting its prey.

Natural Selection

Darwin described the outcome of this struggle for existence among varying plants and animals effectively in Chapter 3 of the *Origin of Species:*

> How is it that varieties, which I have called incipient species, become ultimately converted into good and distinct species, which in most cases

obviously differ from each other far more than do the varieties in the same species? How do those groups of species, which constitute what are called distinct genera and which differ from each other more than do the species of the same genus, arise? All these results . . . follow from the struggle for life. Owing to this struggle, variations, however slight and from whatever cause proceeding, if they be in any degree profitable to the individuals of a species, in their infinitely complex relations to other organic beings and to their physical conditions of life, will tend to the preservation of such individuals, and will generally be inherited by the offspring. The offspring, also, will thus have a better chance of surviving, for, of the many individuals of any species which are periodically born, but a small number can survive. I have called this principle, by which each slight variation, if useful, is preserved, by the term natural selection. But the expression often used by Mr. Spencer, of the Survival of the Fittest, is more accurate, and is sometimes equally convenient. We have seen that man by selection can certainly produce great results, and can adapt organic beings to his own uses, through the accumulation of slight but useful variations, given him by the hand of Nature. But Natural Selection, we shall hereafter see, is a power incessantly ready for action, and is as immeasurably superior to man's feeble efforts as the works of Nature are to those of Art.

BIOGRAPHICAL SKETCH OF DARWIN

We may obtain some idea of the comprehensive basis upon which Darwin proposed his theory from a review of his life, taking the autobiography which he wrote for his children as a guide. Charles Robert Darwin was born on February 12, 1809, the fifth of six children born to Robert Waring Darwin and Susannah Wedgwood Darwin. Darwin obtained most of his elementary schooling at a boarding school in Shrewsbury, England, where his father practiced medicine with notable success. The curriculum was almost entirely classical, and Darwin professes to have found it exceedingly dull and profitless. Although he did not distinguish himself scholastically, he developed a great love for dogs, for collecting all manner of things, and for hunting birds. His father once said to him, "You care for nothing but shooting, dogs, and rat-catching, and you will be a disgrace to yourself and all of your family." But Darwin adds, "My father, who was the kindest man I ever knew and whose memory I love with all my heart, must have been angry and somewhat unjust when he used such words."

Although Darwin found his formal schooling rather fruitless, he did enjoy some cultural avocations during these years. He was fond of poetry, particularly of the historical plays of Shakespeare. He collected minerals and insects with great zeal, but, he says, rather unscientifically. He took much pleasure in watching the habits of birds, and he took some notes on his observations. One of his greatest pleasures was to assist his older brother Erasmus in the latter's chemical experiments, yet his schoolmaster publicly rebuked him for this on the grounds that it was a useless pursuit.

In the fall of 1825, Darwin was sent to the medical school at Edinburgh. His account of the two years at Edinburgh made them seem utterly futile. Instruction was exclusively by means of lectures, which he describes as "incredibly dull." He felt very little motivation to come to grips with his medical studies. "I became convinced from various small circumstances that my father would leave me property enough to subsist on with some comfort, though I never imagined that I should be so rich a man as I am; but my belief was sufficient to check any strenuous efforts to learn medicine." But his achievements at Edinburgh could not have been as mediocre as he himself indicated, for he gained the friendship and respect of well-established scientists, such as Dr. Ainsworth, a geologist, Dr. Coldstream and Dr. Grant, zoologists, and Mr. Macgillivray, an ornithologist who was also curator of the museum. Although he took no courses under these men, he enjoyed their company and learned much natural history from them. Also he joined a students' scientific society before which he read papers on some small research problems that he had undertaken.

In any event, Darwin did not complete his medical education. His father learned that he did not want to be a physician and suggested Darwin prepare himself to be a clergyman of the Church of England. Darwin said that the life of a country clergyman appealed to him, and after some study he was convinced of the truth of the creed of the Church of England. In order to achieve this goal, he needed a degree from an English university, and so Darwin enrolled at Cambridge in January, 1828, and was graduated in January, 1831. His Cambridge years, Darwin said, were "wasted, as far as the academical studies were concerned, as completely as at Edinburgh and at school." The only things he enjoyed in his studies at Cambridge were geometry and the works of Paley, a distinguished eighteenth-century theologian whose beautiful logic and clear expression Darwin admired. He felt that these were the only things in his formal education that contributed to the development of his mind.

Again, Darwin's own estimate of his achievements at Cambridge must have been unduly harsh. He said that he wasted his time with a crowd of sporting men, including some who were dissipated and low-minded. But he also developed a taste for the fine arts and made friends among the more cultured students at Cambridge. And as at Edinburgh, he attracted the friendship and respect of distinguished men of science, who must have seen in this youth something far better than the dilettante he pictured himself to have been. Most important among these were Dr. Henslow, a botanist, and Dr. Sedgwick, a geologist, through whom the young Darwin met many of the most-distinguished men of that time. But Darwin's major interest during his Cambridge years was the collecting of beetles, a study that he pursued with great energy and with some distinction.

Voyage of the *Beagle*

Darwin got the major opportunity of his life through Henslow. The British Admiralty planned a voyage of exploration on H.M.S. *Beagle*, with the preparation of nautical maps of the shore of South America as a primary objective. Henslow was asked to nominate a young naturalist for the voyage, and he urged Darwin to accept this appointment. Darwin's father objected because he felt that this would simply delay the establishment of his son in the clergy. But he added, "If you can find any man of common sense who advises you to go, I will give my consent." Darwin's uncle, Josiah Wedgwood (of the "China" family), whom the elder Darwin had always regarded as one of the most sensible men in the world, kindly fulfilled this condition. The *Beagle* was originally scheduled to sail in September, 1831, but did not actually get under way until December 27, 1831.

As the *Beagle* plied its way back and forth along the coast of South America (Captain Fitzroy was a perfectionist), Darwin availed himself of every opportunity to travel ashore. He marveled at the tropical rain forest; he traversed the pampas with gauchos; he learned of the extinct mammalian fauna of Patagonia by collecting samples in the rich fossil beds there; he climbed the Andes in Chile, where he also experienced the terror of a devastating earthquake. He spent five of the most important weeks of his life on the Galápagos Islands, six hundred miles off the coast of Ecuador, observing the remarkable variation of similar kinds of plants and animals from one island to another.

The *Beagle* continued its voyage across the South Pacific, where Darwin made important geological observations on coral atolls, to Australia, around southern Africa, and finally back to England on October 2, 1836, after a voyage of nearly five years. Throughout the voyage, Darwin took voluminous notes on the geology, botany, and zoology of the regions visited. He also sent a steady stream of biological and geological specimens back to Henslow at Cambridge. Darwin's notes and specimens formed the basis for several books and made valuable contributions to his major work.

Darwin's Publications

Back in England again, Darwin at once set to work upon his *Journal of Researches,* which was based on the journal that he had kept during the voyage of the *Beagle*. It was published in 1839 and was an immediate success. Darwin said that the success of this book, his first literary child, always pleased him more than that of any of his other books. Also, in 1839, he married his cousin Emma Wedgwood. Ten children were born, of whom two daughters and five sons reached maturity. They lived in London until September of 1842. During this time, Darwin (Figure 1–1) was active in scientific society, being secretary of the Geological Society from 1838 to 1841. His closest associate during this time was his old friend Lyell, who perhaps contributed

Figure 1–1.   *Charles Robert Darwin at age 31, from a chalk drawing by George Richmond.*
Courtesy of George P. Darwin.

more than any other man to the modernization of geology; it was he who introduced Darwin (and the rest of the scientific community) to the immensity of geological time.

Darwin's health became progressively worse. Since he was unable to bear much excitement, the Darwins moved to Down, a country residence, in 1842. It was here that Darwin did most of his life's work. As his health forced him to remain in seclusion, the remainder of his biography becomes largely a catalog of his books. Most writers have described Darwin's illness as psychosomatic, but he was exposed to Chagas' disease (a form of trypanosomiasis, which was not understood at the time) while in South America, and he was treated for it with arsenicals for many years. Either of these, the disease or the "cure," may have been sufficient cause for his long illness.

In 1842, Darwin published the first of his major geological works resulting from the voyage of the *Beagle, The Structure and Distribution of Coral Reefs.* In this book he presented a theory of the structure and mode of formation of coral reefs that was very different from the one then generally accepted. Darwin's keen observations and accurate thinking on the subject won support, and his theory is even now generally accepted among geologists. This work was followed in 1844 by *Geological Observations on Volcanic Islands* and in 1846 by *Geological Observations on South America.*

In 1846, Darwin began work on a study of the Cirripedia, or barnacles. His research began with the study of an aberrant barnacle that burrows into the shells of other species, which he collected when the *Beagle* visited the coast of Chile. In order to understand the structure of this new species, he had to dissect more typical forms. Gradually the scope of the study broadened until it included descriptions of all known species of barnacles, living and fossil. This great monograph on the Cirripedia was published in four volumes. The Ray Society published two volumes on the living Cirripedia in 1851 and 1854, respectively, and the Palaeontological Society published the two volumes on fossil species in the same years. Of this work, Darwin said:

> I do not doubt that Sir E. Lytton Bulwer had me in his mind when he introduced in one of his novels a Professor Long, who had written two huge volumes on limpets. . . . My work on the Cirripedia possesses, I think, considerable value, as besides describing several new . . . forms, I made out the homologies of the various parts . . . and proved the existence in certain genera of minute males. . . . The Cirripedia form a highly varying and difficult group of species to class; and my work was of considerable use to me, when I had to discuss in the "Origin of Species" the principles of natural classification. Nevertheless, I doubt whether the work was worth the consumption of so much time.

Yet Sir Joseph Hooker, a distinguished botanist, wrote to one of Darwin's sons, "Your father recognized three stages in his career as a biologist: the mere collector at Cambridge; the collector and observer in the *Beagle,* and for some years afterwards; and the trained naturalist after, and only after, the Cirripede work." T.H. Huxley seems to have concurred in this opinion.

Origin of Species

No other books followed until the *Origin of Species* in 1859. Yet this work had really been in the making for more than twenty years. During the voyage of the *Beagle,* various facts of paleontology and biogeography, which Darwin observed, had suggested to him the possibility that species might not be immutable. But he had no theory to work on. Lyell had attacked geological problems by accumulating all applicable data in the absence of a working theory, in the hope that the sheer weight of facts might throw some light on his problems. As Darwin greatly admired the geological work of Lyell, he determined to apply the same method to the species problem. Accordingly, in July, 1837, he began his first notebook on variation in plants and animals, both under domestication and in nature. He overlooked no possible source of information: personal observations and experiments, published papers of other biologists, conversations with breeders and gardeners, correspondence with biologists at home and abroad, all were represented. As a result of this, Darwin soon saw that human success in producing useful varieties of plants and animals depended on selection of desired variations for breeding stock. But he did not see how selection could be applicable to nature.

In October 1838, Darwin happened to read, for pleasure, Malthus's "Essay on the Principle of Population." Malthus maintained that, as human population tends to increase geometrically and the means of sustenance tends to increase arithmetically, population should inevitably outstrip its resources, and hence poverty and struggle for inadequate resources are normal. It struck Darwin at once that these ideas of human economy and sociology might be extended to the entire living world—that the struggle for existence among plants and animals offered a basis for *natural selection* of those variants that were best fitted to compete. But it was only in 1842, four years later and after the collection of a great deal more data, that he wrote out the first outline of his theory, a pencil draft of thirty-five pages. In 1844, he enlarged this outline to 230 pages. From the time of the completion of the cirripede work in 1854, Darwin devoted all of his time to the study and organization of his notes and to further experiments on transmutation of species.

Early in 1856, Lyell advised Darwin to write out a full account of his ideas on the origin of species. Darwin began this work on a much larger scale than that which finally appeared in the *Origin of Species*. Then, early in the summer of 1858, when this work was perhaps half completed, Alfred Russel Wallace, a young and little-known English naturalist then working at Ternate in the Dutch East Indies, sent Darwin a short essay entitled "On the Tendency of Varieties to Depart Indefinitely from the Original Type." Wallace asked Darwin, if he thought well of this essay, to send it to Lyell for his criticism. Darwin thought very well of it, for he recognized his own theory, and he felt that he ought to withhold his own publication in favor of Wallace. However, Lyell and Hooker had for years been familiar with Darwin's work on the transmutation of species, and Lyell had read Darwin's outline of 1842. These men therefore suggested that Darwin write a short abstract of his theory, and that it be published jointly with Wallace's paper in the *Journal of the Proceedings of the Linnean Society*. These papers appeared in that journal in 1859, together with portions of a letter that Darwin had written to Asa Gray, the great American botanist, in September, 1857, in which he set forth his views on natural selection and the origin of species.

Following this, Lyell and Hooker urged Darwin to prepare for early publication a book on transmutation of species. Accordingly, he condensed the manuscript that he had begun in 1856 to about one-third of its original size and then completed the work on the same reduced scale. The *Origin of Species* was finally published in November, 1859. With regard to the great success of this work, Darwin wrote:

> The success of the "Origin" may, I think, be attributed in large part to my having long before written two condensed sketches, and to my having finally abstracted a much larger manuscript, which was itself an abstract. By this means I was enabled to select the more striking facts and conclusions. I had, also, during many years followed a golden rule, namely, that whenever a published fact, a new observation or thought came across me, which was opposed to my general results, to make a memorandum of it without fail and at once; for I had found by experience that such facts and thoughts were far more apt to escape from the memory than favourable ones. Owing to this habit, very few objections were raised against my views which I had not at least noticed and attempted to answer.

The publication of the *Origin* had a stunning effect on the complacent Victorian mind, even though Darwin had explicitly avoided discussion of human evolution. Characteristically, Darwin declined to take any part in the animated public debate that ensued during the following decade. The outstanding protagonist for the cause of evolution was Thomas Henry Huxley (1825–1895), who earned the epithet of "Darwin's bulldog." Interestingly, when Darwin summoned the courage to address the question of human evolution directly in 1871, the expected public outcry failed to materialize. Public indignation had spent itself.

Most of Darwin's succeeding books presented more fully data and viewpoints that he summarized tersely in the *Origin* or that were otherwise supplementary to his great work. These include *The Fertilization of Orchids,* 1862; *The Variation of Plants and Animals under Domestication,* 1868; *The Descent of Man and Selection in Relation to Sex,* 1871; *The Expression of the Emotions in Men and Animals,* 1872; *Insectivorous Plants,* 1875; *The Effects of Cross- and Self-Fertilization in the Vegetable Kingdom,* 1876; *Different Forms of Flowers on Plants of the Same Species,* 1877; *The Power of Movement in Plants,* 1880; and, finally, *The Formation of Vegetable Mould through the Action of Worms,* 1881. In addition to this immense program of publication, he also brought out revised editions of many of his books, including five revisions of the *Origin.* His autobiography, although written for his children only, was published after his death.

This, then, is the scientific background of the man who wrote the *Origin of Species.* It can be equalled by very few either for breadth or for depth.

Darwin's Mental Qualities

Before concluding these biographical notes, we may be interested in Darwin's estimate of his own mental qualities. His writing is remarkably clear and persuasive, and his style has a charm seldom found in scientific works. Yet he says that "there seems to be a sort of fatality in my mind leading me to put at first my statement or proposition in a wrong or awkward form." Again, he appears to be his own harshest critic. His letters are also very effectively written, and it seems unlikely that he carefully planned and revised these as he did his books. The general manner in which the *Origin* was developed, through a series of outlines based upon a large series of notes, has already been described in detail. He followed this general plan of work for all of his larger books, although he did none quite so thoroughly and over so long a period of years as the *Origin*.

As a young man, Darwin enjoyed poetry, particularly that of Shakespeare, Milton, Byron, Wordsworth, and Shelley. While at Cambridge, he developed a taste for fine paintings and music. But in later years this taste for the fine arts was lost. "I have tried lately to read Shakespeare, and found it so intolerably dull that it nauseated me." The only artistic taste that remained was for novels. He says, "I often bless all novelists. . . . A novel, according to my taste, does not come into the first class unless it contains some person whom one can thoroughly love, and if a pretty woman all the better." He regarded his loss of taste for the arts in general as a personal defect and said that he would cultivate such tastes every week if he had his life to live over. "My mind seems to have become a kind of machine for grinding general laws out of large collections of facts, but why this should have caused the atrophy of that part of the brain alone, on which the higher tastes depend, I cannot conceive."

Darwin (Figure 1–2) regarded himself as rather slow of apprehension and as being incapable of following for long a purely abstract train of thought. But against the charge of some of his critics that he had no powers of reasoning, he defended himself. He pointed out that the *Origin* is one long argument and that it convinced many able people, and he felt justified in saying that this could not have been done by a man without some powers of reasoning. But he felt that he did not exceed in this respect the average successful doctor or lawyer. However, he believed that his powers of observation and his love of natural science were superior. His mind was kept open, and indeed he exercised unusual care in recording any data contrary to his hypotheses: ". . . with the single exception of the Coral Reefs, I cannot remember a single first-formed hypothesis which had not after a time to be given up or greatly modified."

He concludes his autobiography with the statement that "my success as a man of science, whatever this may have amounted to, has been determined . . . by complex and diversified mental qualities . . . the love of science— unbounded patience in long reflecting over any subject—industry in observing and collecting facts—and a fair share of invention as well as of common sense. With such moderate abilities as I possess, it is truly surprising that I should have influenced to a considerable extent the belief of scientific men on some important points."

J like the Photograph very much better Than any other which has been Taken of me.

Ch. Darwin

Figure 1–2. *Darwin in later life, photographed by Margaret Julia Cameron. Darwin wrote, "I like this photograph very much better than any other which has been taken of me." His handwritten note is reproduced here.*

Courtesy of Professor R.D. Keynes and the Physiological Laboratory of Cambridge University.

Darwin died on April 19, 1882, at the age of seventy-three and was buried in Westminster Abbey near the grave of Newton.

ALFRED RUSSEL WALLACE—CO-DISCOVERER

Alfred Russel Wallace (Figure 1–3) was born on January 8, 1823. As a young man, he explored the Amazon Valley with H.W. Bates, a distinguished entomologist. This experience served as the basis for his book *Travels on the Amazon and Rio Negro,* which he published in 1853. In 1854, he began a zoological exploration of the Malay Archipelago, a work that occupied him until 1862 and that resulted in a book entitled *The Malay Archipelago,* which he published in 1869. While on the island of Ternate in February, 1858, he was stricken with intermittent fever. During an attack of the fever, he happened to think of Malthus's "Essay on Population," and "suddenly there flashed

Figure 1–3. Alfred Russel Wallace (1823–1913).
Drawn by Peter Fortey from a photograph taken in the
mid-1860s.

upon me the idea of the survival of the fittest." He thought out the theory
during the rest of the ague fit, wrote it out roughly the same evening, and
then wrote it out in full in the two succeeding evenings. He sent the resulting
paper to Darwin, with whom he was somewhat acquainted. We already know
the rest of this story.

Wallace also was active in the further development of evolutionary literature.
His major contribution was in the field of biogeography, and his most important
work in this field was *Geographical Distribution of Animals,* which was published
in 1876. As he had hoped, this book did for the biogeographical chapters of
the *Origin* what Darwin's book on *The Variation of Plants and Animals under
Domestication* had done for the corresponding chapters of the *Origin.* It was
the most outstanding classical treatment of biogeography. A second book
relating to geographical problems of evolution, *Island Life,* appeared in 1880.
He also published *Contributions to the Theory of Natural Selection* in 1870, *Tropical
Nature and Other Essays* in 1878, and *Darwinism* in 1889. Although Wallace
shares with Darwin the honor of first publication of the theory of the origin
of species by natural selection, he always generously (and properly, for it was
the *Origin* that convinced scientists) gave Darwin full credit for the theory,
as indicated by the title of the last book mentioned above. Wallace died on
November 7, 1913.

SUMMARY

Evolution, "descent with modification," is the process by which related populations diverge, giving rise to new species and higher groups, and by which their adaptations to diverse conditions are refined. Darwin succeeded because he proposed a reasonable theory to explain evolution, and because he amassed much evidence that it had occurred.

Darwin's theory can be summarized in five points:

(1) Nature is prodigal, all species reproducing in excess of the numbers that can survive.

(2) Adult populations, however, are relatively constant.

(3) Therefore, there must be a severe struggle for the means of survival.

(4) All species vary in many characteristics, and some of the variants confer an advantage or disadvantage in the struggle for life.

(5) The result is a natural selection favoring survival and reproduction of the more advantageous variants and elimination of the less advantageous variants.

Although the data base for this theory was provided by Darwin's observations during the voyage of the *Beagle,* it was only in October 1838, when he read Malthus's essay on population, that the above explanation occurred to him. He developed his theory and assembled data in support of it during the next twenty years. Then in 1858 Wallace, during a fever, also thought of Malthus's essay and his many years of experience as a field zoologist, and suddenly he too conceived the theory of natural selection. He wrote a brief paper and sent it to Darwin, who published it along with a short paper of his own. Thus, Wallace ranks as co-discoverer. In the following year, Darwin published the *Origin of Species.* Both men devoted the balance of their lives to the development of the theory of evolution by means of natural selection.

REFERENCES

Barrett, P.H., ed. 1977. *The Collected Papers of Charles Darwin,* vols. 1 and 2. University of Chicago Press, Chicago.
A very valuable source.

Bell, P.R., ed. 1959. *Darwin's Biological Work: Some Aspects Reconsidered.* Cambridge University Press, London and New York.
A collection of essays evaluating Darwin's contributions to biology. Some of the essays add substantial new knowledge. They are of uneven value, but the best are excellent.

Colp, R., Jr. 1977. *To Be an Invalid: The Illness of Charles Darwin.* University of Chicago Press, Chicago.
The author reviews all available data on Darwin's illness and considers many possible diagnoses. After all of this, he favors psychological distress as the most probable hypothesis.

Darwin, C. 1859. *On the Origin of Species by Means of Natural Selection, or the Preservation of Favoured Races in the Struggle for Life.* John Murray, London. Reprinted from the 6th ed. (1872) by Modern Library (Giants Series), New York.
Still the most basic book on evolution; it should be read by every serious student of this subject.

Darwin, C. 1897. *The Life and Letters of Charles Darwin: Including an Autobiographical Chapter.* Edited by his son, F. Darwin. John Murray, London. Reprinted in 1897 by D. Appleton, New York.

Darwin, C. 1958. *The Autobiography of Charles Darwin.* Edited with appendix and notes by his granddaughter, Lady Nora Barlow. Collins, London.

Darwin, C. 1979. *The Illustrated Origin of Species: An Abridged Edition of the Sixth Edition.* Abridged and introduced by R.E. Leakey. Faber and Faber, London.
Leakey has condensed the Origin *considerably, modernized the language in some places, and added some good illustrations. Some students will find the resulting book easier to read than the original.*

Ghiselin, M.T. 1969. *The Triumph of the Darwinian Method.* University of California Press, Berkeley and Los Angeles.
A brilliant reexamination of Darwin's work. With respect to Darwin's study of barnacles, he concludes that "the completed work has nothing less than a rigorous and sweeping critical test for a comprehensive theory of evolutionary biology" and thus that it was the essential prelude to the Origin.

McKinney, L. 1972. *Wallace and Natural Selection.* Yale University Press, New Haven, Conn.
Good reading, and good scholarship relative to the codiscoverer.

Schweber, S.S. 1978. The genesis of natural selection—1838: Some further insights. *BioScience* 28:321–326.
An illuminating study on the sources of Darwin's thought.

Sears, P.B. 1950. *Charles Darwin, the Naturalist as a Cultural Force.* Scribner's, New York.
This book demonstrates the great influence of Darwin on the culture of his time, general as well as scientific.

Stone, I. 1980. *The Origin: A Biographical Novel of Charles Darwin.* Doubleday, Garden City, N.Y.
The master of the biographical novel has written a fascinating, and largely accurate, novel based on our subject.

Wallace, A.R. 1859. On the tendency of varieties to depart indefinitely from the original type. Linnean Society, *Journal of the Proceedings* 3:53–62.

CHAPTER

2

Biogeography

The evidence of evolution has come particularly from the fields of bio-geography, taxonomy, physiology, comparative anatomy, embryology, and paleontology. Some of the more important evidence from each of these fields is summarized in this and the following chapters. Genetics (including molecular genetics), a seventh field of evidence, occupies a large part of the current literature on evolution; it is discussed in Part 2. To think of any aspect of modern life sciences that does not either contribute to or draw from the concept of evolution is difficult. Why, for instance, are primates so valuable in medical research? It is, of course, because of their close genetic (evolutionary) relationship to humans. As the late T. Dobzhansky said, "Nothing in biology makes sense except in the light of evolution."

The study of biogeography, or the geographical distribution of plants and animals, is of particular interest because this is the field that first directed Darwin's attention, while he was on the *Beagle,* to the possibility of the origin of species by means of evolution. He regarded the voyage of the *Beagle* as the most important event in his life. In his autobiography he states:

> During the voyage . . . I had been deeply impressed by discovering in the Pampean formation great fossil animals covered with armour like that on the existing armadillos; secondly, by the manner in which closely allied

animals replace one another in proceeding southwards over the continent; and thirdly, by the South American character of most of the productions of the Galápagos archipelago, and more especially by the manner in which they differ slightly on each island of the group; none of the islands appearing to be very ancient in a geological sense.

It was evident that such facts as these, as well as many others, could only be explained on the supposition that species become modified; and the subject haunted me. But it was equally evident that neither the action of the surrounding conditions, nor the will of the organisms (especially in the case of plants) could account for the innumerable cases in which organisms of every kind are beautifully adapted to their habits of life—for instance, a woodpecker or a tree frog for climbing trees, or a seed for dispersal by hooks or plumes. I had always been much struck by such adaptations, and until these could be explained it seemed to me almost useless to endeavour to prove by indirect evidence that species have been modified. . . .

In October 1838, that is, fifteen months after I had begun my systematic enquiry, I happened to read for amusement "Malthus on Population," and being well prepared to appreciate the struggle for existence which everywhere goes on from long-continued observation of the habits of animals and plants, it at once struck me that under these circumstances favourable variations would tend to be preserved, and unfavourable ones to be destroyed. The result of this would be the formation of new species. Here then I at last got a theory by which to work; but I was so anxious to avoid prejudice, that I determined not for some time to write even the briefest sketch of it.

Thus biogeographical observation prepared Darwin (and Wallace also) to recognize in Malthus's sociology a principle applicable to the entire living world.

The actual distribution of many organisms presents three problems that are difficult to understand if we assume that, in the words of Linnaeus, "There are just so many species as in the beginning the Infinite Being created" and if their present distributions correspond to their places of origin. First, the same or closely similar species sometimes exist in widely separated places, with no representatives in the intermediate territory. For example, alpine species are frequently the same as or closely similar to species much farther north. Second, organisms separated by great physical barriers are usually quite different, even though their physical surroundings may be much the same. Yet the recent fossil organisms of a particular area are usually similar to living organisms in the same area. Finally, the inhabitants of oceanic islands are usually few in number of species, but a large proportion of these are peculiar to each island. They are similar to the inhabitants of the nearest continent, but amphibians and terrestrial mammals are not usually found among them.

DISCONTINUOUS DISTRIBUTION

Nyssa, the black gum and several related trees, offers a good example of the first problem (see Figure 2–1). At present they occur naturally only in

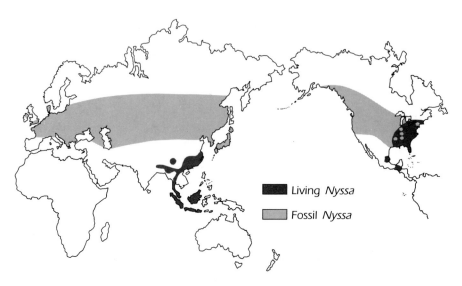

Figure 2–1.   *Distribution of living (dark gray) and fossil (light gray) members of the genus* Nyssa. *This genus is now restricted to Southeast Asia and southeastern North America (with two small populations in Central America), but the genus is known from fossil fruits, pollen, and wood found at more than 150 sites throughout the Holarctic region.*

Southeast Asia and eastern North America (with two small areas in Mexico and Central America). The physical and biotic environments of these areas are similar. In the vast expanse that separates these two populations, *Nyssa* does not occur naturally. The predecessors of Darwin interpreted such facts as indicating that one and the same species had been created independently in more than one place. Many species, however, have very wide ranges, and Darwin pointed out that, if a wide-ranging species were to become extinct in the intermediate portions of its range, the result would be widely separated populations of the same species. Such a distribution is said to be *vicariant*. Actually, fossil evidence indicates that, during a much warmer age, *Nyssa* (and the associated flora and fauna) were distributed continuously over much of the northern hemisphere. During the glacial ages, the climate became too severe for these plants over most of this range, with the result that they became extinct except in the mildest parts.

One of the most striking examples of widely separated populations of the same or closely related species is the common case of the inhabitants of high mountains, which may be identical even though they are separated by great expanses of lowland in which alpine species could not possibly survive. Or again, such mountains may have a flora and fauna closely similar to that of lowlands far to the north (Figure 2–2). Thus the ptarmigans and the varying hare are found in the higher mountains of western United States and in the

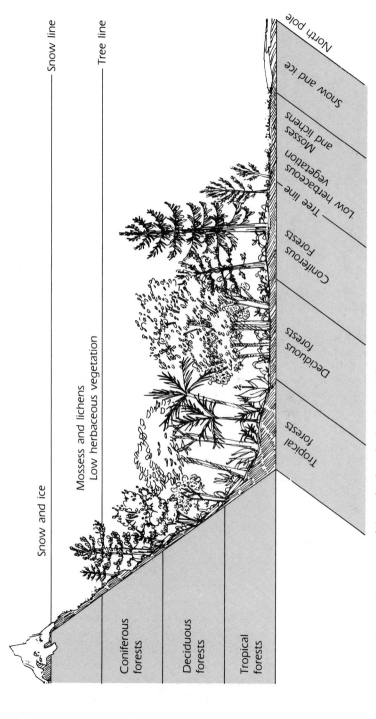

Figure 2–2. *Comparison of latitudinal and altitudinal life zones of plants in North America.*

arctic and subarctic lowlands of Canada and Alaska. In Europe, the mountain hare is found in mountains from the west coast eastward to the Caucasus and Ural mountains and also in the arctic lowlands, but it does not inhabit the intervening lowlands. Darwin pointed out that the plants of the White Mountains of New England are the same as those of Labrador and similar to those of the highest peaks of Europe. This similarity was difficult to explain until Darwin postulated a single origin for each plant species, with subsequent migration and modification. As the glacial ages advanced, arctic plants would progress even further southward, replacing the temperate plants, which, in turn, would migrate closer to the tropics. At the height of the glaciation, an essentially arctic flora would prevail over the entire northern United States, lowlands as well as mountains. As the glaciers retreated, arctic plants would again move northward in the lowlands and upward on the mountains, and the temperate plants would move out from their southern refuges to recapture their former territories. Thus, mountain plants would come to be the same as those of the lowlands to the north. The close similarity of the alpine plants of Europe and America would be understandable because the circumpolar plants from which they all derive are quite uniform.

BIOGEOGRAPHICAL REGIONS

Biogeography divides the modern world into six distinct biogeographical regions (Figure 2–3); biological explorers may feel that they are entering an entirely different world when they go from one region to another. These regions were originally defined on the basis of avian faunas, but their validity holds for terrestrial organisms generally.

The *Holarctic region* includes all of Europe, Asia north of the great mountain ranges, the Himalayan and Nan Ling, Africa north of the Sahara Desert, and North America north of the Mexican Plateau. Typical mammals of this great region include the caribou and the elk, foxes of the genus *Vulpes*, bears, and the marmot tribe. This Holarctic region is often broken up into the Palearctic region (Old World) and the Nearctic region (North America) because the two present characteristic differences, though commonly only on the specific or generic levels.

The *Ethiopian region*, comprising Africa south of the Sahara Desert, is marked by such mammals as the gorilla, giraffe, lion, and hippopotamus. The *Oriental region* includes the portions of Asia south of the Himalayas and the Nan Lings and is marked by tarsiers, gibbons, the orangutan, the Indian elephant, and flying "foxes" (frugivorous bats). The *Neotropical region,* South

Figure 2–3. *Biogeographical regions of the world. Each region has its typical flora and fauna. One representative of each is shown here: Palearctic region—spruce and giant panda; Nearctic region—sagebrush and pronghorn; Neotropical region—rain forest and spider monkey; Australian region—*Eucalyptus *and kangaroo; Ethiopian region—flat-topped* Acacia *and giraffe.*

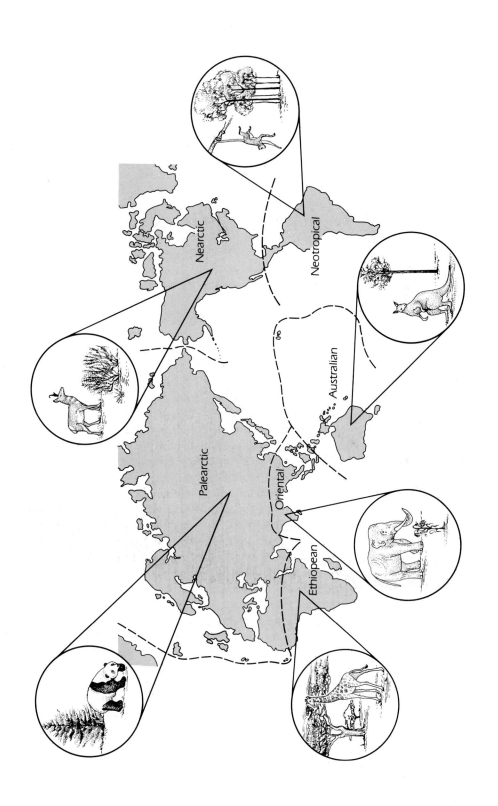

and Central America, has tapirs, sloths, prehensile-tailed monkeys, and vampire bats. Finally, the *Australian region* includes Australia and (like all of the above regions) the associated islands. It is marked by a predominantly marsupial mammalian fauna, together with many other relic forms, and by the complete absence of any native placental mammals other than bats and some rodents.

All of these biogeographical regions are separated from one another by formidable barriers of sea, desert, or mountain, or by climatic zones, and these barriers are geologically ancient. Yet, permanent as they seem, the surface of the earth has changed profoundly through its history; mountain ranges have towered up, then been reduced to low hills, and seas have swept across the continents, only to drain away again. Not surprisingly, past and present distributions of organisms also differ markedly. Thus, the Palearctic and Nearctic regions are separated by the Atlantic and Pacific oceans. But the North Pacific is shallow in the vicinity of Alaska, and animals in the past walked on dry land where the waters of the Bering Sea flow today. Accordingly, it is desirable for some purposes to describe the two regions as a single Holarctic region. The southern continents are now widely separated by the great ocean basins, but there is dramatic evidence of different continental configurations in the past: in the late 1960s, fossil amphibians and reptiles were discovered in 200 million-year-old coal swamps in Antarctica. These fossils were almost identical with those of similar age in South Africa, India, and South America! The Ethiopian region is separated from the Palearctic by the Sahara Desert, a formidable barrier indeed to any organism adapted to temperate conditions or to cold (but not to migrating birds!). Yet even within historic times, the Sahara has been considerably wetter and more temperate than it is today. South America is at present connected to North America by the Isthmus of Panama. The Isthmus, however, is rather recent geologically, and it was submerged during much of the Age of Mammals, so that South America was completely isolated from all other land masses. Even at present, climatic factors prevent many plants and animals from using this connection between the regions. The Oriental region is separated from the Palearctic by the most lofty mountain chain in the world—the Himalayas and the Nan Lings—yet these mountains did not exist 65 million years ago.

A glance at the map shows that the Malay Archipelago extends down from southern Asia and approaches Australia. Thus there is a broken link between the Oriental and Australian regions, and we might imagine that they were connected at one time. However, the channels between some of the islands are very deep, a fact that favors permanence or at least long duration. Wallace found an abrupt discontinuity in the mammals and birds, with typical Oriental forms on Borneo and Bali, typical Australian forms on Lombok and the Celebes islands. A line through the narrow channel that separates them is known as Wallace's line. Indeed, much geological evidence suggests that continuity has never existed here.

Geologists now envision the earth's crust as comprising a number of great plates upon which the continents and the associated islands are located. These plates are separated by mid-oceanic fracture zones. Slowly, convection currents in the underlying layer of the earth tend to move the plates (with the continents), and upwelling in the fracture zones adds new crust at the edges of the plates. The resulting movement of the continents has, over geologic ages, effected major geographic changes. (This subject of *plate tectonics* will be taken up further in Chapter 15.) The Australian plate was long near Antarctica, but it has been moving northward for many millions of years. As the Australian plate has approached the Oriental region, however, there has been limited migration between them, so that Wallace's line actually marks a zone of transition rather than an absolute discontinuity.

The physical conditions within one region may be so closely similar to those of another as to be indistinguishable, so long as biotic factors are neglected. Thus the climate and physiography are much alike over large areas of South America and Africa. Each presents habitats suitable for the plants and animals of the other. Yet they have few organisms in common, and those they do have may be isolated survivors from once-worldwide groups. Thus the Dipnoi (lungfishes) are now represented by only three living genera, *Neoceratodus* in Australia, *Protopterus* in Africa, and *Lepidosiren* in South America. The African and South American forms belong to the same family; the Australian form is the sole member of its family. If we consider only the living species, the fishes of the southern continents appear to have a special relationship, despite the great ocean barriers that separate them. The fossil record shows that the lungfishes were once of worldwide distribution. They have, however, long since become extinct in the face of competition with better-adapted forms in most parts of the world. The southern continents have been a last refuge for these and for so many other primitive forms.

On the basis of any theory of origin, we expect the flora and fauna within any one region to show a certain consistency. Increasing populations that disperse throughout the available territory account for it. But the fact that many plants and animals are *excluded* from lands for which they are eminently well suited is difficult to explain on any theory other than the evolutionary one. This difficulty is exemplified by the problem of widely separated populations of the same or closely similar species, which we have already discussed. Why, if these represent independent creations of the same species, are they not placed in similar parts of *different* regions? Why, for example, should both populations of *Nyssa* be located in the Holarctic region when both the Oriental and the Neotropical regions present eminently satisfactory habitats? And why should generally similar organisms be grouped together in such regions when suitable habitats for almost any organism can be found outside its own region?

Ecological Zones in the Ocean

Distinct biotic regions are not limited to the land masses of the world. Although there is a degree of physical continuity among the oceans, they offer several distinct kinds of habitat, and so ecological factors produce barriers within the oceans (Figure 2–4). On every shore, there is a narrow strand that is alternately covered and exposed by the tides. Beyond this *intertidal,* or *littoral,* zone, is the broad, gently sloping continental shelf, the higher portions of which form the continental islands. The seas over the continental shelves are generally shallow, not over 100 fathoms (600 feet or 183 meters) deep, and they comprise the *neritic* zone. At the edge of the continental shelf, the ocean floor drops off rapidly to great depths. In this great expanse of open sea, there are several depth zones. The surface waters, to a depth of 100 fathoms, comprise the *pelagic* zone, which is inhabited by widely ranging fishes. The water here is subject to wave action and is well oxygenated and lighted. The deeper water, down to 1,000 fathoms, comprises the *bathyal* zone. Here the water is always quiet and poorly lighted. It grows progressively colder with increasing depth and is scantily inhabited. Below this is the *abyssal* zone, into which the sunlight never penetrates. In this zone the water is always cold and quiet. The living forms to be found there are profoundly modified for life in the abyss.

The littoral and neritic zones are by far the most richly inhabited. The deep ocean basins form a barrier to the dispersal of these inhabitants of the continental shelves, with the result that different marine floras and faunas may be quite as isolated from one another as are those of the different continents. Thus Darwin pointed out that the organisms of the east and west coasts of the Americas are quite different, because they are separated by a great land mass. Yet about 30 percent of the fishes on opposite coasts of Panama are identical, correlating with the geological fact that the Isthmus of Panama was submerged during much of the Tertiary period. However, westward of the

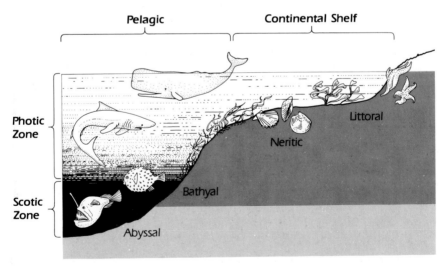

Figure 2–4.   Ecological zones in the ocean. Each zone presents conditions that act as barriers to organisms of other zones.

continental shelf of the west coast lies a great expanse of open sea, which continues until the islands of the Orient and the South Pacific are reached. The organisms found here are utterly different from those of the American continental shelf because the open sea has been a formidable barrier to the littoral and neritic flora and fauna. By contrast, from the islands of the Orient to Africa, a much greater expanse, there is an almost continuous chain of islands or of continental coast, and the flora and fauna are rather uniform throughout this great region.

Darwin's Explanation

Darwin maintained that we could understand these and other puzzling problems of the distribution of plants and animals if we assumed that all organisms of a particular species or higher group had migrated from a common place of origin, with subsequent modification. On this basis, we would expect the floras and faunas of those areas that had been isolated from one another longest (that is, the biogeographical regions) to be most sharply differentiated. Thus the inhabitants of different parts of the same region (for example, the mountains and plains of South America) should resemble one another more closely than the inhabitants of similar parts of different regions (for example, the mountains of South America and of Africa). That is, when the mountains in any region were elevated during geological ages, the new mountains would have been colonized by the inhabitants of the surrounding lowlands. Some of these would have been altogether unfit for the mountain environment; others would have been adaptable to the lower but not to the higher altitudes; and a few could have invaded the highest ranges; and so the proportions of the various organisms would be different from those that characterized the surrounding lowlands. As a result, not only the physical surroundings but also the biotic environment of the mountain colonizers would be different, and both would favor modification of the colonists. Still, these should retain within their structure and habits evidences of their close relationship to their lowland progenitors. On the other hand, the mountain colonists of different regions would be different because of the long isolation of their ancestors. Similar considerations would apply to the colonization of any new territory whatever.

Finally, the fossils in any region should resemble the living organisms of the same region, more closely in the case of recent fossils, less closely in the case of ancient fossils. This is required by the obvious fact that the present inhabitants of any region must have been descended from the past inhabitants. Great migrations of the past may modify the truth of this proposition in some instances, but they cannot take away its general validity. Darwin mentions one example, the fossil armored mammals of South America, which closely resemble the present armadillos of the same continent. Further discussion of this important topic may be found in Chapter 5. All of the above facts follow logically from the Darwinian hypothesis, yet each is anomalous if we assume that each species has been independently created in its present range.

Distribution of Freshwater Organisms

Darwin regarded freshwater organisms as the most noteworthy exception to the principle that organisms separated by a barrier are quite different. River systems and lakes are, of course, separated from one another by barriers of land. Although many freshwater systems frequently empty into the same ocean, salt water is a barrier no less formidable than land to most freshwater organisms. Hence we might expect an unusual degree of differentiation in freshwater floras and faunas, but the opposite is the case. Great similarity exists between freshwater organisms throughout the world, and many individual species are worldwide in distribution. For example, cosmopolitan species are common in such diverse freshwater invertebrates as protozoans, rotifers, oligochaetes, and crustaceans. Darwin believed that this could be accounted for by the fact that most freshwater organisms, in order to survive, must be adapted for frequent short migrations from pond to pond or from stream to stream within a limited locality. These migrations will occasionally cover long distances. Given time on a geological scale, they should result in widespread species.

Darwin gave much attention to what might be called accidental means of transport of freshwater organisms. A common phenomenon is the joining of different rivers and lakes by their floodwaters in the spring. This joining should permit an extensive exchange of their inhabitants. More selective and certainly less common is the transport of fishes and other small organisms by whirlwinds and tornados. A whirlwind, when passing over water, may pick up the surface waters together with any small organisms that happen to be near the surface. Later, when the force of the wind abates, the water and its contents will be dropped. If the water should happen to be dropped over another body of water, the organisms so transported might then multiply and become established in the new locality. This phenomenon is the basis of the rains of fishes that are occasionally reported. Although such reports are usually received with well-justified skepticism, E.W. Gudger examined all such reports critically, and he believed that at least seventy-eight recorded rains of fishes were valid. Shorebirds and waterfowl may also act as agents of dispersal for freshwater organisms. As the birds arise from the water, small organisms, eggs, larvae, and mud containing seeds are likely to cling to the feet of the birds. Because the flight is likely to terminate in a comparable body of fresh water, and as these birds are among the most wide-ranging, much dispersal of small organisms probably occurs in this way. Also, the seeds of many plants retain their viability after passing through the digestive tract of a bird. Thus seeds that are eaten in one body of water may be discharged in a quite distant one, there to germinate.

All of this is not to say that there is a single, worldwide freshwater flora and fauna. Discontinuities do exist among the inhabitants of freshwater systems. But they are less marked than might at first be expected; and when they occur, they correspond to the most ancient and imposing geographical barriers.

ISLAND LIFE

The final category of geographical evidence, and the one that had the greatest effect upon the thinking of Darwin, is that of oceanic islands and their living inhabitants. Darwin observed that such islands beyond the continental shelf are typically poor in numbers of species present, although the success of animals and plants introduced by humans has proven that these islands are well suited to support a much greater variety of organisms. He reasoned that, if all organisms had been created in their present localities, oceanic islands should be as richly inhabited as comparable areas of the continents. Yet the facts are readily understandable based upon his theory of migration from a common place of origin for all members of any group, with subsequent modification, for relatively few species could cross the great water barrier separating oceanic islands from the continental centers of origin.

Of the few species found on oceanic islands, a large number are endemic, that is, found nowhere else. Darwin found twenty-six species of land birds in the Galápagos Islands (Figure 2–5). Of these, twenty-one and possibly twenty-three are endemic. But of the eleven species of marine birds, only two are endemic. This fact is again just what we would expect in accordance with Darwin's theory. The occasional immigrants from the distant mainland of South America, upon arrival in their new environment, would compete with quite different species and in a different ecological setting from their cousins on the mainland, and so their descendants would be modified, eventually reaching the status of new and distinct species. The great water barrier then would greatly reduce the probability of these new species spreading to other localities. For marine birds, however, such a barrier is less formidable, and hence the smaller proportion of endemics is not surprising. (Lest thirty-seven species of birds for a small group of islands sound like a large number, the number of species within a small continental area may be given for comparison. The current checklist of birds on the campus of the University of California at Berkeley lists one hundred ten regular residents or seasonal migrants and thirty-five species as occasional visitors.)

The Amphibia and terrestrial mammals, though not the bats, are usually entirely absent from oceanic islands. When they have been introduced, they frequently have multiplied so greatly as to become a nuisance. The toad *Bufo marinus,* for example, was introduced into Hawaii in the hope that it would aid in the control of insects, but the toads themselves have now become a nuisance on the islands. These groups are unable to cross large water barriers (or saltwater barriers, in the case of the Amphibia); but a barrier across which a mouse, for example, could not swim, might be flown by a bat. Had all species been created in the places in which they now exist, then Amphibia and terrestrial mammals would be as frequent on oceanic islands as on comparable continental areas. Certainly, terrestrial mammals would have been created on these islands as frequently as were bats. But bats are the very mammals that

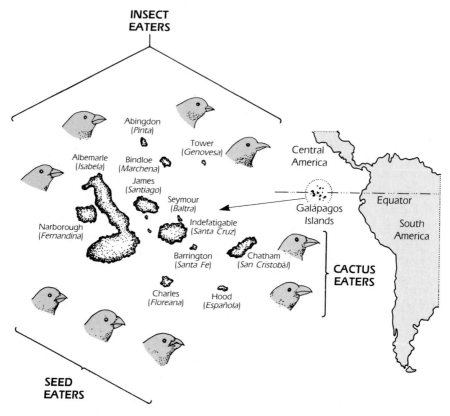

Figure 2–5. *Location of the Galápagos Islands and an enlarged map of the islands. Arranged around the islands are the heads of birds of the subfamily Geospizinae, showing the unusual range of bill structure, which is correlated with feeding habits.*

would have reached the islands most readily if all mammals arose first on the continental land masses and then subsequently invaded such territories as they could.

Finally, the inhabitants of the islands of an archipelago are commonly specifically distinct yet closely related, all of them showing some relationship to the inhabitants of the nearest mainland. Thus, when the *Beagle* visited the Galápagos Archipelago, located between 500 and 600 miles west of South America, Darwin felt that he was stepping on American soil because of the obvious similarity of the plants and animals of these islands to those of the South American continent. The Galápagos Islands include 436 species of flowering plants. Of these, 223 species, more than half of the total, are endemic; and many species are restricted to one or a few islands in the archipelago. Yet all of these plants show a close relationship to South American plants. The climate and the geological character of the islands are utterly different from those of South America; hence the relationship of their plants cannot be understood on the basis of creation of similar plants for similar lands, but only on the basis of migration of plants from the continent to the outlying islands, with subsequent differentiation. Likewise, on the Bermuda Islands, located about 700 miles off the coast of North Carolina, the inhabitants are all North American in character. Many terrestrial vertebrates have been successfully introduced into the islands, but only one, a lizard, is native there. It belongs to a North American genus, but the species is endemic. Land birds are represented by many species, but none are endemic (although there is one endemic subspecies) because Bermuda is near one of the major migration routes for North American birds; hence it is not at all isolated from the viewpoint of the birds. Bats are also common to the mainland and the islands because these flying mammals can readily cross the water barrier. But the land molluscs include a high proportion of endemics, no doubt because of the rarity of a successful crossing of the water barrier.

The only understandable basis for the facts of oceanic islands is Darwin's hypothesis that the islands were colonized from the mainland, with the colonists becoming modified subsequently. As they spread to the various islands of the archipelago, each isolated population was modified independently, thereby forming groups of closely related, endemic species. The connection, then, between the various similar species of an archipelago and of the nearest continent is simply heredity.

Continental islands, which arise from the continental shelf and are readily connected to the continent by moderate changes of sea level, must be distinguished from oceanic islands, which arise from the floor of ocean depths and have never been connected to a continent. The significance of oceanic islands may be highlighted by a biological comparison of two islands that are alike in many other respects, the one oceanic, the other continental. Great Britain and New Zealand lie at comparable distances north and south of the equator, respectively. They are not too different in size and topography. Great Britain, however, is separated from Europe by a mere twenty miles. The English Channel today is as little as 250 feet (75 meters) deep, and it did not even exist until the retreat of continental glaciers. By contrast, New Zealand is more than 1,000 miles from Australia, and the intervening ocean depths reach 15,000 feet (4,500 meters)! Thus, the former is a typical continental island, the latter a typical oceanic island. The flora and fauna of Great Britain are substantially identical with those of continental Europe. There are, of course, fewer species, and subspecies may be different, but there are no sharp contrasts. New Zealand, however, has many endemics. Native mammals are represented only by two species of bats, although a wide variety of mammals thrives on introduction. There is a rich bird fauna, comprising about two hundred fifty species that represent nineteen orders and fifty-four families. Of these, thirty-seven species, like the Canada goose, were recently introduced. A few more are cosmopolitan species, like the marsh hawk. Around half of the rest are endemics, and only about fifty are shared with Australia. Although there is a rich snake fauna in Australia, there are no native snakes in New Zealand, yet the archaic reptile *Sphenodon* is endemic to New Zealand. Amphibians are represented in New Zealand only by an endemic genus of frogs, *Leiopelma*, with two or three species. Thus, there is substantial unity between continent and continental island and sharp contrast between oceanic island and the nearest continent.

SUMMARY

We can readily understand the actual *geographical distribution* of organisms if we assume that each group originated in one of the major regions of the world, then spread to occupy as much space as it could in the face of physical and climatic barriers and of competition from other organisms (including pathogens). *Natural selection,* deriving from geographical, ecological, and biotic factors, has resulted in *adaptation* of the group to a wide variety of ecological niches, and that is evolution. Much of the data of distribution would be anomalous on any other basis. It is little wonder that first-hand experience with so impressive and persuasive a series of facts should have suggested to Darwin the possibility that species are mutable.

REFERENCES

Barlow, Lady Nora, ed. 1946. *Charles Darwin and the Voyage of the Beagle.* Philosophical Library, New York.
> *Darwin's granddaughter here presents selections from his letters to his family and from his notebooks on the voyage.*

Bowman, R.I., ed. 1966. *The Galápagos.* University of California Press, Berkeley and Los Angeles.
> *Essays by many specialists on various aspects of the biology of the islands.*

Carlquist, S. 1965. *Island Life.* Natural History Press, Garden City, N.Y.
> *A beautifully illustrated general account of life on islands.*

Darlington, P.J. 1957. *Zoogeography: The Geographical Distribution of Animals.* Wiley, New York.
> *A readable, thoughtful, and thought-provoking rethinking of the entire field—the first since Wallace's. Unfortunately, it has not been revised since the advent of plate tectonics, but it is still valuable for its comprehensive data on the distribution of organisms.*

Darwin, C. 1845. *Journal of Researches,* 2nd ed. Appleton & Co., New York and London.
> *The original report of the voyage of the* Beagle, *and Darwin's "favorite literary child."*

Grant, P.R. 1981. Speciation and the adaptive radiation of Darwin's finches. *American Scientist* 69:653–663.
> *Population biology and ecology are the bases for research on this important group of island birds.*

Pielou, E.C. 1979. *Biogeography*. Wiley, New York.
  A good, readable introduction to biogeography.
Wallace, A.R. 1876. *The Geographical Distribution of Animals*. Macmillan, New York.
  After a century, this is still a fundamental work in its field.

3

Taxonomy, Comparative Anatomy, and Embryology

The second major category of evidence for evolution is taxonomy, the science of the classification of organisms. Classification would be absolutely necessary because of the sheer numbers of species, even if no other purpose beyond that of facilitating study were to be served. More than 1,000,000 species of animals and 300,000 species of plants are described in biological literature. Large numbers of these may live even in very restricted localities. For example, in Lake Maxinkukee, Indiana, D.S. Jordan identified 64 species of fish, 18 species of amphibians, and 130 species of molluscs. Other groups, not cited, may be as liberally represented. No extensive and orderly study of the living world would be possible unless it were divided into categories about which generalizations could be made. Because taxonomy was originally based mainly on comparative anatomy and comparative embryology, it is appropriate to discuss evidence from these three disciplines together.

EVIDENCE FROM TAXONOMY

Linnaeus and Biological Nomenclature

Modern classification is based on the work of Carolus Linnaeus (1707–1778), a Swedish botanist who undertook the classification of the entire living world. Previously, scientific names of organisms had been short or ponderous descriptions in Latin. Thus, Mark Catesby in 1754 referred to the red-headed woodpecker as *Picus capite toto rubro* and to the red-winged blackbird as *Sturnus niger alis superne rubentibus*. Linnaeus introduced the practice of giving each organism a binomial name, the first of which, the generic name, is shared with other closely similar species, and the second, the specific name, differentiates the species from other members of the same genus. This binomial system of nomenclature is now universally accepted. In it, the red-headed woodpecker becomes *Melanerpes erythrocephalus* and the red-winged blackbird becomes simply *Agelaius phoeniceus*.

The Species Concept.   Basic to this system is the *species concept,* the idea that definite kinds of plants and animals occur. The individuals of any one kind differ from each other only in minor traits (except sex); they are sharply separated in some traits from all other species and are mutually fertile but at least partially sterile when crossed to other species. In the viewpoint of Linnaeus, such species were absolute: "There are just so many species as in the beginning the Infinite Being created." But many present-day biologists think that the species is to some extent an artificial category—that is, that the boundaries between closely related species are arbitrary rather than natural.

Taxonomic Categories.   Linnaeus recognized that several species may have so much in common that they can be grouped together as a *genus,* distinct from other species clusters (other genera). The current edition of the American Ornithologists' Union *Checklist of North American Birds* lists 6 species of *Melanerpes* and 3 of *Agelaius*. Genera can be much larger: the *Checklist* gives 21 species of the genus *Dendroica* (wood warblers). There are a few very large genera. G.L. Stebbins cites *Senecio* (a composite, including the garden flower cineraria) and *Astragalus* (a member of the pea family) with 1,200 species each; *Ficus* (fig) with 600; *Eucalyptus* and *Erica* (heath) with 450 each; *Cassia* (shrubs from which cinnamon and various drugs are obtained) with 400; and *Panicum* (a genus of grasses) with a mere 300 species!

The genera, Linnaeus found, also fall naturally into larger clusters, based on similarity in rather fundamental characters. These groups of like genera he called *orders*. The genus *Melanerpes* belongs to the order Piciformes, along with nearly thirty other genera of woodpeckers and allied birds. *Agelaius* belongs to the great order Passeriformes, including all songbirds. Finally, he grouped the orders into classes, the diverse members of which share only very fundamental characteristics. Thus, all of the orders of birds comprise the single class Aves.

Linnaeus grouped the classes into the two kingdoms, Plantae and Animalia, but he did not feel the need of any category intermediate between class and kingdom. Ernst Haeckel, in the immediate post-Darwinian era, introduced the term *phylum* to include related classes. Phylum means, etymologically, a line of descent, and the term was chosen especially for its appropriateness to the new study of evolution. For a similar reason, Haeckel introduced the *family* as a category intermediate between genus and order.

Thus the modern hierarchy of essential taxonomic categories is species, genus, family, order, class, phylum, and kingdom. Every organism that is described must be assigned to each of these categories, either overtly or implicitly. To demonstrate this system three familiar organisms are classified:

Taxonomic Category	*Red-headed Woodpecker*	*American Lobster*	*Sugar Maple*
Kingdom	Animalia	Animalia	Plantae
Phylum	Chordata	Arthropoda	Tracheophyta
Class	Aves	Malacostraca	Angiospermae
Order	Piciformes	Decapoda	Sapindales
Family	Picidae	Nephropidae	Aceraceae
Genus	*Melanerpes*	*Homarus*	*Acer*
Species	*erythrocephalus*	*americanus*	*saccharum*
Scientific name	*Melanerpes erythrocephalus*	*Homarus americanus*	*Acer saccharum*

All of the taxonomic categories are commonly subdivided for purposes of detailed study. The prefixes "sub-" and "super-" may be added to any of the standard categories to indicate subdivisions or larger assemblages. Other intermediate categories are sometimes used, but they do not have official status in zoology (as determined by the International Congress of Zoology).

Significance of the Taxonomic Hierarchy

T.H. Huxley wrote, "That it is possible to arrange all the varied forms of animals into groups, having this sort of singular subordination one to the other, is a most remarkable circumstance." It is indeed. Linnaeus accounted for it by the *theory of archetypes,* the theory that the Creator worked from a series of plans of limited number. These archetypes, like the plans in an architect's folio, were not all equally distinct, but they fell into definite, classifiable categories. Each class would correspond to a major archetype, the various orders within a class to lesser archetypes, and so on down the hierarchy. Thus, Linnaeus attributed the similarity between species of a genus not to their descent from a common ancestor but to their more or less imperfect copying of rather similar plans of the Creator. Late in his career, he modified this judgment and conceded that species of a genus might be genetically related.

Darwin's explanation of this "remarkable circumstance" is quite different. For him the taxonomic categories simply represented degrees of genetic relationship. Thus, all members of the phylum Chordata have common ancestors, but they are exceedingly remote, and only the most fundamental chordate characters are held in common by extreme members of the phylum. Within any class, however, the degree of relationship is much closer. Thus, more numerous and less fundamental characters are held in common. All birds, for example, share many characters in common. As one goes down the taxonomic scale this trend becomes stronger until finally members of a single species share a common inheritance except for the most minor characters. We would find it difficult to study any group of organisms in detail without feeling that this argument is cogent.

Tree of Life

Taxonomists have always tried to summarize their studies with diagrams. Linnaeus experimented with maplike diagrams, but he found no arrangement that would always place similar forms together and would always separate dissimilar forms. Later, following the *scala naturae* of Aristotle, Lamarck and others tried to arrange living forms on a ladderlike diagram. They reasoned that the adaptation of living forms is progressive, so that any particular animal should be preceded by one somewhat lower in the scale of life and followed by another one somewhat higher. For some large groupings, this ladder works. We may easily concede that amphibians are more advanced than fishes and that reptiles are more advanced than amphibians. But we cannot continue this series up through birds and then mammals. Although one mammal has advanced so far beyond all other animals that he alone studies the world in which he lives, nonetheless the majority of birds are in every respect quite as "high" as are the majority of mammals. And so two rungs are required at the same level on the ladder. Not only is this type of dilemma repeated frequently among the major groups of organisms, but it becomes much more frequent when the study is extended to the lower levels of classification.

Long before Darwin, as soon as naturalists saw the need for parallel rungs at many levels, they generally accepted the tree as a much better diagram. Everyone understands that the parts of a real tree are related to one another in a branching fashion because the whole organism is the product of growth from a single seed—growth that is accompanied by branching and differentiation. Independent creation and secondary union of the many parts would be unthinkable. The taxonomic tree is not strictly comparable to an actual tree, for the processes of branching and differentiation among organisms are not generally amenable to direct observation. But the fact that no other type of diagram can symbolize the data of taxonomy so readily as a tree strongly suggests that, like a real tree, the tree of life owes its branching character to organic growth and differentiation—in other words, to evolution. Pre-Darwinian biologists were unable to understand why classification seemed to fall into a treelike pattern, yet they agreed that it did so. Long awareness of this fact helped to prepare the way for the acceptance of Darwinism.

A final characteristic of the tree of life deserves special consideration. In any group, some members are likely to be simpler and less specialized than the others. These may be known only as fossils, but many such archaic or primitive species are still living. In this case, the archaic species generally has its closest affinities with fossil rather than with living members of its group. Such primitive species generally resemble members of *other* groups more than they resemble the more specialized members of the same group. To paraphrase, species that are placed near the points of branching in the tree of life show special resemblances to other species in both branches. If each species were created independently of all others, this fact would be inexplicable. Common characters would then be distributed among various groups without regard to their level of specialization. Yet, if the evolutionary theory is correct, then the most primitive members of related groups, having diverged the least from their common ancestry, would illuminate the relationships between groups.

Still another aspect of taxonomy suggests evolution. Although members of a species recognize one another, and mismatings between species are rare, human observers often distinguish between closely related species with difficulty. In not a few cases, the ranges of variation of related species overlap. For example, in the genus *Parus* (chickadees and tits), it is possible to set up a series that seems to blend from one species to another. In Figure 3–1, five species show this progression. If each species were represented by numerous specimens, the gradations would be even finer because of individual variation. In a still more extreme case, the fruit flies *Drosophila pseudoobscura* and *D. persimilis* are so similar that statistical analysis of populations is needed to distinguish them. Such situations strongly suggest descent from rather recent common ancestors.

EVIDENCE FROM COMPARATIVE ANATOMY

Comparative anatomy, the field from which inferences of relationship among animals are most commonly drawn, is an especially important source of evidence for evolution. A study of any particular organ system in diverse representatives of a single phylum yields the impression that the system is based upon a prototype that is varied simply from class to class (with finer variations within each class). We will discuss several such instances.

Vertebral Column

The vertebral column supports the mass of the body and provides the rigidity necessary to resist static and moving stresses. These supporting structures for all vertebrates originate as condensed mesenchymal masses within the

Figure 3–1. Five species of Parus. *(Left to right) they are* P. rufescens, *the chestnut-backed chickadee;* P. hudsonicus, *the boreal chickadee;* P. gambeli, *the mountain chickadee;* P. carolinensis, *the Carolina chickadee; and* P. atricapillus, *the black-capped chickadee.* Specimens photographed in the Peabody Museum of Yale University.

myotomes in each somite, or body segment (Figure 3–2). From these simple beginnings, the typical tetrapod vertebra, with its centrum, neural arch, and spinous, transverse, and articular processes, is formed. The support function of the column is much less important for fishes than for land dwellers, so vertebral development among fishes is modest and variable, with articular processes between adjacent segments lacking. In agnaths, the jawless fishes, vertebral bodies are unknown, and the notochord is unrestricted. In lampreys, the nerve cord is flanked by two pairs of spikelike, cartilaginous arcualia (arch formers) in each segment, but hagfishes lack even these. As A.S. Romer (Figure 3–3) noted, agnaths are classed as vertebrates only by courtesy. In the Chondrichthyes (sharks and rays), much more extensive development of cartilage occurs (Figure 3–4), so that the nerve cord is completely arched over by a double series of plates. In modern sharks, which first appeared in the Jurassic, some 150 million years ago, vertebral centra form partly from the notochordal sheath and partly from the bases of the arches and the surrounding mesenchyme. The centra may be invaded by calcium salts so that they grossly resemble the bony vertebrae of higher fishes. However, the notochord, although greatly restricted, is still continuous.

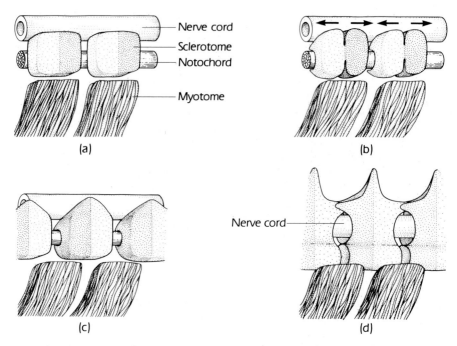

Figure 3–2. *Development of vertebrae. (a) In the early embryo each body segment has a pair of sclerotomes from which the vertebrae will develop. Below each sclerotome is shown part of the myotome (muscle rudiment) of the same segment. (b) The sclerotomes split into anterior and posterior halves, which move apart. (c) Then the posterior part of one fuses with the anterior part of the one behind it. (d) From these secondarily formed blocks of mesenchyme, the vertebrae are differentiated, with the result that a vertebra straddles two muscle segments rather than coinciding with one.*

The vertebrae of the Osteichthyes, including the modern bony fishes, are good, bony vertebrae, unlike those of sharks. However, the development of the centrum and the obliteration of the notochord among ray-finned fishes did not occur extensively until the Jurassic, and then only in the most-advanced members, the teleosts (some advanced holosteans had ossified vertebrae, but most did not). A peculiarity of some bony fishes is the development of two centra per body segment, as in *Amia,* the bowfin. One group of bony fishes, the rhipidistian crossopterygians, developed ossified vertebrae in the Devonian, some 350 million years ago, and it is within this group that the ancestors of the amphibians are to be found.

Vertebral evolution (Figure 3–5) reached a higher level in early amphibians than in any group of fishes. Early amphibian vertebrae were of a complex nature, with two elements forming the centrum: an anterior intercentrum and a posterior pleurocentrum. In the temnospondyl labyrinthodonts, primitive

Figure 3–3.   *Alfred Sherwood Romer (1894–1973), distinguished American paleontologist and vertebrate zoologist, whose books fired the evolutionary interests of several generations of students.*
Courtesy of Mrs. Ruth Romer.

amphibians that lived from mid-Paleozoic until early Mesozoic, the intercentrum became dominant, and the pleurocentrum was eventually eliminated. On the other hand, in anthracosaurs, the amphibians that led to reptiles, the pleurocentrum achieved dominance at the expense of the intercentrum. The single centrum of reptiles, birds, and mammals, and apparently of living amphibians too, is thus a pleurocentrum. The intercentrum persisted as an inconspicuous ventral element in primitive reptiles (as in *Sphenodon*, one of the most-primitive

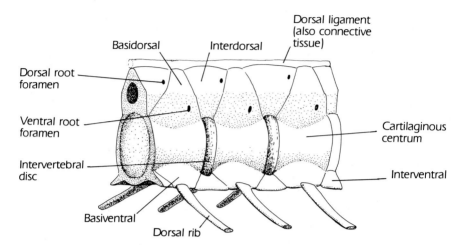

Figure 3–4. *Shark vertebrae. The cartilaginous centra are separated by intervertebral discs. The neural arches, consisting of basidorsals rising above each centrum and interdorsals filling in between the basidorsals, form a roof over the spinal cord.*

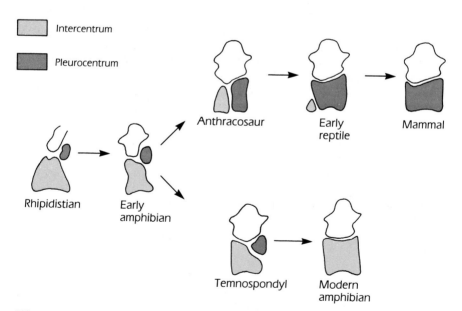

Figure 3–5. *Scheme of vertebral evolution showing the derivation of the simple vertebrae of living mammals and amphibians from the complex vertebrae of rhipidistian fishes and labyrinthodont amphibians. In the line of descent leading to reptiles and mammals, the pleurocentrum became dominant; in the line leading to the living amphibians, the intercentrum dominated.*

The crustacean appendage typically bears a basal *protopodite* of two segments. Distal to it are two parallel structures, a medial *endopodite* and a lateral *exopodite*, each consisting of a series of segments. The appendages of various body segments have been modified in the most diverse ways, adaptive to widely different uses (Figure 3–9). The crayfish is a familiar example. The first two pairs of appendages have been greatly modified to form antennae. Whether the eyestalk also represents a greatly reduced appendage is still subject to debate among specialists. The heavy, biting mandibles are formed by the shortened protopodite and endopodite of the third segment; the exopodite has been eliminated altogether. The fourth pair of appendages forms the first maxilla, in which the protopodite and the endopodite form a flattened plate that is used for handling food. The fifth pair is the second maxillae; the exopodite is present here, and together with a dorsal outgrowth called the epipodite, it forms a vane to pass a current of water over the gills. This is the last of the appendages of the head region.

The first three pairs of thoracic appendages are maxillipeds, the protopodites of which are flattened and serve for the handling of food, just as do the maxillae. The endopodite and exopodite are present on these appendages, but they are not very large and they may be sensory in function. The first maxilliped bears an epipodite; the second and third bear gills that extend dorsally under the carapace. The remaining five pairs of thoracic appendages are specialized as walking legs, with the exopodite absent and the endopodite of the first three terminating in a pincer. The first pair of walking legs forms the major pincers (or chelae) of the organism and serve as defensive and food-procuring organs. All but the last pair of walking legs bear gills.

There are six pairs of abdominal appendages. In females the first is often much reduced or even absent; in males it is modified, together with the second abdominal appendage, to form a copulatory organ. In females the second appendage forms a swimmeret or pleopod in which a short protopodite is followed by a more or less equally developed, filamentous exopodite and endopodite. In both sexes the third, fourth, and fifth abdominal appendages follow this pattern, which appears to be the primitive appendage pattern. The final appendage consists of a short protopodite and a broad, flat endopodite and exopodite, forming the telson, or terminal fan, of the crayfish.

Thus, within a single organism, the basic crustacean appendage is modified to serve no fewer than six to ten different functions (depending on how we classify them). If all of these appendages had originally been created for the function they now serve, then why are they built on the same pattern as are the legs? Most of these functions are subserved in other groups by organs having nothing to do with appendages. Antennae of molluscs, for example, are constructed on a completely different plan, yet there is no reason to suppose that they function less efficiently than those of the Crustacea. Nor is there any intrinsic reason why mouth parts, copulatory organs, and gills should be

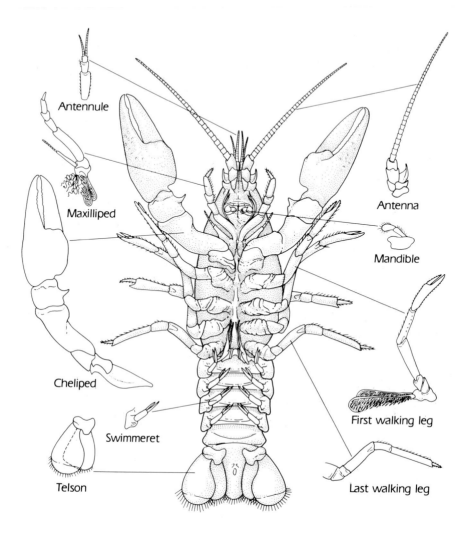

Figure 3–9. *Types of appendages of the crayfish. Antennule, with subequal endopodite and exopodite extending from the protopodite. Antenna, consisting of a heavy protopodite and a long, slender endopodite; the exopodite is reduced to a scale seen at the left of the protopodite. Mandible, with a massive protopodite paralleled by a threadlike endopodite. Maxilliped, one of a series of food-handling appendages; dorsal to the protopodite is a gill, and ventral to it are the slender exopodite and the heavier endopodite. Chiloped, a powerful limb for grasping and tearing food; from the short protopodite a gill extends dorsally and a large endopodite ventrally. Typical walking leg, similar to the chiloped, but much slenderer. Last walking leg has a simple claw rather than a pincer. Swimmeret, with very primitive biramous structure; it has a proximal protopodite and a parallel endopodite and exopodite, the former somewhat larger than the latter. Telson, with the same parts modified into broad swimming parts.*

based on a leglike structure. Yet it is so in the Crustacea. These facts, so puzzling on the basis of any other theory, are easily understandable on the basis of the evolutionary concept. Given a primitive crustacean in which all of the appendages are in a simple condition, somewhat like the swimmerets of the crayfish, natural selection should favor diversification of function. There is a general tendency among the higher phyla for centralization of sensory functions in the head region; hence, the sensory functions of the anteriormost appendages become intensified at the expense of other functions, and these appendages become antennae and possibly eyes. Those appendages nearest the mouth naturally are used for feeding, and so they become specialized for chewing, biting, and food handling. Those appendages nearest the reproductive organs are, in the male, modified for the transfer of sperm to the female. Other appendages continue to serve the original locomotor function in various ways.

If the Crustacea as a whole are considered, then the range of adaptations of these appendages is still greater. In some of the Crustacea, mouth parts are more numerous than in the crayfish. In those with more numerous mouth parts, there are fewer legs. Such a relationship is utterly inexplicable except on the theory that both the mouth parts and the legs have been derived from primitive appendages by adaptive modification. In the barnacles most of the appendages have been suppressed, and the thoracic appendages have been specialized as plumelike cirri that sweep a current of food-bearing water toward the mouth. In the lobster, the abdominal appendages are flattened to form broad, oarlike plates, effective as swimming organs. In the crabs, which normally hold the abdomen recurved against the ventral part of the thorax, the abdominal appendages are much reduced or missing entirely, except the first two, which serve as copulatory organs. But through all of this great range of variation, a single pattern is discernible, a fact that bespeaks an organic relationship of all of the Crustacea.

Adaptive Radiation in the Forelimbs of Mammals

The same principle holds true within each class. The forelimb of the mammals is a clear example. There is always a single long bone, the humerus, in the upper arm. In the forearm there are two parallel bones, the ulna and the radius. In the wrist there are typically eight carpal bones arranged as two rows of four. Five parallel metacarpals form the skeleton of the palm of the hand, and rows of three phalanges each form the skeleton of the digits, excepting the first digit, which has only two phalanges.

The tenrecs (order Insectivora, family Tenrecidae) show a primitive arm structure (Figure 3–10). Their relatives, the moles (family Talpidae), are highly modified for digging. All of the bones of the limbs are short and broad and

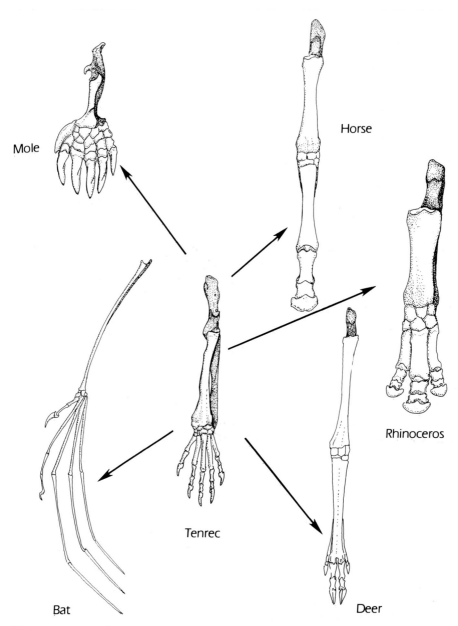

Figure 3–10.   *Adaptive radiation in the forelimbs of mammals. The forelimb of a tenrec approximates the primitive form. Portions of the forelimbs of deer, rhinoceros, horse, mole, and bat show critical modification by changes of proportion, fusion of parts, or loss of parts. (Not drawn to the same scale.)*

give the limb a shovellike appearance. Thus adaptation* is attained by mutual fitting of structure (the shovellike limb), function (digging), and environment (the subterranean habitat). In the bats (order Chiroptera), the humerus, radius and ulna, and four of the digits are greatly elongated to support the wing membrane. In the ungulates, the humerus is short and heavy. The remaining bones of the forelimb are generally elongated, and the digits are reduced in number. Fusion of bones is common in adults, but in the embryos the primitive centers of ossification can be identified. The details naturally differ considerably among the various families of ungulates.

Examples are numerous, but the principle remains the same throughout. Within any taxonomic category, all of the members appear to be built on a common plan, with variations among the members that adapt each to its mode of life. The higher the category examined, the greater the scope of variation, but the common plan is always discernible. To some of the predecessors of Darwin, supernal archetypes accounted for this fact. But since Darwin's time, the great majority of biologists have been convinced that close anatomical similarity must be based on close genetic relationship, and more remote resemblances are based on more remote genetic relationship.

Homology and Analogy Contrasted

Another highly suggestive aspect of comparative anatomy is the comparison of *homologous structures,* which, within a single group, are used for quite different purposes, and *analogous structures,* which, although quite different morphologically and developed in different groups, have nonetheless a certain similarity that is based on adaptation to the same function. If each structure had been created for the purpose for which it is now used, analogy should be far more pervasive and important than homology. Two examples will show that this is not the case.

A classic example is provided by wings, which have been developed independently by insects, reptiles, birds, and bats (Figure 3–11). All the wings are analogous because they are adapted to the same function. The insect wing shares nothing but the planing surface with the others, for it is simply a membrane supported by chitinous veins. All of the vertebrate wings are

---

* Adaptation has been a major concept of evolutionary biology. A quite recent proposal restricts the term to structures fitted to their original function and suggests that *exaptation* (Latin, *ex-*, "by reason of"; and *aptus,* "fitness for") be used whenever a secondary or derived function is concerned. There are both linguistic and biological objections to this proposal. Linguistically, *ex-* rarely means "by reason of." More commonly, it means "from," "out of," "free from" or "former"; any of which, combined with *aptus,* seems to suggest unfitness. Biologically, because of the long history of life, there is seldom assurance that any structure retains its original function. Hence, to replace a useful and universally understood word by one of ambiguous derivation seems futile.

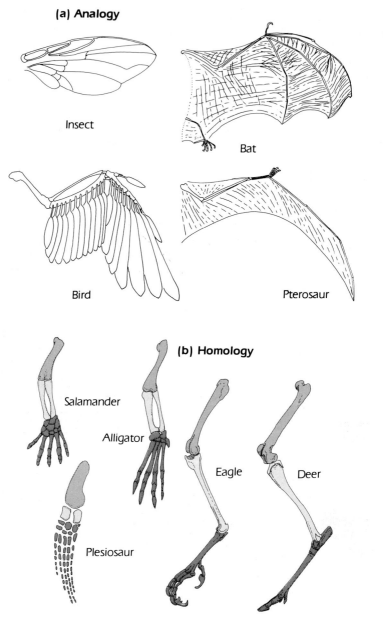

(a) Analogy

Insect

Bat

Bird

Pterosaur

(b) Homology

Salamander

Alligator

Eagle

Deer

Plesiosaur

Figure 3–11. (*a*) *Analogy: wings of an insect, a bird, a bat, and a pterosaur. In each, the planing surface is formed from different materials, and the resemblances are only analogical.* (*b*) *Homology: hind limbs of a deer, an alligator, an eagle, a plesiosaur, and a salamander. Bones with the same intensity of shading are homologous throughout, although they are modified in size, in details of shape, by reduction, or even by fusion of bones (as in the deer and the eagle). Identical materials have been modified to serve the needs of quite different animals.*

constructed from the typical bones of the forelimb of a tetrapod (land vertebrate), and to this extent they may be considered homologous. But the several groups of flying vertebrates are not descended from a common ancestor that flew; flight has developed independently in at least three different lines of descent. The wings of pterosaurs, extinct flying reptiles, were formed by a fold of skin stretched between the body, the posterior surface of the arm, and an immensely elongated fourth digit; the other digits remained free. In birds the feathers that form the planing surface are inserted on all of the three major segments of the appendage. The first digit, which bears feathers, is somewhat independent of the others and can be moved separately by some birds. The second and third digits are fused to form the major skeletal basis of the distal part of the wing. The fourth and the fifth digits are missing altogether. In bats a fold of skin extends from the body to the arm to form the planing surface. This time only the first digit is free and of typical size; the other four are elongated to form the major supports of the wing. Therefore, although the bones of the wings of all of these flying vertebrates are homologous, the planing surfaces are made of different materials, and they are only analogous.

Another case in which analogy and homology contrast significantly is that of eyes. The vertebrate eye is one of the most complex and efficient visual organs in the animal kingdom. Minor variations occur among the various vertebrate classes: identical parts differ in proportions and may function somewhat differently. Vertebrates vary widely in their abilities to see color or to see at night. The muscles of accommodation in mammals vary the tension on the suspensory ligament of the lens, thus controlling its curvature. In birds, more rapid accommodation is achieved when the same muscles contract, thus pulling the lens closer to the retina, or relax, thus permitting it to move farther away. But in all vertebrates, the eye is homologous, constructed of identical materials that are used in similar ways.

The most nearly similar eye outside of the Chordata is that of the Cephalopoda (squids, octopi, and their allies). Superficially, the cephalopod eye bears a close resemblance to that of the vertebrates, but detailed examination shows that in every part different materials have been used in different ways. Embryologically, the cephalopod eye develops from the skin, whereas the vertebrate eye develops from the brain, with the exception of the lens, which is a skin derivative. Yet even the lenses differ fundamentally: the vertebrate lens is cellular and the cephalopod lens is a crystalline secretion of skin cells. Vertebrate and cephalopod eyes are only analogous.

Vestigial Organs

A final contribution of comparative anatomy concerns *vestigial*, or *rudimentary*, organs (see Figure 9–17). These dwarfed and apparently useless organs are found in many plants and animals, relatives of which may have the same organ in a fully developed and functional condition. The great significance of vestigial organs is simply that life has a history—organisms are not ideal creations, but instead they possess structures that are evidently modified from preexisting structures. In many cases the reduced organ has assumed a new function completely unlike that for which it was originally adapted; in other cases it is apparently useless. Perhaps the most widely known example is the human vermiform appendix (Figure 3–12). This small structure, the constricted terminal portion of the cecum, is notorious as a seat of disease. In other primates, this organ is considerably larger. In mammals that have a coarse diet with considerable amounts of cellulose, the cecum forms a large sac in which mixtures of food and enzymes can react for long periods of time, and a constricted appendix may be lacking altogether. Why a useless and disease-ridden structure should have been created especially to plague humans is inexplicable (although the appendix may be important in the development of the immunological functions of the lymphatic system). But to view the human appendix as a degenerating legacy from ancestors with a much coarser diet is easily understandable.

Humans show a number of other vestigial features—a few of which we mention here. In the inner corner of the eye of all tetrapods is a transparent membranous fold, the nictitating membrane. In most vertebrates, this "third eyelid" can sweep clear across the eyeball to cleanse it, much like the blinking of a mammal. In birds, this membrane is particularly well developed. Its use can be easily observed if a captive owl is watched by daylight. In some mammals, horses for instance, it is well developed and fully functional; in many other mammals, as in humans, it forms a mere crescentic fold at the inner corner of each eye, with no known function. It seems best understood as a degenerating structure inherited from an ancestor to which it was actually useful, as it still is to the majority of vertebrates.

The human external ear muscles present a similar situation. Many mammals move the external ear freely in order to detect sounds efficiently. The complete

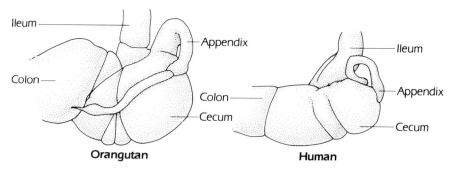

Figure 3–12.   *The appendix in orangutans and in humans.*

muscular apparatus for these movements is present in humans, but it is vestigial. Although children sometimes pride themselves on their ability to wiggle their ears, the ability has no real usefulness; and even this limited ability is not shared by everyone. The presence of these muscles, then, suggests descent from an ancestor to which they were really useful.

Similarly, most mammals have a well-developed tail, but all of the higher primates lack a tail. It is represented in them by vestigial caudal vertebrae, usually three to five in humans. Rarely, a fleshy tail extends a few inches beyond the caudal vertebrae. Whether or not an external tail is present, the muscles that move the tails of other mammals are also present in the primates.

A final human example is the wisdom teeth. The wisdom teeth, or third molars, are the posterior-most teeth, as well as the last to erupt. In other primates, these teeth are as sound and as fully developed as the rest of the dentition. In humans they are far more variable than are the other teeth with respect to size and time of eruption. Frequently, they fail to erupt altogether. When they are present, they are far more subject to all types of dental defect than are the other teeth. Thus, these teeth are probably vestigial, and in view of the frequency with which they fail to erupt they may in time be completely lost.

Many examples of vestigial characters may be found among other animals. Whales, like terrestrial mammals, have ears, but they have lost the pinna, and the auditory canal is nearly blocked. Also among whales, the hind limbs are completely missing, yet in some species rudiments of the pelvic girdle still remain, without connection to the vertebral column. In ungulates (horse, deer, and other hoofed animals), the smaller bone of the lower rear leg, the fibula, has been reduced to a mere splint on the larger bone, the tibia. A similar reduction of the fibula has occurred in the birds. Perhaps no feature of the anatomy of snakes is so generally known as their leglessness. No snake shows any vestige of the forelimbs, but some (pythons and boas) have small, ineffective rudiments of the hindlimbs. (See Figure 9–16.) These are capped by claws that show externally, but they are so reduced that the claws appear at a glance to be scarcely more than raised scales. Males may now use the clawed spurs, mere vestiges of once-useful hindlimbs, to prepare females for copulation.

Many animals, both vertebrates and invertebrates, have become adapted to life in deep caves, where the light of the sun never reaches. Since they exist in perpetual darkness, there is no selective advantage for them to maintain eyesight, and in fact blindness is a general characteristic of such cave dwellers. They exhibit all degrees of eye degeneration, from just short of the typical functional condition to complete absence of the eyes. Examples include the blind, cave-dwelling salamander of central Europe, *Proteus anguineus;* the many species of cave-dwelling fishes of the United States as well as other parts of the world; and the blind crayfishes. The latter have eyestalks that do not bear eyes. Although we can understand the existence of such degenerated eyes easily on the basis of the theory of descent from ancestors with functional eyes, their presence is inexplicable, indeed it is contradictory, on the basis of any other theory.

Typical beetles are strong fliers. Madeira, a wind-swept island about 600 miles off the coast of Portugal and 400 miles to the west of North Africa, has a rich beetle fauna. But a large proportion of these Madeiran species either are wingless or have much reduced wings of no use for flight. If we assume that the reduction or loss of these wings is a result of natural selection, then we may surmise the selective force: beetles in flight would be quite likely to be blown out to sea and lost. Thus a strong selective force should favor any variations toward reduction of wings. Those species that have been on the island and subject to this selection longest should be most numerous in the list of flightless species. In fact, almost all of the endemic species are flightless; most of the flying species are also represented in Europe or Africa or have close relatives on these continents. Only the evolutionary explanation fits these facts.

When structures undergo a reduction in size together with a loss of their typical function, that is, when they become vestigial, they are commonly considered to be degenerate and functionless. G.G. Simpson pointed out that this view need not be true at all; the loss of the original function may be accompanied by specialization for a new function. For example, the wing of penguins has been reduced to a point that will not permit flight, but at the same time it has become a highly efficient paddle for swimming. The wings of rheas, ostriches, and other running birds are also much reduced and have been described as "at the most still used for display of the decorative wing feathers." But Simpson has observed that the rheas, when running, spread their wings and use them as balancers, especially when turning rapidly. This explanation probably applies to the running birds generally.

A comparable case is afforded by the pineal gland. This small structure grows dorsally from the forebrain, and its histological structure shows a mixture of glandular and nervous characteristics. In the lampreys and in many reptiles, including the primitive, lizardlike *Sphenodon* of New Zealand, this structure forms a third eye, located on the dorsal surface of the skull. Fossil evidence shows that this feature was present in the earliest vertebrates, the jawless Agnatha; in the Crossopterygii, the fishes that gave rise to the tetrapods; in virtually all fossil amphibians; and in many early reptiles. Since possession of a pineal eye was a primitive character for the vertebrates, the pineal gland of most extant vertebrates seems to be a vestigial eyestalk. However, it may well have an important new function because it is commonly regarded as an endocrine gland with an important influence upon diurnal and seasonal rhythms that prepare animals for breeding in the spring and migrating or hibernating in the fall.

EVIDENCE FROM EMBRYOLOGY

Comparative embryology, a specialized branch of anatomy, furnishes evidence for evolution regarded by Darwin as "second to none in importance." Ernst Haeckel brought this field into prominence in the immediate post-Darwinian period with his *biogenetic law,* which states that "ontogeny recapitulates phylogeny." He believed that embryonic stages correspond to ancestral adults and hence provide direct evidence of lines of descent. Recapitulations do occur, but not as Haeckel thought. Resemblances are chiefly among embryos, not among embryos and adults; and embryos too have adaptive problems. Embryos evolve, just as do the adults that develop from them.

A striking example occurs among crustaceans. A series of six larval types (Figure 3–13) strongly resemble adults in a sequence from primitive to advanced. The larvae pass through these stages at successive molts, with larvae of primitive crustaceans stopping early in the series and larvae of advanced crustaceans going through most of the stages, as indicated in Figure 3–13. Failure of the *Cypris* larva to appear in the development of higher crustaceans simply means that *Cypris* and barnacles represent a side branch of crustacean evolution.

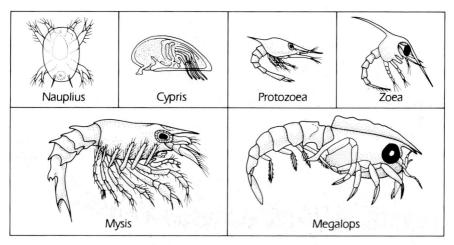

Naupulius | Cypris | Protozoea | Zoea

Mysis | Megalops

Figure 3–13. *Developmental stages of crustaceans. Larval stages often resemble adults of groups with corresponding names. If each larval type is represented by its initial, with m_1 and m_2 representing mysis and megalops larvae respectively, then the developmental histories of some major groups of crustaceans are as follows (note that italic letters indicate stages completed in the egg, all others are free-swimming larvae):*

n→adult *Cypris*
n→c→adult barnacle
n→p→z→adult *Mysis*

n→p→z→m$_1$→adult *Penaeus*
n→*p*→*z*→m$_1$→adult lobster
n→*p*→*z*→m$_1$→m$_2$→adult crab

Examples among Vertebrates

Human ontogeny, considered biogenetically, indicates a long, complicated history. The fertilized egg corresponds to a protozoan ancestor, but it soon becomes multicellular, thus indicating a primitive metazoan grade. Gastrulation changes the embryo to a coelenteratelike form, which soon changes to a triploblastic form, like that of a flatworm. Fundamental chordate characters (dorsal nerve tube, notochord, and pharynx specialized for respiration) are then developed. Fishlike characters, such as gill pouches and aortic arches, appear, followed by tetrapod characters, such as the pentadactyl limb and metanephric kidney. Finally mammalian, then primate, and at last specifically human characters appear (Figure 3–14).

The details of the development of specific systems are impressive. The kidneys of vertebrates are all developed from the *nephrotome,* a segmented mass of mesoderm lying on either side of the somites, from which so many of the serially homologous structures of the body are developed. Yet there are three distinct types of kidneys among the vertebrates. All vertebrate embryos first develop a pronephric type of kidney, which utilizes only the anterior-most part of the nephrotome. Only the hagfishes and a few of the bony fishes (and those only in part) retain this structure as the functional kidney of the adult. In all other vertebrates (including all of the bony fishes), a mesonephric type of kidney is developed posterior to the pronephros, and the pronephros either degenerates or is partly incorporated into the mesonephros. This long, ribbonlike mesonephros is the functional kidney of adult fishes and amphibians and of the embryos of reptiles, birds, and mammals. Finally, in all of the last-mentioned classes, a third type of kidney, the compact metanephros, is formed posterior to the mesonephros, to serve as the functional kidney of the adult organism.

Similarly, all vertebrate embryos develop a series (most commonly six) of aortic arches, each of which runs unbroken from the ventral aorta to the dorsal aorta (Figure 3–15), much as in adult amphioxus. In the fishes, these arches are modified in several ways, all of which involve the separation of each aortic arch into a ventral afferent branchial artery and a dorsal efferent branchial artery, connected by a capillary network in the gill filaments. In the Sarcopterygii, the group of fishes most closely related to the Amphibia, the first arch drops out; it is largely missing in the adult, but its ventral and dorsal roots, together with new growths from them, form the major arteries of the head (the external and internal carotids). The sixth arch has given rise to a pulmonary branch, which supplies the lungs.

The tendency for parts to drop out after having been formed in the embryo and for the remaining parts to be diverted to completely different functions from the original purely respiratory function is the principal feature of the embryology of this part of the circulatory system of all tetrapods. Among the urodeles, the main portions of the first and second arches drop out, so that

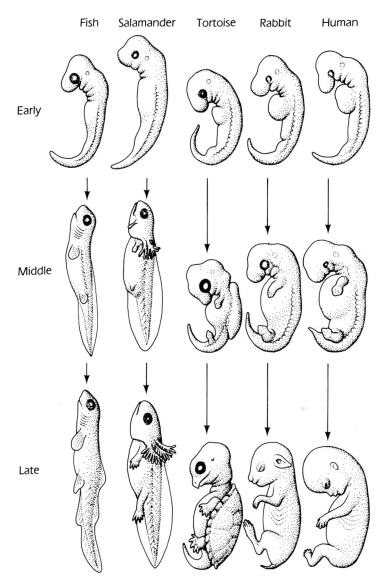

Figure 3–14.   *Embryos of a series of vertebrates compared at three stages in development. This series has been used as evidence for recapitulation. It clearly illustrates von Baer's principle that general characters develop first, then the less general, and finally the special characters.*

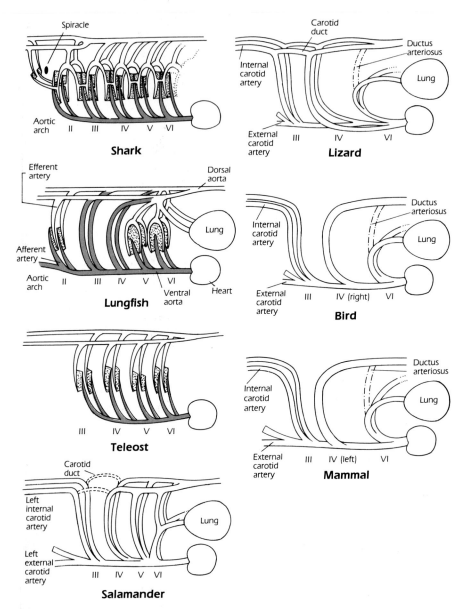

Figure 3–15.   *Diagrams of the aortic arches and their derivatives in a series of vertebrates. Roman numerals indicate aortic arches.*

the carotids arise from the third arch. The third arch is broken by a capillary network early in development, but it soon becomes continuous again. The dorsal connection between arches three and four disappears so that the ventral connections appear as common carotid arteries on either side, with the two fourth arches supplying the major circulation to the body. The fifth arch becomes reduced in size and may be lost altogether, and the sixth arch again gives rise to a pulmonary branch.

In the Anura and in the reptiles, this process goes a little farther, with the fifth arch being lost completely and with the dorsal part of the sixth arch also being lost, so that all of the blood entering the sixth arch must go into the pulmonary artery. The birds have essentially the reptilian system; however, the left fourth aortic arch degenerates, leaving the right fourth arch to carry the entire systemic circulation. In the mammals, it is the right fourth arch that degenerates and the left one that persists. Thus, of the six original pairs of aortic arches, only three persist in the tetrapods. Arch three serves the head region, arch four (but only one of the pair) serves the systemic circulation, and arch six serves the lungs. Yet all six pairs are present in the embryos of birds and mammals.

We could relate a similar story with respect to almost any organ system in any major group. The details differ, but the general facts are the same. When, in the course of embryonic development, a new organ system is formed, its structure is closely similar even in the most widely dissimilar species of the same class, or even phylum in many instances. As differentiation proceeds, detectable differences first become apparent between the embryos of those species of which the adults are most widely different. For instance, the very early embryos of fishes and of mammals are similar, but they soon become recognizably differentiated. As development proceeds, the similarities of the embryos become progressively restricted to smaller and smaller taxonomic groups until finally the characters that distinguish the adults of closely related species are formed. In many instances, this process is not completed until after birth (or hatching). Thus juvenile robins have the speckled breast that is typical of the adults of most species of the thrush family, to which the robin belongs.

Many examples indicate relationship within as well as between classes. The whalebone whales, which feed by straining minute organisms out of the sea water, have no teeth in the adult stage, yet their embryos have a set of tooth buds that are resorbed without ever erupting. Whale embryos also have a coat of hair that is entirely lost to the adults. The absence of hens' teeth is proverbial, yet E.J. Kollar and C. Fisher have shown that epithelium from the first two pharyngeal arches of the chick is capable of inducing tooth formation by mouse mesenchyme. Such facts are readily understandable on the principle that the basic factors of embryology are determined by heredity, and so are common to related species and groups. Both whales and birds are descended from ancestors which had teeth. The hereditary factors which initiate

CAMROSE LUTHERAN COLLEGE
LIBRARY

tooth development are still present and active. But an additional hereditary change (mutation) that acts later in development has been independently acquired in each group since its origin. This change causes the tooth buds to abort.

When much more radical changes in developmental pattern occur, only the early embryonic stages give a clue to the true relationships of the organisms involved. Some extreme examples occur among the cirripedes (barnacles). Typical cirripedes show many crustacean characters, justifying their placement in this subphylum. Nonetheless, they are sufficiently aberrant to have deceived so competent a zoologist as Cuvier, who treated them as a class of the Mollusca. Yet the larvae of all cirripedes are unmistakably crustacean larvae, and it was this fact that finally established the correct taxonomic position of the barnacles. A much more extreme example is presented by *Sacculina*, a very aberrant barnacle that parasitizes crabs. This organism goes through all of the typical developmental stages of a barnacle until it settles down on the abdomen of its host. At this point, other barnacles would develop the usual adult structures of the group, but *Sacculina* undergoes a degenerative development. Appendages are lost, the body of the parasite becomes saclike, and nutrient rhizoids invade and ramify through the tissues of the host, withdrawing nutrients. The only organ system that is well developed is the reproductive system. At this point, the parasite resembles no other organism, and it defied classification until the larvae were studied. The larvae showed clearly that the parasite could be nothing else but a much-degenerated barnacle. The case of the enteroxenid snails that parasitize echinoderms is similar. Externally, they are wormlike. Internally, they have only a much reduced gut and an hermaphroditic reproductive system. Yet they produce typical snail larvae.

Examples among Plants

The embryology of plants is generally simpler than that of animals; hence the recapitulation principle is not so abundantly exemplified in it. Yet there are good examples available in the plant kingdom. The *Acacia* trees, for example, have highly compound leaves (Figure 3–16), yet their seedlings have simple leaves like those of their ancestors. Similarly, the leaves of the adult *Eucalyptus* trees are narrow blades, turned to present the thin edge to the sun, but the leaves of the seedling are broad and oriented like those of more familiar trees (Figure 3–17). Adult cactus plants have no typical leaves at all, although these may be represented by the needles, but the seedlings have readily recognizable leaves. The live oaks of southern United States, which retain their foliage year-round, are considered to be more primitive than the northern species, which are deciduous. However, the saplings of northern oaks commonly retain their leaves during the winter, recapitulating what appears to have been an ancestral character. If we go to a much simpler plant, a moss spore first develops the algalike protonema, then the leafy-stemmed gametophyte grows up from the protonema.

Figure 3–16. *A mature* Acacia *shows the extreme subdivision of its compound leaf. The leaves of the seedling* Acacia *are simple and undivided.*

An interesting and possibly comparable phenomenon occurs in conifers. If conifers are injured and the wound is allowed to heal, the new growth may differ histologically from normal tissue. In such instances, the new tissue shows a type of structure that is well known from fossil conifers of the Mesozoic era.

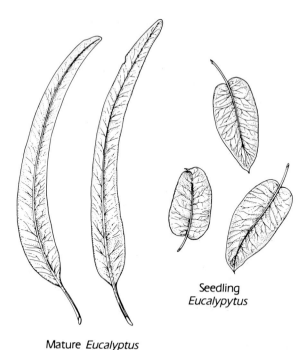

Seedling
Eucalypytus

Mature *Eucalyptus*

Figure 3–17. Eucalyptus *leaves. The mature plant exhibits long, narrow leaves. The seedling has leaves similar to those of less-specialized trees.*

Difficulties of the Biogenetic Law

Because of numerous examples like those described, many biologists of Haeckel's time thought that embryology, when sufficiently known, would be a golden key to problems of phylogeny. Yet there was much unsound biology associated with the biogenetic law, and few aspects of evolutionary science have been so heavily attacked in recent years. The reasons are simple. The recapitulation theory assumes that embryos need only repeat the past, condensing some stages, eliminating others, without adapting to the embryonic mode of life. Actually, the embryo must cope with a hostile environment, even as do adults. All pelagic larvae are subject to heavy predation. The first chapter in embryology texts frequently deals with differences in cleavage patterns, which are correlated with amounts of yolk in the eggs, an adaptive trait of fundamental biological importance. Fetal membranes of amniotes are an obvious series of embryonic adaptations. J. Needham has described an interesting series of adaptations of eggs to fresh water, salt water, and terrestrial habitats. Each requires different excretory physiology, and the last also requires a cleidoic egg (egg encased in a shell) and other adaptations for water conservation. Thus mutations can affect embryonic as well as adult stages; and these mutations are also subject to natural selection, so that embryonic adaptations become part of the normal pattern of development.

The beautifully simple embryology of the echinoderms played an important role in the establishment of the biogenetic law. Yet the comprehensive study of echinoderm embryology by H.B. Fell reveals that extensive differences among various groups of echinoderms are referable to embryonic adaptations. Fell even cast doubt on the echinoderm-chordate relationship because the hemichordate larva did not fit into the scheme of larval relationships that he had worked out. Many biologists, however, feel that this discrepancy is a result of extensive modifications of larvae by natural selection since the separation of their ancestors.

Another difficulty of the biogenetic law was Haeckel's emphasis on resemblances of embryos to *adult* ancestors. Subsequent study has shown that the resemblance is primarily between embryos of related animals; only incidentally do they sometimes suggest adults. Thus ontogeny does recapitulate phylogeny in a significant way, but it is ancestral embryonic, not adult, stages that are repeated, and even these may be drastically modified by adaptive mutations that are favored by natural selection.

von Baer's Principles

In formulating his biogenetic law, Haeckel started from von Baer's principles of embryonic differentiation, and these are perhaps a sounder, better guide to the embryological evidence for evolution. These principles are the following: (1) general characters appear in development before special characters; (2) the characters progress from the more general, to the less general, and finally to the special characters; (3) an animal during development departs progressively from the form of other animals; (4) young stages of an animal are like young, or embryonic, stages of lower animals, but not like adults of those animals. The attack on the biogenetic law has never produced evidence that the findings of embryology do not support the theory of evolution; it has merely shown that Haeckel and his followers read into it more than the data will support. A return to the principles of von Baer makes possible a reasonable evolutionary interpretation of the facts of comparative embryology without straining the evidence.

SUMMARY

We have reviewed the significance of three major lines of evidence for evolution: taxonomy, comparative anatomy, and embryology, with taxonomy depending heavily on comparative anatomy and embryology. It became evident to Linnaeus, the great eighteenth-century naturalist, that the categories of the living world of plants and animals are arranged hierarchically. Great assemblages of organisms (phyla) are grouped together on the basis of only a few characters in common; and successively lower taxa (classes, orders, families) contain progressively fewer organisms with ever-more characters in common. At low taxonomic levels (genera and species), many characters are held in common, and differences are only minor. Darwin inferred that this pattern strongly suggests the degree of genetic relatedness: pairs of species are closely similar to each other because of recent modification from a common ancestor; phyla are very dissimilar one from another because of the remoteness of their ancestry. The arrangement of living organisms into taxonomic trees is an eloquent expression of the tendency of organisms to diversify during evolution.

Comparative anatomy is the tracing of organ systems across taxonomic series of organisms. The morphology of a particular organ or organ system—the pattern of the bones of the skull or of the vertebrae, for instance—varies from species to species, but the range of variation is relatively small. The typical pattern of variation strongly suggests modification of a common pattern, not separate creation of the parts of individual species. The discovery of *homologies,* characters that result from modification of structures found in a common ancestor, is an important goal of comparative anatomy and taxonomy. Homologies can be identified even when an organ undergoes striking modification due to the demands of a particular mode of life. For example, the bones in the wing of a bird, the paddle of a seal, and the arm of a human can readily be recognized as homologous. Evolution does not create perfect structures out of nothing; but rather, out of preexisting structures, it creates structures that work. The imperfection of anatomical structures argues for the process of evolution, particularly with regard to the existence of *vestigial organs,* or structures that are dwarfed in size and reduced or modified in function compared with their state in related organisms. Wisdom teeth (third molars), a vermiform appendix, and a tailbone (coccyx) are examples in humans. The existence of such vestiges is strong evidence for the operation of evolution.

Embryology testifies that living things have a history. A century ago Haeckel proclaimed that ontogeny recapitulates phylogeny. He overstated the case by failing to recognize that embryonic stages of organisms resemble other embryos rather than other adults and that embryos must themselves adapt to the requirements of embryonic life. Nonetheless, embryonic development of invertebrates, vertebrates, and plants all demonstrate strikingly important common features in the order of appearance of characters. General characters (e.g., those that define only phyla or classes) appear first and are followed by progressively more specific characters. For example, early stages of mammal embryos resemble early stages of reptiles or even fishes; intermediate stages of mammals resemble one another; only at late embryonic stages do species-specific patterns appear. These observations are meaningless if each species is a special creation with no relationship to any other species; but they are readily understood on the basis of descent with modification.

REFERENCES

Davis, D.D. 1949. Comparative anatomy and the evolution of the vertebrates. In *Genetics, Paleontology, and Evolution,* edited by Jepson, Mayr, and Simpson. Princeton University Press, Princeton, N.J.
A careful analysis of the role of comparative anatomy in studies on evolution.

De Beer, G.R. 1958. *Embryos and Ancestors,* 3rd ed. Oxford University Press, New York.
A critical review, on a broad basis.

Diamond, J.M. 1981. Flightlessness and fear of flying in island species. *Nature* 293: 507–508.
Discusses many examples.

Eichler, V.B. 1978. *Atlas of Comparative Embryology.* Mosby, St. Louis, Mo.
Includes invertebrates as well as vertebrates.

Gould, S.J. 1977. *Ontogeny and Phylogeny.* Harvard University Press, Cambridge, Mass.
A paleontologist's viewpoint on the embryological evidence.

Hennig, W. 1966. *Phylogenetic Systematics.* University of Illinois Press, Champaign, Ill.
A strong plea for phylogenetic relationship as the primary factor in taxonomy.

Hildebrand, M. 1982. *Analysis of Vertebrate Structure,* 2nd ed. Wiley, New York.
A lively presentation of comparative anatomy stressing biomechanics and function.

Kollar, E.J., and C. Fisher. 1980. Tooth induction in chick epithelium: Expression of quiescent genes for enamel synthesis. *Science* 207:993–995.
The report of the experiment in which "hen's teeth" were actually synthesized.

Linnaeus, Carolus. 1758. *Systema Naturae,* 10th ed. Leyden.
The starting point of modern taxonomy.

Lutzen, J. 1979. Studies on the life history of *Enteroxenus* Bonnevie, a gastropod endoparasitic in aspidochirote holothurians. *Ophelia* 18:1–51.
A thorough review of these extraordinarily degenerate parasitic snails, which are recognizable as snails (or even as animals!) only by their embryos.

Mayr, E. 1969. *Principles of Systematic Zoology.* McGraw-Hill, New York.
An excellent treatment of the subject, with discussion of its evolutionary significance.

Oppenheimer, J. 1959. Embryology and evolution: Nineteenth century hopes and twentieth century realities. *Quarterly Review of Biology* 34:271–277.
An excellent summary and analysis of the controversies that have raged around the relationship of embryology to evolution.

Raff, R.A., and T.C. Kaufman. 1983. *Embryos, Genes, and Evolution: The Developmental-Genetic Basis of Evolutionary Change.* Macmillan, New York, and Collier Macmillan, London.
This valuable book is a partially successful attempt to integrate modern thinking in embryology and evolution.

Romer, A.S., and T.S. Parsons. 1977. *The Vertebrate Body,* 5th ed. Saunders, Philadelphia.
An excellent evolutionary treatment of comparative anatomy.

Sporne, K.R. 1956. The phylogenetic classification of the Angiosperms. *Biological Reviews* 31:1–29.
Includes a critical review of the biogenetic law as applied to plants.

4

Comparative Physiology and Biochemistry

For more than two centuries before Darwin, the great investigative impetus derived from the Renaissance resulted in the accumulation of a large store of biological knowledge. On the whole, however, it was a disorganized array of data, a burgeoning chaos, for there was no basic biological principle to integrate the whole field. The publication of the *Origin* caused a revolution in biological thinking because it provided just such a principle, a rational framework upon which the grand scope of biology could be organized. This achievement is one of the strong arguments for evolution. The biology of Darwin's time was, however, almost exclusively morphological: physiology was in its infancy, and biochemistry did not yet exist. Consequently, these fields were the last to be influenced by the Darwinian revolution in biological thinking. Yet many biologists regard the morphological traits that we have been analyzing as simply the more obvious expressions of specific chemical compounds and processes. In short, they feel that evolution is basically a biochemical phenomenon. That an array of cogent evidence for evolution has emerged from these fields is, therefore, significant.

EVIDENCE FROM BASIC PHYSIOLOGY

On the most basic level, protoplasm appears to be one substance, varied in minor ways from species to species, throughout the living world. It contains nearly the same elements, compounded into roughly the same proportions of proteins, carbohydrates, fats, water, and supplementary substances. The most basic functions of protoplasm are similar, with few exceptions, throughout the living world. This fact is impressive and strongly suggests community of origin, with the most fundamental properties of living things remaining rather constant, while variation in less essential respects has produced the immensely varied forms of the living world.

Much the same is the story with respect to the chemistry of the chromosomes, the physical basis of heredity, incompletely though it is known. Fish sperm comprised the experimental material for the first studies on the chemical constitution of the chromosomes, but subsequent studies on such diverse materials as yeast and mammalian liver cells have all led to similar conclusions. Throughout the living world, the chromosomes appear to consist of nucleic acid, usually combined with basic proteins. Histone and protamine, the simplest types of proteins, predominate, but globulin and some incompletely identified proteins are also present. These more-complex proteins may contribute more to the functioning of the genetic system than their quantity would indicate. The nucleic acids are also rather uniform. The basic unit of structure, or nucleotide, consists of phosphoric acid, a pentose (a sugar based on a five-carbon chain), and a purine or pyrimidine base. These units are highly polymerized (like molecules joined together to form much larger molecules) to form long, double-spiral chains joined to each other by weak bonds between the bases (see Figure 7–9). Nucleic acids differ from one another chiefly in the sequence of base pairs that join together the nucleotide chains, and much evidence now indicates that the specificity of the gene (the unit of inheritance: see Chapter 7) depends upon this trait also. Indeed, we may define a gene as a sequence of nucleotides in a DNA molecule that specifies a protein (or a polypeptide). In view of the great diversity of organisms, all of which owe their attributes to the chromosomes that they possess, that the chromosomes themselves should have so uniform a constitution is astonishing. Unity of the chromosomes, like that of protoplasm, strongly suggests community of origin and constancy of basic properties of life.

Molecular Phylogeny

The gene is essentially a sequence of base pairs in the DNA. More specifically, a single nucleotide may include any one of four bases: adenine (A), cytosine (C), guanine (G), or thymine (T). In the double helix of the DNA molecule, the two chains are joined by hydrogen bonds between the bases of the paired nucleotides. For stereochemical reasons, however, A can pair only with T, and C only with G. Because the pairs may be oriented in either direction, four types of bonds are possible: A-T, T-A, C-G, and G-C. Thus, each of the two strands of a DNA molecule is complementary to the other. The sequence of the four kinds of bonds in the enormous DNA macromolecule determines the information content of the gene, much as the sequence of dots and dashes determines the content of a message in Morse code. Because evolution is based upon changes in the genes, comparing the base sequences of the DNAs of the members of phylogenetic series is of great interest.

This goal seemed remote until an experimental comparison of DNAs of such a series was achieved by B.H. Hoyer, B.J. McCarthy, and E.T. Bolton in 1964. They extracted DNA from a test species A (H. sapiens, for example) and separated the pentose-phosphate chains by heating. They then trapped these chains in a gel so that the complementary chains could not rejoin upon cooling. They prepared DNA similarly from another species, B, which they had given radioactive phosphorus to mark its DNA. They sheared the DNA of species B into fragments small enough to diffuse through a gel. They then incubated blocks of gel with the trapped single strands of DNA from the test species A in a solution of the sheared fragments of radioactive DNA from species B. If the two species had any genes in common, B fragments diffusing through the gel could meet and react with the complementary portions of the trapped strands, thereby reestablishing double helices. They then washed the gel blocks to remove uncombined fragments and measured the radioactivity remaining in the block. This measurement should be proportional to the degree of genetic relationship between the two species.

When both samples of DNA are taken from the same species (A tested against A), only about 20 to 25 percent of the radioactivity is recovered; hence, results using different species must be expressed as a fraction of this amount. Losses occur because fragment may rejoin fragment and because complementary strands, even when present and uncombined, may by chance fail to meet. Measured in this way, members of the same order show about 20 percent relationship, and members of the same phylum show around 5 percent relationship. More distant relationships fail to give positive results. Even such distantly related vertebrates as fish and mammals showed a significant amount of common genetic material. This common material may represent the genetic basis of the most-fundamental vertebrate characters, such as the notochord, the dorsal nerve tube, a pharynx modified for respiration, and essential enzyme systems.

The proportionality of common segments of DNA in two species to their degree of relationship as determined by other methods is easily explained on the basis of the theory of descent with modification. It is unnecessary and anomalous on the basis of any other theory.

Comparative Biochemistry of Plants

Comparative biochemistry of plants is replete with evidence of evolution. Chlorophyll, the green pigment that taps solar energy and thus is the driving force behind all biological processes, including evolution, is a complex compound based upon the porphyrin ring in combination with magnesium. It occurs in several different forms, but chlorophyll *a* predominates throughout the plant kingdom. Variants include bacteriochlorophyll and chlorophylls *b*, *c*, and *d*. Although chlorophylls do not differ widely in chemical structure, bacterio-chlorophyll is the most atypical. It uses the energy of sunlight to split H_2S, thereby releasing sulfur. Chlorophyll *b* is found in association with chlorophyll *a* in blue-green bacteria, in most eucaryotic algae, and in all higher plants. It differs from chlorophyll *a* only by the substitution of an aldehyde group for a methyl group. Similar small modifications of the molecule result in chlorophylls *c* and *d*, but these chlorophylls occur only in the Phaeophyta or brown algae (*c*) and the Rhodophyta or red algae (*d*).

In terms of evolution, the origin of bacteriochlorophyll made solar energy available to bacteria, but the origin of chlorophyll *a* in blue-green bacteria made oxidative metabolism possible through the splitting of water molecules and the subsequent release of oxygen. In the eucaryotic algae, natural experiments tested the combinations chlorophyll *a-b*, *a-c*, and *a-d*, but the first combination must have been strongly favored because it is nearly universal in the plant kingdom.

Cytochrome c.   Perhaps the best example is that of cytochrome *c*, a respiratory pigment found in all cells. The sequence of amino acids in this protein has been determined for a sampling of species widely scattered through the world of life, including several fungi, the gymnospermous tree *Ginkgo*, and some flowering plants.

Ginkgo and the fungi represent widely separated divisions of the plant kingdom, and they differ in about 30 of the 112 amino acids in the chain. The *Ginkgo* and the flowering plants represent, in one taxonomy, different classes of the same division; and in another, different but related divisions. They are separated by around 20 amino acid replacements. Finally, the various flowering plants tested differ by a much smaller number of amino acid replacements, the average being about 8. Thus, groups that are widely separated taxonomically are most unlike chemically, and those that are taxonomically close are also chemically close, just as we would expect on the basis of the evolutionary hypothesis.

Ferredoxin, an iron-containing protein, is easily sequenced because it is a small protein. That of the bacterium *Clostridium* has only 55 amino acid residues. Among the first 28, 12 are repeated in identical positions in the second half of the molecule, thus suggesting that an original short gene was duplicated to make a longer one. Subsequent mutations differentiated the two parts of the molecule. Ferredoxin of the bacterium *Chromatidium* has 81 residues; those of blue-green bacteria have 97 to 105, thus suggesting triplication and quadruplication respectively.

Sequences have been determined for several plants, including *Scenedesmus* (a green alga), spinach, alfalfa, and *Leucaena* (a tree). All have a chain of 97 amino acids. As we might expect, *Scenedesmus,* which is remote from the others taxonomically, is also most different biochemically. If we take *Leucaena* as standard, then the alga differs from it at 38 of the 97 positions, spinach differs at only 21, and alfalfa differs at 28. Again, the biochemical data are roughly parallel to the taxonomic positions as determined on other grounds. If we assume that the number of amino acid substitutions in these plants is roughly proportional to the time elapsed since the separation of their ancestors, then the data make good evolutionary sense.

Flavonoid Pigments.   Flavonoids include some of the most important plant pigments, such as anthocyanins, flavones, and flavonols. Chemically, they are based on the phenolic ring. The simpler flavonoids have only three rings, with small side chains such as the methyl group. The more complex pigments may include as many as six rings, and side chains may be large and complex. Among the algae, there are no flavonoids; only a few of the simpler ones occur in mosses. A wide variety of simple flavonoids occurs among primitive vascular plants such as horsetails, lycopods, and ferns. Among gymnosperms, the whole range of simple flavonoids occurs. Among the angiosperms, the most-advanced plants, we find the full range of complexity of the flavonoids. Thus, a parallel exists between the taxonomy of the plants and the variety and structural complexity of the pigments.

Enzymes and Hormones

Closely similar or identical enzymes and hormones are common to large groups of animals. Some of the digestive enzymes are especially widespread. Trypsin, the protein-splitting enzyme, is found in many groups of animals ranging from the Protozoa to the Mammalia. Amylase, the starch-splitting enzyme, is found from sponges to humans. The thyroid hormone is found in all vertebrates and has been proved to be interchangeable among them. The successful use of beef thyroid in the treatment of human thyroid deficiencies is well known. This hormone is also essential for the metamorphosis of frogs. If a frog's thyroid gland is removed surgically, the frog will not metamorphose. Feeding mammalian thyroid tissue to such a frog will correct the deficiency and permit metamorphosis.

Even more striking is the case of the melanophore-expanding hormone of amphibians. This pituitary hormone causes the pigmented cells of the skin to expand, thus darkening the color of the animal. It has no known effect in mammals, yet an extract of mammalian pituitary glands is just as effective in amphibians as is their own hormone. Thus the melanophore-expanding hormone of mammals might be regarded as a "vestigial" hormone, the presence of which is understandable only on the basis of descent from an ancestor to which the hormone was useful.

Of course, vertebrate hormones are variable. We have long known that the pituitary growth hormone is not interchangeable among mammals. Beef hormone, for example, is not effective in treating deficiencies of growth hormone in children. Analysis of growth hormones, isolated and purified from various mammals, has revealed small but functionally significant differences, comparable to the differences of structure with which comparative anatomy deals.

HEMATOLOGICAL CHARACTERISTICS

If a drop of blood from a species that uses hemoglobin as its respiratory pigment is treated in the proper way, we can obtain crystals of hematin. The details of crystal structure differ from species to species, paralleling classification in a remarkable way. Crystals obtained from all of the species of a genus share many characteristics; crystals from members of different classes have characteristics that are mutually exclusive. Thus, on the basis of hematin crystals alone, we can distinguish among the various classes of vertebrates.

Comparative Serology

The most impressive physiological evidence is drawn from the field of comparative serology. If a small amount of the blood serum of any animal is injected into a guinea pig (or other test animal), the foreign blood acts as an *antigen*, that is, it causes the production in the serum of the guinea pig of *antibodies* that will precipitate and destroy the antigen if a second inoculation should occur. The guinea pig is then said to be *immunized* against the kind of blood that was injected. The precipitation reaction will occur in a test tube as well as in the bloodstream. Thus, if we prepare an antiserum (serum rich in specific antibodies) from an immunized animal and add to it a few drops of antigenic serum, a precipitate will be formed.

This precipitate can be measured by two methods. The first of these is the ring test method. A small quantity of undiluted antiserum is placed in a test tube, and diluted antigenic serum is carefully layered over it. A ring of precipitate then forms at the interface between the two sera. The greatest dilution of the antigenic serum at which a ring is obtained gives a measure of the strength of the reaction, with a high dilution corresponding to a strong reaction. In the second method, the two sera are mixed; the precipitate makes the solution turbid, and the photometric measurement of the absorption of light gives an excellent measurement of the strength of the reaction.

Such antigen-antibody reactions are highly specific. An antibody that precipitates the blood of one species is generally ineffective against the bloods of other species. Yet the specificity is not complete, for serum immunized against the blood of any species will precipitate the bloods of related species, but in ever-decreasing degrees as the relationship grows more distant. For example, a guinea pig is immunized with blood from the common salamander *Necturus*. A sample of serum from the immunized guinea pig is divided among four test tubes, a few drops of serum is then added from *Necturus* and from three other species of salamanders—*Amphiuma, Siren,* and *Cryptobranchus*. The greatest amount of precipitate will occur in the tube to which the serum from *Necturus* has been added. Sera from *Amphiuma* and *Siren* will give about equal precipitates, but rather less than that from *Necturus*. The serum from *Cryptobranchus* will give only a slight precipitate. These findings corroborate evidence from other fields (especially comparative anatomy); it is generally agreed that the first three genera are fairly closely related, while *Cryptobranchus* is a' much more primitive salamander.

Similarly, if serum from an animal immunized against human blood were divided among five tubes, and serum added from a human, an anthropoid ape, an old-world monkey, a new-world monkey, and a lemur, the amount of precipitate formed would also decrease in that order. Thus, the results of serological tests support the theories of relationship that were originally based on comparative morphology. They indicate that the serum proteins show varying degrees of homology, just as do gross structures. That this should be just a coincidence is not imaginable; it is exactly what would be expected on the basis of Darwin's theory that similar species have been formed by descent with modification from a common ancestor.

V.M. Sarich and A.C. Wilson have used the microcomplement fixation method (see biochemistry texts for details) to test the homologies of serum albumins. The method is sensitive to very small differences (single amino acid substitutions) and only very small quantities of antiserum are required. Using this method on organisms whose ancestors are known to have separated at definite times in the past, they concluded that, in a given series of proteins, amino acid substitutions proceed at a constant rate. Hence, the degree of difference of serum albumins, if it can be determined, should provide a key to the time of separation of these organisms' ancestors. Applying this information to the primates (Figure 4–1), they estimated the time of the last common ancestor of *H. sapiens,* chimpanzee, and gorilla at five million years ago; of these three and the orangutan, eight million years ago; of these four and the gibbon and its close cousin, the siamang, ten million years ago; and finally of all of these hominoids and the old-world monkeys, thirty million years ago. At the time, paleontologists were thinking in terms of much greater intervals, but subsequent developments in paleontology have tended to reduce the gap. The molecular clock of Sarich and Wilson may yet be proven accurate.

We have discussed examples taken from among the vertebrates, the group that has been most thoroughly investigated. Extensive studies have also been

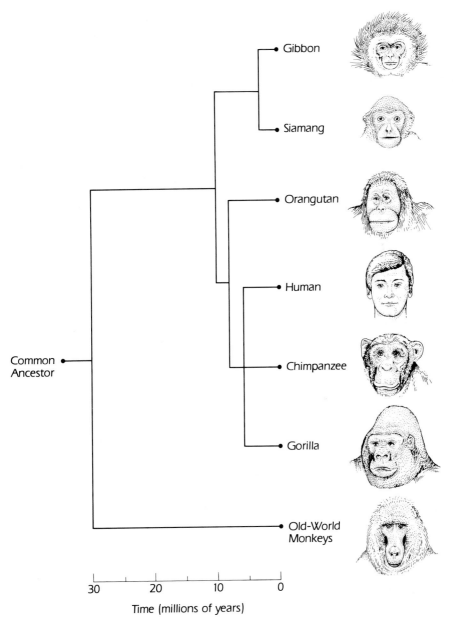

Figure 4–1.   *Times of divergence between the various hominoids, as estimated from immunological data. The time of divergence of hominoids and Old World monkeys is assumed to be 30 million years ago. Figures not drawn to scale.*

made on the serological systematics of the Crustacea, Insecta, and Mollusca. Everywhere, the same fundamental result is obtained: animals regarded as closely related on morphological grounds also show close serological affinity. In general, species of a single genus show very close serological similarity, genera of the same family show moderate serological similarity, and families of the same order show slight but detectable similarity. Frequently, different orders of the same class show too little similarity to permit useful comparisons. There are, however, exceptions, especially among the birds, which have diverged less serologically than have other classes of vertebrates.

In general, serology has verified the taxonomy that was worked out on other bases, but in some cases, serological data have been decisive in resolving difficult problems. Thus the horseshoe crab, *Xiphosura*, was long regarded as a crustacean. Serological tests, however, showed that it has little affinity with typical crustaceans but a strong affinity with spiders. Subsequent studies demonstrated that these aberrant animals share with spiders the fundamental arachnid morphology, whereas their resemblance to crustaceans is a result of superficial convergence.

Although we might reasonably suppose that only those groups characterized by a blood circulatory system could be investigated by the serological method, C. Mez successfully applied the method to problems in plant taxonomy. He injected proteins from a plant into a rabbit, thus immunizing the rabbit against the proteins of the plant species used. He then prepared antibody-containing serum and divided it among a series of test tubes. To each he added a few drops of a solution of proteins from various plants related to the original test plant. Those that gave a precipitate when added in very dilute solution were regarded as being closely related to the test plant; those that gave a precipitate when added only in concentrated solution were regarded as being only distantly related. The results led to a classification that compares very favorably with that which has been worked out by plant morphologists. Thus the method of comparative serology has unexpectedly wide applications, and Mez's method could probably be applied profitably to any group of organisms.

We should note that these biochemical evidences of evolution are *quantitative*. In chapters 10 and 14 the quantitative aspects of evolution are the focus of attention, rather than the simple demonstration of the fact of evolution.

Blood Groups

Like all blood groups, the well-known human A-B-O blood groups are based on antigenic proteins of the red blood cells. Red cells may carry antigen A, antigen B, both (AB), or neither (O), and a person's blood group is named accordingly. Any person's serum contains antibodies capable of agglutinating and destroying cells carrying those antigens not present in his or her own blood. This immune system is inherited on a simple basis (multiple alleles:

Chapter 7). Comparable series of blood groups have been found in many other animals; but only other primates, which are morphologically closest to *H. sapiens,* share the A-B-O groups. Chimpanzees are predominantly group A, but group O also occurs among them. Gorillas and orangutans are known to possess groups A, B, and AB. Among various species of *Macaca* (the rhesus group), all four blood groups are known. In other primates, the same antigenic proteins can be demonstrated in the saliva (as also in humans) but not in the blood cells. The inference of relationship is unavoidable.

BIOCHEMISTRY AND RECAPITULATION

Phosphagens

Few chapters in physiology have been so thoroughly investigated as muscle contraction. Energy-rich phosphate compounds play a key role. Adenosine triphosphate (ATP) breaks down to yield energy for the contraction. Then a second energy-rich compound, called a phosphagen, breaks down and releases energy for the resynthesis of ATP. These reactions are anaerobic, but the cycle is completed by the oxidation of glucose to provide energy for resynthesis of the phosphagen.

In the muscle of vertebrates, the phosphagen is always a specific compound, creatine phosphate; in most invertebrates it is arginine phosphate. It is important to determine which characterizes the most primitive chordates and which characterizes those groups from which the chordates may have arisen. Actually, the Hemichordata, the most primitive group allied to the chordates, has *both* phosphagens, a condition found elsewhere only in certain echinoderms, allies of the sea stars. On embryological grounds, echinoderms were already considered as probably close to the ancestry of the chordates.

This picture is not uncomplicated, for, although most invertebrates have only arginine phosphate, annelids lack it; they have instead a substance that is similar to creatine phosphate and may be identical with it. However, serological evidence affirms the relationship of echinoderms and protochordates and fails to show relationship of the annelids to either of these groups.

Arginine and creatine are closely related chemically, and the former is actually used by vertebrates in the synthesis of the latter. In the embryos of sharks, arginine is abundant, but its occurrence in adults is more restricted. Thus, the creatine metabolism of vertebrates is possibly a biochemical recapitulation, comparable to the embryological recapitulations discussed in the preceding chapter.

Visual Pigments

Vision among vertebrates depends upon one or the other of two chemical systems in the rods of the retina. Freshwater fishes have visual purple, a porphyropsin-vitamin A_2 system; marine fishes and land vertebrates have visual red, a rhodopsin-vitamin A_1 system. That marine and land vertebrates should both contrast with freshwater fishes is surprising until we recall that vertebrates probably arose in fresh water, then migrated from it in the two directions.

However, some fishes, like the salmon, are anadromous: they live principally in the sea but return to fresh water to breed. Others, like the eel, are catadromous: they live much of their life cycle in fresh water but return to the sea to breed. Amphibians, of course, may live much of their adult lives on land, but they return to fresh water to breed. Anadromous fishes and amphibians are both hatched in fresh water, and they undergo larval development there. After a metamorphosis involving profound anatomical and physiological changes, they migrate to salt water and to land, respectively, where they live much of their adult lives. Finally, after changes that are partly a reversal of those occurring at metamorphosis and that G. Wald has called a second metamorphosis, they return to fresh water to spawn.

Wald and his collaborators have studied the visual pigments in these animals, and the results are most illuminating. Actually tadpoles and young anadromous fishes have porphyropsin in their rods. At metamorphosis, that is, when they are preparing to migrate to land or to the sea, the morphological changes are accompanied by a change to a predominantly rhodopsin visual system. Later, when the mature animals are ready to return to fresh water for breeding, they once again revert to a porphyropsin system. In catadromous fishes, the facts are similar, but the sequence of changes is reversed.

From other evidence scientists had concluded that the vertebrates had originated in fresh water and that the ancestral fishes had differentiated there, thus giving rise to freshwater and marine fishes and to amphibians (Figure 4–2). If this conclusion is correct, then the sequence of changes in visual pigments is best interpreted as a *recapitulation*—a condensed repetition of ancestral history.

Similar sequences have been worked out for some aspects of blood physiology and excretory physiology in amphibians. Thus, the principle of recapitulation can give meaning to many otherwise inexplicable phenomena of biochemistry and physiology. According to Wald, "without the rationalizations of phylogeny, comparative biochemistry is little more than a catalogue." Evolutionary (phylogenetic) considerations do give meaning to this great and fundamental body of data, and this is cogent evidence for evolution.

SUMMARY

The evidence from comparative physiology and biochemistry is, thus, similar to that from comparative anatomy and embryology. Protoplasm is much the same in all organisms, bespeaking the unity of the world of life. Similar physiological pathways, catalyzed by similar enzymes, occur throughout the world of life.

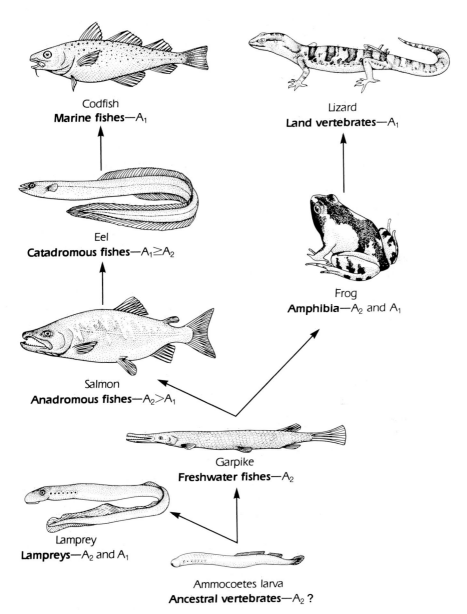

Codfish
Marine fishes—A_1

Lizard
Land vertebrates—A_1

Eel
Catadromous fishes—$A_1 \geq A_2$

Frog
Amphibia—A_2 and A_1

Salmon
Anadromous fishes—$A_2 > A_1$

Garpike
Freshwater fishes—A_2

Lamprey
Lampreys—A_2 and A_1

Ammocoetes larva
Ancestral vertebrates—A_2 ?

Figure 4–2. Biochemical recapitulation. The ancestral vertebrate (approximated by the ammocoetes larva) probably had a porphyropsin–vitamin A_2 system, as do all freshwater fishes today. When their descendants invaded the seas and the land (as amphibians), they developed the rhodopsin–vitamin A_1 system. Amphibians and anadromous fishes, which alternate between freshwater and land or marine habitats, begin life with the A_2 system, then switch to A_1 at metamorphosis. When they return to freshwater to breed, they undergo a second metamorphosis that includes reversion to the A_2 system. Catadromous fishes follow a similar but reversed sequence. Figures not drawn to scale.

At the biochemical level, there are *homologous molecular series* in which the degree of molecular differentiation is roughly proportional to the degree of taxonomic separation of the species in which the compounds occur. For example, all plants have cytochrome *c*, comprising a chain of 112 amino acids. Fungi and gymnosperms are very widely separated taxonomically, and they have about 30 amino acid differences in the chain; gymnosperms and angiosperms are much closer, although in different classes (and perhaps different divisions), and they differ in about 20 amino acids; finally, different species of angiosperms average about 8 amino acid differences. Such biochemical substitutions may proceed at a regular rate, making it possible to correlate biochemical and paleontological data and establish a *"molecular clock"* for evolutionary events.

Finally, even such phenomena as *recapitulation* (visual systems of vertebrates) and *vestigial compounds* (melanophore-expanding hormone of mammals) have been demonstrated. All of these things fit harmoniously into the evolutionary picture, whereas they are anomalous and unnecessary in any other frame of reference.

REFERENCES

Ayala, F.J. 1977. *Molecular Evolution*. Sinauer Associates, Sunderland, Mass.
   A thorough and clear treatment of all of the material in this chapter and, of course, much more.
Barker, W.C., and M.O. Dayhoff. 1980. Evolutionary and functional relationships of homologous physiological mechanisms. *BioScience* 30:593–600.
   A valuable analysis of a variety of homologous protein series among vertebrates.
Barrington, E.J.W. 1979. *Hormones and Evolution*, vols. 1 and 2. Academic Press, London and New York.
Doolittle, R.F. 1981. Similar amino acid sequences: Chance or common ancestry? *Science* 214:149–159.
   The author concludes that some similar sequences may result from chance, but that their regular occurrence in a series of related animals results from common ancestry.
Ferguson, A. 1979. *Biochemical Systematics and Evolution*. Blackie, Glascow, Scotland.
   Best for electrophoretic studies of proteins.
Florkin, M., and B.T. Scheer, eds. 1967–1979. *Chemical Zoology*, vols. 1–11. Academic Press, London and New York.
   A virtual encyclopedia of biochemical aspects of evolution in animals.
Hoyer, B.H., B.J. McCarthy, and E.T. Bolton. 1964. A molecular approach in the systematics of higher organisms. *Science* 144:959–967.
   A seminal paper.

5

Paleontology

The evidence discussed so far indicates that the species now populating the earth must have been produced by inheritance of changes from common ancestors, that is, by evolution. However, the only possible direct evidence for a specific line of descent is a series of fossils leading stepwise from an ancestral to a descended species. Hence the science of paleontology, the study of the remains of ancient organisms, has unique importance for evolution.

GEOLOGICAL TIME SCALE

One of the outstanding accomplishments of James Hutton and Charles Lyell, the founders of modern geology, was to establish the great antiquity of the earth. If the earth were no older than the 6,000 years inferred from Biblical genealogies, Darwin's studies on evolution would have been pointless, for the time necessary for evolution simply would not have been available. How do we determine the age of rocks? Geologists use two methods to date rocks: one is relative dating by position in the strata, or successive layers of rock, and the other is so-called absolute dating by geochemical analysis of radioactive decay, a technique of the twentieth century.

Relative Dating

As sediments slowly settle in the oceans and other large bodies of water, carrying with them the remains of plants and animals, they are gradually compacted into layer upon layer of rock. Of these sedimentary rocks, the deepest strata should be the oldest, the most superficial the most recent. Thick, or fine-grained, strata represent long-continued deposition, and thin, or coarse-grained, strata represent short periods of deposition. Thus, the strata of rock give some insight into the ages of rock deposits, but because rates of sedimentation are variable, only relative estimates are possible.

The principles of relative dating were worked out by William Smith, a civil engineer and builder of canals in England in the last years of the eighteenth and the first of the nineteenth century. An astute observer of the rocks through which he dug, he noticed that different sets of strata were characterized by different sets of fossils. Older strata lying below and younger strata lying above a given horizon had different fossils. He discovered that he could recognize the age of a rock many miles removed from familiar localities and even when the rock type changed, because the fossils were the same. He thus realized that particular fossils are never repeated in the earth's history: once an organism is extinct, it never returns. Smith did not understand the biological meaning of fossils, but he did understand their usefulness for dating.

As the importance of Smith's discovery became apparent, geologists mapped rock units of various parts of the British Isles and of the Continent and characterized them by their fossils. They formalized systems of rocks in type areas (a *type area* is the first region for which a particular system is described); they worked out their temporal relations with other systems, and the geological time scale began to emerge. Thus, the Cambrian period represents the *time* during which rocks of the Cambrian system were deposited in the type area of Wales, and by implication, everywhere else in the world where characteristic Cambrian fossils are found. The Permian period represents the time during which rocks of the Permian system were deposited in the type section of the Ural Mountains of Russia, and so on. By 1879, when the Ordovician system was separated from the older Cambrian and the younger Silurian systems, the geological time table as we know it today (Table 5–1) had crystallized.

The geological time scale is based on the time-specific, nonrepeated nature of fossils, and on the observed physical superposition of strata. Its development largely predates the general acceptance of evolution (Lyell was never fully convinced by his close friend Darwin) and is in no way dependent upon evolutionary assumptions. We now know of course that the evolutionary processes of change and extinction lie at its base.

The grossest division of geological time is into Cryptozoic (Greek for "hidden life"), or Precambrian, and Phanerozoic (Greek for "visible life"). Although the Precambrian covers 85 percent of the 4.6 billion years of earth history, the scarcity of fossils during this time makes subdivision difficult.

The Phanerozoic is divided into three *eras* on the basis of the evolutionary succession of dominant forms of life. The Paleozoic era (Greek for "ancient life") was characterized by fishes, archaic amphibians, trilobites, brachiopods, corals, and giant ferns. The Mesozoic era (Greek for "middle life") hosted clams, ammonites, and crinoids in the seas, and reptiles, gymnosperms, and cycads on land. In the Cenozoic era (Greek for "recent life"), mammals and angiosperms dominated the land, and clams, snails, and foraminiferans teemed in the seas. Earlier life forms sometimes survived into later eras, but they were no longer dominant.

The eras are immensely long periods of time, and they may encompass quite different physical and biotic conditions. They are divided into *periods* of shorter but still very long duration. Thus, the Paleozoic era, lasting more than 300 million years, is subdivided into seven periods varying from 30 million to 75 million years. The periods are divided into either two or three parts, designated as Lower, Middle, and Upper if the emphasis is on the rocks themselves or on the fossils they contain, or as Early, Middle, and Late if the reference is to the segment of time. An exception is the Tertiary period. Because so many sediments have been preserved and because they document so many organisms evolving rapidly, stratigraphers have divided the Tertiary period into five *epochs*. This division is based on a very interesting concept. Studying assemblages of fossil molluscs from Italy, Lyell defined the epochs on the basis of the percentage of living species found in the fossil assemblage. Thus in the Eocene a mere 3 percent of living species are found; in the Miocene, 17 percent (today we insert the Oligocene between the first two epochs); in the Pliocene, 50 to 67 percent; and in the Pleistocene (of the succeeding Quaternary period), 90 percent. Later the Paleocene was added before the Eocene. Although the five epochs average about 12 million years in duration, they range from 3.2 million (Pliocene) to 17.5 million (Eocene):

Pliocene begins 5 million years before present.

Miocene begins 22.5 million years before present.

Oligocene begins 37.5 million years before present.

Eocene begins 55 million years before present.

Paleocene begins 65 million years before present.

Still finer stratigraphic divisions are the *stages*, of which there are typically half a dozen or more per period. In North America seventeen stages during the Tertiary average roughly 3.5 million years in length. Finally in certain favorable cases, stages are divided into *zones* that represent a mere million years of time. Zones are defined by the presence of one or several index fossils, animals that were abundant, widely distributed, and rapidly evolving so that any one species was of short duration. Different groups serve as index fossils for various times during the earth's history: trilobites in the Cambrian, graptolites

Table 5–1. Life through Geological Time

MYBP*	Era	Period	Epoch	Physical Environment	Biological Events
0.011	CENOZOIC	Quaternary	Recent	Glaciers retreat; climate milder.	Age of humans; rise of herbaceous plants.
1.8	CENOZOIC	Quaternary	Pleistocene	Glaciation in northern hemisphere; cool and wet in southern hemisphere.	Humans emerge; large mammals decline, small mammals flourish.
5 / 22.5 / 37.5 / 55 / 65	CENOZOIC	Tertiary	Pliocene / Miocene / Oligocene / Eocene / Paleocene	Elevation of the great mountains of the world: Alps, Himalayas, Rockies, Andes; separation of North America from Europe and its joining to South America; climate begins warm, ends cool.	Dominance of mammals and flowering plants on land; forests decline, grasslands spread; birds and modern bony fishes radiate; snails, bivalves, and protozoans abundant in seas.
141	MESOZOIC	Cretaceous		Vast flooding of continents; warm humid climate; separation of India from Africa, Antarctica.	First flowering plants, marsupial and placental mammals; dinosaurs peak, decline, disappear; great extinctions of vertebrates and invertebrates on land and sea; modern sharks thrive in seas.
195	MESOZOIC	Jurassic		Warm climate, less arid than Triassic; incipient opening of North and South Atlantic.	Cycads and broad-leafed conifers dominant on land; dinosaurs flourish; first birds and salamanders; coiled and bivalved molluscs and protozoans important in seas.
230	MESOZOIC	Triassic		Land surface high, climate arid and cool, but warming through period; initial breakup of supercontinent.	Extinction of seed ferns and archaic amphibians; peak then decline of mammallike reptiles; first appearance of frogs, turtles, crocodiles, mammals, flying reptiles, dinosaurs.
280	PALEOZOIC	Permian		Land surface high; definitive assembly of continents into Pangaea; southern continent glaciation, centered in South Africa; aridity.	Great marine life extinctions: trilobites disappear, brachiopods decline, many groups reduced; ginkos appear, mammallike reptiles flourish.

MYBP*	Era	Period	Physical conditions	Life
310	PALEOZOIC	Pennsylvanian (Upper Carboniferous)	Southern continent glaciation, centered in South Africa; widespread development of coal swamps.	Dominance of sphenopsids, lycopsids, ferns, and seed ferns; giant insects on land; foraminiferans flourish in seas; first reptiles appear, amphibians thrive.
345		Mississippian (Lower Carboniferous)	Land surface low; warm, lime-rich seas extensive, but coal measures already in Europe.	Echinoderms abundant in seas; nautiloids and trilobites in decline, terrestrial amphibians rare.
395		Devonian	Land surface high; mountain building in North America and Europe.	The age of fishes: first bony fishes (both ray fins and fleshy fins), sharks; archaic jawless and jawed fishes peak, decline, disappear; first gymnosperms, amphibians; graptolites become extinct.
435		Silurian	Land higher, seas less extensive; evaporites formed at margins of seas.	Earliest known land plants; fishes become important; invertebrates of Ordovician continue.
500		Ordovician	Vast flooding of continents; northern continents warm, glaciation centered in North Africa.	Trilobites, graptolites, nautiloids, corals, brachiopods, molluscs important in seas; first reefs; vertebrates remain rare, fragmentary.
570		Cambrian	First flooding of long-emergent continents.	Sudden appearance of organisms with hard parts; all major phyla known; trilobites and brachiopods characteristic; fragmentary vertebrates appear.
4600	PRECAMBRIAN		At different times, volcanic activity, sedimentation, erosion, glaciation.	Late Precambrian, wide variety of algae; protozoans; Ediacaran fauna of sponges, coelenterates, echinoderms, molluscs, annelids, arthropods, others of uncertain affinities. Early Precambrian, bacteria, especially blue-greens, often as stromatolites.

* MYBP = millions of years before present

in the Silurian, ammonites in the Mesozoic, foraminiferans in the Cenozoic. The common denominator of all index fossils is that they were free-living marine animals, a necessary condition to achieve wide distribution.

Radiometric Dating and Chronology of Life on Earth

Although stratigraphic paleontologists have a high level of confidence in their relative dates, absolute dates would still be superior for an understanding of the rates of geological and evolutionary processes. Fortunately, methods for absolute dating are available.

In 1907, Bertram Boltwood introduced a method for dating geological strata based on radioactive elements. Because the new method indicated that the earth was vastly older than had been generally believed, geologists received it at first with skepticism. It has since become the standard by which the accuracy of other methods of dating is judged. The method is based on the fact that uranium-238 slowly disintegrates to produce helium and lead, the latter with an atomic weight of 206. The rate of disintegration is calculable: with any definite amount of uranium, half of the molecules will break down, thus forming lead and helium, in the course of 4.5 billion years. As this figure is independent of the actual quantity of uranium originally present, it is called the *half-life* of the element. If we find a uranium-bearing rock, we can determine the ratio of uranium to lead-206; and from this information, utilizing the half-life, we can calculate the interval since the formation of the rock.

Although the lead method is now universally accepted, it has serious limitations. Uranium is not a common element, and it is often found in geological formations that are not readily fitted into the stratigraphic column. However, additional geochemical methods have been developed. Potassium-40 disintegrates to yield calcium-40 and argon-40; rubidium-87 yields strontium-87; thorium-232 yields lead-208; and uranium-235 yields lead-207. Each of these parent elements has its characteristic half-life, ranging from 12.6×10^8 to as much as 6×10^{10} years. Some of these elements, especially the potassium-argon pair, are more abundant than uranium and are, therefore, more widely useful.

The oldest dated rocks range from about 3.6 to 3.9 billion years in age, and come from localities in Minnesota, Greenland, and South Africa. Apparent blue-green algae, bacteria, and *stromatolites,* the latter consisting of laminae of sediments trapped in mats of blue-green algae, come from South African rocks dated at 3.2 to 3.4 billion years old. A more diverse assemblage of the same sorts of organisms comes from the Gunflint chert of northern Ontario (Figure 5–1). These 2-billion-year-old rocks, which rim part of the northwestern shore of Lake Superior, were the first sediments anywhere to yield Precambrian fossils. E.S. Barghoorn and S.A. Tyler reported this find in 1954 and thus ended a century-long mystery of the apparent absence of simple organisms

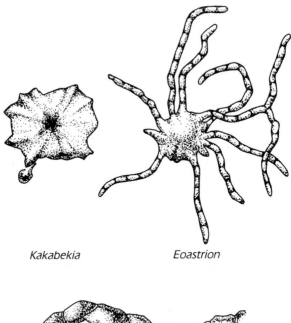

Kakabekia   Eoastrion

Gunflintia

Figure 5–1. *Microfossils from the 1.9-billion-year-old Gunflint Chert of Ontario that contains the first undoubted Precambrian microfossils ever discovered.* Kakabekia, Eoastrion, *and* Gunflintia *were probably iron/manganese-oxidizing bacteria.*

from ancient rocks. This find was an important vindication of Darwin's prediction of the diversification of life from simple antecedents. About 700 million years ago, soft-bodied invertebrates appeared in the fossil record at a number of localities in Russia, China, Europe, Africa, and North America. The most-famous locality in the Ediacara Hills of South Australia has given its name to the recently proposed Ediacaran period of latest Precambrian time. At the beginning of the Cambrian, 570 million years ago, fossils with preservable hard parts appeared and became abundant, thereby making it possible for geologists to utilize the biostratigraphic system of relative dating. Although radiometric dates are always desirable and in fact needed, they are not a practical tool for the field paleontologist. Fossils can often provide an accurate stratigraphic date right in the field; radiometric dating requires careful sampling followed by sophisticated and expensive geochemistry that is performed by only a few specialists.

The Radiocarbon Method

Geochronologists have developed another radioactive method for shorter range determinations of age, up to a maximum of a hundred thousand years. Living organisms utilize a small, but constant, proportion of their organic carbon in the radioactive form. The half-life of radioactive carbon is $5,760 \pm 30$ years. Hence, remnants of bone, wood, or other carbon-containing remains of dead organisms can be assayed for their radiocarbon content. The difference between the average amount in fresh tissue and in the fossil may then be attributed to radioactive disintegration, and the age of the fossil calculated from the half-life. This method has proved useful in the study of late Pleistocene and recent remains. It has been checked against historical objects of known age, such as wood from Egyptian tombs, and has been found reliable.

One crucial part of the stratigraphic column is still difficult to date— the span between about a hundred thousand years and about a million years (which covers a crucial part of human evolution). Because the half-life of potassium 40 is so long, it has not usually been considered practical for age determinations of less than a million years, and because the half-life of carbon 14 is so short (geologically speaking!), a hundred thousand years (or even less) is its maximum span. J.F. Evernden, however, claims that the potassium 40 method has been refined enough to bridge that gap, at least for potassium-rich deposits. Moreover, J.L. Bada and R.R.R. Protsch have shown that amino acids, which are always levo-rotatory in living organisms, are slowly racemized after death. At a given temperature, the rate of racemization is a constant that can be expressed as a half-life of racemization. The rate for aspartic acid, with which they worked, is on the order of fifteen thousand to twenty thousand years at $20°$ C. The details of the method are too complex for description here, but using this method with aspartic acid, these researchers have dated human fossils up to seventy thousand years of age. Each amino acid has its own characteristic half-life of racemization, and Bada and Protsch believe that dating fossils up to several hundred thousand years old may be possible if the appropriate amino acid is selected for measurement of racemization. The process however is sensitive to the presence of water and to temperature and thus seems to require extraordinary stability of environmental conditions—conditions probably too rare to have any general utility for fossils. This technique and the date of seventy thousand years for humans in the New World remain controversial.

Other techniques of dating offer promise. Geologically young marine sediments are now successfully dated by the use of intermediate products of uranium decay. As we have seen, uranium breaks down to lead over an immense period of time. Fortunately, however, intermediate nuclides are formed in a twinkling, geologically speaking. Uranium-238 in solution yields thorium-230, which has a half-life of eighty thousand years; and uranium-235 yields protactinium-231, which has a half-life of only thirty-four thousand years.

Yet another recently developed technique is that of *fission tracks,* which has been used to date materials from a few centuries old (glass) to a few billion years old (moon rocks). Fission tracks record the decay of uranium trapped in crystalline minerals such as zircon or mica. As uranium decays, nuclear particles are shot off at high speed and disrupt the crystal lattice of the adjacent mineral. With the aid of an electron microscope, we can see and count these disruptions, called fission tracks. The older the rock, the greater the number of fission tracks. With a variety of techniques for dating rocks now available, we can be optimistic about the prospects for dating rocks and fossils that lie in the previously troublesome range of a hundred thousand to a million years of age.

FOSSILIZATION

A fossil is a record of an organism that lived in the past, whose remains have come into comparative equilibrium with the sediments in which it was buried. The organism itself may be preserved; it may be dissolved, leaving a natural mold; or the mold may be filled, forming a natural cast. The fossil may be preserved only as a black, carbonized film or impression on the rock. Footprints, animal trails, leaf prints, or plant rootlets may become useful fossils. The remains may be very large—the largest dinosaur remains approach 100 feet (32 m) in length—or they may be microscopic foraminiferans or pollen grains. As a general rule, hard parts are necessary for preservation: teeth and bones of vertebrates, shells and spicules of invertebrates, woody parts of plants. Rarely, fine-grained sediments preserve the form of soft parts: thus, feather impressions of *Archaeopteryx,* the first bird, have survived, and the soft parts of 400-million-year-old trilobites have been revealed by X rays! Even fossil feces, called coprolites, may yield important information about the food habits of extinct animals.

G.G. Simpson estimated that only about one percent of organisms are potentially preservable as fossils, but even of those species that are potentially preservable, the vast majority of individuals are not fossilized after death. Decay and destruction ordinarily await the dying organism. Predators and scavengers may not only eat away the soft parts of the body, but they may also break up skeletal structures beyond all hope of preservation or recognition. It is only the unusual instance in which the organism is rapidly buried or in some other way protected from scavengers and from oxidation that a fossil may be formed. Thus the fossil record, even if completely known, would be fragmentary, because the first step toward fossil formation never happens to most organisms.

The most common method of fossilization is burial in the sediments that are continually deposited on the floor of the oceans or on the bottom of lagoons, lakes, and rivers. When aquatic organisms die, they may fall into deep sedimentary deposits in which the bodies are protected from scavengers and from oxidation. The soft parts of the body gradually decay and are carried away by the seepage of water. Bones and other hard parts may remain, or they may be replaced particle by particle by minerals in the water. As this process continues, the layer of sediment that is being deposited grows ever thicker, and its lower portions gradually harden into rock—the sedimentary, stratified rock that is characteristic of the beds of aquatic deposits everywhere and that is the hallmark of ancient seas in areas that are now dry land.

However, organisms may be buried by other means. Dust storms can have the same effect in causing fossilization of terrestrial organisms. Again, volcanic ash may also rapidly bury organisms and thus preserve them as fossils. Pompeii, which was buried by volcanic ash from Mount Vesuvius in A.D. 79, has been extensively investigated in modern times. Whole families, together with their domestic animals, have been preserved as cavities in the ash, from which casts can be made. Desert forms may be dried out by the hot, dry, desert winds, then buried under the shifting sands. Other organic remains are preserved intact in caves.

Some special methods of burial are also occasionally effective. Where a petroleum spring comes up, evaporation of the more volatile oils produces first a pool of sticky tar and then one of viscous asphalt. During the Pleistocene epoch at Rancho La Brea in southern California, many Pleistocene and recent mammals and birds were trapped in such pools, and they are among the best preserved fossils (Figure 5–2). The process probably worked in the following ways. Small mammals, herbivores, and birds tried to reach the rain pools on the surface of the asphalt. In so doing, they became stuck in the soft asphalt, and predators were then ensnared while attempting to catch them. Water birds alighted on the water pools and then became entrapped in the asphalt around the edges. For this reason, Rancho La Brea is one of the richest known sources of well-preserved fossils of recent mammals and birds. Because the city of Los Angeles has grown up around it, it no longer entraps the wild fauna of the region, but the Los Angeles fire department occasionally had to rescue children from the asphalt until the pits were securely fenced in.

Other unusual methods of burial include the entrapment of insects in pine pitch, which is itself then fossilized to form amber (Figure 5–3). Such fossils are sometimes preserved almost perfectly, so that even histological details are comparable to those of freshly fixed specimens. Also, Pleistocene mammals have been preserved in ice or in frozen mud.

Lastly, organisms may be petrified—that is, their actual tissues may be replaced, particle by particle, by minerals in solution in the local waters. The principal minerals utilized in this type of fossilization are iron pyrite, silica, calcium carbonate, and other carbonates. A widely known example of petrifaction is afforded by the petrified forests of southwestern United States; however, animal remains may also be petrified. Generally, this process preserves only the hard parts of the body, but occasionally soft parts are so well preserved

Figure 5–2.   *One of the larger tar pits at Rancho La Brea. The mammoth that is struggling to escape is, of course, a fiberglass model of a specimen that was trapped in the pits long ago.* Photo courtesy Char Giardina and Vickie Chamberlin.

Figure 5–3.   *A 26-million-year-old insect,* Strepsiptera, *in amber. Upper Oligocene of the Dominican Republic. Note that details of structure of antennae (flabellate), eyes ("raspberry"), and wings can be seen.*
Photo courtesy of Dr. George O. Poinar, Jr., University of California, Berkeley.

that even fine details of cells can be made out in thin sections. Most of the fossils from sedimentary rocks are of this type: the original material has been replaced by minerals from the surrounding medium.

EVALUATION OF THE FOSSIL RECORD

The fossil record is necessarily incomplete. Most species are not fossilizable, and of those that are, most individuals are not buried as is necessary for preservation. Organisms not living in areas of sedimentation are unlikely to be preserved, and this fact rules out most areas on land, except those in or near rivers, lakes, or floodplains.

Animals living in mountains are poorly represented in the fossil record, as are those native to tropical forests, whose skeletons are typically dissolved by the acids in the leaf litter beneath the trees. Marine deposits tend to contain a more complete record of life, distributed over a wider area, than do continental deposits.

The record is also biased by the fact that generally only the hard parts of the body are fossilized. In some instances, such parts are taxonomically useful; in other instances they are not. Among vertebrates, the skeletal parts are the most commonly preserved; specific parts, for instance teeth, skulls, and feet, are of immense value taxonomically. Skeletal remains may give diversified information about a vertebrate. Obviously, a complete skeleton demonstrates the size of the animal; but in the hands of a competent anatomist, even a single bone or a mere fragment of an important bone may offer a basis for a reasonable estimate of size. From the scars of muscle attachments on the bones, the sizes and contours of the muscles can be determined. With this information, it is an easy step to arrive at the general appearance of the animal and its characteristic gait and speed. The skulls give an indication of relative intelligence. The teeth indicate the type of diet. Thus vertebrate skeletons are among the most satisfactory fossils.

But preservation of hard parts does not always lead to so fortunate a result. Among plants, woody parts are most commonly preserved. But these are of secondary importance taxonomically. The flowers, which are of great importance for plant taxonomy, are rarely preserved. For many groups, fossils are rare because there are no hard parts at all.

Even size of an organism affects chance of preservation. Protistans, typically smaller than the grain size of sediments, may be preserved in huge numbers if chemical conditions permit. For vertebrates, on the other hand, a deer-sized animal is more likely to be preserved (and found) than a rodent-sized one; and elephant-sized animals may actually be overrepresented relative to their abundance as living animals.

Even when fossils are preserved, they may be subsequently destroyed by the mighty forces within the earth, which have thrust rocks formed on the sea floor thousands of feet up into mountain peaks. Too, processes of erosion by wind, water, and ice destroy countless specimens before they are ever discovered. Not all periods in earth history have been equally favorable for the formation and preservation of sedimentary rocks. The levels of the continents have risen and fallen through geological time. During the Ordovician and the Cretaceous periods, the continents were submerged under shallow seas; during Devonian and Triassic periods they emerged. The record of life in the seas is at its best for the former periods, and the record of life on land is relatively good for the latter periods.

Chances of preservation of a fossil of a given age depend on the overall volume of sediment of that age that survives to the present. For instance, large volumes of Tertiary age sediment still exist, and we know more kinds of organisms from this period than from any other; conversely, Cambrian sediments are comparatively rare, and we know fewer kinds of organisms from the Cambrian than from any other period.

A whole branch of paleontology has emerged in the past decade that deals with biases in the fossil record. *Taphonomy* is the study of the processes of burial and of the subsequent events that blur the information content of fossils.

In order to appreciate the incompleteness of the fossil record at the intuitive level, we might consider the flora and fauna in our own area. How many different species live there? How many individuals die each year? Of those that die, how many have we actually seen? Of those that we have seen, how many are likely to become fossils, and how many are likely to decay and disappear without trace? Or let us consider a drive across the United States or Canada from coast to coast. How many kinds of mammals would we see on our trip? We may count domestic animals and may even detour to drive through the national parks and spend a few hours there. In fact, suppose we stopped to examine road kills too. Could we reconstruct the extant mammalian fauna of North America, which consists of some 135 genera of land mammals (including bats) on the basis of sightings from a car window plus examination of road kills? Of course not! The fossil record presents paleontologists with similar sorts of problems. For some groups, such as Cenozoic mammals of western North America, we are confident that our knowledge is relatively complete; for other groups, for instance Paleozoic annelids, we are equally confident that our knowledge is extremely incomplete.

In documenting the change of organisms through time, the fossil record provides the strongest evidence imaginable for the fact of evolution. It also uniquely documents rates of evolution and patterns of diversity through time. The fossil record is more begrudging in its readiness to yield information on particular lines of descent, or phylogenies. Because the record is so incomplete and because events of speciation often take place in areas removed from the center of a species' range (*allopatric speciation*), an apparent descendant species may, when carefully examined, turn out to be collateral. Because successive position in the rock strata is not enough to establish direct descent, the fossil record can be frustrating. As Vincent Sarich put it, all living animals have ancestors, but any given fossil need have no descendant. Because the unraveling of phylogenies is a highly interpretative activity, we must use great caution when we analyze the fossil record for this purpose.

Representation in the Fossil Record

With the biases of the fossil record in mind, let us evaluate the record for various groups. The Protozoa (unicellular, or better, acellular animals) generally have not been fossilized, but those that have calcareous or siliceous shells have been fossilized in immense numbers. These are chiefly the Foraminifera, Heliozoa, and Radiolaria, all of them orders of the subphylum Rhizopoda, which is best known to the beginning student for the amoeba. At the end of the Paleozoic era, foraminiferans were evolving rapidly and are important index fossils; in the Cretaceous period chalk deposits were formed of tiny calcareous particles of foraminiferans and marine algae. They again became important index fossils in marine deposits of the Cenozoic era, and comparison with closely related living forms may make them useful indicators of the environment. Porifera (sponges) are most commonly represented in the fossil record by their spicules, either calcareous or siliceous; however, they tell little about the age, the environment, or the organism itself, so they are of little importance.

Soft-bodied Coelenterata (hydroids, jellyfish, and their allies) have left a scant record in the rocks (although they dominated the rare Ediacaran faunas of late Precambrian times); in contrast, corals have been important fossils since early in the Paleozoic. They almost became extinct in the Permian, but modern types appeared in the Triassic and diversified in the Cenozoic. The Annelida and other wormlike phyla have been fossilized so rarely that such fossils as are available have little value for tracing the history of these groups. Yet a few of the known annelid fossils are surprisingly complete. Because of the importance of several of the wormlike phyla among living animals, this deficiency is especially serious in the fossil record. Marine Arthropoda are abundantly represented in the fossil record, but insects are rather scantily represented. Trilobites (Figure 5–4) were present from the base of the Cambrian, were dominant members of benthic communities during the first half of the Paleozoic, but were extinct by the end of the era. Sea scorpions (eurypterids), related to spiders, were formidable predators that reached a length of 5 feet (1.5 m) or more during the Silurian! The oldest fossils of *Limulus,* the horseshoe crab (Figure 5–5), date from the Ordovician. This animal, so common on the East Coast today, is, therefore, truly a living fossil. The bivalved Brachiopoda (Figure 5–6), superficially clamlike, are a minor phylum of only a few species today. They, too, date from the base of the Cambrian, and they were important members of Paleozoic benthic communities. They declined in importance during the Mesozoic. Many of them are useful for dating strata.

Figure 5–4.   Phacops rana, *a trilobite of Devonian age.*

The Mollusca, including such animals as sea cradles, octopuses, clams, and snails today, have left an excellent record. Clams date from the mid-Cambrian, but they became more important in the Mesozoic and Cenozoic. The conical nautiloids approached a maximum length of 20 feet (6 m) during the Ordovician. As the animal grew, it secreted an expanded end on its shell as well as a septum to close the narrower chamber behind it. Later, the shells were coiled, and the sutures of septa to shell became complex. These nautiloids

Figure 5–5.   Limulus, *the horseshoe crab, a living fossil.*

Figure 5–6.   A bedding plane with many specimens of the Devonian brachiopod, Spirifer.

evolved rapidly during the Mesozoic, for which they are important index fossils. Snails were present throughout the Paleozoic, but they became important only in the Cenozoic. Echinoderms are of ancient origin, dating from the Cambrian or before, and they have been abundant in the seas at various times. Certain limestones of Mississippian age are composed largely of crinoid ossicles, and an extinct group, the blastoids, were important at this time too. Echinoids evolved rapidly during the Cretaceous, and they are useful fossils in the English chalk. Finally, the vertebrates are present in the fossil record in good numbers, and they are often quite complete and readily interpretable. The prochordates, however, lacking hard parts, are unknown as fossils, unless the recently described conodontochordates qualify as prochordates.

Fragmentary though the fossil record is, it gives striking testimony to the fact of evolution, and considerable detail can be worked out for many lines of descent. The most ancient animal fossils include only invertebrates. Then primitive fishlike vertebrates appear, and these gradually give way to fishes of more modern aspect. Later, amphibians and reptiles appear in the fossil record, and birds and mammals finally appear quite late. (Chapter 20 treats the succession of vertebrates in detail.) In most major groups (order, class, and phylum), there is marked change from one geological period to the next, but always a particular fauna resembles that of another period near it in time more closely than it does that of any other period remote from it in time. Finally, the fossils of recent organisms blend into our present living flora and fauna, with often the same genera and even the same species being represented.

Catastrophism versus Evolution

Before the time of Darwin, it was customary to explain the fossil record by assuming that all life had been destroyed by catastrophes from time to time, each catastrophe being followed by a new creation. As more and more paleontological information accumulated, however, it became apparent that the number of catastrophes necessary to account for the known succession of floras and faunas was absurdly large and, further, that extinction of different contemporaneous groups was not simultaneous as would be necessary under the theory of catastrophism. Under Darwin's theory we do not need to assume any catastrophes. Species simply change continuously under the influence of natural selection. The inevitable result is a changed aspect of the flora and fauna from one period to the next, with the difference increasing throughout time. Evolution does not require that the rate of change in different groups or in different members of the same group be the same.

FOSSILS AND PHYLOGENY

The phylogeny of horses was the first to be deduced from the fossil record, an early triumph for the fledgling science of evolution. By 1876, T.H. Huxley and V. Kovalesky, each independently studying European fossils, determined that a continuous series linked the Eocene *Palaeotherium*, the Miocene *Anchitherium*, the Pliocene *Hipparion*, to the Pleistocene and Recent *Equus*—horses, zebras, and asses. However, it very quickly became apparent that the European fossils represented offshoots from the direct line of horse ancestry. O.C. Marsh at Yale demonstrated that the rich fossil record of the western United States documents a continuous record of horses through time, almost without gap. Palaeotheres are found only in Europe and are not included in modern accounts of the phylogeny of horses; *Anchitherium* and *Hipparion* appeared first in North America, then migrated to the Old World.

To speak of the "mainline" of horse descent is perhaps misleading. As T.S. Westoll put it, horses did not run the evolutionary race with orthogenetic blinkers. During the Miocene and Pliocene great diversity characterized horses, and three toes were the rule, not the exception. In a sense, single-toed *Equus* was a late, divergent, surviving offshoot whose existence today, far from being inevitable, could not possibly have been predicted from the Miocene or even Pliocene history of horses.

The fossil record lavishly demonstrates evolutionary trends affecting nearly every part of the skeleton. The record begins with little *Hyracotherium* (formerly *Eohippus*) of the latest Paleocene and Early Eocene. Some species stood less than 1 foot (30 cm) high; others towered to 20 inches (50 cm), and their short faces (jaws) and simple teeth show them to have been browsers in forest underbrush. The front feet had four toes and a splint (vestigial toe), the hind feet had three toes and two splints. In both fore and hind feet, the third or central digit was the most prominent (Figure 5–7). Through the Eocene, horses (*Orohippus, Epihippus*) remained small, but the teeth became more complex. In the Oligocene (*Mesohippus, Miohippus*), size increased to 24 inches (61 cm), the fourth toe on the front foot disappeared, and premolar teeth came to resemble molars (enhancing their grinding action). In the Miocene, size increased to as much as 40 inches (1 m), and a great radiation of types occurred. Browsing horses still existed; *Anchitherium* migrated to Europe and flourished there. By the end of the Miocene, three-toed, grazing horses with high-crowned, long-lasting teeth were dominant. *Merychippus* was most characteristic and was very recognizably like a modern horse in overall appearance of head and feet (the side toes being reduced to *dew toes*, or toes that did not touch the ground when the animal was at rest). Half a dozen other genera occurred, varying in stature, body proportions, and fine details of tooth structure; of these, *Hipparion* and *Cormohipparion* were the most widespread since they migrated from North America via Alaska to Siberia and Eurasia, and their descendants lasted into the Pleistocene in Africa. Single-toed horses (*Pliohippus*) appeared in the Pliocene; and several different forms, among them *Astrohippus* and *Dinohippus,* compete in recent schemes for the ancestry of *Equus,* the genus that dominated the Pleistocene in Europe, Asia, Africa, and North America and includes all living equids.

This, then, is a brief sketch of a superb fossil record, which covers 60 million years and includes five continents. It documents evolutionary trends and patterns of diversity through time, patterns of origin and extinction, even migrations between continents. Some samples record dynamics of individual populations (birthrates, death rates, life spans). In some cases we may even determine longevities of genera. Although the genera we have described are fairly distinctive, advanced species of one genus closely resemble primitive species of the succeeding genus. In practice, therefore, distinctions are difficult and somewhat arbitrary. Is the record a perfect one? Of course not! Does it record phylogeny—the taxon by taxon pattern of relationships? In part. When we approach it at the generic level, as we did with horses, we have a pretty clear idea of the phylogeny. At the species level, where more precision is required, the picture is considerably less clear. First, there is the practical problem of species recognition. Even for living organisms, the species problem is a real one (see Chapter 11). We do not even have a consensus on whether the number of species of living horses is six, seven, or eight! For paleontologists the time dimension is a further complication. Only when taxonomic deter-minations are made can a phylogeny, or hypothesis of relationships, be con-structed. It is thus not surprising that, with so many fossils available for study, workers are reaching conflicting conclusions about the phylogeny of

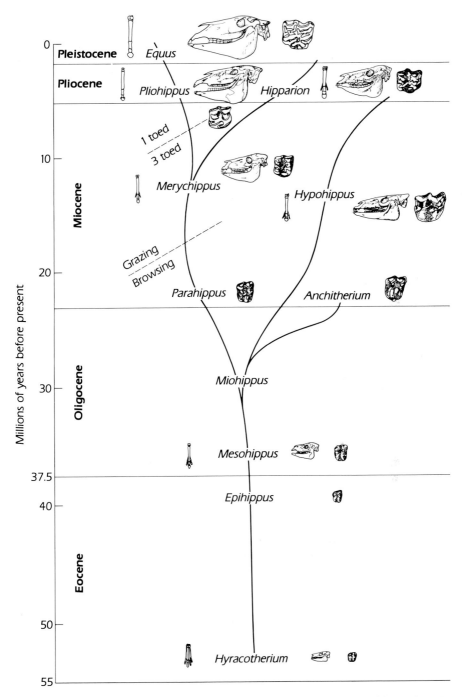

Figure 5-7. *Schematic drawing showing major trends in the evolution of horses from the Eocene to the present, a span of nearly 60 million years. Note the diversity of horses during the Miocene and Pliocene and the extinction of all except the single-toed species by the end of the Pliocene.*

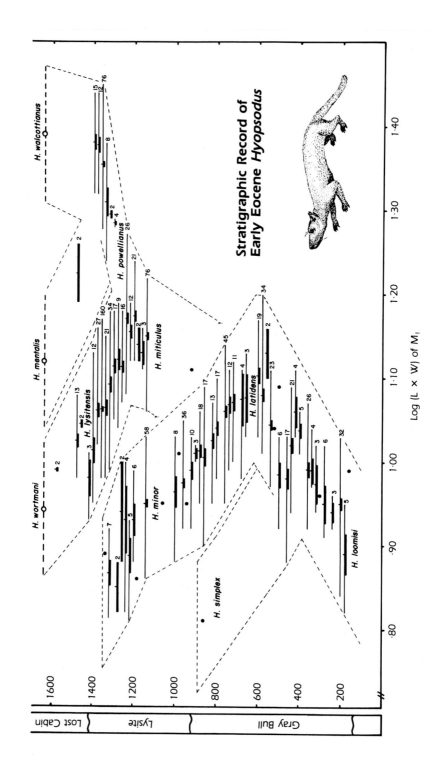

Stratigraphic Record of
Early Eocene *Hyopsodus*

Log (L × W) of M₁

the species of modern *Equus* and its Plio-Pleistocene antecedents. We can be optimistic that our understanding will improve, not because new fossils will be found (as certainly they will be), but because students of horses are becoming increasingly sophisticated in their techniques of analysis.

FINE RESOLUTION: LIMITS TO THE FOSSIL RECORD

The fossil records of many organisms show major evolutionary trends over a period of time without specifying, at our present level of knowledge, a species-level phylogeny. Included among the examples are the transition from reptiles to mammals; the origin and early radiation of flowering plants; and many segments of the fossil record of ammonites, coiled molluscs of the Mesozoic. Species-level phylogenies do exist and have been documented, but they are much fewer in number. Part of the reason for this scarcity is that very few fossil successions have been studied in the required detail; for T.J.M. Schopf the required detail means 20 specimens or more per stratigraphic level, 50 or so levels to cover an adequate time span (up to 10 million years), and several localities to control for geographic factors—all told, 5,000 to 10,000 specimens. Studies that approach this desideratum include those of the Early Eocene mammals of the Bighorn Basin of Wyoming (Figure 5–8) by P.D. Gingerich and his colleagues; Pliocene snails from Lake Turkana by P.G. Williamson; and Antarctic radiolarans by D.E. Kellogg.

When displayed in such detail, records show impressive patterns of species-level relationships; they also show, just as Darwin inferred, a recurring pattern of gaps. In the Bighorn Basin, 2,000 meters of fossil-rich sediment span 10 million years of early Cenozoic time; and in one part of the basin, a 750-meter section spans 2 million years, probably one of the most favorable fossil sequences for any kind of organism at any time in earth history. Here, 1 meter of sediment has an average temporal value of 2,700 years. But how much time can actually be resolved? Gingerich followed calculations based on the technique of P.M. Sadler and concluded that for this highly favorable sedimentary regime, the chance of any single 1-year time interval being

Figure 5–8.   Stratigraphic record of the early Eocene condylarth Hyopsodus *from the Bighorn Basin of north central Wyoming. Each horizontal line represents the range of a measurement, log (length × width) of the first molar, in specimens from its stratographic level. The heavier line on each bar marks the mean ± the standard deviation. One of the most-detailed fossil collections ever made shows the gradual divergence of the species of this genus over a period of 2.5 million years.*

Reproduced by permission of the *American Journal of Science* from P.D. Gingerich. 1976. Paleontology and phylogeny: Patterns of evolution at the species level in early Tertiary mammals. *Amer. J. Sci.* 276:1–28. Courtesy of Dr. P.D. Gingerich, University of Michigan.

recorded was one in seven hundred! For a 5,000-year interval, the odds favoring preservation improve to one in four, and for a 50,000-year interval, the chances of preservation increase to nearly one in one. The data are extremely important in suggesting the limits of resolution of paleontological sampling. The fossil record is incomplete because the processes of sedimentation are discontinuous: rivers don't flood every year, only some years.

Can a good, well-sampled fossil record yield a reasonable phylogeny? For P.D. Gingerich, the pattern of divergence read from the fossil record *is* a phylogeny; this reading divergence is his *stratophenetic* technique. For C. Patterson, the information content of fossils is too low to rely upon for the reconstruction of phylogenies; according to him, fossils are to be "plugged in" to phylogenies based on living animals. This viewpoint, generally associated with the school of taxonomy called *cladism* (see Chapter 22), states that the order of appearance of fossils in the geological record bears no necessary relationship to the evolutionary state (i.e., primitive or advanced) of a series of fossils. These views are the extremes between which the majority of paleontologists labor. But it remains true that the reconstruction of phylogeny is a difficult task, with or without fossils.

SUMMARY

The fossil record documents the progression of life through time. Darwin believed that his theory would stand or fall based on the pattern of change that the rocks show. The vigorous development of geology during the first half of the nineteenth century set the stage for the theory of evolution. Hutton, Playfair, and Lyell demonstrated that the earth is old, immensely older than recorded human history, and that processes of erosion and deposition by running water were responsible for building up the stratigraphic record.

The age of rocks and of the fossils they contain is determined either by *relative dating* or by *absolute dating*. Geologists obtain relative dates by assigning rocks to levels in the layers of the earth's crust: the oldest rocks were laid down first, and successively younger rocks were laid down on top of preceding ones. Strata of different ages contain distinctive sets of fossils, and the succession of fossils holds true over great distances, even when rock types change. The position of strata with given sets of fossils relative to other strata with different sets of fossils became the basis for the stratigraphic system of relative dating.

Beginning early in this century, geochemists have developed techniques for absolute dating that depend on the decay of radioactive isotopes. Uranium-238 decays to lead-206 radioactively, with a half-life for uranium-238 of 4.51 billion years; and carbon-14 decays to carbon-12 in a mere 5,760 years. Geochronologists measure the proportion of the parent isotope to the decay product in unaltered rocks, using these and many other pairs of isotopes, and have, thereby, built up a useful series of absolute dates—for example, the earth dates to 4.6 billion years; the first fossils to 3.4 billion years; the first metazoans, to 670 million years; the appearance of humans, to 2 million years; the rise of Neanderthals, to 250,000 years; and the dawn of civilization in the Near East, to 10,000 years.

The fossil record is incomplete. Only organisms with hard parts and living in areas where sedimentation was taking place were likely to be preserved. Preservation in the sea was more likely than on land, and preservation in lakes and rivers was far more likely than in forests or on mountains. The fossil record documents the beginning of life in the sea; the appearance and radiation of hard-shelled invertebrates in the Cambrian; the rise of sea fishes in the Ordovician and Silurian; the establishment of primitive land plants early in the Devonian; the appearance of amphibians at the end of the Devonian; the diversification of plants, insects, amphibians, and reptiles in the Pennsylvanian and Permian; the appearance of mammals and dinosaurs in the Triassic; the appearance of birds in the Jurassic; the rise of flowering plants, the extinction of dinosaurs, and the radiation of mammals at the end of the Cretaceous; the rise and fall of archaic mammalian types through the Cenozoic; and the development of our own species during the Pleistocene.

The fossil record presents a grand panorama of the succession of life over an immense period of time. We know almost nothing about the evolutionary history of some organisms; whereas the record documents the detailed history of major trends over a significant period of time for other organisms. Excellent records exist for parts of the histories of foraminifera, ammonites, and flowering plants; for the transition from reptiles to mammals; and for many segments of the record of mammals.

The fossil record of horses, one of the finest of any group of organisms, spans almost 60 million years and covers five continents; it documents origin, diversification, extinctions, trends, and rates. Not only does the record enlighten the history of horses themselves, but it also provides insight into the process of evolution. The horse record, however, also documents the limitation of any fossil record: many details of the transition from one species to another are lacking. In fact, very few species-level phylogenies have been documented for any organism because such documentation would require the careful collection of literally thousands of specimens. The fossil record may faithfully document evolutionary events lasting 100,000 years or longer, but it is unlikely to record events lasting 10,000 years or shorter.

REFERENCES

Barghoorn, E.S. 1978. The oldest fossils. In L. LaPorte, ed., *Evolution and the Fossil Record.* Freeman, San Francisco, pp. 44–56.
  A review of ancient fossils from Africa, Australia, and Canada.

Berry, W.B.N. 1968. *Growth of a Prehistoric Time Scale.* Freeman, San Francisco.
A very readable essay about how the stratigraphic time scale that geologists use came into being.

Cloud, P., and M.F. Glaessner. 1982. The Ediacaran period and system: Metazoa inherit the earth. *Science* 217:783–792.
A review of latest Precambrian faunas from around the world, with an interesting stratigraphic proposal.

Darwin, C.R. 1859. On the imperfections of the fossil record. *Origin,* Chapter X.
This is still a very perceptive and valuable essay.

Glaessner, M.F. 1978. Pre-Cambrian animals. In L. LaPorte, ed., *Evolution and the Fossil Record.* Freeman, San Francisco, pp. 57–63.
A reprint of a popular 1961 account of the Ediacaran fauna of South Australia.

Gould, S.J. 1980. G.G. Simpson, paleontology, and the modern synthesis. In *The Evolutionary Synthesis,* E. Mayr and W.B. Provine, eds. Harvard University Press, Cambridge, Mass.
This essay points out the surprising fact that until the publication of Simpson's book in 1944, most paleontologists did not support the concept of evolution by natural selection.

Hallam, A. 1977. *Patterns of Evolution.* Elsevier, Amsterdam.
In this book, experts on various major groups of organisms review the fossil records, rates of evolution, and patterns of diversity through time of their own groups.

Hallam, A. 1981. *Facies Interpretation and the Stratigraphic Record.* Freeman, San Francisco.
A detailed look at the methodologies and results of studies of the fossil record.

Harland, W.B., et al. 1982. *A Geologic Time Scale.* Cambridge University Press, New York.
A timely synthesis of two decades of geochronological activity.

Levin, H.L. 1978. *The Earth through Time.* Saunders, Philadelphia.
A well-presented text that illustrates both fossils and rocks through time.

McKerrow, W.S., ed. 1978. *The Ecology of Fossils: An Illustrated Guide.* MIT Press, Cambridge, Mass.
Although this book concentrates on British marine invertebrates, its applicability is broader.

Patterson, C. 1981. Significance of fossils in determining evolutionary relationships. *Ann. Rev. System. Ecol.* 12:195–223.
A scathing critique on the practice of relying on fossils for phylogenetic interpretations.

Raup, D.M., and S.M. Stanley. 1978. *Principles of Paleontology,* 2nd ed. Freeman, San Francisco.
A stimulating book, by two of the leaders in the field, about the kinds of information that can be obtained from fossils.

Romer, A.S. 1966. *Vertebrate Paleontology,* 3rd ed. University of Chicago Press, Chicago.
Still authoritative on fossil vertebrates, although revision is needed.

Schindel, D.E. 1982. Resolution analysis: A new approach to gaps in the fossil record. *Paleobiology* 8:340–353.
This paper elaborates on the question of time resolution.

Sepkoski, J.J., Jr., et al. 1981. Phanerozoic marine diversity and the fossil record. *Nature* 293:435–437.
After a comprehensive review of the fossil record of marine invertebrates, the authors conclude that the increase in diversity through time is genuine and reflects real evolutionary phenomena.

Shipman, P. 1981. *Life History of a Fossil: An Introduction to Taphonomy and Paleoecology.* Harvard University Press, Cambridge, Mass.

This book probes the conditions under which fossils are formed and the methods by which ecological information is extracted from fossil assemblages. The book is based on the author's experience with African fossils and recent mammals.

Simpson, G.G. 1951. *Horses.* Oxford University Press, New York.

This classic is still in print and is still a valuable overview of horses, including their fossil record.

Simpson, G.G. 1976. The compleat paleontologist? *Ann. Rev. Earth and Planetary Science* 4:1–13.

Reflections of this distinguished paleontologist.

van Eysinga, F.W.B. 1976. *Geological Time Table,* 3rd ed. Elsevier, Amsterdam.

Wetherill, G.W. 1982. Dating very old objects. *Natural History* 91(9):14–20.

A very lucid exposition of a variety of new radiometric dating techniques.

Woodburne, M.O., and B.J. MacFadden. 1982. A reappraisal of the systematics, biogeography, and evolution of fossil horses. *Paleobiology* 8:315–327.

An update, with fresh details, of the horse story, complementing the older work.

6

Development of Evolutionary Thought

The history of evolutionary thought begins in ancient Greek times. We shall touch on this period only lightly, but a brief summary is worthwhile.

EVOLUTIONARY THOUGHT IN CLASSICAL TIMES

The concept of evolution is first found in the writings of Anaximander (611–547 B.C.). He proposed that the earth first existed in a fluid state and that, as it gradually dried, fishlike humans were formed. Only later did they develop to the point that they could live on land. Anaximander did not speculate on the origin of other animals, but he did believe in spontaneous generation. Xenophanes (576–480 B.C.) studied under Anaximander and was the first to recognize fossils as the remains of animals of the remote past. Heraclitus (535–475 B.C.) emphasized perpetual change in nature, but he was not particularly concerned with organisms.

Empedocles (495–435 B.C.) is sometimes called the father of the theory of evolution. He believed that all things were formed from the four classical elements—fire, water, earth, and air—and were acted upon by the combining force of love and the devisive force of hate. The interaction of the elements led first to the origin of plants, then to a long series of trial-and-error events by which plants gave rise to animal life. The various animal parts arose in fortuitous combinations, including not only the animals that now populate the earth but also the monsters of Greek mythology (chimeras, centaurs, and so on). The unworkable combinations ended in extinction; the more viable types survived. Thus, he envisioned a sort of natural selection, but this selection acted on some highly unnatural organisms. Democritus (450 B.C.–?) suggested the idea of adaptation of individual organs. Cuvier called him the first comparative anatomist.

Aristotle

Aristotle (384–322 B.C.) reached the pinnacle in the knowledge of biology during Greek times. Because he made extensive observations of animals, he is often called the founder of scientific zoology. He taught that observation and experimentation were the basis for knowledge of animals, although he did not always resist the temptations of unfettered imagination and hearsay evidence. Aristotle observed a gradation in nature that ranged from inanimate minerals, to plants, to plantlike animals like sponges and anemones, to animals capable of sensibility and locomotion, and finally to humans, the culmination of a long and continuous development. The basis of this development was metaphysical, the result of a perfecting principle in nature. Aristotle emphatically rejected Empedocles' idea of selection of well-adapted organisms.

Fathers of the Church

Some of the fathers of the church interpreted the Scriptures as being favorable to evolutionary concepts. Gregory of Nyssa (A.D. 331–396) taught that God created matter and endowed it with its fundamental properties, in which all of the qualities of the world we know were potential. The world thus resulted from the gradual development of matter according to its potential (*creatio in rationes seminales*), rather than by a series of explicit acts of creation.

St. Augustine (A.D. 353–430) carried this idea of *creatio in rationes seminales* much further by applying it specifically to animal life. He believed that the Creator would not have deliberately made such disagreeable things as mice and mosquitos, yet He might have permitted their development by the action of natural laws. However, Fr. A.L. Schlitzer, a contemporary theologian who is interested in evolution, said that St. Augustine's most important contribution to evolution was not the concept of *creatio in rationes seminales,* but rather the freedom with which he interpreted the Scriptures.

St. Thomas Aquinas (1225–1274) did not himself contribute to the concept of evolution, but he did quote St. Augustine with approval, and he referred to the idea of direct creation on the six days of Genesis as being favored "by the superficial reading of Scripture." Subsequently, however, a more literal interpretation of Genesis prevailed, and interest in evolution reached a low ebb.

RENAISSANCE AND AFTER

Linnaeus and Taxonomy

One reason for the lack of progress in evolutionary thinking during the late Middle Ages was the lack of progress in biology in general and in taxonomy in particular, a study that was almost nonexistent. During the Renaissance the great intellectual ferment spread to the field of biology and might be said to have culminated there in the publication of the *Origin of Species.*

John Ray (1628–1704) made great progress toward a satisfactory taxonomy, but it was Carolus Linnaeus (1707–1778) who laid the foundation of modern taxonomy. He described every species of plant and animal known at that time, and he introduced binomial nomenclature and hierarchical classification. Although Linnaeus contributed an essential tool to evolutionary science, he believed in the fixity of species—that is, the theory that each species was specially created and that absolute limits separated one species from another— as evidenced by his statement that "there are just so many species as in the beginning the Infinite Being created." As his experience grew, however, he modified this view. He finally concluded that all species of a genus had a common origin and that hybridization was a factor in their diversification.

Natural Philosophers

A group of natural philosophers of the seventeenth and eighteenth centuries contributed to the development of evolutionary thought. Francis Bacon's (1561– 1626) most important contribution was his strong advocacy of observation, experiment, and induction as the method of sound science. He also suggested that species might be mutable.

G. W. Leibnitz, co-inventor of infinitesimal calculus, developed a principle of continuity in nature and applied it to biology. In this principle he stated that "all natural orders of beings present but a single chain, in which the different classes of animals, like so many rings, are so closely united that it is not possible either by observation or imagination to determine where one ends or begins." Citing the ammonites in particular, he said that "it is credible that . . . even the species of animals are often changed."

Emmanuel Kant (1724–1804) studied comparative anatomy, and he was impressed by the common plan found in all systems of whole groups of animals.

He considered that, starting with a simple, typical form, variations of proportions and size might produce a great variety of species, and he considered the "supposition that they have an actual blood-relationship, due to derivation from a common parent."

Naturalists

Numerous naturalists of the eighteenth century contributed in some measure to the development of the concept of evolution. To the name of Linnaeus, we may add the names of Charles Bonnet, G.L.L. Buffon, and Erasmus Darwin.

Charles Bonnet (1720–1793) coined the word *evolution,* yet he used it as an embryological term, equivalent to the modern word *epigenesis.* He was greatly impressed by Leibnitz's law of continuity and postulated the inter-connectedness of all forms of life. Yet he did not see genetic continuity in these life forms.

G.L.L. Buffon (1707–1788) has been called the father of the modern theory of evolution (an overly enthusiastic evaluation). He began as a very conservative special creationist (see discussion in Chapter 23), then swung to an extreme evolutionary position, and finally reverted to an intermediate position. Studies in comparative anatomy, which showed the existence of useless organs whose presence could not otherwise be explained, convinced Buffon that animals must have changed since their creation. He thought of changed animals as being degenerate. Thus, he saw donkeys as degenerate horses and apes as degenerate humans. Like Bonnet, he wrote about the chain of life, or chain of being, extending from zoophytes to human beings, and in some passages he seems to anticipate the theory of natural selection. He appears to have assumed the inheritance of acquired characteristics, and he attributed modification of organisms to the direct action of the environment.

Erasmus Darwin (1731–1802), of particular interest because he was the grandfather of Charles Darwin, was a physician, had broad interests in biology, and was a prolific poet. He clearly expressed the idea that all living things shared a common ancestry and that the reactions of plants as well as animals to their conditions of life were responsible for adaptive modifications. These modifications were then inherited by their offspring. Further, he suggested that "millions of ages" might be required for the process. His evolutionary ideas were so close to those that Lamarck published only a little later that Charles Darwin suspected Lamarck of plagiarism. This suspicion, however, does not seem well-founded. Lamarck generously acknowledged his debt to his many predecessors. Further, Erasmus Darwin published his ideas in medico-scientific works that were unlikely to be noticed by general biologists, particularly those working in another language. Finally, several of his works were written in rather pedantic poetry, which did little to invite study by the scientific community.

LAMARCK

By far the most important pre-Darwinian student of evolution was Jean Baptiste Lamarck (1744–1829), a French biologist (Figure 6–1). He began his career as a botanist, but became a zoologist when he was offered an appointment in zoology at the Jardin des Plantes (an institute of general science in spite of its name). Lamarck's services to general zoology are manifold, although his name is usually associated with an outmoded theory of evolution. His extensive studies of invertebrates resulted in a greatly improved classification, including the recognition of the invertebrates and the vertebrates as distinct sections of the Animal kingdom. He came close to the cell theory thirty-nine years before Schleiden and Schwann formulated it.

Lamarck's systematic studies convinced him that species were not constant but were derived from preexisting species. To account for such development, he devised an elaborate theory, which may be summarized in four propositions: (1) living organisms and their component parts tend continually to increase in size; (2) production of a new organ results from a new need and from the new movement that this need starts and maintains; (3) if an organ is used constantly, it will tend to become highly developed, whereas disuse results in degeneration; and (4) modifications produced according to these three principles during the lifetime of an individual will be inherited by its offspring, and cumulative changes will result over a period of time.

Figure 6–1. Jean Baptiste Pierre Antoine de Monet, chevalier de Lamarck (1744–1829).
Drawn by David B. Weishampel.

Lamarck first published his theory in 1802, and he defended it vigorously until his death. For it, he suffered both social and scientific ostracism, but he had the courage of his convictions. He failed to convince his contemporary scientists, not simply because the temper of the times was opposed to evolution, for many others were skeptical of the fixity of species, but because of the implausibility of his major theses. His first principle, the tendency of organisms and their component parts to increase in size, although illustrated by many actual lines of descent, is far from universally true. Many groups of organisms show no tendency whatever to produce strains leading to gigantism. In not a few groups, size reduction has been a prominent feature of evolution. The second principle, that new organs result from new needs, is manifestly false. In the case of plants, Lamarck believed that the environment acted directly upon the plant, causing the production of such new characters as might adapt the plant to its environment. In the case of animals, Lamarck believed that the environment acted by means of the nervous system; in other words, the desire of the animal leads to the formation of new structures. In its crudest form, this would mean that a person who muses "Birds can fly, so why can't I?" should sprout wings and take to the air!

Lamarck did not present quite as crude an example. He did explain the long neck and high shoulders of the giraffe on a similar basis, however. Giraffes browse the leaves of trees. He contended that their original proportions were like those of typical short-necked mammals but that as they strained to reach ever higher and higher leaves, their shoulders grew higher and their necks longer in response to their need. The increase was cumulative from generation to generation.

His final proposition was that the characters acquired by an individual during its lifetime are inheritable by its offspring. This proposition is necessary if environmentally produced modifications are to have any evolutionary significance. However, every serious experimental study designed to test this principle has discredited it, with two questionable exceptions. The first study was concerned with learning in rats. By means of electric shocks, W. McDougall trained rats to leave a maze via a darkened exit rather than a lighted one. He claimed that the progeny of the trained rats learned more rapidly than did the progeny of the control (untrained) rats and hence that learning, an acquired trait par excellence, was inheritable. McDougall's experiments were severely criticized because the rats were genetically mixed, the intensity of light and shock were variable, the method of selecting animals for breeding was not specified, and even the control rats showed increased speed of learning in later generations. The last fact suggests that the technique of experimentation might have improved as the research progressed. The most important reason for doubting the results of McDougall's study, however, was provided by W.E. Agar and collaborators, who repeated the experiments thoroughly, with negative results.

In the second questionable study, E. Steele and R. Gorczynski claim the existence of Lamarckian inheritance of immunological characters. They induced immunological tolerance to foreign cells in mice, then used the males for breeding. They claim that a significant percentage of these tolerant males transmit their tolerance to their progeny. Efforts to duplicate the experiments in other laboratories have failed. These experiments and their interpretation are still hotly debated, and skepticism appears to be well justified.

In contrast to this paucity of positive evidence for the inheritance of acquired characters, countless experiments have led only to negative results. For example, Jewish boys have been circumcised for thousands of years, yet this has not resulted in any tendency whatever toward reduction of the prepuce among Jews. Examples are abundant, and they all lead to the same conclusion: acquired characters are not inherited.

A molecular explanation now exists for why acquired characters are not inheritable. Genetic information is encoded in the DNA of the chromosomes (see Chapter 7). The gene functions as a template upon which a complementary copy of itself is synthesized. This messenger RNA, carrying the genetic message, passes into the cytoplasm, where it serves as a template on which a specific protein is synthesized. These proteins include the principal structural materials of the living organism as well as the enzymes that control the processes of life. Thus, the functional sequence runs from DNA through RNA to protein. No mechanism for reversal of this sequence is known; but RNA viruses do direct synthesis of DNA in the host, and *prions* (see Chapter 16) add a further note of caution.

DARWIN

So when Darwin, jointly with Wallace, brought forth the theory of the origin of species by natural selection, there was no other evolutionary theory to compete with it. The rapidity with which it achieved worldwide acceptance by the majority of competent scientists is generally known, as is the bitter controversy it produced among the lay public and among some scientists. Its rapid acceptance was said to be due to the fact that evolution was "in the air" at the time. Darwin states in his autobiography that, on the contrary, he had discussed his ideas with many naturalists over a period of twenty years before the publication of the *Origin of Species,* and he had not found any of them seriously inclined to agree with him. It seems more probable that Charles Kingsley was right when he said that "Darwin is conquering everywhere, and is rushing in like a flood by the mere force of truth and fact." Darwin attributed the success of his theory to the fact that the *Origin of Species* was highly condensed from a mass of data that he had compiled and critically studied over a period of twenty years before publication (see Chapter 1). His success was also based in part on the fact that he was the first to propose a plausible cause for evolution and the first to amass a significant array of evidence that it had actually occurred (Chapters 2–5).

Darwin realized that an understanding of heredity was essential for evolutionary studies, but he apparently never came across Mendel's work, since he stated in the last edition of the *Origin* that the fundamental principles of heredity were still unknown. To fill the need of a working hypothesis, he devised the theory of pangenesis. According to this theory, all organs produce *pangenes*, minute particles that are carried away by the bloodstream and segregated out into the gametes. Thus, every mature gamete contains a pangene from every organ of the animal producing it. In the developing zygote, each pangene tends to cause the formation of a duplicate of the organ from which it originally came. This theory plainly provides for the inheritance of acquired characters. Darwin did not suggest that this theory was correct. He proposed it simply as a working hypothesis that could serve as a starting point for investigation. The theory of pangenesis has been universally discarded.

Unknown to Darwin, Gregor Mendel (1822–1884; Figure 6–2) was carrying out breeding experiments on genetically pure strains of peas in a monastery garden in Brünn, Austria (now Brno, Czechoslovakia). Mendel correctly recognized the fundamental properties of genetic "elements," which we now call genes. He published his findings in 1866 in the Proceedings of the Brünn Society for the Study of Natural Science, where they were nearly completely ignored. In 1900 they were simultaneously and independently discovered by Correns, De Vries, and von Tschermak; and Mendel was accorded due recognition posthumously.

The history of evolutionary thought subsequent to the publication of the *Origin of Species* may be divided, following Stebbins (personal communication), into three periods: the Romantic Period, extending from 1860 to about 1903; the Agnostic Period, or Period of Reaction, extending from 1903 to about 1937; and the Period of Modern Synthesis, beginning about 1937 and still in progress.

ROMANTIC PERIOD (1860–1903)

The Romantic Period was characterized by extreme enthusiasm for Darwinism, together with an uncritical acceptance of whatever data were claimed to support Darwinism. Negative evidence was given little weight (in contrast to Darwin's own practice), and absurd extremes of interpretation in order to make observed facts fit Darwinian theory were quite common. Leaders of this group in England included T.H. Huxley, Herbert Spencer, and George Romanes; in the United States David Starr Jordan and Asa Gray were the leaders. As a group, they went to interpretive extremes, reading adaptive significance into every organic structure, even on the most imaginary evidence. They often

Figure 6–2.   Gregor Johann Mendel (1822–1884).
Photo by Carl Pietzner.

cited excellent anatomical and taxonomic evidence, but experiments to test adaptive values were unusual, if not unknown.

We should not think of them as second-rate biologists who were blinded by the brilliance of a great man; on the contrary, they were excellent men in their respective fields. Huxley made brilliant contributions to the development of invertebrate zoology, taxonomy, and vertebrate anatomy. Spencer was one of the leading philosophers of his time. Romanes began his career as an invertebrate neurologist, but he soon became engrossed exclusively in evolutionary problems. Jordan was undoubtedly one of the best ichthyologists who has ever lived. And Gray was a botanist of such stature that his work still has great influence. Nor should we think that they never ventured to differ from Darwin, for these men were independent thinkers. Yet the atmosphere of approbation was extraordinary.

It has been said that evolution was born in England but found its home in Germany. The German evolutionists of the Romantic Period were more strictly Darwinian than their English and American colleagues in the sense that they were generally more thorough and careful collectors of data. The leaders in Germany were Karl Gegenbaur, Ernst Haeckel, and August Weismann.

Gegenbaur (1826–1903) was a comparative anatomist, and undoubtedly one of the greatest and most influential; his students held most of the chairs of anatomy in European universities throughout the Romantic Period. He and his collaborators made exhaustive, detailed studies of all classes of vertebrates and used the data so obtained in support of Darwinian theory. Gegenbaur's influence on ideas of vertebrate phylogeny remained strong far into the present century.

Although Ernst Haeckel (1834–1919) did much less experimental work than did Gegenbaur, he did significant work in anatomy, embryology, and taxonomy. His studies in comparative embryology led him to extend von Baer's principles into the biogenetic law, which he supported by extensive publication. He based extensive phylogenies upon embryological evidence, interpreted biogenetically. Bateson has said that this "law" dominated all of the zoology of the last half of the nineteenth century. Haeckel did all his bona fide scientific work in his youth. In later life he became primarily a controversialist and popularizer, a fact that is said to have earned him the contempt of Gegenbaur.

August Weismann's (1834–1914) first interest was heredity, the aspect of Darwinism that Darwin himself had recognized as weakest. Weismann also seems not to have heard of Mendel's work. He was hampered by progressive blindness, which became complete before he finished his major works. Over much of his career, his graduate students made observations and reported them to him in detail. He drew his data principally from cytology, especially from mitosis. Weismann reasoned that, since the hereditary mechanism must be orderly and since only the chromosomes were divided in an exact and orderly fashion in mitosis, the chromosomes must be the physical basis of heredity. The facts of meiosis were not yet known, but he predicted reduction divisions because otherwise the chromosomes would double in number from one generation to the next, an unstable situation. Because of the small size of the chromosomes, he suggested that the hereditary units were chemical in nature. Beyond these propositions, which have long since been proven, his hypothesis of heredity was purely speculative and has been disproven. Unlike Darwin, Weismann was loath to admit the existence of factors other than natural selection in the origin of species.

Also prominent during the Romantic Period were Karl Pearson and Darwin's cousin, Francis Galton. In many respects, the theories of these men were much more akin to those prevalent in the Period of Modern Synthesis, for Pearson and Galton laid the basis for the new sciences of statistics and biometry, which play so prominent a role in modern evolutionary studies.

AGNOSTIC PERIOD (1903–1937)

Such uncritical enthusiasm could not help but be followed by a wave of skepticism and disillusionment, and so the Agnostic Period set in soon after the turn of the century. Many factors converged to cause this change. In part, it resulted from the palpably false extremes of interpretation of evidence, which were so common during the Romantic Period. A second factor was the rediscovery of Mendel's laws of heredity. Today Mendelism is the foundation of most studies in evolution, but at that time the permanence of the gene seemed to raise formidable obstacles to the origin of new species. Genetics was regarded as a sort of blind alley at the end of which stood the sign: THE GENE, DEAD END.

A third factor was the work of W. Johannsen on inheritance of size in beans. He found that, in a seed stock of variable inheritance, selection is highly effective in increasing or decreasing the size of the beans. If a genetically pure line is obtained, however, then selection no longer has any effect. To illustrate this theory he selected beans weighing 20, 40, and 60 centigrams from a pure line with an average weight of 49.2 centigrams. The average weights of their offspring were 45.9, 49.5, and 48.2 centigrams, respectively. Plainly, selection of the parents had not influenced the average weights of the offspring at all. Johannsen concluded that selection could be effective only in a stock with hereditary variability, but that variations produced by the environment (including nutrition, sunlight, temperature, and moisture) were unimportant for evolution.

An additional factor was the mutation theory of H. De Vries. In his studies of the evening primrose, *Oenothera,* De Vries had discovered sudden changes of considerable magnitude that were inherited like Mendelian genes. He called such sudden hereditary changes *mutations,* and he believed that some of his mutants were actually new species, produced in a single step. Thus, *Oenothera lamarckiana* occurred suddenly in a form much larger than normal, and De Vries described it as a new species, *O. gigas.* De Vries, incidentally, was one of the codiscoverers of Mendel's work. Evolution was then conceived as a series of mutations occurring in pure lines. Practically no room existed for the operation of natural selection.

Finally, much of the work of the Romantic Period had been taxonomic in character, and now taxonomy was in disrepute. *Taxonomist* became a term of reproach, and taxonomists were regarded as merely biological file clerks. Contributing to this contempt was the fact that many taxonomists were Lamarckian in their viewpoints.

Biologists generally still believed in evolution, but they were gravely doubtful that the main causal factors were known or that the necessary clues for the discovery of these factors were at hand. This viewpoint is typified by William Bateson, who began his address before the 1921 convention of the

American Association for the Advancement of Science by remarking, "I may seem behind the times in asking you to devote an hour to the old topic of evolution." Later in the same address, he said

> Discussions of evolution came to an end primarily because it was obvious that no progress was being made. . . . When students of other sciences ask us what is now currently believed about the origin of species, we have no clear answer to give. Faith has given place to agnosticism . . . we have absolute certainty that new forms of life, new orders and new species have arisen on earth. That is proven by the paleontological record . . . our faith in evolution stands unshaken.

This, then, was the tenor of evolutionary thinking during the Agnostic Period. Bateson's speech was printed in *Science* in January, 1922. During the year that followed, *Science* published only a single challenge to Bateson's position, written by H.F. Osborn, an elderly zoologist who had reached the peak of his career during the Romantic Period. Scientific interest in evolution was, indeed, at a low ebb. Yet even at this time, F.B. Sumner was experimenting with selection, anticipating the spirit of the Period of Modern Synthesis.

While evolutionary studies were largely suspended during this time, studies in the many branches of biology that contribute to an understanding of evolution were actively pursued. The result was that the stumbling blocks that caused the reaction were gradually removed, thereby paving the way for the Period of Modern Synthesis. The most important developments occurred in genetics. Researchers found that the large mutations with which De Vries worked were rather rare and that much smaller mutations, comparable to the individual fluctuations of which Darwin wrote, were quite frequent. Further, organisms with large mutations were usually less viable than ones with corresponding normal alleles. Taxonomists, meanwhile, had shown that natural species do not differ from one another in single, striking traits, but rather they differ quantitatively in a large number of traits.

Study of wild species in the laboratory showed that pure lines are rare in nature and are usually found only in self-fertilizing plants; hence the pure line concept could no longer have a serious bearing on evolution. Instead, it appeared not only that wild species are variable, but that they commonly take up latent variability (heterozygous recessive genes) "like a sponge" (Chetverikov, see Chapter 9). Both geneticists and taxonomists undertook the study of variability in wild species by using the statistical methods devised by Galton and extended by his successors.

Finally, a new systematics developed, in which subspecies, species, and to some extent genera were studied by the methods of population genetics, ecology, physiology, and every possible approach in addition to the ecology, physiology, and every possible approach in addition to the classical purely morphological approach, with the goal being to determine the dynamics of the origin of species. On the higher taxonomic levels, scientists began to apply to the problems of phylogeny the knowledge gained at the lower levels.

PERIOD OF MODERN SYNTHESIS (1937–)

Thus the bases of the agnostic reaction were gradually destroyed, and the Period of Modern Synthesis became possible. This period has been marked by confidence that the processes of evolution as well as the fact of evolution are open to study. On the lower taxonomic levels, genetic, ecological, geographical, and morphological studies have all been brought to bear upon the problems of the origin of hereditary variation and the origin of species. On the higher levels, paleontologists, especially, have applied the new knowledge gained on the lower levels to the problems of phylogeny. Yet it is misleading to cite only these fields, for the study of evolution at present is truly the "modern synthesis" of all biological disciplines. To find a major area of biology that does not make important contributions to the study of evolution in the current period would be difficult.

The scientific work that culminated in the Period of Modern Synthesis was done throughout the Agnostic Period, yet meaningful exchange among specialists in different fields, especially geneticists and naturalists, was lacking. Books by outstanding scientists who bridged traditionally separate fields of biology catalyzed the modern synthesis and thus brought about rapprochement of previously conflicting fields.

First among these scientists was Theodosius Dobzhansky (Figure 6–3), who in 1937 published *Genetics and the Origin of Species,* a book that launched the modern synthesis. Dobzhansky started his career as an entomologist, then became a leading *Drosophila* geneticist. He was greatly influenced by S.S. Chetverikov (Figure 6–4), a Soviet biologist whose studies on ecological genetics first demonstrated the great genetic variability of wild populations. Dobzhansky's book is a cornerstone of the neo-Darwinian theory—a theory we examine in later chapters. Dobzhansky built upon foundations laid a few years earlier by R.A. Fisher (Figure 6–5), who also saw the processes of Mendelian genetics as the basis of evolution.

The late R.B. Goldschmidt (1878–1958) was closely associated with the rise of genetics after the rediscovery of Mendelism. He did extensive research in geographic variation, taxonomy, and physiological genetics, yet he championed an alternative to the dominant neo-Darwinian theory. R.A. Fisher, J.B.S. Haldane, and Sewall Wright (Figure 6–6) were foremost in the statistical analysis of populations.

Ernst Mayr (Figure 6–7) is a systematic ornithologist, and his work on geographic variation, isolation, and speciation has been very important. In 1942 he published *Systematics and the Origin of Species,* one of the classics of

Figure 6–3.   Theodosius Dobzhansky (1900–1975).
Courtesy of The Rockefeller University Archives.

Figure 6–4.   Sergei S. Chetverikov (1880–1959).
Courtesy of *Genetics*.

Figure 6–5. *Ronald A. Fisher (1890–1962).* Courtesy of *Caryologia.*

Figure 6–6. *Sewall Wright (1889–   ).* Courtesy of Sewall Wright.

Figure 6–7. Ernst Mayr (1904–   ).
Courtesy of Ernst Mayr.

the modern synthesis. In the same year, J.S. Huxley published *Evolution, the Modern Synthesis,* thus naming the current period. In 1944, G.G. Simpson (Figure 6–8) published *Mode and Tempo in Evolution.* Previously, paleontologists had generally been Lamarckian, but in this book Simpson succeeded in applying population genetics and Darwinian principles to the fossil record. He brought *macroevolution,* the origin of higher taxa, into the modern synthesis. The last

Figure 6–8. George Gaylord Simpson (1902–1984).
Courtesy of G.G. Simpson.

of the great foundation books of the modern synthesis, *Variation and Evolution in Plants,* was published in 1950 by G.L. Stebbins (Figure 6–9). Other botanists of great importance in the modern synthesis include such men as Ernest Babcock (Figure 6–10), E. Anderson, J. Clausen, D.D. Keck, and W.M. Hiesey.

Figure 6–9.   *G. Ledyard Stebbins (1906–   ).* Courtesy of G.L. Stebbins.

Figure 6–10.   *Ernest B. Babcock (1877–1954).* Courtesy of the Department of Genetics, University of California, Berkeley.

Mayr commented that by 1947 there was near-unanimous acceptance of the synthetic theory among evolutionary biologists. In that year, the newly founded Society for the Study of Evolution commenced publication of its journal, *Evolution,* in which much subsequent work from all fields of biology has been published. Collectively, these papers have provided a great deal of data that has reinforced the synthetic, or neo-Darwinian, theory. Thus, the modern synthesis is proceeding on many fronts, and we devote most of the succeeding chapters to its development.

SUMMARY

For almost as long as humans have contemplated the natural world, they have recorded thoughts on the subjects of origins and change. Greek writings on the subject begin with Anaximander in the sixth century B.C. Aristotle in the fourth century B.C. was the founder of scientific zoology; he conceived of a gradation in nature from inanimate to animate, from plant to animal, and from lower animal to higher animal, with humans the culmination of a long period of development. In the early Christian era, St. Augustine taught that God created matter and endowed it with its fundamental properties; the development of the diversity of life, including unpleasant vermin, followed from the inherent properties of matter, not from creative acts of God. St. Thomas Aquinas (thirteenth century) followed St. Augustine in teaching that a literal interpretation of Genesis was not necessary.

The founder of modern taxonomy was Linnaeus, who in 1758 established the system of taxonomy that we still use today. He believed that God created all organisms, although he later admitted the possibility of limited modification of species. In the eighteenth century a concept prevailed of a chain of life, or a chain of being—that is, a somewhat vague sense of continuity among living things. Bonnet coined the term evolution but applied it to embryology. Buffon was a great naturalist of the eighteenth century and had some concept of the change of organisms. He was, however, impressed with the process of degeneration: he thought of donkeys as degenerate horses and apes as degenerate humans. Lamarck in 1802 authored the most important pre-Darwinian theory of evolution. He did not, however, succeed in convincing the scientific community to accept his belief that new organs arise from need and that characters acquired during the lifetime of individuals are passed on to their offspring. These propositions are beguiling and even now occasionally appear in one form or another, but they are without genetic foundation.

Charles Darwin published the *Origin of Species* in 1859; it was an electrifying success, literally a Victorian best-seller. He published at a time when many doubted that species were created in a particular form and were fixed in that form. He was a superb naturalist who was capable of adducing facts from every facet of natural history. He produced massive documentation accumulated over a period of twenty years. Where others before had seen species as ideal forms, he recognized individual variation of members of a species as one of the fundamental facts of the living world. His mechanism of natural selection convinced the scientific community where Lamarck's mechanism had failed. Unfortunately, Darwin did not understand the mechanism of heredity and was unaware that the experiments that he required were even then being carried out by Gregor Mendel in a monastery garden in Austria (modern Czechoslovakia). Although published in 1866, Mendel's work did not become widely known until 1900.

The publication of the *Origin* was followed by the *Romantic Period* (1860–1903), during which there was a great burst of activity by evolutionary biologists—particularly in Britain, Germany, and the United States—in the fields of taxonomy, embryology, and cell biology. Although important advances were made in these fields, lack of critical thinking and excessive zeal in interpreting adaptive structures also resulted in some very bad work. Consequently the *Agnostic Period* (1903–1937) set in. When Mendel's laws of heredity were discovered in 1900, the gene appeared to be a stable structure that prevented evolution. De Vries reported mutations in plants that seemed to create species at a single step, with no need for the operation of natural selection. To some extent, taxonomists were typologists with an ideal concept that failed to take into account the range of variation of individuals. To a greater extent, there was simply no communication between laboratory geneticists and field naturalists.

During the *Period of Modern Synthesis* (1937–   ) geneticists took to the field and applied population approaches to the genetics of natural populations. They demonstrated the harmony of genetics, population biology, taxonomy, and natural selection. The keystone publication was T. Dobzhansky's *Genetics and the Origin of the Species*. Other major contributors were E. Mayr, who integrated taxonomy with the synthesis (1942), G.G. Simpson, who brought paleontology into the fold in 1944, and G.L. Stebbins, who published the major work on plants in 1950. Since that time, the study of evolution has been advancing on many fronts, as we will detail in succeeding chapters.

REFERENCES

Bateson, W. 1922. Evolutionary faith and modern doubts. *Science* 55:55–61.
  A brilliant statement of the agnostic reaction to evolution.
Burkhardt, R.W., Jr. 1977. *The Spirit of System: Lamarck and Evolutionary Biology.* Harvard University Press, Cambridge, Mass.
  An excellent study of Lamarck and Lamarckian biology.
Dobzhansky, T. 1971. *Genetics of the Evolutionary Process.* Columbia University Press, New York.
  A late statement of the views of this major architect of the modern synthesis.

Goldschmidt, R.B. 1956. *Portraits from Memory: Recollections of a Zoologist.* University of Washington Press, Seattle.
In these pages, some of the scientists discussed in this chapter live once more.

Irvine, W. 1955. *Apes, Angels, and Victorians.* McGraw-Hill, New York.
Fine scholarship and lively prose illuminate this account of the lives and times of Darwin, T.H. Huxley and their contemporaries.

Lewin, R. 1981. Lamarck will not lie down. *Science* 213:316–321.
A thorough and readable report on the Steele-Gorczynski experiments in immunology, which have been given a Lamarckian interpretation.

Mayr, E. 1982. *The Growth of Biological Thought.* Belknap Press, Harvard University, Cambridge, Mass.
A remarkably comprehensive analytical survey of the history of the great ideas in biology by one of this century's most important biologists.

Mayr, E., and W.B. Provine, eds. 1980. *The Evolutionary Synthesis.* Harvard University Press, Cambridge, Mass.
A very important study of the factors that resulted in the modern synthesis as seen through the eyes of some of its major participants. Mayr's introductory essay (pp. 1–48) provides excellent background for both the Agnostic Period and the Period of Modern Synthesis as described in this book.

Osborn, H.F. 1894. *From the Greeks to Darwin.* Columbia University Press, New York.
This classic on the earlier history of evolutionary thought can still be read with pleasure and profit nearly a century after it was written.

Ruse, M. 1979. *The Darwinian Revolution: Science Red in Tooth and Claw.* University of Chicago Press, Chicago.
The viewpoint of a historian of science.

2

The Origin of Variation

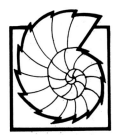

Darwin defined evolution as "descent with modification." Since descent depends on heredity, *hereditary* modifications must be the basic materials for evolution, and the origin of these modifications must be the first major problem of evolutionary science. Lamarck attempted to solve this problem by postulating that the action of the environment on an organism tends to produce adaptive modifications and that these acquired characters are inherited. He failed because both of these theses are easily disprovable.

Darwin sidestepped this problem. He simply accepted without explanation the observed fact that organisms do vary one from another. He did not distinguish between inheritable and noninheritable variations. Darwin's problem was to show the action of natural selection in the formation of new species, given a variable progenitor. The second major problem of evolution is how the varying arrays of organisms become sorted out into species and higher categories. We must analyze the role of selection and of other factors in this process. The differences between the several subspecies of a species may be quite as great as those between species of a genus, but the former interbreed freely and blend from one to another in nature, whereas the latter generally do not. The origin and nature of the barriers that account for this difference are therefore of great importance for evolution.

The two main problems of evolution, therefore, are (1) the origin of variation and (2) the origin of species (and of higher categories). Part 2, comprising Chapters 7 and 8, takes up the first problem, and Part 3 is concerned with the second problem.

CHAPTER

7

Elementary Genetics, Gene Mutation, and the Evolutionary Synthesis

Darwin wrote in the *Origin of Species* that "the fundamental principles of inheritance are still largely unknown"; a good possibility exists that it was the *Origin* that stimulated Mendel's studies of inheritance.* Ironically, when Mendelism was rediscovered in 1900, biologists generally regarded the new science of heredity as a stumbling block in the way of evolution because the apparent permanence of the genes, the units of heredity, seemed to preclude any evolution more fundamental than the reshuffling of a few alternative forms of the genes. The stability of the gene is, indeed, one of its significant properties, but another property has great importance for evolution: *mutation*. A mutation is an inheritable change in a gene; it occurs at a very low but measurable rate. Mutations provide a store of variability upon which natural selection can act. We will, therefore, review the science of genetics as it applies to evolution.

* See *Folia Mendeliana* 6:151–182 (Brno, 1971).

ELEMENTARY MENDELIAN CONCEPTS

Mendelian Laws

The unit of Mendelian heredity is the *gene*. The genes are parts of the chromosomes; hence they ordinarily exist in pairs, just as the chromosomes do. Whenever the two genes of a pair are identical, there can be no doubt what trait they will determine. Whenever the two genes of a pair are unlike, however, several possibilities arise. The effect of one gene may show up to the exclusion of the other, in which case the one that is expressed is referred to as dominant, and its *allele* (alternative gene) is referred to as recessive. (Dominant and recessive genes are symbolized by capital and small letters, respectively.) Or the two genes may collaborate to cause a trait intermediate between the two pure types. Lastly, the two unlike genes could collaborate to produce a character unlike either pure type.

Genes are inherited in a statistically predictable manner. When gametes are formed, the two genes of a pair are separated into sister gametes, so that each gamete contains only one gene for each character (Mendel's first law—the *law of segregation*.) As a result, whenever an organism is heterozygous for a particular gene (that is, the two members of the pair are unlike), two types of gametes will be formed in equal numbers. These unlike genes are not in any way diluted or modified in the direction of an intermediate because of their association in the hybrid. If either gene again becomes homozygous (both members of the pair alike) in a zygote, the original character will reappear unmodified. With the exception of certain special cases, when hybrid organisms interbreed, any type of sperm or pollen has an equal probability of fertilizing any type of egg or ovule (the principle of random fertilization). As a result, the offspring of two hybrids (*Aa*) are 25 percent homozygous dominant (*AA*), 25 percent homozygous recessive (*aa*), and 50 percent heterozygous (*Aa*). As the heterozygotes show the dominant character, 75 percent of the offspring will express the dominant character and 25 percent the recessive—the famous 3 to 1 ratio. Such ratios are usually not obtained exactly; rather, because they depend upon the laws of chance, deviations within the limits of statistical probability occur.

Figure 7–1 offers an example. In the fruit fly, *Drosophila*, normal wing (*Vg*) is dominant over vestigial (*vg*). If homozygous strains of the two are crossed, all of the offspring (F_1 or first filial generation) must be heterozygous (*Vgvg*), and the wings will be normal because of dominance. If these offspring are then interbred, each parent will produce two kinds of gametes (*Vg* and *vg*) in equal numbers. Random fertilization of the two kinds of eggs by the two kinds of sperm results in offspring of which 25 percent are homozygous normal (*VgVg*), 50 percent are heterozygous (*Vgvg*), and 25 percent are homozygous vestigial (*vgvg*). Thus, because of dominance, the F_2 generation consists of 3 normal-winged flies to 1 vestigial-winged fly.

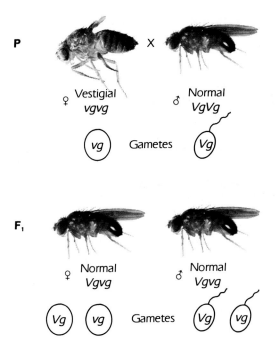

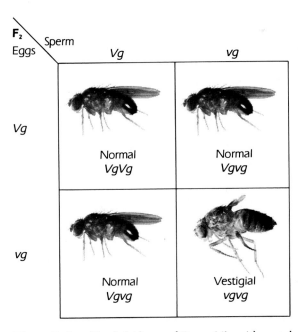

Figure 7–1. *Monohybrid cross of* Drosophila *with normal or with vestigial wings. Note dominance of normal and the 3 : 1 ratio.*

An example without dominance is afforded by flower color in four-o'clocks (*Mirabilis jalapa*, Figure 7–2). If red-flowering plants (*RR*) are crossed with whites (*rr*), then the F_1 plants are all hybrids (*Rr*) and are pink flowering. Again, these offspring produce pollen and ovules of two types (*R* and *r*) in

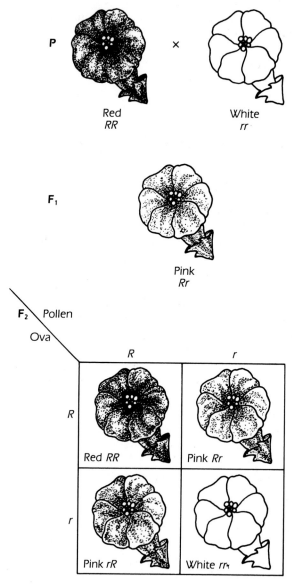

Figure 7–2. Monohybrid cross between red and white flowering four o'clocks, Mirabilis jalapa. Because the heterozygotes produce pink flowers (intermediate), the F_2 ratio is 1 : 2 : 1.

equal numbers. Hence random fertilization leads to an F_2 generation consisting of 25 percent red-flowering (*RR*), 50 percent pink-flowering (*Rr*), and 25 percent white-flowering plants (*rr*).

Such pairs of alleles exist because the original gene has mutated, that is, it has undergone a reproducible change that results in a modified character. Nothing, however, restricts the number of alternative forms of a gene to two; in fact, we know of large series of *multiple alleles*. Thus, the white-eye gene of the fruit fly, *Drosophila*, is represented by at least 14 alleles, and the self-sterility genes of many plants are represented by large numbers of alleles, up to two hundred in some cases. Only *two* alleles of any such series can be represented in any individual (although any number can be present in a population), and these two are inherited in the usual Mendelian fashion, as described above.

If a cross is made between organisms differing in two independent pairs of genes, then each segregates as though the other were not there. This fact is shown by the ratio obtained: 9 dominant for both genes, to 3 dominant for the first but recessive for the second, to 3 recessive for the first but dominant for the second, to 1 recessive for both (9 : 3 : 3 : 1). But this result is simply the algebraic expansion of the binomial $(3 + 1)^2$.

An example of this type of segregation is shown in Figure 7–3. In *Drosophila*, gray body (really black and yellow banding), *E*, is dominant, and ebony, *e*, is recessive. If gray, normal-winged flies are crossed with ebony, vestigial-winged flies, the F_1 are all dihybrids (*Ee Vgvg*), and they are gray with normal, long wings because of dominance. When gametes are formed, one gene of each pair is included in each gamete, but the gametes include all possible combinations in equal numbers. In this example, gametes formed are of types *E Vg, e Vg, E vg,* and *e vg*. Random fertilization yields an F_2 that consists of 9 *E? Vg?* : 3 *E? vgvg* : 3 *ee Vg?* : 1 *ee vgvg* (the question mark after each dominant gene indicates that the second member of the pair could be either dominant or recessive); in other words, the result is 9 gray, normal : 3 gray, vestigial : 3 ebony, normal : 1 ebony, vestigial.

If three pairs of genes, all on different chromosomes, differ in a cross, then the ratio of phenotypes (appearances) obtained is the expansion of $(3 + 1)^3$, or 27 : 9 : 9 : 9 : 3 : 3 : 3 : 1. The same *principle of independent assortment* applies to any number of gene pairs. However, when a cross involves gene differences for two or more pairs of genes located on the same pair of chromosomes, we expect the number of recombinations of genes to be like that of a monohybrid cross. That is, if a cross is made between two homozygous individuals, *AABB* and *aabb*, then one of the chromosomes of the offspring should contain the two dominant genes, *AB*, its homologue should contain the two recessive genes, *ab*, and these two combinations should be maintained indefinitely. This phenomenon, called *linkage*, is a simple consequence of the fact that the genes are more numerous than the chromosomes. But linkage is not absolute; blocks of material may be exchanged between homologous

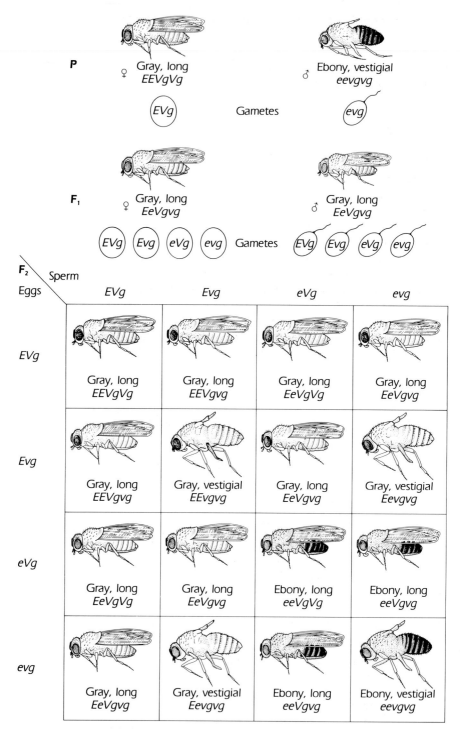

Figure 7–3. *Dihybrid cross between gray-bodied, long-winged and ebony-bodied, vestigial-winged Drosophila. Note that normal wing and gray body are dominant and that the offspring segregate in a ratio of 9 : 3 : 3 : 1. Only females are shown in the F₂ in order to avoid the added complication of sex differences.*

chromosomes. Thus, in the example discussed above, the combination Ab and aB could be formed, though only in a minority of the gametes. This phenomenon is called *crossing-over*.

Sex Linkage

Sex was the first character to be successfully interpreted in terms of the chromosomes. Geneticists found that many species of grasshoppers have 23 chromosomes in males and 24 in females. Thus 11 pairs of chromosomes (autosomes—A) are identical in the two sexes; the twelfth pair is complete in the female but represented by only one chromosome in the male. This is the *sex differential* pair, or the X chromosomes. Consequently, meiosis (maturation divisions) results in eggs all of which are similar, having 11A + X. Sperm, however, are of two classes: 11A + X and 11A + O (no sex differential chromosome). When these two kinds of sperm fertilize the eggs, the parental conditions are reestablished, that is, half of the zygotes have 22A + 2X and become females; the other half have 22A + X + O and become males (Figure 7–4). More commonly, as in *Drosophila* and humans, the X chromosome of

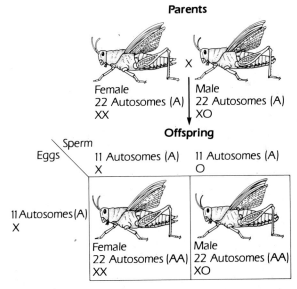

Figure 7–4. *Sex determination in a grasshopper. Females have 22 autosomes (A) and a pair of X chromosomes. Males have 22 autosomes plus an X that has no mate (indicated by 0). All eggs have 11 autosomes (one of each pair) and an X. Sperm are of two types: all have 11 autosomes, but half have an X and half have 0. Fertilization of the eggs by the two types of sperm results in two types of zygotes, 22 autosomes and 2X, which develop as females, and 22 autosomes and an X and 0, which develop as males.*

the male has an unlike mate, the Y chromosome. Thus in *Drosophila* (Figure 7–5), females have 6A + 2X; males have 6A + X + Y. Meiosis results in eggs with 3A + X and sperm of two types, 3A + X and 3A + Y. Fertilization then leads to two equal classes of progeny, 6A + 2X (females) and 6A + X + Y (males).

In general, few mutant genes are known on Y chromosomes. (One such gene in *Drosophila* has a somatic effect and several affect male fertility; in humans, only one Y-linked gene is known with reasonable certainty.) By contrast, X chromosomes have many such genes; in humans more than 100 are known. The pattern of inheritance of such *sex-linked* genes necessarily

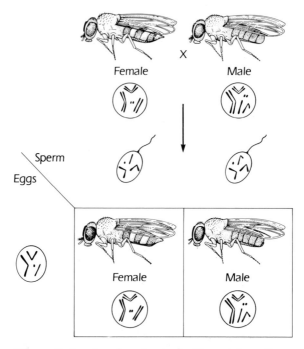

Figure 7–5. Sex determination in Drosophila melano-gaster. *This species has only four pairs of chromosomes. The first, or sex differential pair, consists of a pair of rodlike X chromosomes in the female, an X and a hooked Y in the male. Autosomes (A), as always, are identical in the two sexes: two V-shaped pairs (second and third chromosomes) and the dotlike fourth pair. All eggs contain one chromosome of each pair, including an X. All sperm contain one of each autosomal pair, but half of them have an X, the other half a Y. Accordingly, half of the zygotes should have 6A + 2X and develop as females; the other half have 6A + X + Y and develop as males.*

follows that of the X chromosomes and so is readily identified. Because such genes are unpaired in males, sex-linked recessive genes inherited from the mother show up in male progeny.

Sex-determining mechanisms other than sex-differential chromosomes have also evolved. Mendel confirmed an earlier observation that in the honeybee (and other Hymenoptera, the order of bees, wasps, and ants), males develop from unfertilized eggs (haploid), and females develop from fertilized eggs (diploid). In the parasitic wasp *Habrobracon* (Figure 7–6), P.W. Whiting found a series of multiple alleles for sex determination: X_a, X_b, X_c . . . . Fertilized eggs are usually heterozygous for the sex-determining alleles, as X_bX_g, and they develop as females; whereas unfertilized eggs develop as males. Rarely, however, a fertilized egg is homozygous, as X_eX_e, and these develop as males of very low viability. The selective force that favors multiple alleles is clear: the greater the number of sex-determining alleles in a population, the lower the probability of defective males (or conversely, the higher the probability of normal haplo-diploid sex determination). Diploid males of low viability have since been found in the honeybee also.

Environmental factors, too, may determine sex. Sexual dimorphism is extreme in *Bonellia viridis*, an echiuroid worm (Figure 7–7). (This small group is usually lumped with the annelids, but it is best treated as a minor phylum.) Females have a body about three centimeters long, complex organ systems,

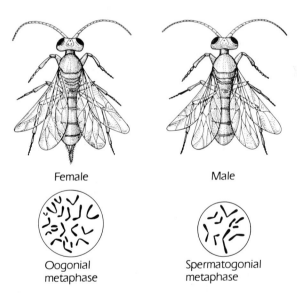

Female   Male

Oogonial metaphase   Spermatogonial metaphase

Figure 7–6.   Habrobracon juglandis. *Female with ovipositor (left); and male (right). Below each is a metaphase plate, an oogonium with 20 chromosomes, and a spermatogonium with only 10 chromosomes.*

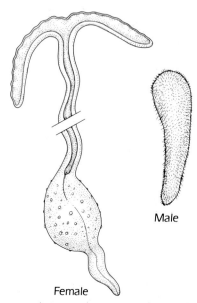

Male

Female

Figure 7–7.   Bonellia viridis. *The body of the female is two to three centimeters long, but the proboscis may approach two meters in length! In contrast, the male is about the size of a large* Paramecium *(up to 0.3 mm) and lives as a parasite of the female.*

and a proboscis for feeding that may approach two meters in length. In contrast, males are comparable in size to *Paramecium* (ca 0.2mm), have rudimentary organ systems, and live as parasites on females. Eggs of *Bonellia* have the genetic capacity to develop as either sex. The larvae soon settle on a substrate and metamorphose into adults. Those that settle on the ocean floor become females; those that settle on the proboscis of an established female become males, probably because the female secretes a substance that controls the development of the newly arrived larva.

Environmental sex determination occurs even among the vertebrates. Sex differential chromosomes are known among snakes, but they do not exist in turtles or crocodilians. Nest temperature during a critical period in development determines sex in these reptiles. M.W.J. Ferguson and T. Joanen studied the alligator and found a critical period from the seventh to the twenty-first day of a sixty-five-day developmental period. Nest temperatures below 30°C resulted in females, nest temperatures 34° or higher resulted in males. Intermediate temperatures gave mixed broods of various proportions. In turtles, J.J. Bull found similar facts, with the middle third of incubation being the critical period. In this group, low temperatures favored males, higher temperatures favored females.

Quantitative Inheritance

We should also give special mention to the inheritance of quantitative traits, which include all of those individual differences that must be defined by measurement rather than qualitatively. Examples include size, proportions of parts, intensity of color, rate of production of a vitamin, concentration of a protein, and many others. The genes that influence these traits are inherited in the usual Mendelian fashion, but swarms of pairs of genes collaborate in the determination of each character. The effect of each such gene is small, so that variation in a heterozygous population appears to be continuous. Also, dominance is often lacking, so that there is a simple additive effect of the genes present. (That is, the effect of two "plus" genes in one pair is the same as that of one "plus" gene in each of two pairs.) Because of these characteristics, quantitative inheritance can only be studied with the tools of statistics.

All of the basic phenomena of inheritance can be understood in terms of these few principles. The reader should review them and the entire field of elementary genetics thoroughly before going further in the study of evolution.

THE GENE THEORY

W. Johannsen named the hereditary unit the *gene* for convenience only, with no theoretical implications. The name soon became closely associated, however, with the theory of T.H. Morgan that the genes are corpuscular bodies in the chromosomes, arranged in linear order like beads on a string, that each gene is separated from all of the others and different from all of the others in substance, and that all of the genes of a particular chromosome are held together by an indifferent substance. This theory was based upon three types of evidence: the fact of mutation; the fact that a linear order of genes in the chromosome can be established by crossover tests (Chapter 8); and the fact that, once this order is established, it can be reshuffled by subsequent chromosomal rearrangements (Chapter 8).

This *morphological* concept of the gene dominated genetic thinking for many years. Its boundaries were soon blurred by position effects (Chapter 8). Current research emphasizes the gene as a biochemical and functional unit, and its morphology no longer excites interest.

Biochemistry of Chromosomes

Since the end of the nineteenth century it has been known that the chromosomes consist of basic proteins and nucleic acid; later, neutral and even acid proteins were reported. The proteins are polymers of amino acids of which there are twenty kinds. These amino acids are joined by peptide linkages, a molecule of water splitting out between the acid group of one and the basic (amino) group of the next to form a saltlike bond (Figure 7–8). This bonding

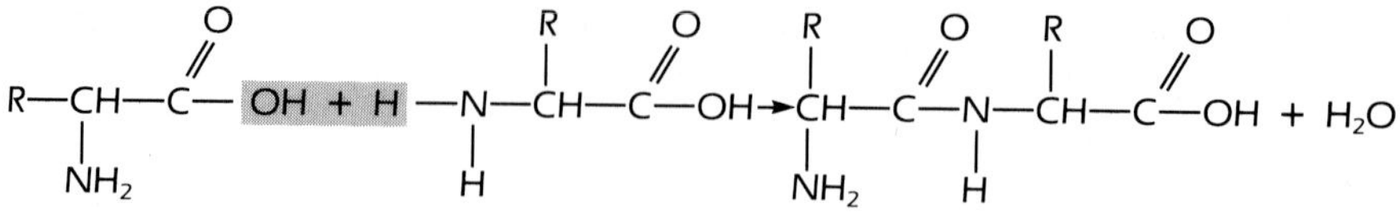

Figure 7-8. *Formation of the peptide bond.*

leaves a free acid radical at one end and a free amino radical at the other, each capable of peptide linkage to yet another amino acid. As a result, protein molecules may be immense polymers; they were considered to be the largest and most complex compounds. In contrast, nucleic acids seemed to be simple and monotonous. The unit of structure is the nucleotide, consisting of a five-carbon sugar conjugated with a phosphate radical and with a basic group, either a purine (adenine or guanine) or a pyrimidine (cytosine or thymine). Researchers at first thought that the nucleic acid molecule was simply a large polymer in which the four nucleotides were repeated indefinitely in much the same sequence. If we let S stand for the sugar, P for phosphate, and A, T, C, and G for the bases, then the structure was envisioned like this:

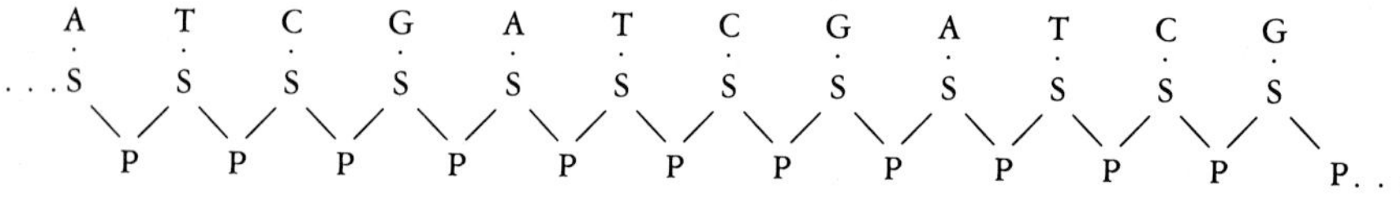

Two kinds of nucleic acid were distinguished, deoxyribonucleic acid (DNA) and ribonucleic acid (RNA), the latter with ribose as its sugar and associated with the cytoplasm; the former with less fully oxidized ribose (hence deoxy) and associated with the nucleus. This supposed rather invariant structure of the nucleic acids (which proved to be erroneous) contrasted with the rich variety of proteins and convinced most geneticists that the gene must be sought in the proteins.

DNA as the Gene

A sequence of events, which began in the late 1920s, changed that picture radically. First, mice were injected with a mixture of live, nonpathogenic and dead, pathogenic *Pneumococcus*. This treatment should have been harmless, but the mice died. Evidently, the pathogenicity of the killed strain was introduced into the live but formerly nonpathogenic strain. This change was called *transformation,* and a long search for the transforming principle began. Various protein fractions, lipids, carbohydrates, trace elements—all failed. But when DNA was tested, it did transform some of the live bacterial cells. This fact strongly suggested that DNA was genetically more active than had been supposed.

Experiments with bacteriophages (viruses that infect bacteria) had similar import. A bacteriophage may attach to the bacterial chromosome, and after the destruction of the bacterial cell, the phage may carry a bit of bacterial DNA into the next cell that it infects. If the two bacterial strains differ genetically, traits of the first may then appear in the second. This process is called viral *transduction*. In both transformation and transduction, a segment of DNA from a donor strain is introduced into the cells of a recipient strain.

Another approach was to separate tobacco mosaic virus into protein and nucleic acid fractions, then attempt to infect tobacco leaves with each. The protein fraction never produced symptoms. It was difficult to produce an infection with the nucleic acid fraction; yet it could be done, and it showed the typical symptoms. Apparently, the protein functioned to facilitate entry of the nucleic acid into the cell. When protein and nucleic acid fractions were mixed, they self-assembled to form fully infective virus particles. The final proof was obtained by the use of radioactively marked bacteriophages. When radio-sulfur, which is incorporated into protein but not into nucleic acid, was used, all of the radioactivity remained *outside* the bacterial cells after infection by the virus. But when radio-phosphorus, which is incorporated into nucleic acid but not into protein, was used, all of the label was found *inside* the bacterial cells after infection. As reproduction of the virus occurs inside the bacterial cells, this outcome shows that only the nucleic acid part of the virus is needed for viral heredity.

Thus, by the late 1940s, DNA was strongly implicated in the gene, but full acceptance was delayed because the structure of DNA, as it was then known, did not provide a basis for the necessary variability nor for replication. Researchers in many laboratories then attacked the problem of the structure of DNA. In 1952 J.D. Watson and F.H.C. Crick, working at Cambridge, achieved success. Their principal data were of three sorts. First, many analyses of DNA had shown that the quantities of adenine and thymine were always equal, as were those of cytosine and guanine. Second, X-ray diffraction spectra of DNA crystals, principally in the laboratory of M.F.H. Wilkins and R. Franklin, showed a repeat every 3.4 Å and a major repeat every 34 Å. Finally, DNA "melts" at a moderate temperature (below 100°C), a fact that suggests the presence of many hydrogen bonds (which are very weak).

The Watson-Crick Model of DNA

Watson and Crick found that these diverse data all fit if it were assumed that the DNA molecule consists of *two* long polymer chains spiraled around one another, with the nucleotides spaced 3.4 Å apart and rotated 36° from one to the next, so that a complete turn of the spiral occurs every ten nucleotides or every 34 Å. The two spirals are held together by hydrogen bonds between the bases, so that base pairs can be visualized as the rungs of a spiraled ladder (Figure 7–9). Of the bases, however, purines are larger than

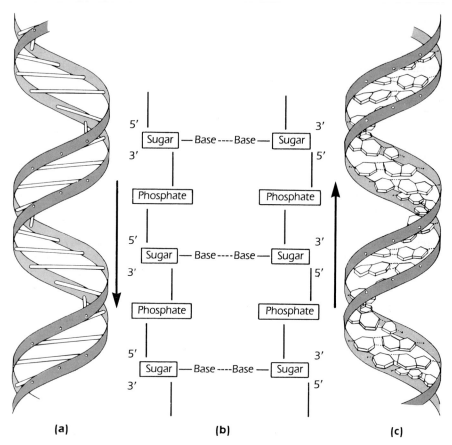

Figure 7–9.   *Watson-Crick model of DNA.* (*a*) *Diagram of the double helix.* (*b*) *Components of a small segment of DNA, including only three base pairs.* (*c*) *Stereochemical relationships of the base pairs.*

pyrimidines, and so the rungs of the ladder will "fit" only when combinations of the same dimensions occur. Actually, adenine always pairs with thymine, cytosine with guanine. Because the molecule is directional, four different rungs are possible: A-T, T-A, C-G, and G-C. This pairing immediately explains the parity of A and T and of C and G in analyses.

This double-helix structure soon suggested a probable method of replication of DNA. If the two strands should separate in a medium containing the necessary nucleotides, then each of the separated strands could serve as a template upon which a complementary strand could be synthesized by attraction of A to T, T to A, C to G, and G to C. Notice that the new strand is not the duplicate of its template, but its *complement*. For example, if a template strand has bases in the sequence AAATGCATG, the strand synthesized upon it will have the sequence TTTACGTAC. The result is the formation of two double helices, each with one old and one new strand, each identical to the other and to the parent double helix—all of the requirements of genetic reproduction. The double-helix structure of DNA and its manner of reproduction have been fully verified experimentally. For this discovery, Watson, Crick, and Wilkins were awarded the Nobel Prize for Medicine or Physiology in 1962.

Coding

Soon after the publication of Watson and Crick, the physicist and cosmologist G. Gamov suggested that sequences of three nucleotides, each with its run of three base pairs, might encode the information for incorporation of the amino acids into proteins. His reasoning was beautifully simple. Natural proteins include twenty amino acids. If a single base pair were the unit of coding (the *codon*), only four amino acids could be coded. If two base pairs comprised the codon, then $4^2 = 16$ amino acids could be coded—still not enough. But if three base pairs comprised the codon, then $4^3 = 64$ codons would be possible—more than enough.

Protein Synthesis

Subsequent research has proven the Gamov hypothesis, and the code has been broken. But if one codon is assigned to each amino acid, what of the other forty-four codons? Most of the amino acids are represented by more than one codon, and three codons serve as punctuation in the code. Only two amino acids, methionine and tryptophan, are encoded by only one codon, and several are encoded by as many as six! Three codons do not specify amino acids but serve to terminate a polypeptide chain. When the gene functions in the synthesis of a protein (Figure 7–10), the two strands of DNA separate in the region which determines that protein, and one of the two strands, called the *sense strand,* serves as a template for the synthesis of messenger RNA (mRNA). In this synthesis, the rule of complementarity is followed, with one difference: where A occurs in the DNA of the sense strand, uracil (U) rather than T is incorporated into the mRNA, which remains single-stranded. By custom, the codons are defined by the sequence in the mRNA. For example, one of the codons for phenylalanine is UUU, which immediately implies AAA in the sense strand and TTT in the anti-sense strand of the DNA. Table 7–1 provides the entire code.

The mRNA then passes into the cytoplasm, where it attaches at one end to one of the numerous ribosomes, double particles of RNA and protein, which are intermediaries in the synthesis of proteins. In the cytoplasm are an abundance of amino acids, each molecule attached to a transfer RNA (tRNA) molecule. Each tRNA molecule is characterized by a triplet called an *anticodon,*

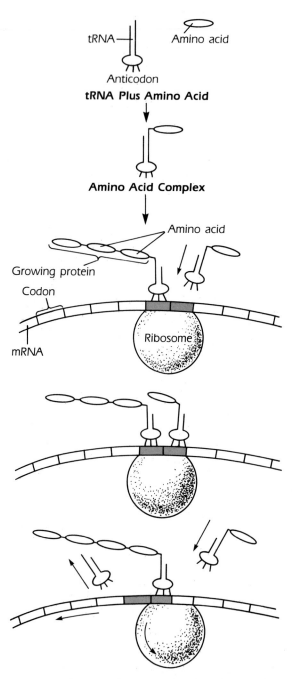

Figure 7–10. *Protein synthesis. Starting from the top, a tRNA molecule couples with an amino acid molecule. Then the tRNA collides with a ribosome and reacts with a codon on an mRNA molecule that is attached to the ribosome. Finally, the amino acid forms a peptide linkage to a polypeptide chain, and the preceding tRNA is released as the next one approaches.*

Table 7–1. Codons and the Corresponding Amino Acids

Base 1	Base 2				Base 3
	U	C	A	G	
U	UUU UUC } Phenylalanine UUA UUG }	UCU UCC UCA UCG } Serine	UAU UAC } Tyrosine UAA Ochre UAG Amber	UGU UGC } Cysteine UGA Stop UGG Tryptophan	U C A G
C	CUU CUC CUA CUG } Leucine	CCU CCC CCA CCG } Proline	CAU CAC } Histidine CAA CAG } Glutamine	CGU CGC CGA CGG } Arginine	U C A G
A	AUU AUC AUA } Isoleucine AUG Methionine	ACU ACC ACA ACG } Threonine	AAU AAC } Asparagine AAA AAG } Lysine	AGU AGC } Serine AGA AGG } Arginine	U C A G
G	GUU GUC GUA GUG } Valine	GCU GCC GCA GCG } Alanine	GAU GAC } Aspartic acid GAA GAG } Glutamic acid	GGU GGC GGA GGG } Glycine	U C A G

The table presents all possible permutations of the four bases of RNA taken in groups of three, together with the amino acid encoded by each. The three italicized letters of each amino acid are commonly used as abbreviations. Note that only two amino acids, methionine and tryptophan, are uniquely determined; all others are represented by two or more codons. Usually, codons for a given amino acid begin with the same two bases. For example, the four codons for alanine all begin with GC. However, three amino acids (leucine, serine, and arginine) have two different initial sequences.

which can form a temporary bond to the corresponding codon for its particular amino acid (for example, AAA to UUU). The tRNAs, with their attached amino acids, are in Brownian movement, and they constantly bump the ribosomes, thereby testing the codon-anticodon reactivity. Thus, if anticodon AAA contacts codon UUU on a ribosome, they bind; and the phenylalanine that is attached to the tRNA at once forms a peptide bond to the amino acid that preceded it. It will then release its tRNA, and the ribosome will move to the next codon in the series. Thus, long protein chains can be assembled, with the amino acids in the same sequence as the codons in the mRNA and in the DNA. A segment of DNA that encodes one polypeptide is called a *cistron,* a term often used synonymously with the term *gene.*

An interesting question is whether the genetic code is universal—that is, the same for all organisms. At first, it appeared to be universal, for tests in a wide variety of organisms gave identical results. Then research showed that a few codons are read differently in the mitochondria of some organisms. Thus, the code is nearly, but not completely, universal.

Genes in Pieces

That the gene must be unitary seems almost self-evident; and for procaryotes this is strictly true. When a eucaryotic gene is transcribed into mRNA, however, segments of the mRNA molecule are often removed enzymatically, and the retained segments are then spliced to form a much shorter mRNA molecule, which serves as a template for protein synthesis. The eliminated segments are called *introns;* the retained segments, *exons*. Nor are the introns restricted to one per gene. There are two introns in the gene for human insulin, and P. Chambon has found as many as nineteen in a gene for albumen! Evidently, the gene can be highly fragmented. How does this fact relate to evolution? One possibility is that this type of structure may facilitate recombination of variants of the exons, thus accelerating evolution. Another is that, even though the introns are eliminated, they may be necessary for translation of the exons into protein. Decisive tests of these hypotheses are still in the future. We return to this subject in Chapter 9.

Molecular Genetics and Morphological Characters

We have examined the fundamental character of the genetic system, which is the functional basis of evolution. Yet in the preceding chapters, discussion centered not on amino acids and proteins, but on patterns of symmetry, length of limbs, acuity of senses, and other traits that are not ordinarily expressed in molecular terms. How can this gap be bridged? In most instances, we do not yet have an adequate answer to that question, but biochemists are confident that facts of morphology and behavior as well as physiology must ultimately be referable to a molecular basis. Because most proteins have enzymatic activity and because biological reactions are always under enzymatic control, morphology as well may have its ultimate explanation and control in the mechanism of protein synthesis.

Gene Number and Gene Size

Attempts have been made to estimate the number of pairs of genes for various organisms. Obviously, a direct estimate of the number of genes is not available for any complex organism, both because a complete study of all hereditary characters of such an organism has never been made and because a particular gene is identifiable by genetic methods only when it is available in more than one form—in other words, when it has mutated to form two or more alleles.

Estimates for the fruit fly, *Drosophila melanogaster,* one of the best-known organisms genetically, range from 5,000 to 15,000. J. Belling estimated about 2,200 genes for *Lilium,* and C. Stern estimated not less than 5,000 nor more than 120,000 genes for humans, with 20,000 a fairly probable intermediate choice. Phage T4 has a chromosome 5.5×10^5 Å in length, or 1.6×10^5 base pairs in length. If we assume an average protein size of 300 to 350 amino acids, which is reasonable, then the average cistron would be about 1,000 base pairs in length, and the chromosome could accommodate 160 such genes. About half of this number of genes has been studied and mapped for T4. Where a specific protein has been fully analyzed, the size of the corresponding gene can be exactly deduced from the parameters of the codons and the DNA. For instance, cytochrome c of wheat germ is 112 amino acids long; hence it is encoded by a cistron 336 base pairs in length, or $336 \times 3.4 = 1142.4$ Å, a little over 0.001 micron. This protein is rather small; the estimate for T4 is more typical.

Estimates of gene size for typical genes of higher organisms have been based on the volume of the chromosomes (especially the euchromatic parts) divided by the estimated number of genes. By this method, J.W. Gowen and H. Gay calculated the average size of the genes of *Drosophila* to be 1×10^{-18} cubic centimeters. D.C. Pease and R.F. Baker, using the electron microscope, observed in the salivary gland chromosomes of *Drosophila* leaf-shaped bodies, which they believed to be the genes. These bodies vary in size by a factor of about 3, but the average is about 1×10^{-17} cubic centimeters. As the salivary gland chromosomes are giant chromosomes to begin with, the mere fact that these bodies are ten times the size estimated by Gowen and Gay cannot be regarded as conflicting with their estimate. But neither is there any proof that these bodies actually are the genes.

Repetitive DNA

Such estimates of gene size are maximal for another reason: the amount of DNA in a typical eucaryotic nucleus appears to be much greater than the amount needed to encode all of the structural and regulatory genes of the organism. Biochemical studies indicate that the typical structural gene sequence is unique. Genes that encode the transfer and ribosomal RNAs may be repeated moderately, as are those genes that encode globins, histones, and some other abundant proteins. But extragenic DNA may be highly repetitive. R. Britten and D. Kohne estimated that some DNA sequences may be repeated hundreds of thousands of times! Only the unique DNA and some of the moderately repeated sequences are transcribed into mRNA and translated into protein. Moderately repetitive sequences may be transcribed into tRNAs and rRNAs. Some of the highly repetitive DNA is used structurally in the centromeric regions of the chromosomes, but it is not transcribed. Possibly some intermediately repetitive sequences are transcribed but not translated. (We return to the repetitive DNA in Chapter 9.)

MUTATION OF GENES

Abundant evidence indicates that the genes mutate to produce permanently inheritable alleles. We must, therefore, study this source of variability for its bearing on evolution. A mutation is a permanent change in a gene. Alleles exist only because the original wild-type gene has mutated at some time. Like the original gene, the mutant is recognized by the character it causes.

Mutation in Nature

Critics of genetics long held that the mutants with which geneticists worked could not be significant because such mutants were not really natural phenomena but a sort of degenerative outcome of the laboratory environment. This idea has been abundantly disproved; many investigators have found, in nature, mutants identical with or similar to those dealt with by laboratory biologists. The original *Drosophila* mutant (white eyes) was, as a matter of record, captured in nature by the entomologist F.E. Lutz at about the same time that it arose in the laboratory stocks of T.H. Morgan. Goldschmidt found many mutants in the wild populations of *Drosophila* near Berlin. S.S. Chetverikov inbred the offspring of 239 wild *Drosophila melanogaster* from southern Russia, and 32 recessive mutants segregated out. N.P. Dubinin collected the same species in several localities in the Caucasian Mountains. On inbreeding them, Dubinin found that the incidence of lethal genes varied from 0 to 21.4 percent in different localities, and visible mutations varied from 3.9 to 33.1 percent. Many of these had minor effects, such as a slight reduction in size of bristles.

E. Baur studied wild snapdragons (*Antirrhinum*) and found that about 10 percent of the plants showed at least one mutant. Dobzhansky showed that in wild populations of *Drosophila pseudoobscura* 75 percent of the chromosomes showed at least one mutant. L.R. Dice found that wild populations of the deer mouse *Peromyscus* always have many mutants. We may then regard as established that mutation is a normal phenomenon in nature.

An enormous amount of genetic variability has been demonstrated by electrophoretic sampling of natural populations, a method introduced by R.C. Lewontin and J.L. Hubby. Proteins are electrically charged molecules, and substitution of one amino acid for another may change the charge. Because of this fact, variants of a given protein, called *polymorphs,* migrate at different rates in an electrical field. Electrophoresis, a very sensitive method for separating polymorphic proteins, exploits this property. In electrophoresis, a sample of a protein solution is placed on a gel between electrodes. After allowing sufficient time for migration, a stain is applied to make the separated protein spots visible.

Lewontin and Hubby studied eighteen different proteins of *Drosophila* and found that more than a third of these were polymorphic. Further, they easily identified heterozygosity because each allele produces its own variant of the protein under study. The average individual was heterozygous for 12 percent of the genes tested. This figure, large as it is, is an underestimate because some amino acid substitutions are electrophoretically "silent"—that is, they do not change the charge of the molecule.

Similar data have been demonstrated for many species, such as fruit flies, crabs, mice, and humans. The extent of protein (and therefore of genic) polymorphism ranges to as high as 86 percent! This extensive polymorphism must have originated through mutation.

Rate of Mutation

The rate at which mutation occurs has also been the subject of many studies. W.P. Spencer has shown that the rate of mutation in *Drosophila* differs from time to time. These differences he attributed to environmental influences. M. Demerec, finding that the rate varied in diverse genetic strains of *Drosophila*, proposed that the rate of mutation is itself controlled by specific genes. He found that the rate of lethal mutation varied from one in every hundred chromosomes to one in every thousand chromosomes. P.T. Ives identified such "mutator" genes in wild populations of *Drosophila*, genes that increase the mutation rate as much as tenfold. Baur found that 5 to 7 percent of the progeny of normal *Antirrhinum majus* showed at least one mutant, whereas he observed none in *A. siculum* during twenty years of breeding. These data were explained on the basis of a great difference in mutation rate in these two closely related species. L.S. Stadler measured the frequency of the appearance of eight mutants in maize. When he noted the number of times each occurred per million gametes, he obtained the following series: 492, 106, 11, 2.4, 2.3, 2.2, 1.2, and 0. It is clear, then, that different genes mutate at different rates even within the same strain. This conclusion is supported by laboratory studies of *Drosophila*. Some mutants have appeared many times in laboratory stocks, some rather rarely, and some are known only from single records.

These mutation rates suggest a major problem in the study: the lower the rate, the larger the sample that must be studied in order to get adequate data. In Stadler's corn, *R* mutated 492 times per million gametes, so that samples of only 10,000 seeds should average nearly five mutations. But what of *Sh*, which showed a rate of only 1.2 per million gametes? Clearly, a sample of 10,000 seeds would be unlikely to include even one mutation. With most eucaryotes, making adequate studies of mutation is difficult and costly. Bacterial cultures, in contrast, produce immense numbers of cells in a short time and at a much more modest cost. Consequently, procaryotes are often preferred for studies on mutation.

We now envision the mode of action of *mutator genes* in terms of the physiology of DNA replication. At replication, the two strands of the double helix separate, and each serves as a template upon which a new strand, complementary to its template, is synthesized. This process is catalyzed by an enzyme, DNA polymerase, which not only catalyzes the polymerization of the new chain of nucleotides but also "proofreads" for the correctness of the complementarity of the new strand as compared to its template. If it finds an error, it removes the "wrong" nucleotide and substitutes the correct one. Mutator genes result from mutations that produce defective DNA polymerases, enzymes that proofread with reduced efficiency, and consequently cause an increased frequency of mutation.

Direction of Mutation

It is commonly said that the direction of mutation is random—that is, that chance alone determines in which of an infinity of possible ways a particular gene will actually mutate. This idea is true in the sense that the environment does not cause the appearance of mutants that are appropriate to it. In all environments, both natural and experimental, the majority of the mutations are disadvantageous: their prospective fate is elimination by natural selection. Yet, a priori, we would expect a limited number of ways in which a particular gene can mutate. For any known chemical substance, certain classes of reaction are possible and certain classes are not.

Studies in biochemical genetics have shown that gene mutation usually consists of the replacement or deletion of a single base pair in a codon. Clearly, in such a one-step modification of a codon, a given amino acid can be replaced only with certain others. This fact may partially explain why certain mutations recur very commonly.

Mutation and Populations

Mutation, then, is a universal phenomenon in the living world. The fate of a newly mutated gene depends both on chance factors and on its selective value. Purely by chance, a potentially valuable mutation may be lost in a polar body, or it may be carried by a sperm that fails to fertilize an egg. Once established in a zygote, dominant genes are subject immediately to the test of natural selection, and a favored new allele may spread in the population. Severely disadvantageous alleles will generally be eliminated quickly, but those that are moderately disadvantageous may be retained in the population for several to many generations. Occasionally, they may even become advantageous under later conditions. Recessive genes will be subject to selection only when they become homozygous, hence such mutations may spread widely in the heterozygous condition. Selection acts not only upon the newly mutated genes but also upon the store of genetic variability, which, in every natural population, has been passed on from earlier generations. Of course, mutation is the ultimate source of all genetic variability.

Early students of the genetics of natural populations expected each species to have a normal genotype, often called the *wild type,* from which any mutations were more or less exceptional deviants. They believed the typical member of a species to be homozygous for most of the wild-type alleles, with perhaps a few mutant alleles included. This was the *classical* theory of the genetic structure of populations.

A long series of researches, beginning with Chetverikov's studies of morphological characters in *Drosophila* and culminating in the electrophoretic study of biochemical mutants in many species, have demonstrated that the wild type is illusory, for the typical wild population is highly heterozygous, with a balance between the origin of new variability by mutation and the elimination of alleles by selection or by chance factors. This view is termed the *balance* theory of the genetic structure of populations, and it has achieved general acceptance.

Experimental Production of Mutation

As early as 1927, H.J. Muller showed that mutations were produced in *Drosophila* at several hundred times the normal frequency if the gonads were X-rayed. He selected sex-linked lethal mutations for special study because of the ease with which they can be identified. Muller and others have since shown that any high-energy radiation will produce mutations. In 1947, Muller was awarded the Nobel Prize in Medicine on the basis of this work. With any type of radiation, the rate of induced mutation is directly proportional to the dosage, but it is independent of the rate of administration of the radiation. Chemical mutagens were sought unsuccessfully for many years until, during World War II, C. Auerbach demonstrated that mustard gas is as effective as radiation. Subsequently, researchers have found mutagenic activity in such diverse substances as urethane, formaldehyde, peroxides, manganese chloride, aluminum chloride, purines, and pyrimidines, and the list may well become indefinitely long.

When Muller's work was first reported, the question was raised whether cosmic radiation or some other naturally occurring radiation was responsible for mutation in nature. Alpine plants tend to differ greatly from one population to the next. As ultraviolet radiation is more intense at high altitudes, the suggestion was made that ultraviolet-induced mutation might explain the diversity of alpine plants. However, even in the absence of a high mutation rate, this phenomenon could be explained by selection of different mutations occurring in isolated areas or by chance fixation of different mutants (genetic drift, see Chapter 9). E.B. Babcock tried raising *Drosophila* in areas of high and low natural radiation. He did not observe significant differences in mutation rate. Hence it must be admitted that studies of experimental mutation have not yet led to the understanding of naturally occurring mutation. Chemical mutagens may be of great importance in nature.

GEOGRAPHIC SUBSPECIES AND NEO-DARWINIAN EVOLUTION

An important aspect of natural species is the fact that the various types of variability occurring within a species are not scattered evenly over its entire range; rather, local populations, more or less isolated from their neighbors, show distinctive patterns of the variable characters of their species, with the result that they may be defined as subspecies. They are fertile in crosses, and intergradations between adjacent subspecies commonly occur wherever their ranges meet. Nevertheless, the several subspecies of any one species do inhabit breeding ranges that are mutually exclusive for the most part. Closely related species may have identical ranges, but subspecies replace one another geographically. The whole series of geographic subspecies for any species is called a *Rassenkreis* (German for "group, or circle, of races").

Wherever subspecies are distributed over an area that presents a progressive change in some physical feature, such as mean annual temperature, some of the differences in the subspecies are likely to show a progressive change also. The occurrence of such *clines,* in which variations of characters parallel progressive change of climate or topography, strongly suggests that the clinal characters are of adaptive value to the various subspecies. A limited amount of exchange of breeding individuals occurs between adjacent local populations, and wherever the ranges of two subspecies meet, there is some interbreeding. As a result, *gene flow* takes place throughout the species; and E. Mayr, especially, has emphasized this as a factor that tends to maintain the integrity of the species as the unit of evolution.

Thus, W.W. Alpatov, for example, found that the average size of honeybees increased progressively from southern to northern Europe, while the lengths of their legs and tongue decreased along the same cline. Similarly, G. Turesson has shown that plants commonly show definite character complexes according to the type of habitat in which they live. Thus in *Primula* (primrose) there are *ecotypes* adapted to alpine habitats, ecotypes adapted to meadowlands, and ecotypes adapted to as many more types of habitat as the plant inhabits naturally (Figure 7–11). We can scarcely doubt the adaptational value of a character complex that is always found in plants that live in a particular type of environment.

The relationship of clines to subspecies is not a simple one. A cline will usually be divided into subspecies only if there are abrupt discontinuities along the cline. W.L. Brown argues that clines are the primary expression of intraspecific variability and that "subspecies" are illusory results of the intersection of clines for different characters. J.A. Endler has simulated clines of *Drosophila* in the laboratory and has included the introduction of measured levels of selection and controlled amounts of gene flow among the experimental demes. He finds that selection is much more important than is gene flow in modifying the phenotype of the experimental population. Similarly, P.R. Ehrlich and P.H. Raven studied a series of natural populations of both plants and animals and concluded that the importance of gene flow had been overestimated.

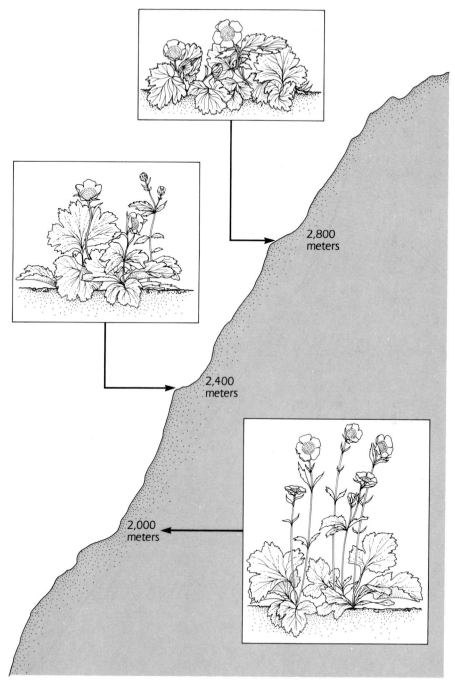

Fi*gure* 7–11.   *Ecotypes of* Geum montanum *at different altitudes in the Tirol.*

Sometimes factors leading to such cline formation can be summarized in concise rules. *Gloger's rule* states that animals in cool, dry regions tend to be lighter in color than those in warm, humid regions. *Bergmann's rule* states that, in a given species or genus, populations in cooler climates have a larger mean body size than do populations in warmer climates. This rule is believed to depend upon the fact that a large body has a smaller surface area per unit volume than does a small body. Thus, conformity to the rule promotes heat conservation in the north and heat conduction in the south. A corollary is *Allen's rule,* which states that extremities (limbs and tails) tend to be smaller in cooler climates than they are in warmer climates. Again, this difference is interpreted in terms of restriction (north) or extension (south) of radiating surfaces.

P.F. Scholander and other comparative physiologists doubt that the observed size variations are sufficient to account for climatic adaptation. Arctic mammals and birds have effective insulation of fur or feathers as well as circulatory control of heat loss, which permit them to maintain the body temperature against severe cold by basal metabolism alone. The less-protected extremities are physiologically adapted to function well even when thoroughly chilled. Scholander believes that these findings invalidate Bergmann's and Allen's rules. W.A. Calder and J.R. King reviewed the literature on ecogeographic variation in birds and found many examples in which correlation of size of body and appendages with climate showed a previously unsuspected degree of precision. They concluded that, although physiological factors may be predominant, nonetheless selection for size commonly acts through the thermoregulatory consequences of size as predicted by Bergmann's (body size) and Allen's (length of appendages) rules.

Subspecies need not occur in clines. That is, if a series of subspecies, A, B, C, D, and E occur in that order over the geographic range of the species, A might resemble B most closely in size, and it might resemble D most closely in color. In a third character, it might resemble C most closely, and so on. In many cases, subspecies can be defined only by a statistical analysis of several variable characters in the various populations of a species. Subspecies of North Sea herring, for example, have been defined by statistical studies of number of vertebrae, number of scales in the lateral line, number of rows of scales above and below the lateral line, and many other variable characters. The resulting statistics show that the averages and standard deviations for each school are typical and differ from all others, thereby justifying their description as subspecies.

Darwin had already spoken of subspecies as "incipient species [which] . . . become ultimately converted into good and distinct species . . . by . . . natural selection." As already mentioned, Darwin realized that the weakest point in his theory was the lack of knowledge of heredity. The dominant school of evolutionary thinking today is based upon the Darwinian principles of the prodigality of nature, variation, and natural selection. But to them is added the interpretation of variation in terms of the theory of the gene, and the study of populations and changes in gene frequency with the tools of statistics, which, like genetics, did not exist in Darwin's time.

In addition, evolutionary processes are strongly influenced by ecological, physiological, structural, and chance factors. This viewpoint has been called neo-Darwinism, the synthetic theory, and even the biological theory. It was founded on the pioneer work of S. Wright, S.S. Chetverikov, R.A. Fisher, and J.B.S. Haldane and was further developed by the zoologists T. Dobzhansky, J. Huxley, E. Mayr, and G.G. Simpson; by the botanists E.B. Babcock and G.L. Stebbins; and by many others. Since the 1940s, this modern synthesis has dominated evolutionary studies.

For the neo-Darwinians, then, subspecies are, at least potentially, incipient species. Each subspecies of a particular species is characterized by a particular complex of gene-determined characters, and these various genes are derived by mutation from an originally more uniform progenitor. The genes that any particular subspecies possesses are not peculiar to it but are found in varying percentages in different subspecies. It is the frequencies of genes that are unique for each subspecies, not the individual genes. Such gene combinations are the materials upon which selection acts. Within such subspecies, mutation continues at random, with the result that different alleles may arise in different subspecies. Subspecies are usually at least partially isolated from each other geographically. Other types of isolation may also arise, for example, physiological or ecological isolation. Within such isolated subspecies, new mutations, as they arise, cannot spread throughout the species. Thus, by the gradual accumulation of mutations, such isolated subspecies may "become ultimately good and distinct species" (Darwin).

The spectrum of mutation extends from gross defects such as lethality or crippling, through moderate effects like color change or change in bodily proportions, to minor changes detectable only by special methods. Neo-Darwinians generally emphasize the last type of change for several reasons. First, most of the larger mutations are plainly abnormalities, which would be eliminated by natural selection. This may be a consequence of the fact that the more-favorable large mutations, which must have recurred many times in the past because of the high natural mutation rate, have already been incorporated into the normal genotype of the species. Second, although unproven, it is generally accepted that a species can harmoniously incorporate into its genotype a series of small mutations more readily than it can a single, equivalent, larger mutation. Most important, however, is the fact that subspecies and closely related species generally differ from each other in a series of quantitative traits, such as size, proportions, intensity of color, or extent of a pigmented area. As explained above, such traits are generally inherited upon the basis of swarms of genes, each with a very small individual effect.

SUMMARY

The modern evolutionary synthesis is plausible in every detail, and it has the advantage of combining the greater part of historical Darwinism with the main aspects of genetics, ecology, and the rest of the biological sciences. Restated from a different point of view, the origin of variability is the first of the major problems of evolution. Gene mutation and chromosomal mutation (see Chapter 8) are the only demonstrated sources of inheritable variability. An average mutation rate is 1/100,000 gametes per gene.

Since some have doubted that so rare a phenomenon could produce the range of diversity that is the product of evolution, let us consider an example in detail. A codfish, a typical vertebrate, has about 20,000 pairs of genes. Although each gene has its own rate of mutation, their separate probabilities may be added together to get the collective probability that *some* gene will mutate. The result is an overall probability of 1 in 5 that any particular gamete will carry some new mutation. A codfish spawns in excess of 10,000,000 eggs; and so about 2,000,000 of these may be expected to carry one or more newly mutated genes. If we make the plausible but gratuitous assumption that only 1 mutation in 10,000 is potentially valuable to the species, the spawn of a single codfish could include as many as 200 eggs in which mutation had produced alleles of value to the species in its evolution—that is, in its continual adaptation to the changing conditions of life. It is plausible, then, that mutation may produce enough hereditary variability to satisfy the requirements of evolution.

The variant genes that are formed by mutation are reshuffled into new combinations by the mechanism of sexual reproduction. For most of the mutants, the prospective fate is elimination by natural selection, but the occasional mutant that has adaptive value will tend to spread in its population, and chance factors may also favor some alleles. Thus, the species tends to approach a state of adaptation to the ever-changing conditions of life.

A typical species, however, is distributed over a wide area characterized by quite different conditions. The deer mouse, *Peromyscus maniculatus,* for example, ranges over most of the United States and considerable parts of Canada and Mexico. In this enormous territory, some populations live on prairies and others in mountains. Some inhabit the deserts of the southwest; others live on the shores of the Great Lakes. Some must cope with the rigors of the Canadian winter; others never know frost. Clearly, the characters that adapt the species to such unlike habitats must be quite different. As a result, natural selection will favor different alleles in different regions, and so recognizable geographic races, the subspecies, are formed.

Even subspecies, however, may be widely distributed and very numerous. Within that range, each is restricted by its ecological requirements to limited areas within its total range. The result is a checkerboard of moderately sized mouse communities, ranging from a few pairs up to several thousand mice.

These local demes are the real breeding populations, within which there is free exchange of genes. A considerable, but more-limited, exchange of genes occurs among neighboring demes because of occasional migration of individuals. Finally, wherever subspecies meet, there is some interbreeding; potentially, gene flow may thus maintain the unity of the species, provided that it is not opposed by natural selection. When small, peripheral populations are isolated from the rest of the species, they may accumulate differences that impose a sterility barrier and thus effectively cause new species to arise. If they were to move back into the territory of the parent species, they would be unable to interbreed with them.

All aspects of this process have been tested experimentally and have been verified at the subspecific level. The assumption that the same processes, continued over long reaches of time, would result in the formation of species, genera, and the higher categories, is most attractive, but it is unproven. In view of the great body of supporting data from all aspects of biology and from other sciences as well, we can hardly wonder at the wide acceptance this theory has achieved. Nonetheless, the acceptance has not been universal, and we will consider an alternative in the following chapter.

REFERENCES

Ayala, F.J., and J.A. Kiger. 1980. *Modern Genetics*. Benjamin-Cummings, Menlo Park, Calif.
: *This introduction to genetics is at its best in the section on evolution.*

Babcock, E.B. 1947. The genus Crepis, Part 1. *Univ. Calif. Publs. Botany* 21:1–199.
: *Old, but a neo-Darwinian classic of great value.*

Blum, H.F. 1969. *Time's Arrow and Evolution*, revised ed. Princeton University Press, Princeton, N.J.
: *The application of the laws of thermodynamics to mutation and evolution is explored here.*

Calder, W.A., and J.R. King. 1974. Thermal and caloric relations of birds. In D.S. Farner and J.R. King, eds. *Avian Biology* 4:260–413. Academic Press, New York and London.
: *This paper includes a recent and quantitative review of the ecogeographic rules.*

Dobzhansky, T. 1982. *Genetics and the Origin of Species*. Columbia University Press, New York.
: *A reprint of the first edition (1937), with an introduction by Stephen Jay Gould. See also the second edition (1942) and the third edition (1951).*

Dobzhansky, T. 1971. *Genetics of the Evolutionary Process*. Columbia University Press, New York.
: *This book states the views of this most important contributor to the modern viewpoint on evolution. It supersedes, but does not entirely replace, his* Genetics and the Origin of Species *(3rd ed., 1951, Columbia University Press). (Auerbach, Demerec, Dice, Dubinin, Ives, Spencer, Stern, and Chetverikov).*

Hayes, William. 1969. *Genetics of Bacteria and Their Viruses*, 2nd ed. Wiley, New York.

Includes a clear presentation of the DNA and the gene in procaryotes.

Lewontin, R.C. 1974. *The Genetic Basis of Evolutionary Change.* Columbia University Press, New York.

A thorough treatment of the subject at a rather advanced level.

Morgan, T.H. 1928. *The Theory of the Gene*, 2nd ed. Yale University Press, New Haven, Conn.

A classic of modern biology. (Belling)

Suzuki, D.T., A.J.F. Griffiths, and R.C. Lewontin. 1981. *An Introduction to Genetic Analysis*, 2nd ed. Freeman, San Francisco.

An excellent introduction to genetics.

Tamarin, R.H. 1982. *Principles of Genetics.* Willard Grant Press, Boston.

An up-to-date and clearly written introduction to genetics.

Watson, J.D. 1969. *The Double Helix.* Atheneum, New York.

A highly personal account of the discovery of the structure of DNA by one of the discoverers.

8

Chromosomal Mutations

One of the major achievements of classical genetics was the demonstration, principally by T.H. Morgan and his associates, that each gene can be assigned, by means of crossover tests, to a definite locus, or position, in the chromosomes. If a wild-type female *Drosophila* is mated to a yellow-bodied, white-eyed male, the F_1 females should all be heterozygous, with the two dominant genes in the X chromosome derived from the mother and the two recessive genes in the X chromosome derived from the father. If no crossing-over occurs and if these F_1 females are backcrossed to the recessive type, then all of the backcross generation should be of one parental type or the other. Actually, these types (gray-bodied, red-eyed, and yellow-bodied, white-eyed) make up 98.5 percent of the offspring; the remaining 1.5 percent comprises flies that are gray-bodied but white-eyed, or yellow-bodied but red-eyed.

LONGITUDINAL DIFFERENTIATION OF THE CHROMOSOMES

Crossing-Over

The exceptional types can be accounted for only by the exchange of genes between the two X chromosomes of the female parent. The frequency of the exchange is typical for any two pairs of genes and is the same in reciprocal crosses. If the cross involves yellow and cut (a wing mutant), the exceptional types will always make up 20 percent of the backcross generation. If white and cut are tested, then the exceptional types comprise 18.5 percent of the progeny. These data give a basis for mapping the relative positions of the genes in the chromosome, if it is assumed that the frequency of crossing-over is a function of the distance between the genes: the farther apart the genes, the more crossing-over occurs. Thus the white and yellow genes, which show only 1.5 percent of crossing-over, should be fairly close together in the chromosome, while white and cut should be twelve times as far apart. Such crossover experiments between sets of three pairs of genes have made it possible to map the chromosomes of the genetically better known plants and animals. They leave no room for doubt that the chromosomes must be differentiated longitudinally.

Specificity of Synapsis

Cytological evidence of the longitudinal differentiation of the chromosomes is also available. The synapsis of homologous chromosomes is specific not only for the whole chromosomes but for the chromomeres, which are the smallest visible components of the leptotene (greatly extended preparatory to synapsis) chromosomes (Figure 8–1). Whenever the chromomeres are individually identifiable because of differences of size or shape, only like, never unlike, chromomeres synapse. If a group of chromomeres has been lost from one chromosome of a pair, then at synapsis the corresponding chromomeres of its mate form an unpaired bulge projecting to one side of the chromosome. Other gross changes in the chromosomes produce changes in the synaptic behavior of the chromosomes, which are understandable only on the principle that synapsis is specific for each point along the length of the chromosome.

Salivary Gland Chromosomes

The most impressive cytological evidence for the longitudinal differentiation of the chromosomes is derived from the study of the salivary gland chromosomes of Diptera, especially *Drosophila,* in which they have been most intensely studied. These giant chromosomes (Figure 8–2), up to half a millimeter in length, would be visible to the naked eye if they were opaque. They are closely synapsed pairs in which reorganization on the molecular level appears to be involved in the great size increase. These chromosomes have a cross-banded appearance, and the order, shape, and intensity of the bands are all perfectly regular, so that the salivary gland chromosomes can be mapped and even small segments can be identified by reference to a standard map.

Appropriate staining methods and chemical treatments can also produce banding patterns in chromosomes of other species. As in the case of the salivary gland chromosomes, the regular pattern of banding is important evidence of the serially differentiated structure of the chromosomes. It is, however, always far less detailed than is the banding of the salivary gland chromosomes.

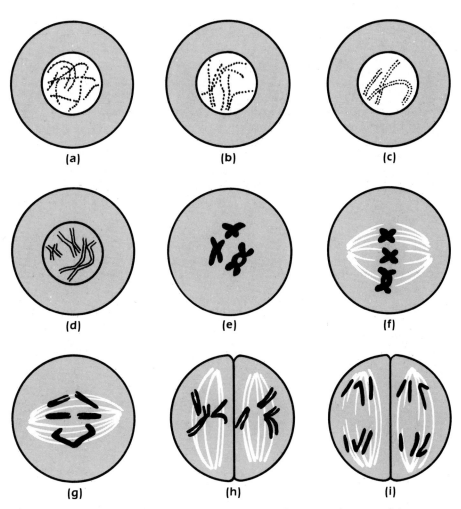

Figure 8–1.   *Meiosis in* Crepis capillaris, *a common weed with only three pairs of chromosomes.* (*a*)–(*e*) *Stages in the prophase of the first meiotic division; notice the serial structure of the chromosomes in* (*a*) *through* (*c*). (*f*) *First metaphase, with three tetrads or bivalents.* (*g*) *First anaphase.* (*h*) *Second metaphase, with three dyads or univalents on each spindle.* (*i*) *Late second anaphase, showing four reduced genomes.*

Fine Structure Analysis

Molecular geneticists have demonstrated this linear differentiation of the chromosomes even at the molecular level. As discussed in Chapter 7, the codons are arranged in linear order in each cistron and the order of codons corresponds exactly to the order of amino acids in the resulting polypeptide. In principle, mutation can occur in any of the many codons of a cistron, and recombination between different intragenic mutants should then occur. (Consult

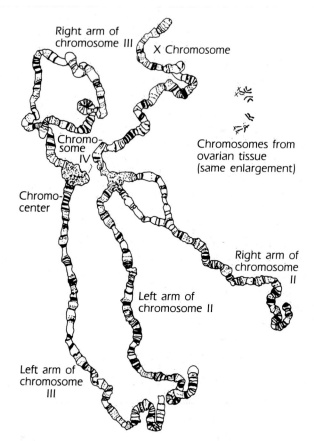

Right arm of
chromosome III

X Chromosome

Chromo-
some
IV

Chromo-
center

Chromosomes from
ovarian tissue
(same enlargement)

Chromo-
center

Left arm of
chromosome II

Right arm of
chromosome
II

Left arm of
chromosome
III

Figure 8–2. *The salivary gland chromosomes of* Drosophila.
After T.S. Painter, *Journal of Heredity* (1934) 25:466.

a genetics text for a more-detailed discussion.) S. Benzer studied the rII (rapid lysis) region of the chromosome of phage T4 and found that the more than 1,600 mutations that he studied were distributed among 251 linearly arranged sites, although about half were located at two "hot spots." He called these studies on the localization of mutations within the gene "fine structure analysis." It is the coup de grâce in the demonstration of the linear differentiation of the chromosomes, for it demonstrates that the individual nucleotide pair is the minimal unit of recombination.

Overlapping Genes

The detailed mapping of the chromosomes of many organisms was one of the great achievements of Mendelian genetics. A secure conclusion stemming from this work seemed to be that each gene was clearly separate from all others. There could be no overlap. When Benzer and others extended the method to the molecular level, this conclusion was confirmed and strengthened. Then, in 1978, F. Sanger determined the complete nucleotide sequence of phage ΦX174. This bacteriophage has only nine genes. Assignment of sequences of codons to specific genes gave unexpected results. In the sequence ...ATGA..., TGA is the first codon of gene C, but ATG is the last codon of gene A! A similar situation occurs where genes C and D meet. Genes D and J share a single base. Even more surprising are the locations of genes B and E, which proved to be entirely included within genes A and D respectively!

These facts have implications for evolution, namely, mutation in one gene of an overlapping set should entail mutation in the other gene. The fact that such simultaneous mutations were not known was a reason for the earlier rejection of the hypothesis that genes might overlap. Because viruses are extremely small particles, some geneticists suggest that overlapping genes may be a special adaptation for maximal compactness. Now that overlapping genes have been demonstrated in a virus, however, we cannot eliminate the possibility that they may also occur in eucaryotes; and in them, so simple an explanation would not be tenable. At present, we must acknowledge overlapping genes as one of the unsolved problems of evolutionary biology.

ARCHITECTURAL CHANGES IN THE CHROMOSOMES

The above data lead to the conclusion that, beneath the homogeneous appearance so often presented by the chromosomes, there is a longitudinal physicochemical differentiation, the "architecture of the chromosomes." Once the standard architecture for any chromosome is established, it is possible to recognize various rearrangements. Four such types of rearrangement are known. A *deletion,* or *deficiency,* constitutes the loss of a segment of a chromosome (Figure 8–3). Such losses can be produced experimentally by high-energy radiation, and they may be caused by radiation in nature also. At synapsis, in an organism that is heterozygous for a deficiency, the unpaired portion of the normal chromosome projects to one side as a loop, whereas those points that are present in both chromosomes of the pair are synapsed normally.

Just the opposite of a deletion is a *duplication,* in which a segment of a chromosome is repeated. In a duplication heterozygote, synapsis looks very much like it does in a deficiency heterozygote, this time because a region which is present only once in the normal chromosome is present twice (or more times) in the modified chromosome. In the salivary gland chromosomes, the difference between a deficiency and a duplication can be easily detected by the pattern of banding. The production of duplications is believed to be based on unequal crossing-over, that is, crossing-over in which the breaks in the homologous chromosomes occur at somewhat different points. This unequal crossing-over would produce, simultaneously, a duplication and a deficiency.

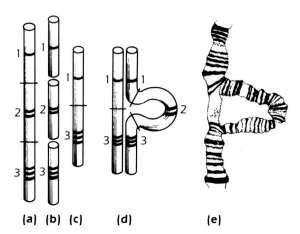

(a) (b) (c)   (d)   (e)

Figure 8–3.   *Heterozygous deficiency in the salivary gland chromosomes. (a) Segment of the chromosome diagrammed with three sections, marked by 1, 2, and 3 bands respectively. (b) Breaks occur at each end of section 2. (c) Sections 1 and 3 heal together to form a deficient chromosome. (d) Synapsis of the deficient chromosome with a normal one is possible only if section 2 of the normal chromosome forms an unpaired loop while sections 1 and 3 are perfectly synapsed, point for point. (e) An actual heterozygous deficiency as it appears in the salivary gland chromosomes.*

A third type of chromosomal rearrangement is an *inversion,* a reversal of a segment of a chromosome. Thus, a chromosome in which the order of parts runs ABCDEFGHIJK might undergo an inversion of the segment D to H, so that the order of parts runs ABCHGFEDIJK. In the heterozygote, homologous point still synapses with homologous point, with the result that one chromosome of the pair must form a twisted loop, while the other continues around it without twisting (Figure 8–4). An inversion can be produced only if a chromosome is broken at two points, and if it reheals in the reversed relationship. Inversions are produced frequently in radiation experiments. Naturally occurring radiation probably plays a role in the production of naturally occurring inversions; in any case, inversions are rather common in nature.

In the fourth and final type of architectural rearrangement of the chromosomes, called a *translocation,* a segment of one chromosome has been transferred to a nonhomologous chromosome (Figure 8–5). Typically, translocations are reciprocal—that is, segments are exchanged between two nonhomologous chromosomes, so that we may speak of "illegitimate crossing-over." In synapsis of the heterozygote, the exchanged portions retain their original synaptic specificity, with the result that two different tetrads are bound together.

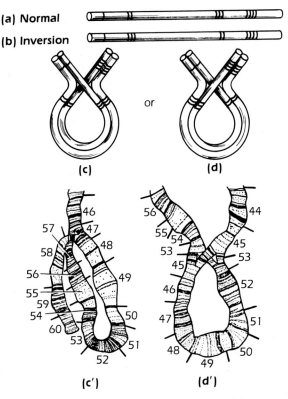

(a) Normal

(b) Inversion

or

(c)   (d)

(c')   (d')

Figure 8–4. Heterozygous inversion. (a) Normal chromosome, with serial structure indicated diagrammatically. (b) Its homologue, with the central portion inverted. (c) and (d) Synapsis, with one of the pair in a twisted loop and the other looped around the first without twisting so that synapsis is still homologous point for homologous point. (c') and (d') Salivary chromosomes with a heterozygous inversion. Numbered sections refer to a standard map of the salivary gland chromosomes. Using this map, any cytogeneticist can identify the chromosomal region depicted.
After T. Dobzhansky and D. Socolov, *Journal of Heredity* (1939) 30:9.

Position Effects

In accordance with the classic gene theory (Chapter 7), we might expect that such chromosomal rearrangements would alter linkage relations but would never actually affect phenotypes. In fact there are many examples in which each type of chromosomal rearrangement behaves as though it were a gene

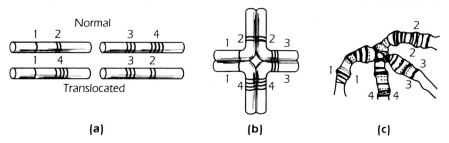

Figure 8–5. *Translocation heterozygote. (a) Above, nonhomologous chromosomes, the ends of one designated as 1 and 2, those of the other as 3 and 4; below, their translocated homologues, designated 1 and 4 and 3 and 2 respectively. (b) At synapsis the two tetrads are bound together because homologous point still attracts homologous point. (c) Such a translocation in the salivary gland chromosomes.*
After 0. Mackensen, *Journal of Heredity* (1935) 26:170.

mutation—that is, the rearrangement causes a definite phenotype and is inherited according to the usual principles of Mendelian heredity. Many rearrangements are lethal in the homozygous state, but some are not; and the fact that closely related species can often be shown to differ by a few homozygous rearrangements indicates that these can play a definite role in evolution.

An obvious explanation of the mutational effect of rearrangements would be that the same force that causes the breaking and the rearrangement of the chromosome also causes mutation of the gene nearest the break. Everyone who has worked in this field agrees that few, if any, of the mutational effects of chromosomal rearrangements can be explained on this basis. N.P. Dubinin, in the course of a study of a reciprocal translocation between the second and third chromosomes of *Drosophila,* obtained the final disproof. The untranslocated third chromosome carried the recessive gene *hairy,* which causes excessive development of the bristles when homozygous. The translocated third chromosome, however, carried the normal allele of the *hairy* gene. When these two genes were interchanged by a crossover, *hairy* became dominant over normal. As crossing-over has no such influence in the absence of the translocation, change of dominance must be a simple consequence of change of neighborhood: *hairy* gene in a normal chromosome is recessive, *hairy* gene in a translocated chromosome is dominant. Because the rearrangement of the genes in the chromosome seems to have a mutant effect independent of the genes, such chromosomal mutations are called *position effects*.

Bar-Eye "Gene"

The Bar-eye mutation of *Drosophila* was the first to be analyzed in terms of position effect. The eyes of normal flies are oval shaped. The Bar gene, however, results in a smaller number of facets than usual, and so the eyes are narrower than those of normal flies. C. Zeleny showed that, in homozygous Bar-eyed stocks, about one fly in 1,500 mutated to the more extreme Doublebar, while an equal number mutated back to normal. By experiments involving crossing-over between other genes near Bar, A.H. Sturtevant and T.H. Morgan showed that this simultaneous mutation from Bar to Doublebar and to normal was based on unequal crossing-over. Thus, although each chromosome of the pair ought to contain the Bar gene, after the unequal crossing-over, one chromosome had *two* Bar genes in tandem, and the other chromosome had none at all. Those zygotes having the chromosome with no Bar gene were normal; those having the chromosome with the two Bar genes showed the Doublebar phenotype. Two Bar genes in the same chromosome appeared to cause a different phenotype than did two Bar genes located one in each of a pair of chromosomes (Doublebar and Bar, respectively). To describe this unexpected phenomenon, Sturtevant and Morgan introduced the term *position effect*.

The analysis of the Bar "gene" could be completed only after the introduction of the salivary gland chromosome technique. In the region of the X chromosome, which is indicated by crossover tests to include this gene, a series of about six bands is present only once in wild-type flies, twice in Bar flies, and three times in Doublebar flies (Figure 8–6). The Bar "gene" appears actually to be

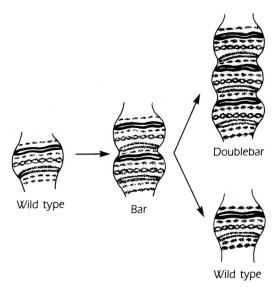

Wild type   Bar

Doublebar

Wild type

Figure 8–6.   The Bar "gene" in the salivary gland chromosomes of Drosophila.
After C. Bridges and E. Sutton in M.J.D. White, *Animal Cytology and Evolution,* 3rd ed. (New York: Cambridge University Press, 1973).

a duplication of a small segment of the X chromosome. If crossing-over is unequal in a fly homozygous for this duplication, then the resulting chromosomes will be one normal one, in which this segment is present only once, and one in which the segment is present three times. The normal allele of Bar now becomes simply the unmodified chromosome.

Chromosomal Rearrangements Differentiating Species

So far, we have discussed two position effects, one resulting from a translocation, the other from a duplication. Both types play a role in evolution. The morphology and synaptic behavior of the chromosomes of many species suggest that duplications may be present. Duplication has been suggested as a possible source of "new" genes, the duplicated ones presumably becoming completely different from the original through repeated mutations. Evidence of this possibility is the common occurrence of *pseudo-alleles*. These are genes of similar effects that are located so close together in the chromosome that they are rarely separated by crossing-over.

The idea that new genes might arise in this way was for a long time hotly controversial, but it is now well established for many proteins. A good example is hemoglobin, the respiratory pigment of the blood of vertebrates. Similarity of the amino acid sequence of hemoglobin in primitive animals, such as the lamprey, to that of myoglobin, the oxygen-binding protein of muscle, suggests that the gene for myoglobin duplicated, with one of the units retaining its original function, while the second mutated to produce its protein in blood cells. In higher fishes, duplication and further mutation of the hemoglobin gene may have resulted in the formation of two hemoglobin chains, the alpha and beta chains, each of which may be differentiated from species to species by replacement of amino acids (reflecting base-pair replacements in the DNA). An alternative interpretation of pseudoalleles is that they indicate that segments of varying size of a chromosome are concerned with a unified function. Although this is undoubtedly true, the two interpretations are not mutually exclusive, and the latter cannot be used, as it once was, as an argument against origin of new genes by duplication and divergence.

Translocations certainly play a role in producing changes of chromosome number. For example, S. Makino studied the chromosomes of the Japanese loaches (teleost fishes of the family Cobitidae) *Misgurnus anguillicaudatus* and *Barbatula oreas*. These closely related species have 26 and 24 pairs of chromosomes, respectively, and these are, for the most part, rod-shaped (Figure 8–7). However, two pairs of the chromosomes of *B. oreas* are V-shaped, and Makino drew the unavoidable conclusion that the reduction in chromosome number had been accomplished by translocations, uniting originally distinct chromosome pairs. Similarly, in *Drosophila* there are six basic chromosome arms, which in different species are combined in different ways to give haploid numbers varying from three to six (Figure 8–8).

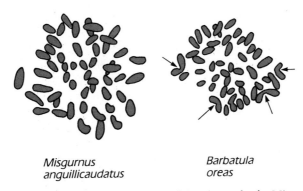

Misgurnus
anguillicaudatus

Barbatula
oreas

Figure 8–7.   *The chromosomes of the Japanese loaches* Misgurnus anguillicaudatus *and* Barbatula oreas. *Note the
two V-shaped pairs in the latter.*

The total pattern of the chromosomes of a species is called its *karyotype.*
Analysis of karyotypes of related species may illuminate their relationships,
as in the two examples illustrated in Figures 8–7 and 8–8. The higher primates
provide another excellent example. J.J. Yunis and O. Prakash have studied
human chromosomes and those of the great apes by high resolution methods
and have compared the patterns of banding in fine detail. Of the twenty-
three pairs (twenty-four in the apes, because the two arms of chromosome 2
are separate), eighteen appear to be virtually identical in all species. The others
are differentiated by a series of inversions and a few translocations.

These researchers conclude that humans and chimpanzees were the last to
separate and that gorillas separated from the common ancestor somewhat
earlier. Orangutans are much more primitive and separated much earlier from
a common ancestor of all of the higher primates. On morphological grounds,

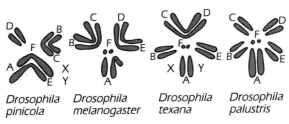

Drosophila
pinicola

Drosophila
melanogaster

Drosophila
texana

Drosophila
palustris

Figure 8–8.   *Chromosome sets of males of* Drosophila
pinicola, D. melanogaster, D. texana, *and* D. palustris.
After M.J.D. White, *Animal Cytology and Evolution,*
3rd ed. (New York: Cambridge University Press, 1973),
pp. 341, 342.

students of the primates have recognized that the families Pongidae (great apes) and Hominidae (humans) comprise the superfamily Hominoidea (along with the Hylobatidae, which Yunis and Prakash did not include in their study). On the basis of the close similarity of their chromosomes and of their apparent derivation from a common karyotype by a few simple rearrangements, Yunis and Prakash suggest that all of these primates belong in *one* family, the Hominidae, with the orangutans comprising the subfamily Ponginae, and gorillas, chimpanzees, and humans comprising the subfamily Homininae. This classification is consistent with biochemical evidence, which shows remarkably little difference among the proteins of the higher primates. The taxonomy of the primates is now in a state of flux, but surely the viewpoint of Yunis and Prakash must be taken into account in any future revision of this important group.

In *Drosophila,* mutations based on all four types of chromosomal rearrangements are known. We have discussed examples of mutant effects of duplications and of translocations. Deficiencies and inversions may also produce position effects. Notch wing is a well-known sex-linked mutation based on a deficiency near the left end of the X chromosome. Curly and dichaete wings are based on inversions of the second and third chromosomes, respectively. R.B. Goldschmidt and others found that about half the classical mutant genes of *Drosophila* are in fact position effects. Of the other half, many have not been adequately examined, and a good portion of them may also prove to be position effects. Are the remaining mutants really bona fide genes in the sense discussed in Chapter 7, or are they also position effects of rearrangements that are too small for detection by present methods? Goldschmidt, from the mid-1930s until his death in 1958, held to the second explanation. Such mutations and other evidence led him to make a concerted attack on the neo-Darwinian, or synthetic, theory and to propose a countertheory.

In the preceding examples, the role of inversions and translocations is emphasized; but in *Sciara ocellaris* and *S. reynoldsi* (fungus gnats) a few small deficiencies and duplications account for the differences between the species. M.J.D. White and others have shown that in many insects for which salivary gland chromosomes are not available, differences between the chromosome complements of related species are nonetheless most easily interpreted in terms of several types of rearrangements. This finding cannot, of course, preclude the possibility that gene mutation might also play a role, even a predominant role, in the differentiation of these species.

A good example among plants is provided by H. Lewis's study of *Clarkia,* a genus of the evening primrose family. *C. biloba* is widely distributed in California; *C. lingulata,* a closely similar species, is found only on the southern edge of the range of the former. *C. lingulata* has been derived from *C. biloba* and differs from it by one translocation, two inversions, and the separation of the two arms of a long chromosome (Figure 8–9). *C. lingulata* survives in a marginal habitat, one consisting of patches of moist land surrounded by

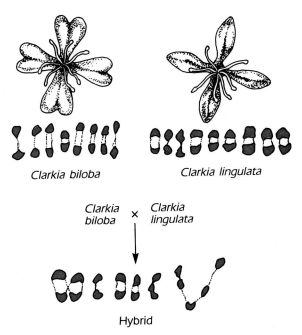

Figure 8–9. *Speciation by chromosomal mutation.* Clarkia biloba *has flowers with bilobate petals;* C. lingulata *has flowers with simple petals. The latter has been derived from the former by a few chromosomal mutations. Below each is its typical set of chromosomes at metaphase or very early anaphase of the first meiotic division, showing 8 bivalents in* C. biloba *and 9 in* C. lingulata. *In the hybrid obtained by crossing these two species, meiosis is modified because of the chromosomal rearrangements. Note the large translocation complex at the right.*

desert. Further examples among plants are provided by the chromosomal races of *Datura* (Chapter 12) and by the many examples of polyploidy (Chapter 13).

Overlapping Inversions and Phylogeny

By far the most complete study of chromosomal differences separating natural races and species is that of T. Dobzhansky and his collaborators on *Drosophila pseudoobscura* and its allies. As originally described, the *pseudoobscura* group comprised *D. pseudoobscura* and *D. miranda,* two closely similar species. But *D. pseudoobscura* was subdivided into races A and B on the basis of a series of differences so refined that they can usually be distinguished only by statistical analysis of populations. These differences, difficult to diagnose, are nonetheless highly consistent, and they include intersterility. After years of detailed study, T. Dobzhansky and C.C. Epling described race B as a distinct species, *D. persimilis,* and reserved the original name for race A. Such closely similar species, differentiated only by trivial morphological characters and yet separated by a sterility barrier, are called *sibling species.* They appear to have diverged from a common ancestor in the recent past.

Phylogeny of the various local populations of this group has been studied by means of the analysis of inversions. When two inversions are present in the same chromosome, they may be independent (ABEDCFGJIHKL), or the second may be included within the first (ABCJIHEFGDKL), or they may overlap (ABGFIHCDEJKL) (Figure 8–10). The last type is of especial interest because we can determine the order of events in that group. If there are three arrangements known for a particular chromosome: (1) ABCDEFGHIJKL, (2) ABGFEDCHIJKL, and (3) ABGFIHCDEJKL, it is obvious that either (1) or (3) could have derived from the other only with (2) as an intermediate step.

Most of the chromosomal variability in the *pseudoobscura* group occurs in chromosome 3. Thirteen different arrangements of this chromosome are known only in *D. pseudoobscura,* seven are known only in *D. persimilis,* and one, called Standard, is found in both species. By detailed analysis of overlapping inversions, starting with the assumption that Standard was the primitive arrangement, Dobzhansky and his collaborators have worked out a nearly complete phylogeny for this group (Figure 8–11). Only a single gap occurs, between Standard and Santa Cruz, a variety of *pseudoobscura.* Even this gap is not entirely unfilled, for the necessary intermediate chromosomal pattern is found in the closely related species *D. miranda.*

Chromosomal Mutation and Evolution

Thus, clearly, chromosomal rearrangements comprise a source of genetic variability that supplements variability produced by mutation at the molecular level (Chapter 7). That chromosomal rearrangements have a role in evolution is demonstrated by the many cases in which closely related species differ by a few rearrangements, and some of these species are sibling species that may be difficult to separate morphologically. As a result, most evolutionary biologists can agree with T. Dobzhansky's evaluation that chromosomal mutation must be one of the mainsprings of evolution, even though its exact role is difficult to assess.

However, one major evolutionary biologist of the early years of the modern synthesis, R.B. Goldschmidt (Figure 8–12), dissented very strongly. He assigned the rearrangements a central role in evolution, and he rejected many of the neo-Darwinian tenets. Let us examine two current developments that are again focusing attention on Goldschmidt's evolutionary biology.

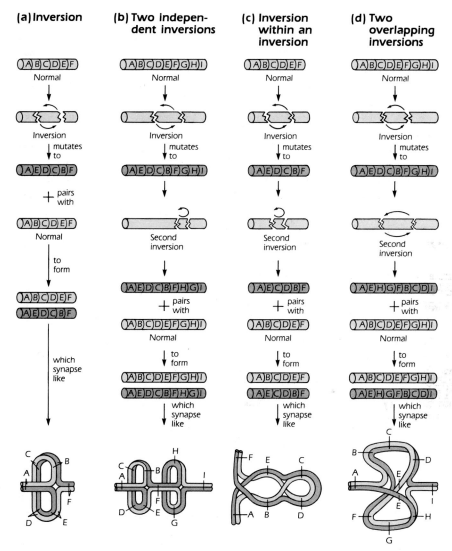

Figure 8–10. *Pairing in salivary gland chromosomes heterozygous for inversions. At the top of each vertical column is an unmodified chromosome with normal banding sequence. Below are diagrams of breakage and rehealing to form new sequences. At the bottom of each column is a diagram of synapsis of the inversion heterozygote (i.e., synapsis of the modified chromosome with the unmodified chromosome). (a) A single inversion, (b) two independent inversions, (c) two inversions, one included within the other, and (d) two overlapping inversions.*

Data from T. Dobzhansky, *Genetics of the Evolutionary Process* (New York: Columbia University Press, 1970), p. 130.

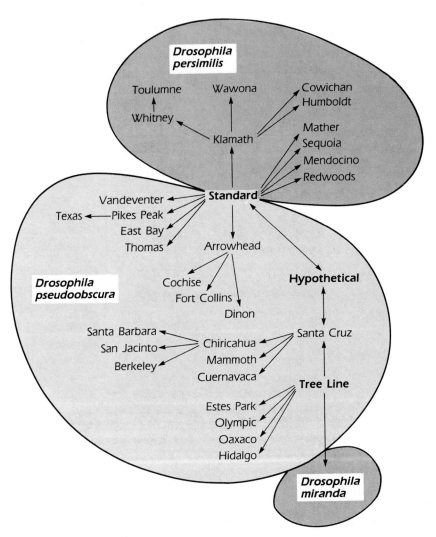

Figure 8–11. *Phylogenetic chart of the* Drosophila pseudoobscura *group as worked out by analysis of overlapping inversions. Race A is* D. pseudoobscura *and race B is* D. persimilis *of the newer classification. The step from standard to Santa Cruz, marked hypothetical, is found in* D. miranda.
Data from T. Dobzhansky, Genetics of the Evolutionary Process (New York: Columbia University Press, 1970).

Figure 8–12.   Richard B. Goldschmidt (1878–1958).
Courtesy of *Caryologia.*

Molecular Drive

G. Dover has investigated the repeated sequences of DNA and has pointed out that they tend to be relatively homogeneous within a species, more dissimilar among species. He attributes the within-species homogeneity to the concerted action of the processes of unequal crossing-over, transposition, and gene conversion. Collectively, he calls these processes *molecular drive,* and he believes that molecular drive causes a cohesive change in populations as a whole, independently of natural selection (see Chapter 9).

Unequal crossing-over, as we noted earlier, causes both deficiencies and duplications. B. McClintock investigated transposition in the late 1940s and found certain genes in corn that moved from place to place in the genome. She called these genes *controlling elements,* because they seemed to activate or turn off other genes near which they were inserted. At the time, such "jumping genes" were not found elsewhere, and her data were given little attention.

In the 1970s plasmids (small, nonessential, accessory chromosomes) came under intensive study in bacteria. Plasmids often include segments that begin and end with identical nucleotide sequences—for example, ABCD . . . DCBA, with each letter representing a series of nucleotides, and the points of ellipsis representing any series of genes, often genes for antibiotic resistance or for virulence. These repeated sequences recognize similar sequences in the bacterial chromosome and synapse with them, so that exchange occurs between bacterial and plasmid chromosomes. The bacterial chromosome usually contains several such recognition sites, so that the *insertion element* of the plasmid may attach at any of several sites in the bacterial chromosome. Observers first noted that genes near the insertion element were turned off—that is, they failed to function. Later, finding examples in which reversed orientation of the insertion element turned on the adjacent genes, researchers suggested that this might be part of the normal mechanism of control of the genetic system. Similar jumping genes have been found in *Drosophila*, and they are probably widespread, although inadequately understood. At first, the transposable elements seemed to offer a means for the spread of antibiotic resistance, but any broader role in evolution was quite conjectural. Dover assigns them a major role in molecular drive.

Gene conversion is the most contentious aspect of molecular drive. When the heterozygote *Aa* replicates at meiosis, we expect the tetrad *AAaa*. In most plants and animals, only a sample of the products of meiosis is recoverable, but the sample data are consistent with the hypothesis that both alleles are replicated. In some fungi, however, it is possible to recover *all* of the products of meiosis. In such species, *AAaa* is indeed commonly produced, but a significant number of tetrads prove to be *AAAa* or *Aaaa*. One allele appears to have been replicated preferentially. Such *gene conversion* has been demonstrated only in a few fungi, and its occurrence in other organisms cannot yet be either confirmed or denied. Dover points out that if conversion were biased in favor of specific nucleotide sequences, it could be a powerful force for the homogenization of the repetitive DNA.

Dover believes that molecular drive results in a considerable degree of uniformity of the repetitive DNA within a species and may also result in speciation independently of natural selection. Two of the three factors in molecular drive—unequal crossing-over and transposition—relate directly to Goldschmidt's conception of the gene. Molecular drive is hotly controversial, and it is still much too early to predict how the controversy will be resolved.

Punctuated Equilibria

N. Eldredge and S.J. Gould base their theory on paleontological data, which show that new groups commonly appear abruptly rather than emerging gradually, as might be expected on the basis of the synthetic theory. The incompleteness of the geological record has usually been cited to explain abrupt appearance, but Eldredge and Gould propose that the discontinuities are real, the result of rapid origin of new species and higher groups. They propose that species and higher groups (except for minor variations that lead nowhere) remain stable over long periods of time (*equilibrium*). Then these groups rapidly, within thousands of years, give rise to new groups (punctuation). The intervals, short enough to be easily missed in the geological record, are followed by another long period of equilibrium.

How is the rapid change effected? Gould especially emphasizes mutation of *regulatory genes*, genes that control the activity of structural genes. Such a mutation, with its primary effect early in embryonic development, might have a cascading effect throughout development and so might lead to a radically changed organism. Gould recognizes Goldschmidt as the forerunner in this line of reasoning. (We return to punctuated equilibria in Chapter 22, after we survey the grand panorama of evolution.)

Systemic Mutation and Evolution

R.B. Goldschmidt belonged to the first generation of geneticists after the rediscovery of Mendelism. He carried out a comprehensive study of the genetics of geographic variation in the nun moth, *Lymantria monacha,* a species ranging across Eurasia from England to Japan. In this study he defined a species in terms of a range of variability rather than of a specific type from which accidental deviations occurred. He soon became convinced that neo-Darwinian evolution serves only to adapt a species to a maximum range of habitats without ever crossing the gap between species. Let us summarize his reasons briefly. At the outset, he rejected the corpuscular gene, which was then closely allied with neo-Darwinism. Many data are identical for point mutations and for position effects, and he concluded that the law of parsimony requires one explanation for all types of mutation. Hence, he regarded point mutations as simply position effects of rearrangements that were too small to be identified by available methods. He conceived of the chromosome as a chemical continuum in which the genes were segments of varying lengths. He even suggested that different genes might overlap—at that time a very bold proposal but one that has now been confirmed (see Chapter 7). Mutation he conceived of as a position effect within a chemically integrated chromosome.

If the accumulation of small mutations under the influence of natural selection cannot go beyond the confines of species, what does? Goldschmidt's answer was *systemic mutations,* mutations with primary effects on early embryonic processes and secondary effects cascading through later development, much like the mutations of regulatory genes to which S.J. Gould and J.W. Valentine currently attach similar importance. Such mutations might produce organisms markedly different from the parents. Consequently, a new species might result from one or a few such systemic mutations, and members of this species would then immediately be put to the test of natural selection. The majority would fail the test, but an occasional "hopeful monster" would successfully establish a new species (or higher group), which would then be "fine-tuned" to a variety of habitats by the neo-Darwinian process.

Thus, in 1940 Goldschmidt proposed on a genetic basis very much what Eldredge and Gould proposed on a paleontological basis some thirty years later. Meanwhile, a different understanding of the gene had developed. The molecular understanding of the gene had not even begun when Goldschmidt first proposed the existence of systemic mutations, and it had not progressed very far by the time of his death in 1958. Systemic mutations still remain generally unverified, and genes are now seen principally as *structural genes,* which encode specific proteins, or as *regulatory genes,* which control the activity of the structural genes. To the regulatory genes, Gould and Valentine ascribe much the same role that Goldschmidt had assigned to systemic mutations— that is, sudden origin of new species without the intervention of the neo-Darwinian process.

Evaluation

The majority of geneticists and other students of evolution rejected some or all of Goldschmidt's theses, yet most agreed that he had amassed a body of highly pertinent evidence that could not be dismissed lightly. His attack on the fashionable theory brought upon him the anger of much of the biological community. Many of his criticisms, however, were justified, and he served a useful function in forcing its defenders to take greater care both with their data and with their reasoning.

To make valid criticisms of the dominant theory is, of course, not enough. A competing theory must be proven on its own merit. Goldschmidt's major points were the invalidity of the corpuscular gene and the necessity of systemic mutations for the formation of new species and higher groups. Let us examine each. The controversy over the theory of the gene seems remote, now that we know that a gene is a specific sequence of nucleotides in the DNA. As we might expect, the gene as it is now understood includes aspects of both of the competing theories of a generation ago. Irreducible units are there, but they are the individual nucleotide pairs, hardly the genes in any sense. The gene is, indeed, a pattern in a chemical continuum—the DNA molecule— but on a scale much finer than Goldschmidt had envisioned. In treating the gene primarily as a functional rather than as a structural unit, Goldschmidt was far ahead of his time.

Is mutation necessarily, as he thought, a matter of architectural rearrangement in the chromosomes? Studies in biochemical genetics have proven that the great majority of mutations are replacements of a single base pair, point mutations par excellence; these studies reenforce the synthetic theory, with its emphasis on point mutations. Position effects are still a well-established phenomenon and undoubtedly have great importance for evolution. Nonetheless, their relative role in evolutionary theory has not increased in the intervening years. If the corpuscular gene has disappeared, neither have systemic mutations been proven, although there are probable examples. The strongest case, the only proven case of systemic mutation, is that of polyploidy in plants (Chapter 13), a case which was well known in Goldschmidt's time; yet even he agreed that this was a special case from which generalization was not justified.

SUMMARY

The chromosomes, comprising the genes, are longitudinally differentiated, as shown genetically by the results of crossover experiments and cytologically by the sequence of chromomeres and the banding of the giant chromosomes of the salivary glands of dipterans. Longitudinal differentiation has even been demonstrated for the sequence of nucleotides within a gene by fine structure analysis. The complete sequence of nucleotides in the chromosome of phage $\Phi X174$ has been determined, with the surprising result that some of the genes overlap.

Once a specific sequence of genes is determined, it can be rearranged in several ways: a part of the chromosome may be removed (deletion) or a part may be repeated (duplication); if the chromosome is broken at two points, a segment may be turned 180° and rehealed into the chromosome in the reverse of the original orientation (inversion); or segments may be exchanged between nonhomologous chromosomes (translocation, often called illegitimate crossing-over). All of these rearrangements are easily identified in the salivary gland chromosomes.

Such chromosomal rearrangements often behave as though they were mutant genes—that is, they cause specific phenotypes. For example, duplication of a small segment of the X chromosome of *Drosophila* causes the Bar-eye phenotype, and triplication of the same segment causes a more extreme phenotype called Doublebar.

We know many examples in which the chromosomes of related species differ by chromosomal rearrangements: the Japanese loaches differ by two translocations; in species of *Drosophila* six basic chromosome arms are combined in different ways to give haploid numbers of 3, 4, 5, or 6; and in the higher primates, eighteen pairs of chromosomes appear to be identical in humans, chimpanzees, and gorillas, and five (humans) or six (apes) pairs are differentiated by a few inversions and translocations. Similarly, *Clarkia lingulata* differs from its parent species, *C. biloba,* by a translocation, two inversions, and the separation of the two arms of a long chromosome. Thus, chromosomal rearrangements clearly play an important role in evolution.

When two inversions overlap, we may be able to trace the order in which they occurred, especially among dipterans, where the salivary gland chromosomes facilitate the research. Dobzhansky has exploited this method to trace the formation of the species and subspecies of the *Drosophila pseudoobscura* group. This example has especial importance because it is fully demonstrated evolution.

R.B. Goldschmidt used the facts of position effect as the basis for a major attack on the theory of the gene and on the neo-Darwinian theory of evolution. He believed that all mutation was basically position effect and that the neo-Darwinian type of evolution served only to adapt species to a variety of ecological niches. The more-fundamental evolution of new species required, according to Goldschmidt, systemic mutations—that is, mutations affecting early embryonic processes, with effects cascading throughout development. The biological community gave this theory a harsh reception, but recent research has refocused attention upon Goldschmidt and his theories.

G. Dover believes that molecular drive maintains the unity of species and may cause the formation of new species without the intervention of Darwinian selection. Molecular drive comprises three processes: unequal crossing-over, transposition, and gene conversion. Dover believes that these processes are particularly effective in determining the makeup of the repetitive DNA, which is so abundant in eucaryotes.

N. Eldredge and S.J. Gould proposed that the sudden appearance of new groups in the fossil record depends not on the incompleteness of the record so much as on punctuated equilibria. They believe that groups remain essentially static for long periods, then rapidly form new species or higher groups. They envision mutations of regulatory genes, with properties much like the systemic mutations of Goldschmidt, as the basis for the rapid formation of new species. Goldschmidt's challenge to the neo-Darwinian, or synthetic, theory of evolution played an important role in the development of the latter; but, more than a generation after his death, it is still the synthetic theory, modified and matured, that dominates and guides evolutionary research. Today, molecular drive and punctuated equilibria have a role similar to that of systemic mutations as a challenge to the dominant theory. How they will be integrated into the modern evolutionary synthesis, or whether they may require a new synthesis, as some enthusiasts believe, are questions that only future research can answer.

REFERENCES

Goldschmidt, R.B. 1982. *The Material Basis of Evolution.* Yale University Press, New Haven, Conn.
  A reproduction of the 1940 edition with an introduction by S.J. Gould.
Goldschmidt, R.B. 1955. *Theoretical Genetics.* University of California Press, Berkeley and Los Angeles.
  The last major work by this author.
King, M.C., and A.C. Wilson. 1975. Evolution at two levels in humans and chimpanzees. *Science* 188:107–116.
  A return to the idea of macroevolution based upon chromosomal rearrangements.
Stebbins, G.L., and F.J. Ayala. 1981. Is a new evolutionary synthesis needed? *Science* 213:967–971.
  This paper presents a vigorous defense of the modern synthesis. The authors conclude that the modern synthesis is generally sound, but that punctuated equilibria, which are consistent with the modern synthesis but not required by it, may be needed as a supplement.
Swanson, C.P., T. Merz, and W.J. Young. 1981. *Cytogenetics: The Chromosome in Division, Inheritance, and Evolution.* 2nd ed. Prentice-Hall, Englewood Cliffs, N.J.
  This introduction to cytogenetics is particularly strong on the evolutionary aspects of cytogenetics.

White, M.J.D. 1973. *Animal Cytology and Evolution*, 3rd ed. Cambridge University Press, New York.

A classic in this field.

White, M.J.D. 1978. *Modes of Speciation.* Freeman, San Francisco.

One of the most experienced of evolutionary cytologists here summarizes his discipline.

Yunis, J.J., and O. Prakash. 1982. The origin of man: A chromosomal pictorial legacy. *Science* 215:1525–1530.

This important paper includes excellent photographs to illustrate homologies among the chromosomes of the higher primates.

PART

3

The Origin of Species and of Higher Categories

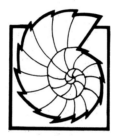

In Part 3, comprising Chapters 9 through 15, we discuss factors that interact with the variations arising from mutation, resulting in the evolution of new species, genera, families, and the rest of the taxonomic hierarchy. This process is, of course, the second major problem of evolution, and we begin it with the study of natural selection.

9

Natural Selection

The early post-Darwinian students of evolution were inclined to think of natural selection as an all-or-none phenomenon: either an organism was favorably endowed and survived to reproductive age, or it was unfavorably endowed and died in the struggle for existence without leaving progeny (Figure 9–1). Among the factors that led to the agnostic reaction at the turn of the century was the fact that no such severe selection had actually been observed as a general phenomenon in nature. Yet it ought to be obvious if it existed. Severe defectives were, of course, eliminated: crippled deer cannot long escape predators. But many minor defectives were observed in nature, and they did leave progeny. Coupled with the new mutation theory of De Vries, according to which a new species might be produced at a single step, these facts seemed to push natural selection out of the picture entirely.

SEVERE VERSUS MILD SELECTIVE FORCES

There is, however, another possibility, which was not considered at the time of the agnostic reaction. If a character confers only a slight disadvantage upon the organisms that exhibit it, those organisms may have a death rate

Figure 9-1.   *The Darwinian view of natural selection.*

somewhat higher than that of their more fortunate cousins, with the result that relatively fewer of them will reach reproductive age, and they will leave somewhat fewer progeny. Conversely, variants that have even a slight advantage will have a somewhat higher survival rate and hence will leave somewhat more descendants. In either case, the net result will be a gradual change in the proportions of each variant in a species.

TYPES OF NATURAL SELECTION

Selection always operates by eliminating the less-favored types, thus indirectly promoting the reproduction of the more favored; but evolutionary biologists sometimes find it convenient to distinguish three types of selection (Figure 9-2), based on the effect on the population selected. *Stabilizing, or normalizing, selection* maintains an already well-adapted condition by eliminating any marked deviations from it. For example, owls prey upon field mice. The normally colored mice blend protectively into their background, whereas mutants that produce more conspicuous colors are taken by the owls in disproportionately large numbers; thus selection tends to maintain the color of the mice within a narrow range.

By contrast, *directional selection* produces a change. It occurs whenever organisms must adapt to changing conditions of life. A good example is the development of industrial melanism, which we discuss later.

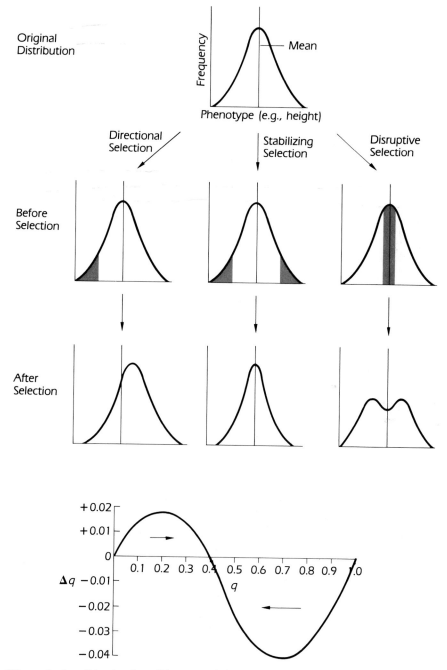

Figure 9–2. *Directional, stabilizing, and disruptive selection. Top, Frequency distribution curve for a hypothetical trait. Middle row, the same curve showing in black the part of the curve subjected to selective elimination for each type of selection. Bottom row, the results of selection.*

Finally, _disruptive selection_ occurs when two or more character states are favored, so that the population tends to break up into subpopulations. In its most extreme form, selection favors variation itself, with many different phenotypes coexisting. For instance, G. Moment reported a study of the brittle star, _Ophiopholis aculeata,_ in which no two among hundreds collected showed identical patterns, and many were radically different. He believes that this massive variation is protective because there is no single pattern that a predator can learn to associate with the prey. Another example is provided by the self-sterility genes of many plants. A pollen grain with a given _S_ allele cannot grow on the style of a plant with the same _S_ allele. As a result, in a population with a small number of _S_ alleles, many sterile pollinations occur, and so there is strong selection for diversification of _S_ alleles. Simpson has called this _fractionating selection._

EXPERIMENTAL DEMONSTRATION OF NATURAL SELECTION

Once researchers realized that selection might effect gradual changes, apparent only on statistical analysis, rather than all-or-none changes, they found it possible to study selective changes in natural populations and even to perform experiments in selection. W. Sukatchew analyzed such changes experimentally in the dandelion, _Taraxacum officinale._ Strains were obtained from the Crimea, Leningrad, and Archangel. As the latitudes, respectively, are 45°, 60°, and 64° (the latitudes of Ottawa, the capital of Canada; Seward, Alaska; and Dawson, the gold rush city of the Yukon), the dandelions were adapted to temperate, subarctic, and arctic conditions, respectively. The three strains were planted in mixed plots in Leningrad at densities of 3 cm and 18 cm between plants. About 60 percent of the Crimean strain survive in the sparsely planted plots, but in the densely planted plots, where competition (and hence selection) was more severe, its survival rate dropped to about 1 percent. At high densities, the Archangel strain had a greater survival rate than did the Leningrad strain (70 percent against 11 percent), but at low densities, the local strain was favored by a small but significant margin (96 percent against 88 percent). These observations are important for two reasons. First, they demonstrate that the differences between populations of the same species result in differential survival when a mixed population is exposed to identical conditions. Second, they demonstrate that selection need not be an all-or-none phenomenon but that it may operate by the statistical transformation of populations.

N.W. Timofeeff-Ressovsky, working in Berlin, similarly tested the relative viability of _Drosophila melanogaster_ and _D. funebris_ at temperatures of 15°, 22°, and 29° C. _D. funebris_ is originally a northern species, and _D. melanogaster_ is originally tropical. His method was to put 150 eggs of each species together in a culture bottle containing insufficient food for 300 larvae. The numbers of adults of each species that emerged were then compared. As *D. melanogaster* usually showed the greater viability, results were expressed by stating the viability of *D. funebris* as a percent of that of *D. melanogaster*.

$$\frac{\text{surviving } D.\ funebris}{\text{surviving } D.\ melanogaster} \times 100$$

Although the survival rate of *D. melanogaster* was almost always higher than that of *D. funebris*, northern races of *D. funebris* did much better relatively, when tested at 15°, than did southern races. In fact, one northern race showed a slightly higher survival rate than did *D. melanogaster*. At the higher temperatures, however, the survival rate of *D. funebris* was never so much as half that of *D. melanogaster*.

Goldschmidt studied adaptation to climatic conditions in the gypsy moth, *Lymantria dispar*. While the conditions of development in this study were determined by climate rather than by experiment, they nonetheless permit an insight into the nature of the selective force. The moths lay their eggs in the fall, but the eggs remain dormant until the following spring. In the spring, when the sum of the daily temperatures has reached a certain minimum, development proceeds and the larvae emerge and feed on the green foliage of certain plants. The hatching time of northern races is quite short; that of Mediterranean races is much longer. Obviously, a more rapid developmental cycle has selective advantage for the northern races, allowing maximal exploitation of the short summer. But if a southern race had a short hatching time, then after a mild winter the larvae would be likely to emerge before foliage appeared on their food-plants, Similarly, Mediterranean plants have a long period of growth of vegetation; in contrast, arctic plants have a very short period, corresponding to the short arctic growing season.

OBSERVED CHANGES IN NATURAL POPULATIONS

The preceding observations leave no room for doubt that selection can actually cause changes in the characters of species. Whether character changes have occurred historically is another question. Fortunately, important changes in response to known selective forces have occurred in some species within historical times.

Industrial Melanism

The best known example is the phenomenon of *industrial melanism,* which occurred in many species of moths during the industrial revolution. Dark (melanistic) forms have been known for hundreds of years as an occasional curiosity; but, beginning with the industrial revolution, they became ever-more numerous until the lighter original forms were rare and prized by collectors. The centers of distribution of the melanistic forms have been the large industrial cities.

Goldschmidt analyzed the melanism of the nun moth, *Lymantria monarcha*, genetically. The original pattern consists of narrow zigzag lines on a light background. Two independent, autosomal mutations cause an increase in the width of the zigzag lines and some pigment deposition between them. A third mutant gene is sex linked and causes an intensification of the pigments. The three pairs of genes, in various combinations, have additive effects and so can cause the wide range of phenotypes that are illustrated in Figure 9–3. Calculation of the mutation rate necessary to cause the observed increase in

Figure 9–3. *Industrial melanism in* Lymantria monacha. *In this species, pigmentation is controlled by two pairs of autosomal genes and one sex-linked pair. The moth at the upper left has the genotype for minimal pigmentation. Genotypes with progressively increasing amounts of pigmentation appear to the right and below, so that the moth at the lower right has the genotype for maximal melanism.*

melanism gave an absurdly high value, and so it was concluded that selection favored melanism.

The British biologist H.B.D. Kettlewell found a simpler situation in the peppered moth, *Biston betularia*. A single dominant gene for melanism exists and its frequency increased from less than 1 percent to more than 98 percent within the short space of fifty generations. The moths fly by night and rest by day on the trunks of trees. They are preyed upon by certain birds. In the absence of industrial pollution, the bark of the trees is light colored and encrusted by lichens that are even lighter, so that a light-colored moth blends into its background when at rest, whereas a dark-colored moth stands out sharply. In industrial regions, smoke blackens the bark and prevents the growth of lichens, so that dark-colored moths blend in and light-colored ones contrast with the background (Figure 9–4). Kettlewell ascertained, through observations

(a)

(b)

(c)

Figure 9–4. Industrial melanism in *Biston* betularia. *(a) Light moth on lichen-covered tree. (b) Dark moth on lichen-covered tree. (c) Both light and dark moths on a dark, sooty tree.*
Courtesy of the National Museum of Natural Sciences, National Museums of Canada.

recorded by motion pictures, that predators take light moths in disproportionately large numbers in industrial areas, and they take dark ones preferentially in unpolluted areas. Clearly predation seems to provide the selective force that favors one color or the other for its protective value.

Other explanations for industrial melanism have been proposed. Early in this century, some biologists favored a Lamarckian explanation. Later, E.B. Ford proposed that melanism was probably secondary to metabolic changes that gave resistance to poisoning by lead salts of industrial smoke. Although Ford may be correct, Kettlewell has proven the value of protective coloration. In preindustrial times, melanism failed to spread because melanistic moths were conspicuous to predators, and possible resistance to lead poisoning had little value. Both traits, however, proved to be preadaptive to the industrial environment; hence, melanism spread rapidly after the onset of the industrial revolution. That this selective trend of well over a century may be reversed is noteworthy. Since the late 1950s a major effort to reduce the levels of pollution has taken place in Britain; and light moths seem to be on the increase once again. This example is important as a well-documented case of transformation of species (at least to the equivalent of a subspecific level) by a known selective force, within historical times.

Additional Examples

Another often-quoted example, reported by W.F.R. Weldon in 1899, is the crab *Carcinus maenus*. The building of a breakwater in an English sound resulted in a higher silt content of the water. During a period of five years, observers noted a decrease in the mean diameter of the carapace of the crabs in the sound. Weldon captured both narrow-carapaced and broad-carapaced crabs and kept them in silty aquaria. The narrow-carapaced crabs survived, the broad-carapaced ones died. Weldon attributed this fact to accumulation of silt on the gills of the broad-carapaced crabs. He assumed that the same thing happened in nature, and so silting constituted a selective force. Although he did not demonstrate the actual silting of the gills, it is plausible.

The relationship of wheat and wheat rust is also illustrative. One of the primary objectives of plant breeders is to develop disease-resistant varieties of commercially valuable plants. To a disease-producing parasite, however, scarcity of susceptible hosts is obviously a severe selective force. The history of rust-resistant wheats is monotonous in its repetition. When a new resistant wheat is introduced, it gives excellent, disease-free crops and is widely adopted. After a few years, however, a few fields show some active rust infection. Then the successful rust spreads rapidly, and a new resistant wheat is again needed. Evidently, mutation has by chance produced a rust variety that is adapted to the new wheat, and it spreads rapidly because it is favored by a strong selection pressure. C.O. Person demonstrated close gene-for-gene correlation in the evolution of host wheat and parasitic rust.

H.J. Quayle reported an interesting case of selection in the scale insects that parasitize citrus trees. The standard method of combating the insects is to cover each tree with a tent and fumigate with hydrocyanic acid. But in each of the three species concerned, cyanide-resistant varieties have appeared and have replaced the original cyanide-sensitive varieties. One of these resistant species has subsequently disappeared for unknown reasons. A similar development occurred with respect to DDT poisoning. When DDT first came into general use late in 1945, it gave promise of being an almost-perfect insecticide for the control of household pests such as flies. But soon DDT-resistant flies began to appear. Under the strong selective force of DDT poisoning campaigns, these resistant strains soon became well established, and in many localities they have largely replaced the original DDT-sensitive flies.

W.L. Brown and E.O. Wilson studied an interesting case in the fire ant, *Solenopsis saevissima*. A large, dark variety of this ant was accidentally introduced into Alabama from South America about 1918. It spread slowly and caused no alarm. During the 1930s, however, a smaller, lighter-colored form appeared, probably as an immigrant from South America. This lighter form proved to be far more aggressive than the darker form and it destroys the nests of the latter. Where both forms have occurred together, the lighter has replaced the darker, evidently as a result of selective advantage in competition. It has become a major pest in the southeastern United States; and large-scale programs of extermination, heavily financed by the Department of Agriculture, have been futile.

Heavy metals in the soil are highly toxic to plants. Thus the area downwind from the nickel mines and smelters at Sudbury, Ontario was so desolate that NASA selected it for training for moon landings. Genotypes for metal tolerance have developed in many plants, and these metal-tolerant genotypes, aided by strong selection pressure, are now actively reinvading the area. Metal-tolerant genotypes are at a disadvantage in nontoxic soils.

Polymorphism

Selection does not lead to uniformity; it leads to high frequencies of those genes that contribute to the most successful genotypes. However, heterozygosity itself has considerable value, for it provides a source of variability, which increases the probability that at least some of the progeny from a mating will be adapted to any environmental changes. Most natural populations of sexually reproducing organisms are highly heterozygous. Whenever this heterozygosity results in obviously different phenotypes, we speak of *polymorphism*. The most-striking cases are those of mimetic polymorphism (see the following discussion).

Specific selective forces may maintain a balanced polymorphism, as in the case of the inversion types of *Drosophila pseudoobscura* (Chapter 8). Dobzhansky has shown that some chromosomal arrangements increase in frequency during the summer; others, during the winter. Hence they must be related to seasonal adaptations. A particularly instructive case is known in humans. Sickle-cell anemia is a fatal disease resulting from homozygosity of a certain gene. The same gene, when heterozygous, causes the harmless sickle-cell trait (which is expressed only under special conditions). Strong selection pressure should tend to eliminate such a gene, but it is nonetheless common in equatorial Africa. This fact was inexplicable until researchers found that heterozygous persons are resistant to malaria, which is prevalent in the same region. Thus selection favors heterozygosity, in spite of the severe liability of homozygosity.

Selection in nature, originally postulated by Darwin as a necessary consequence of the prodigality of nature and of the variability of all species, has been thoroughly vindicated by observations such as these. The following discussion focuses on two special aspects of natural selection: adaptive resemblances or coloration and sexual selection.

ADAPTIVE RESEMBLANCES

The principal types of adaptive coloration are cryptic resemblance, by which the animal blends into its background; warning, or aposematic, coloration, by which an obnoxious or dangerous animal is made obvious to potential predators; and mimicry, by which one species resembles another, presumably taking advantage of aposematism.

Cryptic Coloration

Cryptic coloration (Figure 9–5) is widespread in the animal kingdom. It is useful not only for protecting potential prey but also for permitting predators to avoid detection long enough to make a kill.

Perhaps the simplest form is countershading—that is, dark pigmentation on the dorsal surface and light color on the ventral surface. Its protective function may be well illustrated with a fish. A bird looking down upon the fish sees the dark dorsal surface blending into the darkness of the depths. But a larger fish, looking up from below, sees the light ventral surface blending into the daylight. Although the protective effect of such a pattern is plausible, even probable, it is by no means certain that its occurrence depends upon selective value. Some suggest that the very universality of countershading indicates a simple physical cause. Often development of pigment requires exposure to light. As most organisms are exposed to light much more on the dorsal surface than on the ventral, we would expect countershading to be the rule. The Nile catfish, *Synodontis,* however, normally swims with the ventral surface up; and it is the ventral surface that is darkly pigmented and the dorsal surface that is light. Also, researchers have reversed this pattern experimentally in some fishes by raising them in aquaria lighted from below. But, although such facts demonstrate the *means* of the development of coun-

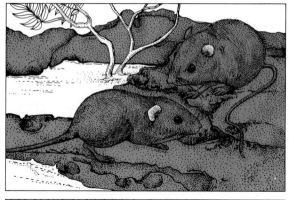

Figure 9–5. *Cryptic coloration of pocket mice from adjoining areas:* Perognathus intermedius ater *on black lava and* P. apache gypsi *on white gypsum sand.*

tershading, they do not disprove its protective value. Neither do they explain why small surface fishes are typically transparent and abyssal fishes are typically dark colored. Both colorations appear to be protective, but both run counter to the light gradient.

Industrial melanism clearly demonstrates the effectiveness of such crypsis against predation. F.B. Sumner's experiments on predation by penguins on mosquito fish (*Gambusia*) are also most illustrative. These fishes slowly change color to match their background. If one group of fish is kept in a black tank and another in a light tank, each group will become adapted to its own background. If fishes from one of the tanks are then transferred to the other, they will contrast with their background until they can again become adapted. Sumner exposed such mixed groups of adapted and unadapted fishes to predation by penguins. Always, both types of fish were taken, but the unadapted fishes were taken in much greater proportion than were the adapted ones (Table

9–1). The differences are highly significant statistically in both types of tank, and thus these data support the protective value of the color changes.

L.R. Dice pioneered experimentation on protective coloration in small mammals. He exposed variously colored races of the white-footed mouse, *Peromyscus,* to predation by owls. When he exposed these variously colored mice on a background nearly matching the color of one race, members of that race were taken less frequently than were the others. Many cases of adaptive resemblance are difficult to interpret in any other way than as examples of cryptic coloration. For example, the arctic hare, the northern weasel, and the ptarmigan molt and grow white coats in the fall, then revert to brown coats in the spring. The protective utility is self-evident. Then there are numerous cases of stick insects (Figure 9–6) and leaf insects (Figure 9–7), which, so long as they do not move, look like parts of the plants on which they are normally found. Many moths have a close resemblance to the bark of the trees on which they usually rest.

In a few instances organisms actively acquire resemblances to the environment. For instance, the masking crabs, of which *Loxorhynchus crispatus* (Figure 9–8) is a good example, actually "plant" upon themselves algae, hydroids, sponges, and other sessile organisms from their environment. If a masking crab is moved from its original locale, it will seek an area in its new environment with the same kinds of organisms that it is carrying. If no such area is available, the crab will remove its riders and replace them with the sessile flora and fauna of the new locality.

T. Eisner and his collaborators have reported a remarkable case among insects. There is a well-known symbiosis between wooly alder aphids, *Prociphilus tesselatus,* which produce a sweet fluid, and ants that shepherd these aphids and "milk" them. In return, the ants protect the aphids against their predators, the larvae of the lacewing, *Chrysopa dossoni.* The wooly appearance of the aphids results from a growth of waxy fibers on their backs. The lacewing larvae pluck the fibers from the aphids and implant them on their own backs. They then

Table 9-1. Predation by Penguins on Adaptively Colored Mosquito Fish

Fish Color	Tank Color	Number of Fish Tested	Number of Fish Eaten	Percent of Fish Eaten
white	light	528	176	39
black	light	528	278	61
white	black	335	217	74
black	black	335	78	26

Data from F.B. Sumner, Evidence for the protective value of changeable coloration in fishes, *American Naturalist* (1935) 69:245–266.

Figure 9–6. Walking stick insect, Diapheromera femorata.
Courtesy of the National Museum of Natural Sciences, National Museums of Canada.

resemble the aphids so closely that human observers found it difficult to identify them. Ants patrolling the aphid colony failed to remove the disguised lacewings, although they quickly removed any unmasked lacewings. If we deny that such complicated behavioral instincts as those of the masking crabs and the lacewing larvae have a protective value, we are left with no explanation for this behavior.

Warning Coloration

Warning, or aposematic, coloration is also common in the animal kingdom. Here the object is just the opposite of that of cryptic coloration: whereas a cryptic pattern tends to render the animal inconspicuous, a warning pattern is obtrusive and advertises the presence of an otherwise well-protected animal. As H.B. Cott has put it, "Their bite is worse than their bark." Such animals have formidable defense mechanisms, produce foul-smelling substances, or

Figure 9–7.   Dead leaf insect, Kallima, is clearly visible with its wings spread, but it tends to disappear when its wings are folded.

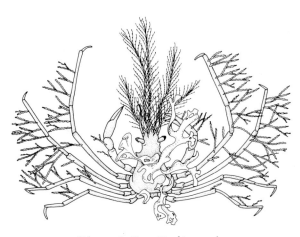

Figure 9–8.   Masking crab.

have a disagreeable taste. The object seems to be for the organism to present an easily recognized appearance and thereby educate potential predators, after an initial bad experience, to avoid that organism. That the method can work is well known to everyone whose dog has attacked a skunk or a porcupine. A second encounter rarely occurs.

The skunk and the porcupine are clear examples of warning coloration among the mammals. Many poisonous snakes are brilliantly colored, and their coloring may be aposematic. In general, the Amphibia are cryptically colored, but some have brilliant aposematic patterns, for example, *Triturus torosus*, the western water dog, which has abundant poisonous skin glands. Other examples are tiny aposematic frogs (Figure 9–9) and dendrobatid frogs of Central and South America, which are brilliantly colored but highly toxic; some supply the poison for the darts of local Indians. Many of the most brilliantly colored insects are believed to taste repugnant (butterflies); others have stings (bees) or emit foul-smelling fluids (coccinellid beetles).

The presumed distastefulness of many aposematic animals has been proven in specific instances. *Oscanius membranaceus*, an aposematic mollusc, secretes dilute sulfuric acid. Most fish will not eat it, and they refuse their usual foods if these foods are treated with the acid. Larvae of the magpie moth, *Abraxas grossulariata*, are aposematic. When these larvae were offered to lizards and frogs, they were seized at once, then dropped and refused thereafter. Subsequently, the predators sat with mouths agape and tongues rolling, as though trying to get rid of an obnoxious taste. To cite a final example, a variety of freshly

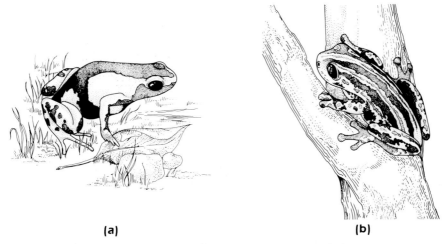

(a)   **(b)**

Figure 9–9.   Aposematic frogs. (a) Mantella cowani *and (b)* Hyperolius marmoratus *have striking patterns in glossy black, red, orange, and yellow. These warning colors, like those of the dendrobatid frogs of Central and South America, are correlated with the secretion of strong toxins by the skin glands.*

killed insects was placed on a bird feeding tray. Birds feeding on them, took most of the cryptically colored insects and left the aposematic ones. Warning coloration, then, does confer protection.

Curiously, crypsis and aposematism are not necessarily antithetical. The bold black-and-white pattern of zebras, for instance, stands out strongly against the tan of the savannah—hardly cryptic. However, when an African hunting dog or a lion surveys a herd of running zebras, it sees a confusing jumble of moving stripes, and the vulnerable individual blends into the protective herd. Another example is that of the coral snake (*Micrurus*), which H.W. Greene recently studied. This small, neotropical snake is poisonous but not usually aggressive. It is boldly ringed in red, yellow, and black—a pattern readily interpreted as aposematic. However, the snake lives under a forest floor litter of yellow and brown leaves, where its pattern, so striking in other surroundings, is highly cryptic. Its pattern may be viewed as cryptic before discovery and as aposematic afterward.

Mimicry

Finally, there are the fascinating cases of mimicry. These are cases of aposematism in which the same pattern (or closely similar patterns) is shared by two or more species. In the original description, H.W. Bates assumed that one of the species, the model, is genuinely aposematic and appropriately marked, while the other species, the mimic, is desirable prey but shares the protection of the model by assuming its cloak. This phenomenon is known as *Batesian mimicry*. F. Müller later described what appeared to be an exceptional case of mimicry in which *both* the model and the mimic appeared to be protected because they were repugnant to predators. This type of mimicry, by which two or more species present a single aposematic pattern for their predators to learn, is called *Müllerian mimicry* and now appears to be much more common than Batesian mimicry. It presumably operates as a kind of double insurance.

R. Mertens described a third type of mimicry, called *Mertensian mimicry*. Mertens studied coral snakes, which are highly poisonous, and false coral snakes, which may be either mildly poisonous or harmless. Mimicry is common among them all. Mertens emphasized that it must be the mildly poisonous snakes that serve as models both for the highly poisonous snakes (Müllerian mimicry) and for the harmless snakes (Batesian mimicry) because a predator never learns caution from the highly poisonous snakes: the first attack is fatal to the predator. H.W. Greene and R.W. McDiarmid have verified these observations and, like Mertens, concluded that Batesian and Müllerian mimicry are adequate to explain all cases.

L.P. Brower and J.V.Z. Brower, in experiments using monarch (model) and viceroy (mimic) butterflies (Figure 9–10) as prey for caged jays, clearly proved the effectiveness of mimicry. Jays that were conditioned as a result of

Figure 9–10. *Mimicry of the monarch butterfly,* Danaus plexippus *(lower right), by the viceroy,* Basilarchus archippus *(upper left). The colors are black and orange, with a small amount of white.*
Courtesy of the National Museum of Natural Sciences, National Museums of Canada.

feeding on monarchs subsequently refused both model and mimic; unconditioned birds, in contrast, took the mimics readily. The Browers showed that the distastefulness of butterflies results from toxins in the food plants of the larvae.

An experiment of Swinnerton shows the effectiveness of mimicry among birds. The African drongos, *Dicrurus afer* and *D. ludwigi,* are black all over and are unpalatable. The flycatcher, *Bradyornis ater,* and the cuckoo-shrike, *Campephaga nigra,* mimic the drongos, but they are themselves edible. The tit, *Parus niger,* resembles the drongos ventrally but has conspicuous white markings dorsally. Swinnerton offered a cat specimens of all five species, turned ventral surface up. The cat refused them all. But when he turned the birds dorsal surface up, the cat quickly took the tit, but continued to refuse the others.

Mimicry may also serve functions other than protection against predators. For example, orchids of the genus *Ophrys* have flowers that mimic, even to the scent, female wasps of the genus *Scolia.* Male wasps try to copulate with the flowers (Figure 9–11), and, in going from one flower to the next, they cross-pollinate them. Another striking case of mimicry in plants is that of *Lithops* (Figure 9–12), the windowed plants of the Namib desert of southern Africa. These plants mimic the gravel in which they grow. The flat-topped surfaces of the leaves are almost flush with the ground. The species are very local, and they mimic the colors of the gravel of each local region. Thus, the plants are cryptic.

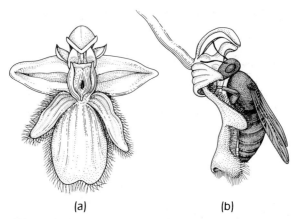

(a) (b)

Figure 9–11. *Pseudocopulation. (a) Flowers of the orchid* Ophrys *mimic female wasps of the genus* Scolia. *(b) Males of* Scolia *attempt to copulate with the flowers and thereby cross-pollinate them.*

Figure 9–12. Lithops ruschiorum, *from the deserts of Namibia, matches its background closely. Arrows indicate some of the plants, can you identify others?* Courtesy of Dr. J.A. Jump.

AGNOSTIC REACTION AGAINST PROTECTIVE RESEMBLANCES

Perhaps no single factor did more to cause the agnostic reaction than did the excessive speculations of nineteenth-century evolutionists on adaptive resemblances. Every possible quirk of nature was interpreted by someone as having adaptive and, therefore, selective value to its possessor.

During the Romantic Period almost every dull color pattern was interpreted as cryptic; almost every brilliant pattern, as aposematic; and almost every resemblance between pairs of species as mimicry. These interpretations involved many gross errors and much uncritical acceptance of scanty and ill-founded data. In subsequent years, the selective value of color was commonly denied altogether or treated as one of the most doubtful aspects of evolutionary theory. Finally, W.L. McAtee published a study, based on the examination of the contents of the stomachs of 80,000 birds, in which he reported that both protectively and nonprotectively colored insects were eaten by North American birds in numbers proportional to their respective populations. Because of its broad experimental basis, this study had great influence in discrediting the whole idea of adaptive coloration.

However, McAtee's study was invalidated because of serious statistical errors. He carefully recorded the number of bird stomachs in which each species of insect was found, but he did not record the number of insects found there, so that if he found one insect of species A and a hundred insects of species B in ten different stomachs, he reported the data identically. Further, he made no distinction among species of predators, yet protection against only some predators may be important. Thus, the porcupine (*Erethizon*), which is aposematically colored and armed with formidable quills, is protected against most carnivores. The fisher (*Martes*), however, successfully takes porcupines by turning them over on their backs, then ripping open the unprotected ventral surface. We would hardly wish to claim that, because a single carnivore takes porcupines, their protection is not effective. If, indeed, their protection were ineffective, many other carnivores would kill porcupines.

Sumner's data on predation by penguins as well as the data on industrial melanism demonstrate that cryptic coloration is effective; the Browers' data prove that warning coloration and mimicry are effective. Collectively, the vast store of data on adaptive resemblances is persuasive testimony to the power of natural selection.

SEXUAL SELECTION

A second special feature of Darwin's theory is sexual selection. Darwin believed that the general theory of natural selection could not account for the color differences that so commonly characterize the sexes nor for other types of ornamentation that differentiate the sexes. He, therefore, proposed the theory of sexual selection to account for such differences. In brief, the theory is that the female selects her mate and that, therefore, any especially attractive male will have improved chances of obtaining a mate and leaving descendants. This leads to the evolution, in the male, of brilliant colors, elaborate combs, ornamentative hooks—in fact, of any secondary sex character that might be regarded as potentially attractive. Darwin also included antlers, tusks, and spurs; these, however, could also be accounted for by the general theory of natural selection since they relate to direct competition between members of the same sex.

The idea of sexual selection requires that either males be more numerous than females or polygamy be the rule, that the ornamentation of the males be attractive to the females, and that the females select their mates. Some positive evidence in favor of the theory has been obtained, but it is scant. Generally speaking, the numbers of males and females seem to be about equal. There are some outstanding examples of polygamous societies among animals. One of the best is the fur seal, which breeds on the Pribilof Islands near Alaska. The males arrive at the breeding grounds before the females; they then engage in mortal combat, with many of the males either being killed or being driven away from the breeding grounds. Upon arrival of the females, each surviving male acquires an extensive "harem."

A reversal of the usual pattern of sex behavior is another phenomenon that has been interpreted as supporting the theory. For example, phalaropes (related to sandpipers) are characterized by brilliant plumage in the female and dull plumage in the male. But the female does the courting, and the male builds the nest and incubates the eggs.

There is much doubt about how much influence ornamentation has upon actual selection of mates. The female seals referred to in the preceding discussion do not appear to have any choice of mates; they simply accept the male that is at hand when they arrive at the breeding grounds. In the deer family, vigorous displays of antlers (Figure 9–13) are directed first and foremost at rival males, and victory in combat may confer on the victor mating privileges with a whole harem of females. In some cases, secondary sex characters appear to excite sexual activity without actually influencing choice of mates. In *Drosophila,* mating is preceded by courtship behavior that includes wing movements by the male. Wingless males can mate with normal females, but it takes much longer. Now Sturtevant has shown that, if a normal pair of flies is put into a bottle along with a wingess male, the female will mate after a normal time lapse, *but she will mate with the wingless male as readily as with the normal one.* It is evident that the courtship of the normal male has hastened receptiveness of the female without influencing her choice of mates.

Much of the literature on sexual selection is speculative, but M. Andersson, a Swedish zoologist, has recently made an experimental study of sexual selection in the African widowbird, *Euplectes progne.* Females are dull colored and have

Figure 9–13.   *Elk,* Cervus canadensis.

short tails. Males have bright wing patches and showy tails that average a half meter in length, and they defend territories within which they keep harems of nesting females. Andersson cut the tail feathers of some of the males to half of the normal length and glued the cuttings to the ends of the tails of other males, thereby creating three classes of males: short tailed, normal tailed, and long tailed. He then released these birds and allowed them to establish territories. The three classes of males were equally successful in defending territories; but the long-tailed birds were the most successful in attracting females to their territories, and the short-tailed birds were least successful. These data suggest that females' choice of the longer-tailed males is a factor in maintaining the sex difference in this species.

R.A. Fisher suggested that if females were to prefer a trait that reduced the fitness of males, extinction could result. The excessive development of the tail, as in the peacock, is a possible example. Presumably, these very conspicuous birds might be more vulnerable to predators than are more modest birds. If females were to select the more extreme males as mates, then the male offspring would be pushed toward still greater extremes, and the female offspring might inherit a preference for these extreme males. Fisher called this *runaway selection.* P.H. Harvey and S.J. Arnold have recently supported this idea.

Finally, most attempts to determine factors in choice of mates have been inconclusive. Courtship behavior in general may well have more to do with sexual excitement than with choice of mates. And simple proximity may well have more influence on choice of mates than does any other factor for most animals. Sexual selection has probably played some role in evolution, but the magnitude of that role has not been successfully assessed, and the general theory of natural selection may include some of the phenomena ascribed to sexual selection. Some Russian biologists, particularly L.S. Davitashvili, have found the theory of sexual selection attractive, and they have elaborated it at length.

SUBSTRATE OF SELECTION

Most biologists, in agreement with Darwin, have treated the *individual organism* as the unit of selection. It is the individual that succeeds and reproduces or that dies without leaving offspring. R. Dawkins, however, promotes the idea that the gene itself is the unit of selection. Indeed, he seems to regard genes as the ultimate organisms, with visible organisms functioning largely as gene carriers, structures produced by the genes to ensure their own reproduction. This idea is reminiscent of Weismann's theory contrasting germplasm and soma. Up to a point, the idea is a truism: selection does favor some alleles over others. It is, however, an oversimplification, if only because the gene is never isolated—that is, whole systems of genes work in concert, and an allele that is favored in one system may be disfavored in another. The creative role of selection is in bringing together such favorable constellations of genes. Only through the reproduction or elimination of whole individuals can the frequency of a gene (or, rather, of a given allele) be changed; and a typical organism carries a constellation of genes that, individually, have quite different selective values. Hence, the organism as a whole, rather than the individual gene, still seems to be the unit of selection.

Others, like V.C. Wynne-Edwards, favor *group selection*. Many organisms, like the marine birds that he has studied, have complex social systems. He argues that the reproduction of the individuals is controlled by the structure of the avian society; that it is the group as a whole that succeeds or fails; and that selection, therefore, works on the group as well as on the individuals. Certainly, group characteristics often influence survival. Herds of bison, for example, work as a unit for protection against predators. Territorial birds, such as song sparrows, divide up suitable habitat early in the breeding season, and supernumerary males are able to reproduce only if they can find an available territory (e.g., a predator eliminates a territory holder). Nonetheless, it is still the individual organism that has the potential for a particular type of social behavior. Cases that were classed as examples of group selection have often been successfully explained in terms of selection of individuals adapted for (or maladapted to) specific types of group behavior. Consequently, most evolutionary biologists are not convinced that the concept of group selection is necessary.

Kin Selection and Sociobiology: The Biology of Cooperation

British biologist W.D. Hamilton brought about the resolution of the group selection problem in his consideration of the question of *altruism*—that is, of how behavior that results in the saving of a group of animals at the risk of the life of the altruist could evolve through natural selection. An example of altruistic behavior would be the giving of a vocal warning signal that draws a predator's attention to the signaler. Surely genes for such behavior would be eliminated, not preserved, by natural selection. The answer, as Hamilton noted, lies in the fact that members of a group are commonly related to each other and thus share common genes—a condition that is certainly true of bees in a hive, termites in a mound, lions in a pride, wolves in a pack, and so on. In this way, the altruist preserves his own genes carried by his siblings and cousins—natural selection at work. J.B.S. Haldane reportedly once declared, "I will lay down my life for two of my brothers or eight of my cousins," thereby inferring precisely how kin selection might work— namely, that the probability of a given gene's being shared by sibs is one-half and by cousins is one-eighth. Substituting kin selection for group selection thus returns the phenomena of group behavior to a strictly Darwinian framework.

The theory of kin selection prepared the way for the development of a new field of social behavior, as heralded by Harvard biologist E.O. Wilson's book, *Sociobiology*. Wilson presented a masterful survey of social behavior within the animal kingdom and pointed out how such behavior is comprehensible on a Darwinian basis. Examples are numerous. A honeybee that defends the hive by stinging an intruder inevitably dies; even though it has not reproduced, it has fostered its own genes by protecting its kin. Contrarily, a lion, entering a new territory and taking control of a pride by ousting the previously dominant male, promptly kills the young cubs. Although heartless by our standards, this behavior is sanctioned by natural selection, whereas expending effort to foster another's genes is not. In promiscuous tribal societies, sociologists have found that a man may provide greater care for his sister's children, who have some of his genes, than for his wife's children, who may or may not carry his genes.

Sociobiology is predicated on the proposition that behavior has a genetic basis and is, therefore, inherited. Because behavioral traits may increase or decrease the chances of survival and reproduction, they are part of the phenotype that is exposed to natural selection. Behavior is thus a character complex that evolves. Cooperation is not a contradiction of natural selection but a consequence of it. In humans, cooperation and competition exist side by side, and both are integral parts of our nature. Sociobiology is most controversial when applied to human society, partly because no specific human behavioral trait has been shown to have a genetic basis. Indeed, behavioral plasticity appears to be an outstanding human trait. (Chapter 23 provides further discussion of this topic.)

Hierarchy of Organization and Selection

S.J. Gould has proposed an interesting variant on the theme of group selection. He notes that a hierarchy of levels of organization exists in the living world and that we can study life at the molecular, cellular, organismic, or community level. He suggests the occurrence of a form of individuality at each level—that is, the gene, the whole organism, the deme (in ecology, an interbreeding population), the species, and the clade (a line of descent). Perhaps we could carry this hierarchy further to include genera and even higher groups. In this hierarchy of levels of organization, Gould suggests that each level may function as an individual for some purposes, and selection might be as effective on these hierarchic individuals as it is on individual organisms. The experiments and observations to test this interesting hypothesis are still in the future.

Strategies of Selection

In general, natural selection works by differential reproductive success. It is not the most-prolific species that dominate the earth. Reproductive success is not simply a matter of numbers of offspring produced; rather, it is the numbers that survive to maturity and reproduce successfully in turn. This fact suggests two contrasting strategies of competition: a species could specialize to produce large enough numbers to offset probable losses, or it could specialize to insure maximum survival of a relatively small number of offspring.

In ecological terms, r represents reproductive capacity, and K represents the carrying capacity of the habitat. The two types of specialization are called r and K strategies, respectively. Many parasites, for example, are r strategists. A sheep liver fluke may produce several hundred thousand eggs in a day. These are passed in the feces of the sheep and they soon hatch as miracidium larvae. The larvae seek a secondary host (usually a pond snail) and give rise to a succession of larval types that reproduce asexually, so that each egg potentially gives rise to hundreds of flukes. Finally, the larvae metamorphose into cercaria larvae, which encyst on grasses along the edge of the pond. If eaten by grazing sheep, the cysts will form mature worms, which migrate to the liver of the sheep. At every stage of this complex life cycle (see texts on parasitology for details), the losses are enormous, so that only a very small fraction of 1 percent of the young ever reach maturity. Dandelions are also r strategists, and they illustrate well two common characteristics of this strategy: ecological opportunism and high potential for dispersal. In general, r strategists resist extinction well because they are prolific enough to survive great losses.

The whooping crane (*Grus americana*) is an extreme example of a K strategist. These magnificent birds, whose total population in the spring of 1982 was only eighty-eight in the wild and twenty-nine in captivity, winter on the Texas Gulf coast. In the spring they migrate to Wood Buffalo National Park in the far north of Canada, where they nest. The female lays just two eggs, but she cares for only one chick. All the parental care is concentrated upon the success of that one chick.

The passenger pigeon (*Ectopistes migratorius*) was also a *K* strategist, for a nesting female laid only one egg. Yet, so successfully did passenger pigeons exploit their ecological niche that they produced an immense population. *K* strategists depend upon precise adaptation to their ecological niche. If the niche is significantly altered, they may not be able to keep pace, and so they are relatively susceptible to extinction.

For the whooping cranes, extinction has been a near miss up to the present time; for the passenger pigeons, it was the final reality. These examples are, of course, extreme. The majority of plant and animal species are intermediate in their strategies, but the concepts of *r* and *K* strategies are useful.

Coevolution

An interesting aspect of natural selection is its coordination of the evolution of ecologically related species, each of which acts as one of the agents of selection for the others. In an earlier discussion, we cited the related genetic systems of wheat and the rusts that parasitize it; and we know other such systems. Parasitism commonly evolves toward lower levels of virulence. Thus, Dutch elm disease was introduced into North America around 1930, and it has devastated the elms there. Yet in Europe, where tree and fungus have been associated for millenia, the disease is mild (although a severe form occurs in England). The mild form in Europe may be a result of the evolution of resistance by the trees or of reduced virulence by the fungus or both. Similarly, mild diseases of Europeans, such as chicken pox, have decimated aboriginal populations elsewhere.

An important theme in the evolution of flowering plants has been the development of systems for dispersal of pollen and seeds. Insects, birds, and bats have coevolved with these plants in mutually beneficial relationships. Before the advent of flowering plants, chancier, more profligate wind dispersal of pollen was the norm. (Some flowering plants still disperse their pollen by wind, as all ragweed-sensitive hayfever sufferers know.) Flowers may attract pollinators with elaborate olfactory, visual, or structural adaptations; and they may reward them with nutritious nectar and pollen. Orchids are a particularly favorable subject for the investigation of evolutionary relationships of plants and their pollinators; indeed, Darwin devoted an entire book to this subject. Although bees may be attracted by the sweet scent of nectar, flies are attracted by putrecine and cadaverine, fetid chemical attractants offered by the lowly skunk cabbage.

Such coevolution is well exemplified by L.W. Macior's researches on pollination of *Pedicularis* (louse wort), a large genus of about 700 species.

These self-sterile plants depend on bumblebees (*Bombus*) for cross-pollination. Of the twenty-two North American species that he has studied, some are vernal (spring flowering), some aestival (summer flowering), and one is autumnal. The vernal species and some of the aestival species produce nectar; other aestival species and the autumnal one lack it. Vernal and early aestival species are visited by queens that forage for nectar, using their long tongues. The queen enters the flower upright (Figure 9–14), her back picking up pollen from the anthers and depositing pollen from a previously visited flower on the stigma. This *nototribic* (Greek, "back rubbing") method of pollination is the simplest and perhaps the most primitive in the genus. It is found in such species as *P. canadensis* and *P. crenulata*. Those early aestival species that produce nectar (e.g., *P. bracteosa*) are also visited by queens, but the rest are visited only by workers, which are smaller and have short tongues. They forage for pollen. Some, like *P. groenlandica*, are also pollinated nototribically. Others are adapted for *sternotribic* (Greek, "chest rubbing") pollination (Figure 9–15). The inverted bee hangs from the flower and picks up pollen on its ventral surface. In the flowers of plants visited sternotribically, the stamens may be concealed within the upper lip of the flower, and the vibration of the wings of the bee shakes the pollen loose. In all cases, a close correspondence between the structure of the flower and the structure and behavior of the bee has evolved.

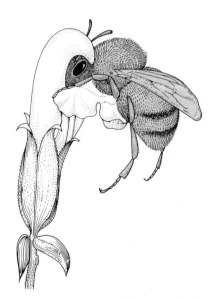

Figure 9–14. Nototribic pollination. The bee enters the flower upright and rubs its back on the stamens and stigma. In passing from flower to flower, cross-pollination is effected.

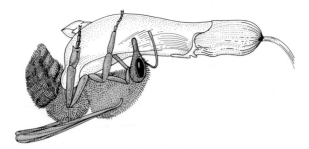

Figure 9–15.   *Sternotribic pollination. The bee hangs beneath the flower and rubs its abdomen over stamens and stigma. It passes from flower to flower and effects cross-fertilization.*

Fruits and nuts are another means of insuring reproduction and survival of flowering plants. Seeds that fall on the ground beneath a parent tree may have little chance of growing to maturity. However, if consumers can be attracted by colorful, fleshy, nutritious fruits, they may eat them, then later defecate the seeds in a more-favorable place, where the seeds may germinate and grow. Indeed, seeds of some species require passage through a gut, which scarifies the seed, in order to germinate.

The dodo (*Raphus cucullatus*), a large, flightless pigeon of the island of Mauritius in the Indian Ocean, rendered this service for *Calvaria major,* a tree endemic to the island. Since the extinction of the dodo some three hundred years ago, no *Calvaria* seedlings have germinated, although surviving trees still set seed today. However, when the nuts are force-fed to turkeys, they germinate after being passed in the feces of the bird. Thus, turkeys are capable of rendering the same service that the dodos once did.

D.H. Janzen and P.S. Martin have noted many trees and shrubs in Costa Rica that have large fruits with tough seeds and that lack agents of dispersal. The fruits fall to the ground and rot, or they are destroyed by seed predators such as bruchid beetles. Introduced cattle eat such fruits avidly and disperse the seeds successfully. In Africa large herbivores (e.g., elephants, rhinos, giraffes) consume similar fruits and disperse the seeds successfully. Janzen and Martin believe that the rich Pleistocene fauna of Central America, which included elephants, camels, giant ground sloths, giant armadillos, horses, bears, bison, and other large mammals, provided agents of dispersal for many of these trees and shrubs. When these animals became extinct some ten thousand years ago, the trees went into a decline, and only a much-reduced remnant survives today.

Another interesting aspect of coevolution is the interaction between herbivores and their food plants. Many plants have evolved defenses against herbivores. The simplest defense is rapid growth, thereby enabling the plant to keep ahead of the herbivores. Range grasses, for example, grow rapidly and spread asexually, by producing runners, as well as sexually, thus keeping ahead of the grazers.

Many plants produce alkaloids, tannins, and other toxins that repel herbivores. Thus, oak leaves contain considerable amounts of tannin, which binds proteins and makes them indigestible. Alkaloids are produced by an array of plants, including the common nightshade family, of which tomatoes and potatoes are useful members. Glycosides are produced by such plants as *Sorghum, Prunus,* and *Digitalis* (the latter is well known for its medicinal use). The list of plant toxins is a very long one.

Nettles and thorns are also best understood as defenses against herbivores. No defense is perfect, however, and many animals have evolved counterdefenses that enable them to feed on the "protected" plants. For example, cucumber leaves contain a toxin against the spider mite, *Tetranychus,* but some mutant strains of the mite are resistant to this toxin. The passion flower vine, *Passiflora,* has evolved toxic compounds against insects, yet the larvae of the butterfly *Heliconius* have not only evolved resistance to the toxins, but they have specialized for feeding on this plant. L.E. Gilbert, however, suggests that at least one species, *P. adenopoda,* may have won this battle againt the butterfly, for it has evolved hooked hairs that immobilize the larvae, thereby causing their death.

Coevolution of predator and prey is widespread. Prey organisms evolve defenses against predators, and predators must evolve means to counter those defenses. The interactions of ants, aphids, and lacewings are excellent examples of complex coevolution involving mutualism, predation, defense, and countering strategies. The case of the porcupine, which is protected against almost all predators by its quills, and the fisher, which has learned to rip open the unprotected belly of the porcupine, is another good example. Even humans and wolves have been described by R.L. Hall and H.S. Sharp as coevolving because of competition for common prey. Is it because of this that many people have so virulent a hatred for the wolf? Yet Huron Indians refer to "brother wolf."

Clearly, coevolution is an enormous subject, the dimensions of which can only be hinted at here.

SOME LIMITATIONS ON NATURAL SELECTION

We have seen that natural selection is a very powerful principle, yet it is not without limitations. No single principle is adequate to explain all aspects of the diversification of life, a point of which Darwin was well aware. Natural selection does not produce perfect organisms, only fit ones. As S.J. Gould has repeatedly emphasized in his writings, it is the *imperfection* of organisms, the recurring tendency to modify existing structures rather than to create new ones *ex nihilo,* that constitutes some of the strongest evidence for evolution. François Jacob used the analogy of natural selection as a tinkerer who fashions parts out of this or that, whatever is available (a wing out of a leg, an ear out of a bit of jaw), rather than as an engineer who creates optimum solutions out of materials designed for one purpose only. Part 4 provides abundant examples of this theme.

Natural selection is constrained by past evolutionary history. Terrestrial vertebrates have two pairs of limbs (or fewer). Is this an adaptive ideal or is it the consequence of the fact that the first vertebrates to crawl out on land had two pairs of limbs? We can easily imagine organisms that might have but did not evolve (e.g., elephantine insects, walking trees, cloven-hoofed horses). Surely mechanical and physiological considerations prevent natural selection from producing the first two, but just as surely the latter represents a path evolutionary history might have but did not follow. The following discussion deals with some factors that limit or appear to limit the role of natural selection in evolution.

Selection and Nonadaptive Characters

One of the objections often raised against the theory of natural selection is that it cannot account for the many cases in which differences between related organisms do not have any evident adaptive value. There are several ways in which such cases may be harmonized with the theory.

First, it is difficult to prove in any particular case that there is no adaptive value. The endocrine glands of vertebrates were thought to be without function as recently as the last century. The demonstration of their manifold functions awaited the development of suitable techniques. The same thing is at least potentially possible with respect to any character that appears to have no adaptive significance when studied by present methods.

Second, an apparently nonadaptive character may have adaptive value during a limited phase of the life cycle. For example, many terrestrial animals return to water for breeding. If we studied them only during their terrestrial phase, their aquatic adaptations would be puzzling. Again, a character may have no selective value under ordinary circumstances, yet be highly valuable in the extremes to which the organism may occasionally be subjected. Thus the plants of the San Francisco Bay region may for many generations never be subjected to any extremes of temperature. Because of this mild climate, a wide variety of plants has been successfully introduced. But when freezing weather does occasionally strike, it wreaks much greater damage among the introduced plants than among the native plants. Studied during "typical" years, the characters that adapt these plants to severe weather are difficult to understand. But the native plants are there not only because they can exploit this usually benign weather, but also because they are capable of withstanding the extremes. The occasional extremes may constitute the most-severe selective force to which these plants must become adapted.

Lastly, a character may actually have no adaptive value. Such characters can be understood on several bases. They may be incidental effects of pleotropic genes—that is, genes with more than one phenotypic effect—so that, while one effect of such a gene is favored by selection, other effects are accidentally preserved along with it. Or the gene for a character of no selective value might be closely linked to one of definite selective value. P.W. Hedrick calls this "genetic hitchhiking," and he has amassed an array of probable examples.

Again, a gene may have become fixed in a species as a result of genetic drift (Chapter 10). Also, a character might be present in vestigial form because it was valuable in the past and has simply not been completely eliminated yet. Examples of such vestigial characters are numerous. Snakes, which have no use for limbs, are descended from typical reptiles with well-developed limbs. Some snakes still have traces of the hind limbs (Figure 9–16). A few of the many vestigial structures in humans are illustrated in Figure 9–17.

Neutral Evolution and the Molecular Clock

M.C. King and T.H. Jukes believe that random fixation of genes by drift, which they call non-Darwinian evolution, plays a major role at the molecular level. They point out that about one-fourth of all possible single-base substitutions in DNA lead to synonymous codons and hence to no change in the proteins. For example, the amino acid serine is encoded by six different codons: UCU, UCA, UCC, UCG, AGU, and AGC. By definition, change from one of these to any other is a mutation, yet selection could not distinguish them because they lead to an identical phenotype. Further, even if amino acid substitutions do occur, groups of amino acids may be nearly equivalent physiologically.

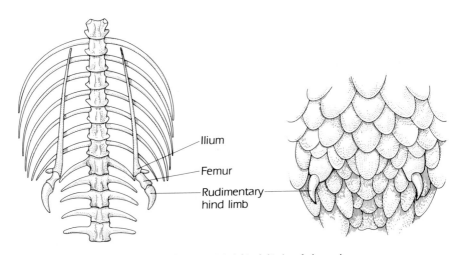

Ilium

Femur

Rudimentary hind limb

Figure 9–16.   Vestigial hind limbs of the python.

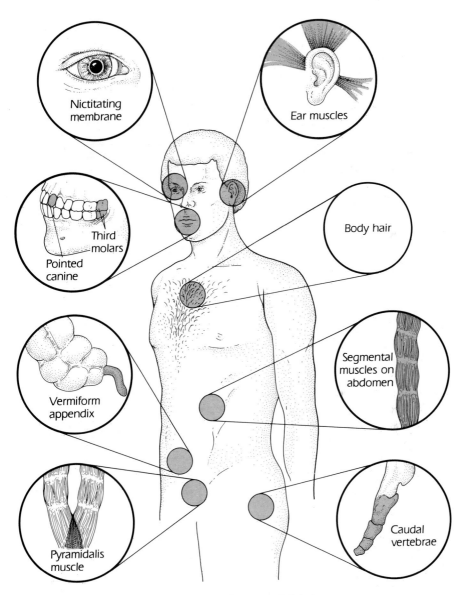

Figure 9–17. Some human vestigial characters.

Thus, at position 66 in the cytochrome *c* molecule, five different amino acids—glutamine, isoleucine, leucine, threonine, and valine—occur in different organisms. As the enzymes function normally in all of these organisms, King and Jukes believe that they must be functionally equivalent, hence that the substitution of amino acids has not been influenced by natural selection. That selection is never effective in such cases has not been clearly proven, yet it is probable that the theory of King and Jukes is widely applicable. Since *non-Darwinian* was often misinterpreted as *anti-Darwinian,* the usual term for this type of evolution is *neutral evolution.*

V.M. Sarich and others have suggested that the rate at which neutral substitutions occur may be approximately constant, and hence that it comprises a *molecular clock.* Comparison of the substitutions in a given protein in a series of organisms may, therefore, provide a basis for estimating the time since the separation of the ancestors of these organisms. Thus, Sarich and A.C. Wilson studied substitutions in serum albumen of primates, and they devised a time scale (see Figure 4–1) that placed the separation of various primates from their common ancestors much more recently than paleontologists had thought. Although this concept was initially very controversial, recent discoveries in paleontology have tended to validate the molecular clock (see Chapter 21).

Some suggest that neutral evolution is simply a modern restatement of the classical theory of population structure (see Chapter 7). This suggestion, however, is an error. The classical theory envisaged a rather uniform genotype and phenotype, the wild type, for each species, with mutants occurring only rarely. In contrast, neutral evolution envisages extensive heterozygosity as a result of random drift, unaided by natural selection.

Molecular Drive and Evolution

Molecular drive, as introduced in Chapter 8, comprises a series of processes—unequal crossing-over, transposition, and gene conversion—that drive the repeated sequences of DNA toward homogeneity within any species. The observed within-species homogeneity has been called *concerted evolution.* G. Dover believes that concerted evolution cannot be based upon natural selection, for the latter works by differential reproductive success of the favored genotypes. In concerted evolution, the species as a whole seems to move toward homogeneity, so that differential success of a part of the population can have nothing to do with it. If this view is correct, then molecular drive is a third means of evolutionary change, in addition to natural selection and genetic drift.

In contrast to the within-species homogeneity of repeated sequences, these may be different between species. Dover suggests that, if separate populations of a species should begin to diverge, then molecular drive might force them apart. The divergence of their repeated sequences might then constitute a barrier that would result in the separation of species. Because this speciation would have nothing to do with natural selection, he calls it *accidental speciation.*

Introns, Exons, and Natural Selection

In Chapter 7 we saw that a eucaryotic gene typically consists of exons, which are transcribed into mRNA and then translated into protein, and introns, which are sequences of nucleotides intervening between the exons and of no known function. A priori, we might expect that the exons would be subject to selection, hence that there would be limits to their variability, and that the introns would be much more variable because of the absence of selection.

M. Kreitman determined the sequence of nucleotides in a gene for a particular enzyme of *Drosophila melanogaster*. The rate of polymorphism (nucleotide substitution) was about 7 percent in both exons and introns. In the exons, however, all of the substitutions were "silent" in the sense that they did not result in amino acid substitutions. The fact that about three-quarters of all random changes in the codons do result in amino acid substitutions can only mean that changes in the enzyme were quickly eliminated by natural selection, probably because of lethality.

SUMMARY

Natural selection may be stabilizing, directional, or disruptive; and its effectiveness has been demonstrated both in experiment and in nature. Most species are highly heterozygous, so that the shifting balance between variable genotypes and natural selection can allow them to exploit a variety of ecological niches. This has led to a great range of adaptations, including such striking phenomena as *crypsis, aposematic coloration*, and *mimicry*, the effectiveness of which now rests upon a firm experimental basis. *Sexual selection*, whereby competition for mates results in sex differential ornamentation or weaponry, has been the subject of much speculation, but there is now some experimental foundation for it.

In general, the individual organism is the substrate of selection; but *kin selection* is also important, particularly in social animals. An interesting suggestion is that demes, species, and clades may also have genuine individuality and may function as units of selection. Whatever the unit of selection, there are two broadly different strategies of selection. Some species are *r* strategists, reproducing in numbers large enough to offset the losses due to natural selection. Others are *K* strategists, producing small numbers of offspring that are carefully nurtured and protected by adaptations that tend to minimize losses.

All species live in complex communities in which they interact with an extensive array of producers (plants), consumers (animals), and decomposers (fungi and bacteria). For any species, many of these other organisms will act as selective agents. As a result, complex *coevolution* may occur.

Finally, as powerful as natural selection is, there are limitations on it. Every species has some characters that appear to be neutral, unexplainable by natural selection. Although such characters are misleading in some cases, in others the neutrality appears to be genuine. Such characters may be "*hitchhikers*," carried along with selectively favored characters. Some are fixed by *genetic drift*. An important aspect of genetic drift is the regular, chance substitution of nucleotides, which forms the basis for the molecular clock. Finally, *molecular drive* is a very recently proposed mode of evolutionary change that is independent of natural selection. Natural selection remains the mainspring of evolution, yet it cannot explain the immense diversity of life without the concurrence of other factors.

REFERENCES

Brower, L.P., and J.V.Z. Brower. 1964. Birds, butterflies, and plant poisons: A study in ecological chemistry. *Zoologica* 49:137–159.
Summarizes and draws conclusions from an excellent series of experiments on mimicry. (Bates, Ford, Müller)

Campbell, B., ed. 1972. *Sexual Selection and the Descent of Man.* Aldine, Chicago.
A centennial volume, commemorating the publication of Darwin's double volume on these subjects. It brings the subject of sexual selection up to date.

Cott, H.B. 1940. *Adaptive Coloration in Animals.* Methuen, London.
An extraordinarily comprehensive review. (Swinnerton, Sumner)

Dawkins, R. 1976. *The Selfish Gene.* Oxford University Press, New York.
A highly readable account of sociobiology with abundant examples from nature, but one predicated on the questionable assumption that the gene, rather than the organism, is the unit of selection.

Dobzhansky, T. 1982. *Genetics and the Origin of Species.* Columbia University Press, New York.
A reprint of the 1937 classic, with an introduction by S.J. Gould. See also the editions of 1941 and 1951. (Dice, Stakman, Sukatchew)

Dobzhansky, T. 1970. *Genetics of the Evolutionary Process.* Columbia University Press, New York.
This book started out as a fourth edition of the above, but the progress of the field made it a really new book. (Quayle, Timofeeff-Ressovsky)

Dover, G. 1982. A molecular drive through evolution. *BioScience* 32:526–533.
A relatively simple and clear presentation of molecular drive.

Dover, G. 1982. Molecular drive: A cohesive mode of species evolution. *Nature* 299:111–117.
A more-advanced and rigorous presentation.

Ford, E.B. 1965. *Genetic Polymorphism.* MIT Press, Cambridge, Mass.
An excellent work, which elaborates on many of the examples of this chapter.

Futuyma, D.J., and M. Slatkin, eds. 1983. *Coevolution.* Sinauer Associates, Sunderland, Mass.
Many researchers report their work on all aspects of coevolution.

Gould, S.J. 1979. *Ever Since Darwin.* Norton, New York.
Collected essays on a variety of evolutionary topics, especially including natural selection and macroevolution, from an on-going column in Natural History.

Gould, S.J., and R.C. Lewontin. 1979. The spandrels of San Marco and the Panglossian paradigm: A critique of the adaptationist programme. *Proceedings of the Royal Society London, B* 205:581–598.
A penetrating analysis of the limits of natural selection.

Hedrick, P.W. 1982. Genetic hitchhiking: A new factor in evolution? *BioScience* 32:845–853.

Neutral or somewhat unfavorable genes may be preserved by close linkage to favored genes. This article makes a strong case for the importance of such "genetic hitchhiking" for evolution.

Janzen, D.H., and P.S. Martin. 1982. Neotropical anachronisms: The fruit the gomphotheres ate. *Science 215:19–27.*

A fascinating example of the evolutionary interactions between plants and animals.

Macior, L.W. 1982. Plant community and pollinator dynamics in the evolution of pollination mechanisms in *Pedicularis* (Scrophulariaceae). In *Pollination and Evolution,* ed. J.A. Armstrong, J.M. Powell, and A.J. Richards. Royal Botanic Gardens, Sydney, Australia.

A review of the author's studies on pollination in twenty-two species.

Regal, P.J. 1977. Ecology and evolution of flowering plant dominance. *Science* 196:622–629.

Explores the reasons for the dominance of flowering plants in the modern world in terms of coevolution with pollen-dispersing animals, especially birds.

Turner, J.R.G. 1977. Butterfly mimicry: the genetical evolution of an adaptation. In *Evolutionary Biology,* ed. M.K. Hecht, W.G. Steere, and B. Wallace, vol. 10, 163–206. Plenum Publishing, New York.

This paper reviews mimicry much more broadly than its title indicates. References include more than twenty papers by the Browers, from 1958 into the 1970s.

Wickler, W. 1968. *Mimicry in Plants and Animals.* World University Library, McGraw-Hill, New York.

A colorfully illustrated survey.

Wilson, E.O. 1971. *Insect Societies.* Harvard University Press, Cambridge, Mass.

A superlative survey of the biology and social life of insects, with special reference to the genetic systems from which these derive.

Wilson, E.O. 1975. *Sociobiology.* Harvard University Press, Cambridge, Mass.

An extremely interesting overview of social behavior in animals and the genetic systems that underlie these behaviors. Very controversial when applied to the study of humans!

CHAPTER

10

Population Genetics

Once biologists realized that selection need not be an all-or-none force but can operate through gradual changes in the frequency of certain characters or character combinations in a species, they had to analyze statistically the changes in the genetic composition of species under various conditions of mutation, selection, and population structure. This problem was attacked mathematically by R.A. Fisher, J.B.S. Haldane, and S. Wright in the early years of the modern synthesis, and more recently by J.F. Crow, M. Kimura, R.C. Lewontin, and others. The results of their calculations are one of the major achievements of the modern synthesis. In the first edition of his brilliant book, Dobzhansky prefaced the discussion of the statistical analysis of variation in populations with the following statement*:

> Only in recent years, a number of investigators . . . have undertaken
> a mathematical analysis of these processes, deducing their regularities from
> the known properties of the Mendelian mechanism of inheritance. The ex-
> perimental work that should test these mathematical deductions is still in

* T. Dobzhansky, *Genetics and the Origin of Species* (New York: Columbia University Press, 1937).

the future, and the data that are necessary for the determination of even the most important constants in this field are wholly lacking.

In the intervening years, many population geneticists have tested the hypotheses of evolutionary mathematics in experimental populations; they frequently use *Drosophila* raised in population cages, large cages to which food can be added without disturbing the flies; and they have developed computer simulations of evolving populations. While the results are impressive, investigation of evolutionary mathematics in nature has not kept pace. It is still possible that the mathematical analysis of evolutionary phenomena, as it is now generally accepted, violates the dictum of Johannsen that "Biology must be handled with mathematics, but not as mathematics." Some of the simplest and most important mathematical work is presented in this chapter.

EVOLUTIONARY MATHEMATICS

The Hardy-Weinberg Law

Evolutionary mathematics begins with the *Hardy-Weinberg Law* (1908), which states that, if alternative forms of a gene are present in a population in a definite proportion, and if random mating and equal viability of all genotypes obtain, then the original proportions will be maintained in all subsequent generations, unless they are upset by some other factor, such as mutation or selection. The following discussion assumes that mutation and selection do not occur (although they normally coexist), and that the population sample is large enough to give statistically valid results. In mathematical terms, the proportion of one allele, A, may be taken as p, and that of the other, a, as $1 - p = q$, making the sum of their proportions 1. Then in the F_2 and in all subsequent generations, the proportions of the possible genotypes will be

$$p^2AA : 2pqAa : q^2aa$$

and the proportions of the genes will remain p in the case of A and q in the case of a. Let us substitute figures now for a monohybrid cross. In the cross $AA \times aa$, p will equal 0.5, and q will also equal 0.5; in other words, the alleles will be present in equal numbers in the experimental population. The expansion of the binomial $(p + q)^2$ gives the F_2 coordinates

$$p^2 + 2pq + q^2$$

as stated above. Substituting numbers and genotypes, we obtain

$$(0.5A + 0.5a)^2 = 0.25AA + 0.50Aa + 0.25\,aa$$

As this result is the familiar $1:2:1$ genotypic ratio of the F_2 of a monohybrid Mendelian cross, it is clear that the elementary Mendelian ratios comprise special applications of the Hardy-Weinberg law.

The Hardy-Weinberg law is just as applicable to problems involving initial allelic ratios not bearing any special relationship to standard Mendelian ratios. If the frequency of A is 0.8 and that of a is 0.2, the expanded formula would read

$$(0.8A + 0.2a)^2 = 0.64AA + 0.32Aa + 0.04aa$$

The sum of the frequencies of the several genotypes still equals 1, which verifies the calculation. Now if this calculation represented fifty organisms, it would include 100 genes. Of these, 80 (all 64 alleles of the 32 AA organisms, plus 16 of the 32 alleles of the 16 Aa organisms) would be A, and 20 (the other 16 alleles of the Aa organisms plus all 4 alleles of the 2 aa organisms) would be a. Thus the values $q = 0.8$ and $p = 0.2$ are maintained. The data are summarized in Figure 10–1.

If the assumptions are changed now by adding only nonrandom breeding, such as self-fertilization, or preferential breeding so that organisms of similar phenotype tend to mate, the result will be an increase of the genotypes AA and aa at the expense of Aa. But the relative proportions of the alleles will remain unchanged.

We can use this formula to determine allelic proportions if character proportions are known. For example, M–N blood groups of 140 Pueblo Indians were tested, with the following results: 83 M, 46 MN, and 11 N. Persons of groups M and N are homozygous for alleles L^M and L^N respectively, while those of group MN are heterozygous, $L^M L^N$. Accordingly, this sample of 140 persons includes 212 L^M alleles (all of the 166 alleles of the group M persons plus 46 of the 92 alleles of the MN persons) out of the total of 280, or 0.757

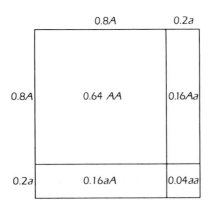

Figure 10–1.   *Diagram of a cross in which the proportions of two alleles are unequal, 0.8A and 0.2a. The lines on the sides of the square are proportional to the frequencies of the alleles, and the areas of the subsquares are proportional to the frequencies of the genotypes derived from these gametes.*

of all alleles present. Similarly, the number of L^N alleles is 68 (the other 46 from the MN persons plus all of the 22 alleles of the 11 N persons), or 0.243 of the total. Is this population sample in Hardy-Weinberg equilibrium? (The three genotypes are selectively neutral.) To test this, let us substitute the gene frequencies in the Hardy-Weinberg equation:

$$(0.757 \, L^M + 0.243 \, L^N)^2 = 0.573 \, L^M L^M + 0.369 \, L^M L^N + 0.059 \, L^N L^N$$

The results are, of course, given in proportions of genotypes. In order to compare it with the original data, we must multiply each of these fractions by 140, the size of the sample:

Calculated	Observed
$140 \times 0.573 = 80.22$	83
$140 \times 0.369 = 51.66$	46
$140 \times 0.059 = 8.26$	11

Intuitively, these deviations seem to be rather minor, and a chi-square test confirms that the results do not differ significantly from expectation on the basis of the Hardy-Weinberg law. This formula is a most effective tool for analyzing the genetic composition of populations.

The consequences of the Hardy-Weinberg law are threefold. First, allelic frequencies do not change from generation to generation. The values of p and q remain constant, and their sum is always 1. Second, the frequencies of the genotypes also remain constant, and they are described by $p^2 AA + 2 \, pqAa + q^2 aa$. Finally, equilibrium is reached after one generation of random breeding. The Hardy-Weinberg law, then, represents a tendency to maintain the status quo, a conservative factor in evolution. In order to apply the formula to evolutionary problems, we must take into account the factors that might upset the equilibrium and cause a change in the relative frequencies of the alleles. The principal calculable factors are mutation and selection. As the mathematics of mutation and selection is rather complicated, we will not present the details here. Full details may be found in the references listed at the end of this chapter.

Selection Pressure and Rates of Evolution

Calculations of the effect of selection are most easily made where selection favors an allele with complete dominance (that is, selection against the recessive). If 1,000 AA or Aa survive for every 999 aa, a selection pressure of 0.001 is said to favor the dominant allele or to oppose the recessive allele. If the relative fitness of the most productive genotypes, AA and Aa, is defined as 1, then the relative fitness of *aa* is $1 - s$. To determine the composition of the next generation, the usual members of the Hardy-Weinberg equation must be multiplied by their relative fitnesses. This gives

$$p^2 + 2pq + q^2(1 - s)$$

The sum of these is no longer 1, because $q^2(1 - s)$ is clearly less than q^2. Unity can again be established by dividing each member of the equation by the new total, which is $1 - sq^2$. Finally, the amount of change per generation in the frequency of *a* will be given by

$$\triangle q = \frac{- spq^2}{1 - sq^2}$$

and of course there is an equivalent increase in the frequency of A, p.

One consequence of this formula is that, for a given value of s, change is most rapid when $p = 0.33$ and $q = 0.67$, and it falls rapidly as q becomes either very large or very small. This is because the product pq^2 is maximal when $q = 0.67$. It falls off in both directions because the square of a number less than 1 is smaller than the number itself. What might the rate of change be? In *Biston betularia*, the dominant allele for melanism rose from less than 1 percent to more than 98 percent within a half century. Kettlewell estimated s at nearly 0.5. Values of s from 0.01 to 0.5 appear to be common, and much higher values occur against severely disadvantageous genes, with a value of 1 against a lethal gene being the limiting case.

B. Kurtén has reviewed the history of a pair of alleles over the past million years. In bears, the growth of the first upper molar is *allometric;* that is, growth in height is more rapid than growth in length. The degree of allometry is genetically controlled, being quite marked in some bears, rather moderate in others. The larger the bear, the higher the crown of the molar in relation to its length (Figure 10–2). The more-extreme allometry is found in modern bears, *Ursus arctos;* a more-moderate degree characterized the Late Pleistocene cave bear *U. spelaeus,* now extinct. Both types of allometry are found in *U. etruscus,* an Early Pleistocene bear that was ancestral to both species. Kurtén has made the reasonable, but unproved, assumption that genes governing the two kinds of allometry are alleles, A_a for the arctoid, more extreme type, and A_s for the spelioid, more moderate type. Both alleles, then, were present in the ancestral *U. etruscus.* A sample from the Mid-Pleistocene bear, which is probably ancestral to *U. spelaeus* and possibly to *U. arctos,* shows about 67 percent A_a and 33 percent A_s. The Hardy-Weinberg formula leads to an expectation of 4 A_aA_a:4 A_aA_s:1 A_sA_s. In this sample, many molars are intermediate and hence probably based upon heterozygotes. The actual sample count was 42 arctoid, 50 intermediate, and 8 spelioid, not a significant deviation from the mathematical expectation of $44.45 : 44.45 : 11.1$ ($= 100$, the sample size). In *U. spelaeus,* which appeared later, the allele A_a disappeared altogether; in *U. arctos,* both alleles have been retained, but A_s is less abundant,

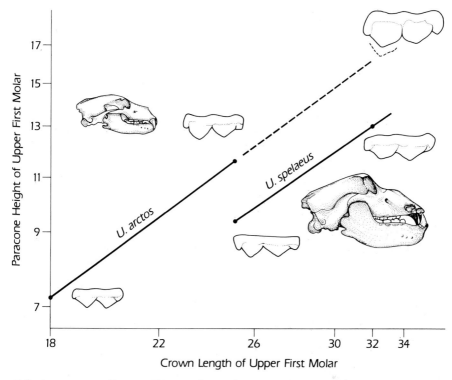

Figure 10–2. *Allometry of bear molars. The prominent cusp on the left of each tooth is the paracone; that on the right, the metacone. On the logarithmic graph, height of the paracone in mm is plotted against length of the crown of the first upper molar. A and B, the smallest and largest molars respectively for* Ursus arctos, *show marked allometry; C and D, the extremes for* U. spelaeus, *a much larger bear, show moderate allometry. The dashed extension of the line for* U. arctos *represents the potential results if arctoid allometry were to continue to the total size of* U. spelaeus.

making up only 23 percent of the gene pool of a recent Finnish population of bears. Thus Kurtén has supplied an outline history of a pair of alleles over the past million years.

Some probable selective forces can be inferred, *Ursus spelaeus* was a large bear, and arctoid allometry would have produced a very high tooth, jutting out of the tooth row and not working harmoniously with the others. The spelioid pattern in *U. arctos,* however, leads to a low-crowned tooth, which must wear down more quickly than its neighbors.

In the early years of the modern synthesis, the usual selection pressures were generally believed to be very small. This belief was based on many studies of quantitative differences among related species. Quantitative traits are determined by swarms of genes that, individually, have very small effects (see

Chapter 7). Further, there was general agreement that a series of small mutations could be integrated into the gene pool of a species more readily than could a single, large mutation. Haldane calculated the results of selection pressure of 0.001 favoring a dominant allele at various values of p and q. He found that it would require 11,739 generations to increase p from 0.000 001 to 0.000 002 (1 in a million to 2 in a million). Yet the change from 0.000 01 to 0.01 would require only 6,920 generations; that from 0.01 to 0.5, only 4,819 generations; but that the change from 0.990 to 0.999 99 would require 309,780 generations (Figure 10–3). Evidently, such mild selection pressure, although effective in the intermediate range, can establish a new gene or bring a well-established one to 100 percent ("fix" it) only with great difficulty.

Even Dobzhansky acknowledged that these figures cast doubt on the adequacy of natural selection alone to achieve the observed results of evolution. Partly because of these calculations, Goldschmidt believed that the neo-Darwinian theory placed upon natural selection a greater burden than it could bear. Accordingly, he proposed the systemic mutations to diminish that burden. Now that it is known that much stronger selection pressures are common, this objection has lost cogency.

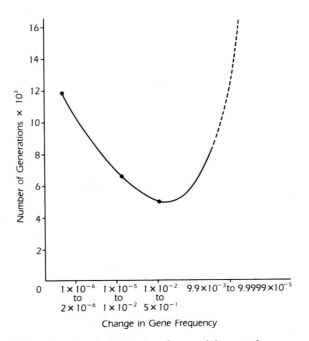

Figure 10–3. *Haldane's data for rate of change in frequency of a dominant allele favored by selection, with s = 0.001. Change is extremely slow at very low or very high frequencies of the allele, but it is moderately rapid at intermediate frequencies.*

Selection against a dominant allele may also occur. Here, the change per generation is given by

$$\Delta p = \frac{- spq/2}{1 - sq}$$

Finally, there are the interesting cases in which selection against both homozygous genotypes occurs, so that the heterozygotes are favored. If selection against allele A is s and that against allele a is t, then the change per generation is given by

$$\Delta q = \frac{pq(sp - tq)}{1 - sp^2 - tq^2}$$

An interesting human example concerns hemoglobin. The allele Hb^A encodes normal hemoglobin, whereas Hb^S encodes a variant that is nearly identical but in which valine replaces glutamic acid at position 6 in the beta chain of the hemoglobin molecule. Small as the difference is, it is important, because persons homozygous for Hb^S have sickle-cell anemia (Figure 10–4), a disease that is usually fatal in the absence of medical intervention. In one tribe studied, the Hardy-Weinberg equation gave an expectation of 0.64 Hb^AHb^A:0.32 Hb^AHb^S:0.04 Hb^SHb^S. Selection against the homozygous Hb^S individuals is severe, as high as 0.87. Such strong selection should eliminate the disfavored allele rapidly, yet in some tribes as much as 40 percent of the people may be heterozygous. The reason is that the heterozygotes, Hb^AHb^S, are resistant to malaria, a prevalent disease in the regions in which the Hb^S allele is common. Accordingly, *balanced selection* against Hb^AHb^A because of susceptibility to malaria and against Hb^SHb^S because of sickle-cell anemia maintains the heterozygote Hb^AHb^S at high frequency.

Mutation Pressure and Genetic Equilibrium

Now let us add the effect of mutation to the mathematics. In any particular case, mutation might proceed only in one direction, $A \rightarrow a$, or it might proceed in both directions, $A \rightleftarrows a$. In the former case, even a slight mutation pressure would eventually lead to a species completely homozygous for the mutant gene, unless a selective disadvantage of the mutant were to prevent it. In the latter case, if all of the genotypes had equal selective value, then the frequencies of the two alleles would reach an equilibrium, $\hat{q}$ (read q-hat) for a or $\hat{p}$ for A, the numerical values of which would depend upon the actual magnitude of the mutation rates in the two directions.

The equilibrium point is related to the two mutation rates in a simple way. If rate $A \rightarrow a = u$ and $a \rightarrow A = v$, the frequency of a at equilibrium is

$$\hat{q} = \frac{u}{u + v}$$

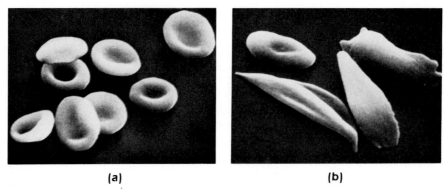

(a)   **(b)**

Figure 10–4. *(a) Normal and (b) sickled red blood cells photographed under the scanning electron microscope.*
Courtesy of Dr. Marion I. Barnhart, Wayne State University Medical School.

and the frequency of A at equilibrium is

$$\hat{p} = \frac{v}{u + v}$$

For example, if the rates are equal, then

$$\hat{q} = \frac{1}{1 + 1} = \frac{1}{2}$$

or equilibrium will occur when the frequencies of A and a are equal. If $u = 2v$, then a is being formed twice as rapidly as A, and

$$\hat{q} = \frac{2}{2 + 1} = \frac{2}{3}$$

or equilibrium will be reached when two-thirds of the genes are a. If $u = 4v$, then 80 percent of the genes will be a at equilibrium.

In nature, neither mutation nor selection will ordinarily occur alone; the two will act simultaneously, perhaps in the same direction, perhaps in opposite directions, to upset the Hardy-Weinberg equilibrium. Most frequently, selection will work against mutation, as the majority of possible mutations are deleterious. This opposition will result in very slow change, if any. But if a particular mutation is favored by selection and if its mutation rate is appreciable, the combined action of mutation and selection might well cause a rapid change.

The frequency at equilibrium of a recessive gene subject both to selection and to replacement by mutation is given by the equation

$$\hat{q} = \sqrt{\frac{u}{s}}$$

where u is the mutation rate and s is the coefficient of selection. If the coefficient of selection is close to 1, as in the case of a lethal or a sublethal gene, then

$$u = \hat{q}^2$$

If selection is against a dominant gene that is also replenished by mutation, then the equilibrium frequency is given by

$$\hat{p} = \frac{u}{s}$$

The difference between the equations for dominant and recessive genes reflects the fact that a dominant gene is continually exposed to selection, whereas a recessive gene is exposed only when it is homozygous.

Migration

Migration, which is gene flow from one population to another, may also disturb the Hardy-Weinberg equilibrium. If the allelic frequencies of the two populations are the same, migration will have no effect; but if the frequencies differ, then migration, like mutation, will cause a shift of allelic frequency of the host population in the direction of the immigrant population. If migration occurs at a regular rate and if the allelic frequencies for a given gene are known in both populations, then the results of migration can be calculated. Let the proportions of migrants in a host population be m, the initial frequency of an allele A in the host population be p_0, and the frequency of A in the migrant population be P. Then, after one generation, the frequency of A in the host population will be

$$p_1 = (1 - m)p_0 + mP = p_0 - m(p_0 - P)$$

and after t generations, it will be

$$p_t = (1 - m)^t (p_0 - P) + P$$

Let us consider a historical example. Blood group O is much the most common among American Indians, and group A is moderately frequent. Groups B and AB are so rare that their occurrence is evidence of white ancestry. Lianne Lacroix tested 184 Cree Indians at Fort George, Quebec, with the following results:

Group	Number	Percent	Allelic Frequency
0	152	82.61	0.909
A	31	16.85	0.088
B	1	0.54	0.003
AB	0	0.0	

This far northern community was, until very recently, extremely isolated, and interracial marriages were rare. Accordingly, these data are probably representative of the aboriginal gene frequency. On the other hand, the Micmac Indians of Nova Scotia were on the main route of the French and English migrations into North America, and intermarriage was common. It is probable that the distribution of blood groups among the Micmacs was originally similar to that among the Crees, but a recent study gave the following results:

Group	Percent	Allelic Frequency
O	48.0	0.669
A	46.66	0.276
B	3.17	0.027
AB	2.17	

For comparison, the same figures from France are as follows:

Group	Percent	Allelic Frequency
O	39.8	0.635
A	42.3	0.277
B	11.8	0.088
AB	6.1	

and from England:

Group	Percent	Allelic Frequency
O	47.9	0.669
A	42.4	0.251
B	8.3	0.050
AB	1.4	

Evidently, migration of Europeans into the homeland of the Micmacs and extensive intermarriage have changed the blood-group distribution of the Micmacs from the aboriginal to an essentially European pattern.

Population Size and the Effectiveness of Selection

So far we have assumed that mutation and selection proceed in a population of indefinitely large size. Wright showed that the effectiveness of selection is greatly affected by the size of the population concerned. The mathematical basis for this proposition is complex, but the results are simple: mild selection pressures are relatively ineffective both in very small and in very large populations. Selection has its maximum effectiveness in moderate-sized populations. What these relative population sizes mean in numerical terms is much less clear than might be desired. We should point out, however, that it is the actual breeding population rather than the total population that is important here.

Opponents of this idea point out that populations of natural species are, in most cases, immense, though good estimates are available only for a few species with small populations. Its advocates, however, point out that such immense natural species are divided into more-or-less isolated subspecies, that these are further subdivided into local populations, and that these local populations are the significant breeding units. Migration from one unit to another will tend to blur the lines that separate them; but, generally speaking, the factors that tend to differentiate local populations and subspecies will be stronger than the migration pressure. This type of breeding population structure is generally acknowledged; the basic questions that remain are, how large is large, and how small is small, from the viewpoint of population dynamics?

Wright believes that evolutionary changes might occur with "explosive" rapidity if favored by a combination of mutation, selection for a character (or combination of characters), and optimum population structure, as described above, all in the presence of an open ecological niche. This may be the answer to the question whether some new principle, in addition to mutation and selection, is necessary to account for the observed results of evolution. Geneticists have pointed out that this may meet the demands of punctuated equilibria.

Genetic Drift, the Sewall Wright Effect

Yet another factor that tends to upset the Hardy-Weinberg equilibrium is genetic drift. Fisher originally described the phenomenon, but he later disavowed its value completely, and it is Wright who gets (and deserves) the major credit for the development of this concept. Genetic drift refers to the accidental fluctuations in the proportion of a particular allele, which depend upon the fact that the assortment of genes into gametes and the combination of gametes to form zygotes are random processes. Drift is important in small, but not in large, populations. Such accidental deviations from the theoretically expected assortment of alleles and from their expected recombinations at fertilization are well known to be responsible for the fact that Mendelian ratios are rarely obtained exactly. But because the deviations in each experiment are random, they tend to cancel out, and hence the validity of Mendelian principles is demonstrable if we add the results of many experiments or perform large-scale experiments.

In nature, the large-scale experiment is already provided. Such accidents of sampling do not have an important effect on large breeding populations, because an accidental increase of *A* in one part of a population will generally be counterbalanced by an accidental increase of *a* in another part of the same population. Or the genetic drift of one season will be reversed in the next season. In small populations, the situation is very different. If there are only a hundred individuals in a particular population (and smaller breeding populations do exist, for example, the whooping crane) and if a particular allele is present only once, then an accident of sampling might easily, in a single generation, remove it irrevocably from the population or increase it manyfold, say to 10 percent. The result is that, in small, isolated populations, genes may be completely lost or completely fixed by genetic drift, without reference to selective value. Genetic drift is thus a force working apart from selection: drift tends to preserve or to destroy genes without distinction, whether favorable, neutral, or unfavorable; whereas selection tends to preserve those that confer some adaptive value and to destroy those that impair the adaptive value of a species. Severe selective forces will, of course, destroy disadvantageous genes irrespective of the population size.

Wright's viewpoint on the significance of genetic drift has been as much misunderstood by his proponents as by his opponents. It has often been said that Wright regards small, isolated populations as optimum for rapid evolution, but this is not the case. Because genetic drift predominates over selection in such populations, he believes that they will show a higher degree of homozygosity than will more typical populations and that they will, on the whole, tend to be poorly adapted. As a result, they may well become evolutionary blind alleys.

A corollary to genetic drift is what Stebbins has called the "bottleneck" phenomenon. It is often said that the numbers of a species tend to remain approximately constant in any locality. But every field biologist knows that a species that is abundant in one year may be difficult to find in another year, only to be on the increase again the following year. In years of scarcity, small populations assume an especial importance, for they are the only source from which the species can again be built up; hence, the term "bottleneck." Accidental changes in the genetic makeup of such bottleneck populations will, therefore, determine changes in the larger populations to be derived from them, and these changes will generally be nonadaptive in character. For example, the arctic hare periodically reaches a peak of population at which disease decimates the population. A "no-rabbit year" follows and then come several years of recovery. Accidental changes in gene frequencies during the years of scarcity must have a profound effect upon the populations of years of abundance. The same thing must be true of lemmings, which are small, microtine, Scandinavian rodents. Every few years its population reaches a prodigious level, and epidemics do great damage. The lemmings then migrate in great numbers, with much loss of life, and the population crashes. The population is reestablished by the few that remain in the mountains. Similarly, those many species that have small overwintering populations must show bottleneck effects.

Very similar is the "founder principle" of E. Mayr. Populations on oceanic islands or other highly isolated places may be established by a very small sample from a continent or from another island. As a minimum, this founding sample may be a single pregnant female animal, or a single seed of a self-compatible plant. In any case, the founders can include only a small part of the genetic variability of their species, and this limitation must influence the populations to which they give rise, even if the latter be very large.

GENETIC IDENTITY AND GENETIC DISTANCE

A final topic for this chapter is that of genetic identity and genetic distance, related concepts that are useful in the evaluation of evolutionary relationships. They are estimates of the extent to which two populations share the same alleles (I) or differ consistently (D). Thus, if at locus A, populations X and Y have the same alleles in the same proportion (i.e., $0.75 \ A_1$ and $0.25 \ A_2$), I would equal 1; if at locus B, X always has B_1 and Y always has B_2, I would equal 0; and if at locus C, X has $0.20 \ C_1$ and $0.80 \ C_2$, while Y has these in reversed proportions, then I for this locus is intermediate, 0.69. To calculate the degree of genetic identity of the two populations, we average the figures for a considerable number of gene loci, typically 20 to 40. The general formula is

$$I = \frac{\sum x_i \, y_i}{\sum x_i^2 \, \sum y_i^2}$$

where x_i and y_i represent the ith gene in populations X and Y, respectively. Genetic distance (D) is based on the number of allelic substitutions required to account for the degree of difference between populations. The formula is

$$D = - \log_e I$$

We might expect this, too, to range from 0 (identity) to 1 (all loci different); however, because a single gene may be substituted repeatedly, values greater than 1 can be obtained.

The results of such calculations are illuminating. F.J. Ayala has investigated *I* and *D* in the *Drosophila willistoni* group, a South American complex that shows many levels of divergence. Members of a single deme have *I* values close to 1 and *D* values just above 0 (e.g., $I = 0.97$ and $D = 0.03$). Members of different subspecies are appreciably differentiated, with *I* on the order of 0.8 and *D* on the order of 0.2 or more. Incipient species—groups that are partially isolated reproductively—have values of the same order as do the subspecies, but for sibling species both *I* and *D* are between 0.5 and 0.6. Finally, morphologically differentiated species of the group have *I* on the order of 0.35, and *D* on the order of 1.05. These values are typical for many other groups that have been investigated. Numerical details differ, but the principle usually holds: *I* is proportional to the degree of relationship that is inferred on other grounds, and *D* is proportional to the degree of separation that is inferred on other grounds.

GENETIC DISTANCE AND ELECTROPHORESIS

To estimate genetic distance, a random sample of gene loci is needed. To obtain a random sample is very difficult with traditional morphological characters, because the genes are identified by their variability. An invariable gene is invisible in Mendelian experiments. If, however, we select a given enzyme or other protein for study by electrophoresis (see Chapter 7), we do not know until *after* the electrophoretic test which proteins are polymorphic and which are invariable. Accordingly, any selection of several proteins should be a random sample with respect to genetic variability and so should provide a valid basis for statistical analysis of the genotypes that specify the proteins. Thus, electrophoresis has made possible an extensive survey of genetic variability in all groups of organisms. It has become a standard tool of the modern taxonomist.

One very important result was the demonstration that, as in *D. willistoni*, the genetic distance of any two populations is roughly proportional to their taxonomic distance. Another important result is that, as might be expected intuitively, morphological and biochemical differentiation are not necessarily closely correlated. In the *D. willistoni* case, the groups that were most distant morphologically were also most distant electrophoretically. In contrast, humans and chimpanzees differ in almost all morphological details, yet M.C. King and A.C. Wilson found that the human and chimpanzee proteins that they tested were nearly identical: an average of only 11 base pairs per 1,000 were different! Similarly, the cichlid fishes (a family of perchlike fishes of Africa, South America, and a few other places) may be morphologically and ecologically differentiated to the generic level, while retaining close similarity of their proteins, as revealed by electrophoresis.

SUMMARY

The *Hardy-Weinberg law* describes the regularities of Mendelian inheritance as it occurs in populations. This law is a conservative principle in evolution, for it states that both allelic and genotypic frequencies remain unchanged from generation to generation, unless some external force disturbs the equilibrium. The principal forces that disturb the Hardy-Weinberg equilibrium are *natural selection*, *mutation*, and *migration*. All of these can be quantified and analyzed mathematically. And population geneticists have done so very successfully in experimental populations and computer simulations. There are also some successes in natural populations, but the problems may be very difficult, and so progress has been much slower.

A fourth disturbing factor is *genetic drift*, the effect of sampling errors that occur because the basic processes of genetics are chance processes. Drift is important in small populations, where it may lead to loss or fixation of alleles without regard to the selective value of these alleles.

Special cases of genetic drift are founder effects, in which a new population may be established from a very small sample of another population that includes little genetic variability, and the bottleneck effect, which occurs when a large population is reduced to a minimum, with only a little of the original variability, and then expands again.

Finally, measures of *genetic identity* and of *genetic distance* can be calculated, particularly on the basis of electrophoretic data. Again, these data verify numerically the intuitive expectation that genetic and taxonomic evolution should be roughly proportional.

REFERENCES

Ayala, F.I. 1982. *Population and Evolutionary Genetics: A Primer.* Benjamin-Cummings, Menlo Park, Calif.
  Excellent quantitative genetics by a leading evolutionary geneticist.
Crow, J.F., and M. Kimura. 1970. *Introduction to Population Genetics Theory.* Burgess Publishing, Minneapolis, Minn.
  A clear presentation by two leaders in the field.
Dobzhansky, T. 1970. *Genetics of the Evolutionary Process.* Columbia University Press, New York.
  Includes an unusually clear presentation of the main features of population genetics.

Falconer, D.S. 1981. *Introduction to Quantitative Genetics*, 2nd ed. Longman, New York.

Excellent and readable coverage of the material of this chapter.

Wallace, B. 1981. *Basic Population Genetics*. Columbia University Press, New York.

For students who already have a good command of elementary genetics, this book is an excellent introduction to its evolutionary applications.

Wright, S. 1968–1978. *Evolution and the Genetics of Populations*. vols. 1–4. University of Chicago Press, Chicago.

An authoritative and encyclopedic treatment of the subject, but difficult reading.

11

Continuous versus Discontinuous Variation: The Species Problem

Nothing about the living world is more obvious than variation everywhere. On the grandest scale, moneran, plant, and animal kingdoms are differentiated. Almost as obviously, the major types, phyla, and classes are differentiated within each kingdom. As we descend the taxonomic hierarchy, variation is demonstrable within each category, although we need more refined methods of observation to establish the fact at the lowest levels. Equally important is the fact that the varying organisms are arranged into discontinuous clusters, more specifically, into a hierarchy of discontinuous clusters. This is the basic fact of taxonomy.

BASES OF DISCONTINUOUS VARIABILITY

Nonetheless, it is often maintained that the observed discontinuity in our present-day flora and fauna is illusory because it depends upon the extinction of intermediates that existed in the past; this, essentially, was Darwin's position. According to this viewpoint, if representatives of every species that has ever existed could be arranged in order from the most primitive to the most specialized, an almost imperceptible transition would be observed, with abrupt discontinuities corresponding only to single gene differences. In other words, the entire living world would show no discontinuities greater than those that actually separate members of the same species. Admittedly, no such assemblage could ever be made: the amount of extinction that has marked the history of life on this earth is too immense, and the fossil record is too incomplete. But the modern viewpoint favors the possibility that an approximately continuous series may have existed, though never at a single time level.

The alternative is, of course, that the larger steps in evolution may have been achieved through systemic mutations or through some other type of macroevolutionary change. If this second viewpoint is correct, then at least part of the deficiency of the fossil record is due to the fact that some of the supposed intermediates (or missing links) never existed, as the paleontologists N. Eldredge, S.J. Gould, and S.M. Stanley strongly argue.

Wright approached the problem of continuous and discontinuous variation from the viewpoint of statistical analysis of permutations of gene combinations. He began with the very modest assumptions that every organism might have 1,000 pairs of genes and that each of these might form a series of ten multiple alleles. The number of possible recombinations of these alleles would then be 10^{1000}. If this entire array of genotypes could be formed and if the resulting organisms were so arranged that each differed from its neighbors only by a single gene, they would undoubtedly form a smooth, continuous series from one end to the other. But it would be patently impossible to form this whole series, because the estimated number of electrons in the visible universe is only 10^{79}. Most of the potential genotypes, if formed, would be monstrosities that would be destroyed by natural selection. Of those genotypes that actually formed, those that survived would cluster around *adaptive peaks*—that is, character combinations that are physiologically harmonious and sufficiently adapted to the demands of the environment in which the organism must face the test of natural selection. Separating these adaptive peaks are *adaptive valleys,* which represent disharmonious, or unworkable, character combinations.

In such a system, a group of closely associated peaks represents a species, with each peak corresponding to a subspecies. Small ranges represent genera, and larger ranges represent families and higher groups (Figure 11–1). Discontinuities depend partly upon the impossibility of forming the entire series of genotypes and partly upon the elimination by natural selection of many of those that are formed. We must realize that this ingenious explanation is a hypothetical model, which should not be expected to correspond completely to the facts of nature because this would require that humans and amoebae be conditioned by the same series of allelic genes. How far this model is applicable to nature is debatable: some believe that it is generally applicable; the advocates of punctuated equilibria believe that its applicability is severely limited.

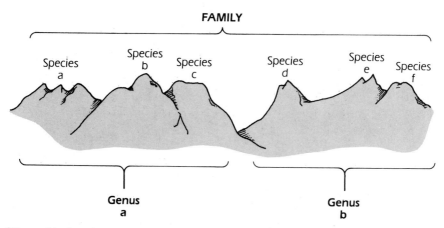

Figure 11–1.   *A taxonomic mountain range. A species consists of a group of closely associated peaks, the subspecies; the species in turn are grouped into genera, and these are grouped into families. The analogy could be carried much further, but the diagram would become increasingly complex.*

SPECIES CONCEPT

In any case, classification is possible because discontinuities do exist between varying series of organisms. We can easily ascertain the limits of the various discontinuous groups on the higher levels, although we may dispute the rank and relationships of a particular group. For example, some systematists treat the Onychophora as a class of Annelida; others, as a class of Arthropoda; and still others, as an independent phylum. Some even unite all three groups in a single phylum Articulata. But advocates of all of these ideas agree which animals are onychophorans and which are not. Again, a "splitter" may treat a group as an order with several families, and a "lumper" may treat it as a single family; but both agree on what genera constitute the group.

To the extent that described groups correspond to actually discontinuous groups in nature (and that is generally the case), the conventional classifications may be said to be natural. To the extent that the rank accorded the various groups is arbitrary, the system of classification itself is arbitrary rather than natural. Taxonomists are generally agreed that, although the discontinuities of the higher groups are real, their assignment to systematic categories is primarily a matter of convenience. Linnaeus himself said this of the higher categories, but he regarded the species as real, each one a specially created unit. The replacement of the archetypal concept by the evolutionary concept has not been accompanied by an abandonment of Linnaeus's idea of the definiteness of species. As Bateson has said, "Though we cannot strictly define species, they yet have properties which varieties have not, and . . . the distinction is not merely a matter of degree."

What, then, is the nature of this unique taxonomic unit, the species? In Chapter 3, we defined species as kinds of plant or animal, the individuals of which differ from each other only in minor traits (except sex); they are sharply separated from all other species in some traits; and they are mutually fertile but at least partially sterile when crossed to other species. We are now in a position to qualify that definition.

Magnitude of Difference

Many other definitions of species have been published, and all of them, including the preceding one, are unsatisfactory in that they do not provide a basis upon which a practical taxonomist can decide whether two similar groups are distinct species or only subspecies. Some have tried to specify the degree of difference necessary to separate good (valid) species; but this attempt is impractical, not only because of the difficulty of formulating a quantitative expression but because the degree of difference between species in some groups seems to be very much greater than in other groups. Moreover, some undoubtedly good species, as *Drosophila pseudoobscura* and *D. persimilis,* show little morphological difference.

Gilia inconspicua, a small flowering plant of the deserts of western North America and South America, is so uniform that it was once treated as a single species throughout that vast range. It has proven to be a complex of more than twenty genetically isolated sibling species. A similar profusion of cryptic species of plethodontid salamanders has been discovered in North America. In contrast, different races of other undoubtedly single species, like *H. sapiens,* show pronounced differences. It appears that the degree of difference is much less important than the constancy of difference—that is, the discontinuity between the groups.

Discontinuity and Interspecific Sterility

But even strictly discontinuous differences need not indicate specific boundaries. Differences conditioned by a single gene will show complete discontinuity if dominance is complete (for either allele), and these differences may be of considerable magnitude. For example, the mutant *tetraptera* in *Drosophila* is characterized by the development of two pairs of wings, as in the dragonflies and many other orders of insects.

The type of discontinuity of greatest interest is that which results from the inability of related species to interbreed. An immense amount of literature exists concerning these phenomena of interspecific sterility and hybrid sterility. Many definitions of species have emphasized this point. It has the merit of being generally true, but it also has some faults. First of all, taxonomists must do most of their work with preserved specimens, and hence they would have difficulty discovering a sterility barrier even if they agreed completely on the validity of the concept. Furthermore, some facts cast doubt on the validity of the concept. It is difficult to distinguish between cases in which organisms cannot interbreed and those in which they can but do not do so for other reasons. For instance, the closely similar species *Drosophila pseudoobscura* and *D. persimilis* do not interbreed in nature because of a genuine sterility barrier. Bison (*Bison bison*) and cattle (*Bos taurus*) do not interbreed naturally, although experimental crosses prove that they are at least partially interfertile. Yet barriers may exist within an undoubtedly single species, as in the case of *Lymantria dispar,* in which racial crosses may result in hybrids that are sterile because of intersexuality. Finally, many cases exist, especially among plants, in which crosses between well-recognized species can be made easily. Frequently, the fertility of such crosses, or of the F_1 from them, is impaired partially or even greatly, but sometimes there is no such problem. Thus, the cross between the horse, *Equus caballus,* and the donkey, *E. asinus,* yields the sterile mule; but the cross between the dog, *Canis familiaris,* and the coyote, *C. latrans,* is fully fertile. Such cases make it difficult to defend sterility as a criterion for the identification of species.

Nonetheless, reproductive isolation remains the common element in most definitions of species. Dobzhansky defined species as a stage in the evolutionary process at which formerly interbreeding arrays become physiologically incapable of interbreeding. For evolutionary considerations, Goldschmidt treated groups that could be crossed as single species; at the same time he acknowledged that, for purposes of formal classification, taxonomists might be justified in treating these groups as separate species. Mayr defined species as groups of interbreeding populations that are reproductively isolated from other such groups. Taking all these definitions into consideration, we may define speciation as the origin of reproductive isolating mechanisms, so that gene flow between the related populations is prevented. (This will be the subject of the following chapter.)

This concept of species based upon gene flow between related populations is often called the *biological species concept.* It has strong logical appeal, especially now that the successes of genetics in analyzing the problems of evolution are so striking. T. Sonneborn, however, has protested against it on two grounds. First, it is difficult to apply it to more than a few of the more thoroughly studied organisms because of the sheer volume of experimental study required for its adequate application. Second, it is not applicable to those many species of Protozoa, lower Metazoa, and plants that reproduce asexually or parthenogenetically. Although some biologists have held that species in these organisms are not comparable to those of sexual organisms, Sonneborn makes a good case for the proposition that species in these groups are also based upon accumulation of genetic differences under the control of natural selection and hence that a species concept that excludes them is unsound.

For all these reasons, many biologists have come to the conclusion that the species is an arbitrary unit in the same way that the higher categories are. R.R. Sokal and T.J. Crovello are among those who have strongly maintained that local populations rather than species are the basic units of evolution. Still others persist in the opinion that the species is a valid natural unit, even though we cannot define it adequately. The quotation from Bateson, with which this discussion was introduced, expresses this viewpoint nicely. Likewise, Dobzhansky* said that

> Some biologists, lacking familiarity with the subject, have, in fact, fallen into this error [that species are arbitrarily determined units]. In reality, no category is arbitrary so long as its limits are made to coincide with those of the discontinuously varying arrays of living forms. Furthermore, the category of species has certain attributes peculiar to itself that restrict the freedom of its usage, and consequently make it methodologically more valuable than the rest.

Goldschmidt found the species he studied to be so sharply, if not necessarily widely, differentiated, that he felt justified in speaking of a "bridgeless gap" separating species from one another. Species have genetic, morphological, ecological, and chronological dimensions. No consensus exists on the reality of species. As noted above, Darwin was not a strong believer in the reality of species. The topic continues to generate controversy.

Rassenkreise and Speciation

Dobzhansky believed that the confusion about the species concept results from the mode of origin of new species. According to the current concept, a widely distributed species should break up into partially isolated subspecies, which become differentiated by selection of different characters and by genetic drift. Thus a *Rassenkreis,* or circle or group of races, is formed, the terminal members of which would eventually become sufficiently differentiated so that the sheer weight of difference would raise a sterility barrier. These would now be good species. With such a gradual origin, there would be no sharp break in morphological characters; hence taxonomists, who must depend mainly upon morphological characters, would have difficulty in determining whether the specific level of differentiation had been attained. This reasoning is based on the assumption that the processes of evolution are the same on the microevolutionary (changes within a species) and macroevolutionary (origin of new species or higher groups) levels. This assumption, which has been strengthened by the data of molecular evolution (Chapter 11), is plausible but unproven.

---

* T. Dobzhansky, *Genetics and the Origin of Species* (New York: Columbia University Press, 1937), p. 307.

Perhaps no more strongly suggestive study of a *Rassenkreis* has been published than that by R.C. Stebbins of *Ensatina eschscholtzii,* a Pacific coast salamander. This species ranges widely in the mountains of California and the Pacific Northwest (Figure 11–2). It is absent, however, from the great central valley of California, a low, hot, arid or semiarid region. Thus the distribution of this species forms an oval, with one segment corresponding to the coastal mountains, the other to the Sierra Nevada. The two are connected at either end. As we might expect, this population is not homogeneous but is broken up into seven subspecies, with intergrading populations usually found wherever

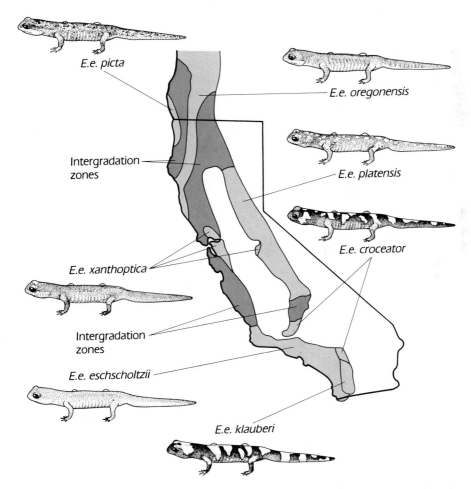

Figure 11–2.   *The* Rassenkreis *of* Ensatina eschscholtzii *on the West Coast. Although borders have been drawn for the range of each subspecies, there is always a broad zone of intergradation (i.e., of natural hybridization) wherever two subspecies meet.* Data from R.C. Stebbins.

the ranges of two subspecies meet. The species probably originated in the northwest, then spread southward along the coastal range and the Sierra Nevada, breaking up into subspecies along the way. The four coastal subspecies are uniformly dark brown to reddish brown dorsally; the three interior subspecies show an increase in orange or yellow spotting from north to south. In southern California, the ends of the chain are brought together, with *Ensatina e. eschscholtzii* representing the coastal division, and *E.e. croceator* and *E.e. klauberi* representing the interior division. The differences among them are marked.

Stebbins's first paper was based upon 203 specimens of *eschscholtzii,* 15 of *croceator,* and 48 of *klauberi.* Inland and coastal forms were not collected in the same locality, and no intergrades were found. Subsequent collections enlarged all groups, and several *croceator-klauberi* intergrades were taken within a few hundred yards of *eschscholtzii* in the same canyon and the same ecological zone, yet no intergrades with *eschscholtzii* were found. Evidence was found for neither ecological isolation nor seasonal differences in breeding. Hence Stebbins believed that these terminal members of the *Ensatina Rassenkreis* met without breeding and that they would be distinct species if the connecting members of the *Rassenkreis* were to become extinct. A later collection, however, included an apparent hybrid, which was intermediate between the two terminal types, and two backcross individuals, one to each of the parental types. These specimens suggest that a sterility barrier had not yet developed. Scant data are available on their breeding behavior, and it is possible that intergrades may be rare for reasons other than intersterility, such as preferential mating, although there is no evidence of this. This is an exceptionally illustrative *Rassenkreis,* the facts of which constitute one of the strongest arguments for the neo-Darwinian theory.

We know of other comparable examples among butterflies and shrews. In situations in which *Rassenkreise* are distributed over long clines rather than in circular patterns, fertility problems are often found when terminal members are brought together experimentally. Thus, *Rana pipiens* (leopard frog) from Quebec and from the coast of the Gulf of Mexico do not produce fertile offspring since they are adapted to very different breeding seasons and developmental temperatures. They are now regarded as the terminal members of a chain of sibling species, the *Rana pipiens* complex.

Allopatric and Sympatric Origin of Species

The Sierra Nevada and coastal subspecies of *Ensatina* occupy quite separate territories and have diverged widely. They are said to be *allopatric,* in contrast to *sympatric* populations, which inhabit the same territory. Allopatric populations obviously have the opportunity to diverge, whereas any tendency to diverge on the part of sympatric populations might be nullified by interbreeding. Accordingly, widespread agreement was achieved early in the period of Modern Synthesis that speciation was normally allopatric. The importance of sympatry was as a test of specific status: if formerly allopatric populations remained distinct after moving into the same territory, then they had achieved specific rank in allopatry.

Mayr, in particular, advocates speciation by small populations isolated on the periphery of the distribution of the parent population. He calls this *peripatric* speciation. He points out that such small, peripheral founder populations are likely to be near ecological niches not occupied by the parent population, and that these founder populations can carry only a small part of the genetic variability of the parent population. He also notes that the genetic reorganization that leads to reproductive isolation would be more readily accomplished in small rather than in large populations.

In Hawaii, there are over 600 species of *Drosophila,* many occupying only a single island. H. Carson's studies indicate that many of these have arisen in small, peripheral isolates. In some cases, the founder population may consist of only a single, fertilized female. Much inbreeding should help to fix the critical elements of the genetic reorganization. Of course, not every peripherally isolated population becomes a new species. Many quickly become extinct; others fail to form reproductive isolating mechanisms; and still others may reunite with the parent population.

G. Bush has reviewed this problem with interesting results. He believes that an array of ecological characters determines which of several modes of speciation are most probable for a given group of organisms. Can speciation occur by the more or less equal subdivision of a large population (allopatric, but not peripatric)? Bush calls this allopatric speciation type *a*, and he believes that it is characteristic of *K* strategists, which are highly mobile, long-lived, and high in competitive ability (as well sharing other ecological characters). The division of the American bison into plains and woodland types offers a good example. Among plants, the separation of the American sycamores into the eastern *Platanus occidentalis* and the western *P. wrightii* and *P. racemosa* might have occurred in this way. Bush believes peripatric speciation, which he designates as allopatric type *b*, to be characteristic of *r* strategists, with short life spans, low competitive ability, and high mobility. Chromosomal rearrangements are often characteristic of peripatrically formed species, as in the species of Hawaiian *Drosophila* cited above.

Both of these types of speciation are generally agreed to be important, but Bush advocates two more types that are more controversial. The first concerns populations that occupy adjacent territories, often with a narrow zone of overlap. This pattern of distribution is called *parapatric* distribution. Parapatric speciation occurs in groups that are *r* strategists of low mobility. Chromosomal rearrangements frequently separate species formed in this way. M.J.D. White has studied wingless grasshoppers in which a complex of chromosomal races are distributed parapatrically (*stasipatrically,* in White's terminology). A nice example occurs in plants that are adapted to withstand the toxic effects of copper, zinc, and lead. These plants, such as the grass

Agrostis, may invade mine tailings and become well established; whereas, only a few feet away, metal-intolerant plants grow. Gene flow between the two is very limited.

Although the importance of parapatric speciation is disputed, there is even less agreement with regard to sympatric speciation. Again, Bush believes that sympatric speciation is characteristic of *r* strategists of varying degrees of mobility. They usually have a narrow ecological specialization and are commonly parasitic. Parasitism is not a severe restriction. As Bush points out, it is very common, with more than a half million species of parasitic insects alone in existence. He believes that reproductive isolation in sympatric speciation is likely to depend on minor mutations of both structural and regulatory genes. The best examples of sympatric speciation are to be found in Bush's studies of Tephritidae, the true fruit flies. A mutation in the fly that results in a change of host plant will have a strong isolating effect; and since the insects commonly mate on the fruit on which they feed, the same mutation will also determine the choice of mates. The hawthorn fly, *Rhagoletis pomonella,* shifted to apples in 1864 and to cherries about a century later (1960). Bush has evidence that a single mutation was responsible for the shift in each case, although further mutations were selected to refine the adaptation. The original and mutant races continue to inhabit the same area, but host preference keeps them apart and provides separate mating sites. Time of emergence has also diverged. These factors keep the original and the mutant races reproductively isolated. Allopatric, and especially peripatric, speciation appears to be most common, but Bush has made a good case for the proposition that other modes of speciation cannot be neglected.

The Species: Class or Individual?

Traditionally, the species has been treated as a class, of which each organism is a member. M.T. Ghiselin, however, has argued that a species—any species—is an individual, not a class, and that organisms of that species are not its members but its parts. His reasoning is rather philosophical and will seem uncongenial to some biologists. He states that the scientific name of a species is a proper name and that only individuals can have proper names. Species, he says, are to evolutionary theory what business firms are to economic theory. Mayr considers Ghiselin's reasoning to be correct, and Gould makes it a key point in his theory of a hierarchy of levels of evolutionary processes. It remains to be seen how biologists generally will receive the idea of species as individuals.

The species question is, then, at once one of the most basic problems of biology and of evolution and one for which no satisfactory answer is available. It is complicated by the difficulty of comparing sexual and asexual species, by cases in which subspecific or specific status is disputed, and by cryptic or sibling species that show only trivial phenotypic differences even though they are reproductively isolated.

Darwin said that a species in any group is whatever a competent specialist on that group says a species is, and it may well be that species are not distinguished by the same criteria in different major groups. Discontinuity rather than degree of difference is likely to play the larger role in the achievement of an answer to the species question. Some aspects of the development of discontinuity will therefore be considered in the next chapter.

SUMMARY

The basic fact of taxonomy is that the varying arrays of organisms are arranged in a *hierarchy of discontinuous clusters,* the species and higher taxonomic groups. S. Wright discussed the origin of taxonomic discontinuities in terms of multiple alleles. If every organism had 1,000 pairs of genes and if each of these formed a series of ten multiple alleles, 10^{1000} genotypes could potentially be formed. Of course, the overwhelming majority of these genotypes have never been formed, but the genotypes that are actually formed cluster around adaptive peaks (positively selected) that are separated by adaptive valleys (negatively selected).

The *species* is the basic unit of classification, but the definition of species has been much disputed. Attempts to define a minimum magnitude of difference between species have failed. Degree of difference is less important than constancy of difference. Many definitions of species have relied upon interspecific sterility, but such definitions are difficult to apply both because there are many exceptions and because taxonomists commonly work with preserved specimens (and, of course, the sterility criterion can never be tested for fossil species). Nonetheless, E. Mayr's biological species concept—that a species is an interbreeding group of populations that is reproductively isolated from other such populations— is now generally accepted.

Widely distributed species commonly form a series of subspecies that are partially isolated from each other. Subspecies differ from one another partly because of selection for differing habitats, partly because of accidental or chance factors. The series of subspecies makes up a *Rassenkreis,* the terminal members of which may approach the status of separate species. It has often been claimed that extinction of the intermediate members of a *Rassenkreis* would leave the terminal members as distinct species.

Related populations living in the same territory are said to be *sympatric,* but those living in different territories are said to be *allopatric.* Mayr especially has argued for the importance of allopatric speciation, particularly in small populations on the periphery of the range of the parent group. He calls this peripatric speciation. Speciation may begin with a founder population, a very small isolate from the parent population. Founders include only a small part of the genetic variability of the parent population. Sympatric speciation has generally been considered to be exceptional, even nonexistent, but G. Bush has studied probable examples among parasitic insects.

Finally, a species has usually been treated as a class or category, of which organisms are members, but M. Ghiselin has proposed that species should be treated as individuals, of which organisms are parts. This idea has met with some approval, but the judgment of the biological community as a whole cannot yet be assessed.

REFERENCES

Brown, C.W., and R.C. Stebbins. 1964. Evidence for hybridization between the blotched and unblotched subspecies of the salamander *Ensatina eschscholtzi*. *Evolution* 18:706.
This paper and the two by Stebbins form the basis of the discussion of the Ensatina Rassenkreis *in this chapter.*

Bush, G. 1975. Modes of animal speciation. *Annual Review of Ecology and Systematics* 6:339–364.
A review of allopatric, sympatric, and other modes of speciation.

Ghiselin, M.T. 1974. A radical solution to the species problem. *Systematic Zoology* 23:536–544.
Species as individuals.

Mayr, E. 1970. The biological meaning of species. *Biological Journal of the Linnaean Society* 1:311–320.
A lucid defense of the reality of species.

Mayr, E. 1982. Speciation and macroevolution. *Evolution* 36:1119–1132.
A discussion of many of the topics of this chapter.

Sokal, R.R., and T.J. Crovello. 1970. The biological species: A critical evaluation. *American Naturalist* 104:127–153.
This paper marshals the arguments against the reality of species.

Stebbins, R.C. 1949. Speciation in salamanders of the plethodontid genus *Ensatina*. *University of California Publications in Zoology* 48:377–526.
In this and the following article, Stebbins describes a particularly illuminating Rassenkreis.

Stebbins, R.C. 1957. Intraspecific sympatry in the lungless salamander *Ensatina eschscholtzii*. *Evolution* 11:265–270.

CHAPTER

12

Isolating Mechanisms and Species Formation

Darwin and his contemporaries gave much thought to the possibility that the variations that form the raw materials of evolution might be "swamped" in crossings to the original type, so that no actual change in the species could occur unless such hybridizing were prevented by isolation of the new variant from the parent stock. This thinking was based upon a pre-Mendelian conception of heredity, the *blending* theory of inheritance, according to which, offspring should always be intermediate between the parents. If this theory were correct, then repeated backcrosses of a hybrid stock to the original type would result in an ever-closer approach to the original type. Moritz Wagner, beginning in 1868, added the concept of *isolation* of variant races as the necessary prerequisite and the inevitable cause of speciation. (The term *speciation*, generally used as a synonym for species formation, is, unfortunately, etymologically incorrect—speci*fic*ation would be more proper—but it seems certain to remain a permanent part of our evolutionary vocabulary.)

MENDELIAN GENETICS, ISOLATION, AND SUBSPECIATION

With the rise of Mendelian genetics, the blending theory of Darwin and Wagner became untenable. A gene-determined character could never be destroyed by crossing to the original type: it could only become heterozygous, with the possibility always present that homozygosity would be reestablished. The gene might thus spread evenly through a species, so that no tendency toward formation of a subspecies could be observed, but this process would not entail loss of the gene nor of the phenotype for which it was responsible. Breeding to the original type, however, could break up combinations of such genes. Selection can only operate on whole organisms, and so particular combinations of genes may have a value that would not be possessed by the separate genes of the genotype. For example, mutations to enhance the sense of smell and to increase running speed (such as longer leg bones) would both be of selective value to a predator that runs down its prey, such as the wolf. The two in combination would be of much greater value than either one alone. Although neither mutant could be destroyed by crossing to the original type, the combination would almost certainly be broken up either by independent assortment or by crossing-over.

As subspeciation is the ordinary prerequisite to speciation in the neo-Darwinian scheme, the formation of breeding populations that have distinct complexes of genes (subspecies) is of great interest. Completely random breeding within a species would result in even distribution of all of the genetic variability of the species, and so no subspecies could be formed. This process, however, would not prevent the mass transformation of a species into a new one if an appropriate selective force were operating, as Wright has pointed out. But complete random breeding probably never occurs, except in very small, endemic species, such as possibly among whooping cranes, which, until recent human intervention, consisted of but a single, small population.

Typically, local populations breed largely among themselves, with relatively little outbreeding. The result is that different populations of single species can build up genotypes that differ consistently in some or many loci. On the basis of the resulting phenotypes, we may classify these as subspecies. Wherever two such subspecies meet, they ordinarily interbreed and the expected Mendelian recombinations occur, with the result that a single, generally intermediate, and highly variable population is formed. Thus any start toward specific status that the subspecies may have made is lost. If, however, a subspecies is sufficiently isolated over a long period of time to prevent interbreeding with its relatives, it may continue to accumulate differences until it is no longer capable of successful interbreeding with its parent species. The subspecies may now be regarded as a new species. As long as the related groups remain geographically separated from each other, we refer to them as *allopatric* species. Whether two geographically separated groups are in fact allopatric species or simply subspecies is frequently in considerable doubt. But if they move into the same territory and fail to interbreed and form intermediates, we regard them as *sympatric* species, and their status is much less in doubt. C.L. Remington has suggested that antihybridization mechanisms may in some cases evolve *after* sympatry has been established, with selection acting to reduce matings between species because of the reduced fitness of the hybrid progeny.

The study of the means by which subspecies, and indeed species themselves, may be isolated from one another has played a major role in modern evolutionary studies. Very few would now care to accept the letter of the dictum of Romanes that "without isolation or the prevention of interbreeding, organic evolution is in no case possible." But, with some tempering, this has been the spirit of many recent studies.

PREMATING MECHANISMS

Mayr classifies the mechanisms by which subspecies and closely related species can be isolated from one another into two main groups, *premating* and *postmating* isolating mechanisms. The former prevent crosses from occurring; the latter reduce or eliminate the success of such crosses as may occur. Mayr defines isolating mechanisms as "biological properties of individuals which prevent the interbreeding of populations that are actually or potentially sympatric." By specifying "biological properties," it was his intention to exclude geographical isolation. Geographical isolation is, nonetheless, worth discussing as a premating mechanism because it often provides the opportunity for more strictly biological mechanisms to develop and because the line between geographical and ecological factors is often blurred. In addition to geographical isolation, we will examine seasonal or habitat isolation, ethological or behavioral isolation, and mechanical isolation.

Geographical Isolation

Distribution maps generally indicate that a particular species is found continuously over broad areas, but, in fact, all species select, in such large areas, those restricted portions that present suitable ecological features. A checkerboard is thus a better model of species distribution than is the typical distribution map. For example, the distribution map (Figure 12–1a) for the American sycamore, or plane tree, *Platanus occidentalis,* indicates a continuous distribution over more than half of the United States, from Texas in the south and Iowa in the north to the Atlantic coast. But to find natural groves of sycamores in this vast area, we must look in rich bottomlands and along the banks of streams. Again, the distribution map (Figure 12–1b) shows that the eastern meadowlark, *Sturnella magna,* is found over most of the United States and southern Canada, east of the Rocky Mountains, but it is found only in open grasslands. The various local populations of most species are similarly

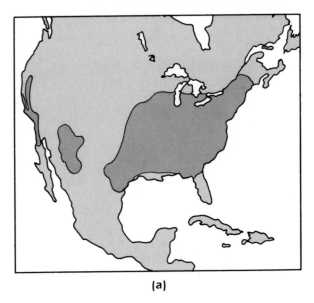

(a)

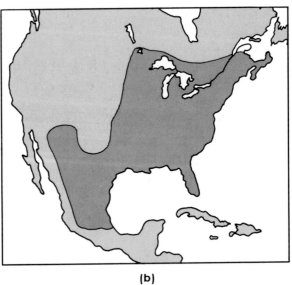

(b)

Figure 12-1. *(a) Distribution of the American sycamore tree,* Platanus: P. occidentalis *in the East,* P. Wrightii *in the Southwest, and* P. racemosa *in the Far West. (b) Distribution of the eastern meadowlark,* Sturnella magna.

separated by barriers of greater or lesser extent of territory that they cannot utilize for ecological reasons. Within the broad areas of mapped distribution, almost every geographical feature may prove to be a barrier to the dispersal of some species.

Even a small amount of salt water is a nearly absolute barrier to amphibians. For this reason, oceanic islands are usually uninhabited by amphibians, except where they have been introduced by humans, as in Hawaii. Salt water also separates many freshwater fishes. For example, on the Pacific coast many freshwater streams follow more or less parallel courses to the ocean. Typically, each stream has its own subspecies or even species. Although the expanse of salt water separating the mouths of neighboring streams may be small, the fishes do not cross it. If the floodwaters of the streams join during the rainy season, then the streams share their fishes. This fact shows that it is actually the salt-water barrier and not a homing instinct or another factor that typically keeps the fish faunas of neighboring streams separate.

Large bodies of water may be effective barriers to land birds. The Amazon River seems to be an absolute barrier to many birds, for in species after species the subspecies on opposite banks are different. This is not true for the Mississippi River, which, great as it is, is a much narrower river. Mayr has pointed out many instances in which neighboring tropical islands are inhabited by different subspecies even though the distances between them are short. When Darwin studied the birds of the Galápagos Islands, he found twenty-six species of land birds and eleven species of marine birds. Of these, twenty-one species of land birds were endemic, and only two species of marine birds were endemic. Islands that are separated only by short distances may have distinct subspecies or species, but, generally, those islands that are most isolated have the highest proportion of endemics.

Many mammals also are stopped by water barriers. Thus, subspecies of mice on opposite sides of major rivers are likely to be different. The zoogeographical realms were defined primarily on the basis of their mammalian fauna, and it is noteworthy that water barriers, the oceans, separate many, though not all, of these realms.

Mountains have sometimes been described as islands in a sea of lowlands. The flora and fauna of mountains are restricted in their dispersal by their inability to cross the lowlands intervening between neighboring mountains or mountain ranges. Mayr studied the mountain birds of New Guinea and found that almost all of them have broken up into subspecies in much the same fashion as have the birds of archipelagos. In some cases, there are even series of distinct altitudinal races, including lowland, midmountain, and alpine races. Many mammals such as mountain goats and bighorn sheep are limited to high mountain habitats. Turel found that a large proportion of alpine plants belonged to endemic races.

Similarly, mountains serve as barriers to lowland organisms. The American opossum, *Didelphis marsupialis,* ranges widely in the eastern United States. The low eastern mountains do not constitute a barrier to it. Introduced into California around 1870 for sport, it has thrived in the low coastal range, but it has been unable to invade the Sierras. The thirteen-lined ground squirrel, *Citellus tridecemlineatus,* is widely distributed in the prairies of north central United States, but it stops short of the Rockies. The cottontail rabbit, *Sylvilagus floridanus,* ranges (Figure 12–2) over most of the United States and southern Canada east of the Rockies, but it has been replaced in the mountains by its cousin, the jackrabbit (*Lepus spp.*). The white-footed mouse, *Peromyscus leucopus,* is similarly distributed, whereas its near relative, *P. maniculatus,* has successfully invaded the mountains. Even simple distance may serve as a barrier. L.R. Dice and P.M. Blossom found that seven species of small mammals were subspecifically distinct at Tucson and at Yuma, yet there seemed to be no barrier to their free dispersal other than distance. As stated at the beginning of this discussion, almost every natural feature may be a barrier to some plant or animal, and the barrier of one is the highway of dispersal of another. The formation of the Panama land bridge in the Pliocene opened a route for interchange between North and South America, but at the same time it separated the marine faunas of the western Caribbean and of the eastern Pacific, allowing speciation to occur.

Ecological and Seasonal Isolation

Geographical isolation blends into ecological isolation because of habitat preferences. Thus, extensive forests serve as barriers to dispersal of grassland organisms, prairies serve as barriers to forest organisms, and more subtle barriers are effective. For example, the red tree mouse, *Phenacomys longicaudus,* lives on a diet of fir needles. It also nests in fir trees and spends most of its life in them. Thus, not only will prairies serve as a barrier to its distribution, but non-fir forests will also be effective barriers.

The well-known homing instinct of many migratory birds also limits random dispersal. The homing instinct has been reported for other vertebrates, including salmon and sea turtles, and even for some invertebrates. The social structure of some species may have a similar effect. Thus, in geese, family groups do not break up at the end of the nesting season, but rather they migrate together and separate only after returning to the nesting ground in the following year. The result is a high degree of inbreeding.

R.H. MacArthur studied five species of *Dendroica* warblers that inhabit the same forests during the breeding season. Their ecological and food preferences are so similar that we might expect intensive competition to result in the elimination of all but the strongest competitor among the five species. In fact, he found that behavioral differences among the birds favored predation upon different species of insects, for the several species of birds forage at

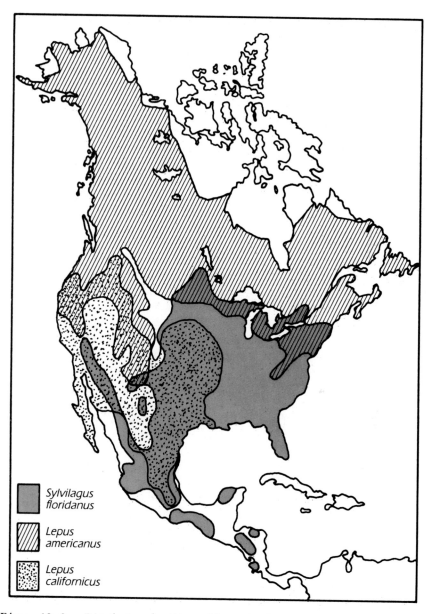

Figure 12–2. *Distribution of rabbits and hares. In the East (gray area), the cottontail,* Sylvilagus floridanus; *to the North and extending southward in the mountains of the West (crosshatched area), the snowshoe hare,* Lepus americanus; *and finally in the West (stippled area), the western jackrabbit (really a hare),* Lepus californicus. *Note the extensive overlaps.*

different levels in the trees and in different ways. Much overlap, however, does occur, especially in predation upon the most abundant and wide-ranging insects. Competition is also reduced by differences in breeding dates and by occupation of different habitats outside of the breeding season. Thus, in spite of their closely similar distribution, these five species of warblers are effectively isolated by ecological factors.

An important ecological difference between subspecies (or species) that may tend to prevent interbreeding is the selection of different habitats, so that potential mates from different populations do not meet even though they exist in the same general area. For example, Dice found that the ranges of two subspecies of *Peromyscus maniculatus* overlap in northern Michigan; and, although they will interbreed in the laboratory, there is no evidence of their interbreeding in nature. One of these races is found principally in forests, the other on sandy beaches. A. Murie has studied a similar case in the same species in Glacier National Park. Here, one of the races is confined to forests, the other to open prairies. The common water snake, *Natrix sipedon,* presents a comparable situation in Florida; freshwater and saltwater races may come close together, but they are kept separate by their habitat preferences.

A. Pictet has described a curious case in Swiss moths. *Nemeophila plantaginis* occurs in different altitudinal races, with a race above 2,700 meters and one below 1,700 meters differing in a single gene. At 2,200 meters, there is a hybrid population in which *all* of the moths are heterozygous. When the two pure races are crossed in the laboratory, the offspring include all three types in the F_1; if these data are correct, the homozygous types must be subject to a severe selective elimination at the 2,200-meter level. Similar data are being gathered for many different groups, both plant and animal, and such differences in habitat requirements are clearly of general importance as barriers to random mating.

A good example is provided by shrubs of the genus *Ceanothus,* of which there are many species in California. *Ceanothus thyrsiflorus* and *C. dentata* are broadly sympatric in the central part of the coastal range; but the former requires good, moist soil, whereas the latter thrives on poor, dry, and often shallow rocky soil. Although they are readily crossable in cultivation, hybrids between them are rare in nature. Their differing ecological requirements effectively isolate them.

Another ecological factor is seasonal isolation, that is, breeding seasons sufficiently different to make interbreeding improbable. For instance, in California five species of cypress trees (*Cupressus*) have been described. These species are subdivided into ten distinct groups that could be called subspecies, all of which have limited distribution and some being represented by only a single grove. These subspecies may grow side by side, forming only a very few hybrid trees or none at all. The factor that prevents interbreeding appears to be simply that the several races shed pollen at somewhat different times and so do not have the opportunity to cross. The occurrence of occasional hybrid trees can easily be explained by the fact that some trees of an early variety may shed pollen later than usual and some of a late variety may shed pollen earlier than usual. E. Anderson has shown that such seasonal differences are important in isolating some species of *Iris*. Researchers have demonstrated the same for some other genera. Seasonal isolation is probably common among plants.

Seasonal isolation is by no means unknown among animals. Generally speaking, the breeding season of warm-blooded vertebrates is long, and the seasons of many members of the same group coincide or overlap broadly. Nonetheless, Mayr cites several instances in which seasonal isolation appears to be effective among birds. Among the cold-blooded vertebrates and among invertebrates, breeding seasons may be restricted, and so seasonal isolation may be highly effective. In the northeastern United States, *Rana clamitans, R. pipiens,* and *R. sylvatica* may all breed in the same pond. But generally *R. sylvatica* will begin breeding before the others appear in the ponds, and *R. clamitans* may not begin breeding until the others are through. Water temperature is the decisive factor in determining when each species begins breeding. *R. sylvatica* begins to call at water temperatures as low as 44° F in southern Michigan, and probably breeds at temperatures only slightly higher. *R. pipiens* begins to breed when the water temperature reaches 55° F or perhaps somewhat higher. Finally, *R. clamitans* does not appear in the ponds until the water temperature exceeds 60° F; the minimum temperature for breeding has not been determined for this species. Elaborate species-specific mating calls probably also prevent mismatings. Actually, crosses among these three species fail in the early embryonic stages, but the seasonal separation may have preceded the sterility barrier and made possible its development.

A similar situation applies to the salamanders *Ambystoma tigrinum* and *A. maculatum* in the same area. *A. tigrinum* begins breeding soon after the spring thaws (late February and early March) and is usually through breeding by the end of March; *A. maculatum* does not begin breeding until late in March or early in April. W.F. Blair has published comparable data for the toads of the genus *Bufo*.

Ethological Isolation

Many instances are known in which behavioral (ethological) differences, particularly in regard to courtship, restrict random mating between members of different species. Among plants, mating is usually based upon more or less mechanical situations, such as wind pollination or insect pollination. But among animals, some preparation usually precedes mating. This preparation is slight for some marine animals, for example, sea urchins, in which discharge of gametes and of some sex-stimulating substance by a single individual may induce all or most of a large colony to shed their sex cells. Gametes are then mixed by the water currents, and fertilization follows. At the other extreme, elaborate courtship, sometimes lasting for many days, may precede mating. Mayr has pointed out that in many birds pair formation may precede copulation by periods of up to several weeks. Among such species, he finds that interspecific hybrids are rare, whereas they may be common between closely related species that do not have such an "engagement" period. He attributes the absence of hybrids among the former to the effect of differences in behavior pattern that may *break* the "engagement" of pairs that are not of the same species.

H.T. Spieth has made a detailed study of the sexual behavior of six species of *Drosophila*. He found that courtship and mating can be divided into six phases, at any one of which incompatible behavior of the potential mates may break off the courtship. Although several hundred attempts were made to obtain interspecific crosses, only once did actual copulation result. In the majority of cases, courtship stopped during the first stage. Yet the observable differences in the courtship patterns of the several species are, to our eyes, rather minor and are concerned with tapping of the female by the male, postural reflexes, and similar movements. In other species, however, differences in courtship patterns may be more pronounced. In crabs of the genus *Uca*, species can be recognized at a distance by their courtship "dances." Courtship dances also play a role in birds, as in the great crested grebe, *Podiceps cristatus* (Figure 12–3), and the mating dances of salamanders and turtles may be striking.

(a)   **(b)**

Figure 12–3. Examples of elaborate species-specific pair bonding and courtship rituals are found among the grebes. (a) Weed-sharing ritual in the European great crested grebe, Podiceps cristatus. (b) Spectacular rushing ceremony in which a pair of North American western grebes, Aechmophorus occidentalis, skitter across the surface of the water, then dive in perfect unison and disappear.

Morphologically, *Rana pipiens*, the leopard frog, seemed to consist of a single species, although it was well known that terminal members (northern and southern end members) from Quebec, Georgia, or the Gulf Coast were incapable of interbreeding. Today, researchers recognize a swarm of species instead of a single one. The distinction among species is more readily made by frogs than by taxonomists, but their characteristic mating songs serve both constituencies. Species of birds that show only minor morphological differences may also be easily differentiated by their songs.

Perhaps scents, songs, and recognition marks all belong here. That scent plays a major role in mating reactions of the Lepidoptera is well established. If a female of a rare species of moth is placed in a screen cage, many males will gather around it soon after the cage is placed outdoors. If the female is instead exposed in a glass container, the males do not assemble. That these scents, resulting from chemical compounds called *pheromones*, are highly specific can be shown by the selective response of males when two closely related species of female are put out in the same area. "Wrong" associations rarely occur. B. Petersen has studied a case in which no fewer than thirty-seven species of a single genus of moths live in a single Scandinavian valley without interbreeding. Visible differences among these species are minor, and Petersen believes that conspecific matings are guaranteed by the scents of the moths. Pheromones of related species may be closely similar, and here temporal isolation may be important. One species may call (release its pheromone) at dusk, whereas another may call only after dark; or other temporal relations may apply.

Songs of birds are well known for their role in mating, and in many instances the sounds produced by insects are known to play the same role. In addition to the recognition of conspecific mates, these sounds may stimulate sexual activity. As pointed out earlier, the wing movements of the male of *Drosophila* hasten the receptiveness of the female, but once excited, she will accept a *wingless* male. Mayr believes that the many phenomena formerly ascribed to sexual selection may be properly understood simply as sex-stimulating mechanisms. This plausible theory, if correct, would immediately bring these phenomena back into accord with the general theory of natural selection; or does it simply mean that sexual selection and natural selection may coincide?

The sedentary character of many animals also prevents the meeting of potential mates. Surprisingly enough, sessile animals such as the sea anemones are not, in effect, the most sedentary, because the small, free-swimming larvae of these marine invertebrates may be widely distributed by ocean currents. Similarly, most plants, although strictly sessile, have means of dispersal of seeds that are at least as efficient as typical means of dispersal for animals, and commonly much more so. But the larger animals, particularly the vertebrates, whose locomotor organs are among their most obvious characteristics, seem to use their powers of locomotion far more to maintain their home range than to expand that range. Extensive experiments in bird banding constitute the best evidence for this, as well as the best-known example, but similar evidence is accumulating for all groups of vertebrates.

Mechanical Isolation

Mechanical factors were once regarded as important isolating mechanisms, but their importance in animals now appears to have been overrated, if not completely unfounded. Among insects, the morphology of the genitalia may be complicated, and it frequently presents the best available taxonomic characters. On this basis, L. Dufour long ago proposed the "lock and key" theory—that is, that there must be an exact correspondence between the morphology of male and female parts to permit copulation. The female genitalia are thus compared to a lock that can be opened by only one key, namely the male genitalia of the same species. This theory is suggestive, but unfortunately there is much more evidence against it than there is for it. On the positive side, a few interspecific crosses in moths have been described in which death resulted from the inability of the copulating pair to separate. Two snails of different species were observed trying to copulate over a period of several hours; they failed, presumably because of mechanical difficulties. On the negative side, there is a great array of evidence. In some cases, extreme differences in the genitalia do not prevent successful copulation. Copulation between insects of strikingly different morphology has been observed. In many genera, the female genitalia may be identical throughout, whereas the males show differences of taxonomic importance. Even extreme size differences may be no bar to copulation: Dachshunds and St. Bernards have been successfully crossed.

In plants, mechanical isolation plays a more important role because of the necessity of morphological compatibility of the flower and the animal—usually an insect, bird, or bat—that cross-pollinates it. In Chapter 9, we discussed the pollination of *Pedicularis* by bumblebees (*Bombus*) and of the orchid *Ophrys* by the wasp *Scolia*. We could just as well have discussed both of these in relation to mechanical isolation. For example, the nectar-producing *Pedicularis* species are pollinated by queen bees, which, unlike the workers of the same species, have tongues long enough to probe for the nectar. Other species of *Pedicularis* and the *Bombus* species that visit them are mutually adapted either to the nototribic or to the sternotribic method of pollen collection and cross-pollination.

Character Displacement

Character displacement is an isolating mechanism first studied by W.L. Brown, Jr. and E.O. Wilson. Closely related species may be strongly differentiated where they are sympatric, yet similar where they are allopatric. Brown and Wilson have cited examples from groups as diverse as birds, fishes, ants, and crabs; others have even cited examples among trilobites. The ant *Lasius flavus* has a wide holarctic distribution, whereas *L. nearcticus* is confined to northeastern United States. In this common area, the two species contrast morphologically and ecologically, yet elsewhere in its range, *L. flavus* is similar to *L. nearcticus*. Again, in rock nuthatches, *Sitta neumayer* ranges from the Balkans eastward into Iran, and *S. tephronota* ranges from Turkestan westward into Armenia. The two species overlap broadly in Iran. Over much of their range, these birds are similar, yet they contrast strongly in the common territory (Figure 12–4).

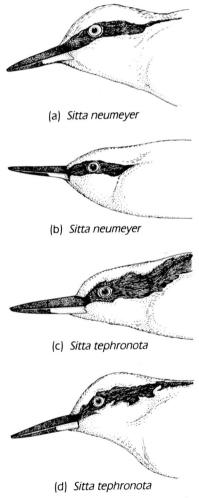

(a) *Sitta neumeyer*

(b) *Sitta neumeyer*

(c) *Sitta tephronota*

(d) *Sitta tephronota*

Figure 12–4. *Character displacement in rock nuthatches. (a)* Sitta neumeyer *from Dalmatia, the western end of its range. (b)* S. neumeyer *from western Iran, where it is sympatric with (c)* S. tephronota. *(d)* S. tephronota *from central Asia. Note that the sympatric birds (b and c) contrast strongly, but the allopatric birds (a and d) are rather similar.*

Brown and Wilson explain character displacement by selection for strong differentiation to avoid mismatings and wastage of gametes. Also, ecological differentiation reduces competition between species, thus permitting a greater total population in a given region. Character displacement may apply to any aspect of the biology of species, including fertility. In the case of the rock nuthatches, however, P.R. Grant believes that the contrasting patterns in Iran are simply the extremes of clines in each species. He cautions that difference in sympatry and similarity in allopatry are not sufficient to prove character displacement; the possibility must be ruled out that clinical variation in each species has not "passively" produced the contrasts in sympatry. This wise proviso may leave character displacement less important than it once appeared.

POSTMATING MECHANISMS

The final category of isolating mechanisms consists of postmating mechanisms, which act through a reduction of fertility. These are ordinarily subdivided into *interspecific sterility,* in which there is a failure to produce an F_1, and *hybrid sterility,* in which a good F_1 is produced, but this hybrid is sterile, and so no F_2 results. Such definitions are overstatements, for the phenomenon is not absolute: fertility may be reduced without producing absolute sterility, and this is commonly the case, particularly among plants. The literature on interspecific sterility and hybrid sterility is immense, going clear back to Aristotle. Although we cannot summarize it all, we will discuss some of the more salient facts.

Interspecific Sterility

Interspecific sterility may be based upon the failure of the pollen or sperm to reach the ovule or egg or upon the production of an inviable zygote. The first type is particularly well known in plants. Frequently, the growth of pollen tubes is slowed down in interspecific crosses or the pollen tube bursts, so that no fertilization is possible. If the species crossed are only remotely related, the pollen tubes may not grow at all. If they are more closely related, they are likely to grow more slowly than normal. As a result, if species A is pollinated both by A and by B pollen, practically all of the fertilizations will be conspecific (A × A) rather than heterospecific (A × B), because A pollen on A styles grow at a normal rate, and B pollen on A styles grow at a reduced rate. Although this phenomenon is well known in interspecific crosses, it is not unknown in intraspecific crosses. Thus in maize, alleles for starchy (*Su*) and sugary (*su*) endosperm are linked to another pair, *Ga ga.* On silks of heterozygous plants (*Su su Ga ga*), pollen *Su Ga* grows more rapidly than does *su ga,* so that there is an excess of starchy plants in the progeny. The two types of pollen tubes have equal growth rates on *sugary* silks.

The bursting of pollen tubes is particularly likely to occur in crosses in which the male parent has a chromosome number greater than that of the female parent. Whenever the chromosome numbers of the species are equal, the ratio of the chromosome number in the style to that in the pollen tube will be 2:1 (diploid to haploid). When the ratio comes closer to 1:1, then pollen tubes commonly burst. This problem is particularly likely to occur in crosses between polyploid species (Chapter 13) and their diploid ancestral species. For example, the commercial tobacco, *Nicotiana tabacum,* with a chromosome number of forty-eight, appears to be derived from *N. sylvestris* and *N. tomentosa,* both of which have twenty-four chromosomes. The cross of either of the latter species to *N. tabacum* can be made easily, provided that *N. tabacum* is used as the female parent. This cross will give a ratio of stylar to pollen tube chromosomes of 4:1. But if the same cross is attempted using *N. tabacum* as the male parent, the ratio is 1:1, and the cross usually fails because of the bursting of pollen tubes. In some cases, the bursting of pollen tubes has been shown to result from a greater osmotic pressure in the pollen tube than in the style. Although this pressure is correlated with chromosome number, the higher osmotic pressure of the pollen tubes is clearly a physiological effect of the genotype rather than a direct effect of the number of chromosomes, for chromosomes never exist in sufficient numbers to have important effects on osmotic pressure. Nothing strictly comparable is known in animals. However, A.S. Serebrovsky has shown that pH and osmotic pressures of vaginal secretions of various domestic mammals have small but consistent differences. Interspecific inseminations generally result in death of the sperm possibly because of osmotic or antigenic incompatibility.

Once a hybrid zygote is formed, there is no assurance that it will reach maturity, for death may occur at any stage during development. Thus, in interfamilial crosses of sea urchins, the paternal chromosomes are extruded and the resulting haploid larva dies at an early stage. In the cross between *Datura stramonium* and *D. metel* (jimsonweeds), development proceeds to the eight-cell stage, then stops. In many plant hybrids, lethality appears to result from an inadequate nutritive relationship between the endosperm and the embryo, for embryos that would otherwise die may be brought to maturity if they are dissected out of the seed and raised on an artificial nutritive medium. The adult plants so obtained may be fully as vigorous as the parent species.

Because little progress has been made in analyzing the genetic basis of interspecific sterility, two cases have especial interest. L. Hollingshead, using the extensive materials of Babcock's laboratory, made many interspecific crosses in the genus *Crepis,* a weed of cosmopolitan distribution. When *C. capillaris* is crossed to *C. tectorum,* the outcome depends upon the strain of the latter that is used. The F_1 from some strains is fully viable; from other strains the F_1 includes fully viable plants and plants that die in the cotyledon stage in a ratio of 1:1; from still other strains all of the progeny die in the cotyledon stage. *C. tectorum* appears then to have a gene that, in crosses to *C. capillaris,* behaves as a dominant lethal. Every effort to find a phenotypic effect of this gene in pure *C. tectorum* has failed, but the gene has also been found to behave as a lethal in crosses with several other species of *Crepis*. This gene evidently functions harmoniously with the *C. tectorum* genome, although in an unknown way, but when it combines with the *C. capillaris* genome, it is so disharmonious that lethality results.

A similar, but better-understood, example occurs in tropical fishes. The moonfish, *Xiphophorus maculatus*, carries a dominant gene, *Sd*, which causes a black spot on the dorsal fin. This spot is made up of macromelanophores, large pigment cells that are capable of tumor formation. Only the recessive allele of this gene occurs in the closely related swordtail, *X. helleri*. The two species can be crossed, and the F₁ is fertile. Hybrids of genotype *Sdsd*, with half of their chromosomes from each parent species, have more heavily pigmented fins than do homozygous *SdSd X. maculatus*, and these heavily pigmented fins may become tumorous (Figure 12–5). If the F₁ hybrids are now backcrossed to *X. helleri* (*sdsd*), the backcross progeny would derive approximately three-fourths of their chromosomes from the swordtail, and half of these zygotes would also be of genotype *Sdsd*. In these backcross progeny, the pigmented tissue always forms lethal melanoma tumors.

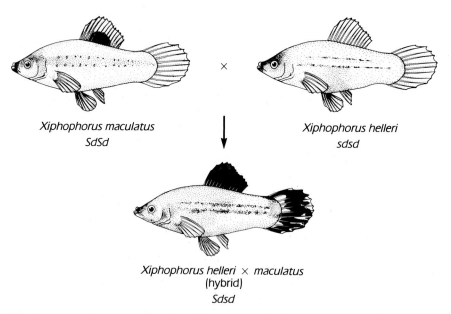

Xiphophorus maculatus
SdSd

×

Xiphophorus helleri
sdsd

Xiphophorus helleri × *maculatus*
(hybrid)
Sdsd

Figure 12–5. *Genetic tumors in tropical fish.* Xiphophorus maculatus *is normally homozygous for a dominant allele,* Sd, *which causes a pigmented spot on the dorsal fin. A closely related fish,* X. helleri, *is homozygous for the recessive allele,* sd, *and does not have a spot. The hybrids between the two species,* Sdsd, *have melanotic tumors that may be lethal. If these hybrids are backcrossed to the recessive parental type, the* Sdsd *progeny always have lethal tumors.*

Another gene pair, $N-n$, for body pigmentation, behaves similarly. Thus the genes Sd and N produce normal color variants in the moonfish; and they produce more extreme variants of the same characters in the F_1 hybrids; but in a genotype closer to that of the swordtail, they produce a still more extreme effect, melanotic tumors, and they have, therefore, become interspecific lethals. In other words, genes that are beneficial or at least harmless in their proper genetic background may become lethal in a different genetic background. Whether these data have any relationship to the fact that pigmented growths in mammals are sometimes precursors of cancer is not at present known.

Hybrid Sterility

The problems of hybrid sterility are not fundamentally different from those of interspecific sterility. In fact, hybrid sterility could be regarded as a special form of interspecific sterility in which the defect is simply delayed for one generation. This delay makes it possible to get more insight into the genetic and cytological mechanisms that are operative. When sterile hybrids are studied cytologically, the most common anomaly found is that the chromosomes fail to synapse. The unsynapsed chromosomes will typically be distributed without division to the two daughter cells at one of the maturation divisions, with the result that the distribution of the chromosomes is random. All of the chromosomes might go to one pole and none to the other, equal numbers might go to each pole, or any intermediate result might occur. The other division is usually equational. Most of the gametes so produced are inviable, but the genomes from the parent species might be accurately separated by random division, with fertile gametes resulting. Authenticated records exist of several mule mares that have produced offspring. In plants, the inclusion of all of the chromosomes of the hybrid within a single gamete may also result in viable offspring, which, however, will be polyploid (see Chapter 13).

Yet the failure of the chromosomes to synapse cannot be the whole explanation of hybrid sterility, for the reproductive systems of species hybrids are often grossly abnormal. In hybrids between *Drosophila melanogaster* and *D. simulans,* the gonads are rudimentary, and meiotic divisions never begin. In other cases, the chromosomes actually do synapse, but the hybrid is sterile anyway. Dobzhansky has described a particularly instructive case in hybrids between *Drosophila pseudoobscura* and *D. persimilis.* If the proper strains of these two species are selected for hybridization, synapsis in the hybrid will be complete, but other strains give partial synapsis or none at all. If failure of synapsis were the essential factor in hybrid sterility, then the first group of hybrids would be fertile, the second partially fertile, and the last highly sterile. In point of fact, the outcome is the same in every case: the first maturation division proceeds normally up to the metaphase, but the development during anaphase is grossly abnormal and results in a single, binucleate cell. The second division does not occur at all, and the giant spermatids degenerate. Hybrid sterility may, evidently, result from a derangement at any point in the long and complicated series of processes that extends from the zygote to the mature gametes that are ready to produce the next generation.

Whenever synapsis of the chromosomes fails completely, no insight into its causes can be obtained. But *partial* synapsis frequently occurs and may permit us to study the factors that prevent its completion. Frequently chromosomal rearrangements are present, and these rearrangements place purely mechanical obstacles in the way of completion of synapsis. I.H. Horton's study of synapsis in hybrids between *Drosophila melanogaster* and *D. simulans* is most illustrative. The meiotic chromosomes of this hybrid are not available for study because the gonads are rudimentary; however, there is every reason to believe that the salivary gland chromosomes, which are well developed in the hybrid, give an accurate picture of the synaptic behavior of the chromosomes of the gonads. Any rearrangements that the chromosomes of the two species show with respect to one another may be assumed to have arisen since their separation, but there is no way of judging which species has the more primitive arrangement.

In general, the salivary gland chromosomes of this hybrid are well synapsed, but in many regions synapsis is irregular or lacking (Figure 12–6). In these regions, ten rearrangements have been identified with certainty. Six of these are inversions; five of them, small (two to twelve bands in length) and one, large (involving nine sections of the third chromosome and including about 250 bands). The small inversions were too short to permit the formation of typical inversion loops, but they did suppress synapsis not only within their own length but also for a variable distance beyond the ends of the inverted sections. Occasionally, a pair of homologous bands did synapse in very short inversions, thereby showing that identical orientation of the bands is not essential for synapsis. As Horton pointed out, this fact raises the possibility that single-band inversions could have a mutational effect. The other four identified rearrangements were at the ends of the chromosomes and were more difficult to analyze; however, the evidence indicates that they were based upon a series of small translocations. In addition, there were fourteen regions in which suppression of synapsis was not associated with definitely identifiable rearrangements. Horton believes that small, cytologically undetectable rearrangements are most probably responsible for the inhibition of synapsis in these regions. He concluded that the differentiation of *Drosophila melanogaster* and *D. simulans* from their common ancestor has involved as many as twenty-four chromosomal rearrangements.

CHROMOSOMAL REARRANGEMENTS AS ISOLATING MECHANISMS

To what extent do such rearrangements, interfering with the normal course of meiosis in the hybrids, constitute genetic barriers between related species? Of course, failure of synapsis cannot be imputed to extensive chromosomal rearrangements in every case, for simple lack of homology will also prevent

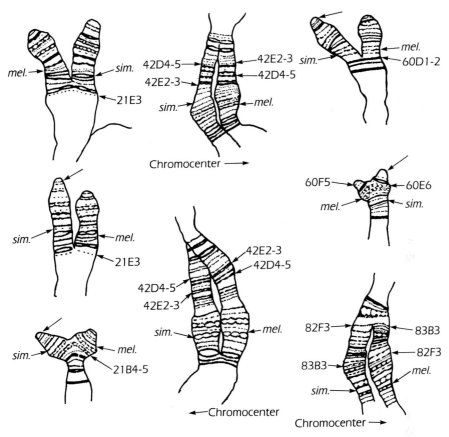

Figure 12–6. *Failure of synapsis because of chromosomal rearrangements in a hybrid between* Drosophila melanogaster (mel.) *and* D. simulans (sim.). *In each figure, the species from which each strand is derived is marked. Numbers identify the bands with respect to the standard map of the salivary gland chromosomes of* D. melanogaster.
Data from I.H. Horton, *Genetics* 24(1939):235–243.

synapsis in a hybrid. But closely related species have a considerable amount of homology between their chromosomes. Moreover, there are many specific cases in which chromosomal rearrangements produce considerable disturbance of the mechanism of meiosis.

Examples from *Drosophila*

Genetic interspecific isolating mechanisms exist when both of the parental types (pure species) are fully fertile and the hybrid is largely or entirely sterile. Generally speaking, chromosomal rearrangements studied in the laboratory do not meet this requirement. Although the heterozygous types do show reduced fertility, the homozygous rearrangements are commonly inviable in *Drosophila* (but often not in plants). That this need not be so is proven by the fact that fully fertile species are known in nature, the chromosomes of which can be shown to be rearranged with respect to one another. In the example above, if *D. melanogaster* be regarded as retaining the ancestral chromosomal pattern (an arbitrary assumption), then *D. simulans* must be homozygous for more than twenty rearrangements, certainly including both inversions and translocations and possibly including duplications and deletions as well. Although the majority of rearrangements are deleterious, as is the case with the majority of gene mutations, there is always a possibility that a particular rearrangement or *combination of rearrangements* may establish a new harmonious genetic system that can be maintained in the homozygous state and that is isolated from the parent type by the sterility of the hybrids. Species pairs such as *D. melanogaster* and *D. simulans* prove that this phenomenon has actually occurred in the formation of species now existing.

Perhaps the most thoroughly studied case in which inversions differentiate the chromosomes of species is the *Drosophila pseudoobscura* group, studied by Dobzhansky and his collaborators (see Chapter 8). A series of overlapping inversions enabled them to reconstruct the exact phylogeny of three species, *D. pseudoobscura, D. persimilis,* and *D. miranda,* together with the details of subspeciation in the first two species. Another excellent example is provided by the rearrangements that differentiate the chromosomes of the higher primates, as studied by Yunis and Prakash (see Chapter 8).

Oenothera and Translocation Complexes

An unusual situation is found in the genus *Oenothera* (Figure 12–7), the evening primroses, the subject of De Vries's studies, during which he rediscovered Mendel's laws. He found some aspects of the genetic behavior of *Oenothera* to be anomalous. A single species, *O. hookeri,* which is native to the Pacific coast of North America, behaves like a typical plant genetically. It has large flowers and is normally cross-fertilized. Other species are found east of the Rocky Mountains. They are difficult to treat taxonomically, and there is no agreement on the number of species, taxonomists inclined to lump ("lumpers") reckoning as few as one at one extreme, and those inclined to split ("splitters") as high as one hundred at the other extreme. All species east of the Rocky Mountains are characterized by small flowers and self-fertilization, and all show unusual genetic behavior.

The distinctive genetic features of most *Oenothera* species are four. First, they show only 50 percent fertility as compared to *O. hookeri,* a result of the formation of defective seed rather than of a small seed set. Second, when most *Oenothera* species are crossed to *O. hookeri,* "twin" hybrids are formed; that is, there are two classes of F_1 plants that differ in many traits. Third, in spite of the high degree of heterozygosity, which is demonstrated by the formation of the twin hybrids, the plants breed true when selfed (their normal method

Figure 12–7.   A flowering stalk of Oenothera biennis,
the subject of many evolutionary researches.

of reproduction). Finally, crossing-over rarely enters into the results of *Oenothera* crosses; but when it does, it always involves large blocks of characters. The results of crossing-over in *Oenothera* are so striking that crossover products were originally interpreted as large mutations.

All of these characteristics are understandable in terms of a series of translocations in the various species of *Oenothera*. To show how, a more thorough discussion of the behavior of heterozygous translocations (Figure 12–8) will be helpful. Let us assume that two pairs of chromosomes are named, respectively, $a \cdot a'$ and $b \cdot b'$, each letter identifying a chromosome end. At synapsis, when normal pair formation occurs, each pair behaves independently of the other. But now let us suppose that a translocation occurs, so that only one normal chromosome of each pair remains, and the other two now have the constitutions $a \cdot b'$ and $b \cdot a'$. Homologous point still synapses with homologous point, so that the translocated chromosomes bind the two tetrads together, forming a cross in the pachytene (a particularly clear phase immediately after completion of synapsis), and a ring of four chromosomes on the metaphase plate. In the division of such a ring, alternate members typically go to the same pole in the anaphasic movement, with the result that half of the gametes formed get both of the normal chromosomes and half get both of the translocated chromosomes.

There is no reason why only two pairs of chromosomes should be involved in a translocation complex. If three chromosomes, designated as $a \cdot a'$, $b \cdot b'$,

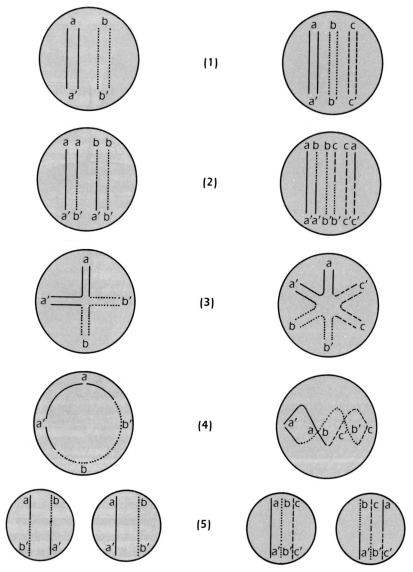

Figure 12–8. *Meiotic behavior of translocation complexes. Sequence at left is for two pairs of translocated chromosomes; and at right, sequence for three pairs. (1) The untranslocated chromosome pairs. (2) The chromosomes after translocation. (3) Configurations of the translocation heterozygotes in the pachytene. (4) (Left) metaphase ring, with translocated chromosomes alternating with untranslocated ones; (right) early anaphase, showing alternate disjunction of the chromosomes. (5) The two types of gametes formed in each case by alternate disjunction.*

and c·c′, are translocated so as to yield, in addition to the normal chromosomes, strands of the constitution b·a′, c·b′, and a·c′, then a ring of six would form, as illustrated at the right of Figure 12–8. Again, alternate disjunction occurs at anaphase, so that half the gametes have only untranslocated chromosomes and the other half have only translocated chromosomes. Theoretically, the limit to the number of chromosomes that may be joined in such a translocation complex is set only by the total number of chromosomes that a plant possesses, and this limit is actually reached in some species of *Oenothera*. *O. hookeri* has seven pairs of chromosomes, all of which behave independently. The other species all show translocation ring formation of some degree; in some, all fourteen chromosomes form a single large ring at the metaphase of the first meiotic division.

Oenothera shows one further genetic peculiarity: because of the alternate distribution of chromosomes of these rings, only two types of gamete are formed. Three types of zygote should, therefore, result—the two homozygous types and the heterozygous type. But only the heterozygous type ever appears in the progeny of selfed plants because each chromosome complex (a group of chromosomes inherited as a unit because of alternate disjunction) contains a recessive lethal gene. It is, however, a different gene in each complex. The result is that neither complex can become homozygous.

The 50 percent fertility that characterizes most *Oenothera* species is, then, a result of the balanced lethals that the several chromosome complexes carry. The twin hybrids are formed when any species is crossed to *O. hookeri* because each complex of the heterozygous parent (*O. hookeri* is largely homozygous) is radically different in genetic content from the other. In other words, most species of *Oenothera* are highly heterozygous. Yet, in spite of their heterozygosity, the plants breed true when selfed. Independent assortment is prevented by the alternate distribution of the chromosomes of the rings in all of the translocated chromosomes. Only parental type zygotes form the progeny, because the homozygous ones are eliminated by the balanced lethals.

The problem of the absence of crossover effects in *Oenothera* crosses is more complex. An inspection reveals that each chromosome consists of two ends that are perfectly paired and a central segment that is mechanically held out of contact with its homologue. Crossing-over should occur normally in the pairing arms of such chromosomes. But the pairing arms appear to be largely homozygous throughout the genus, so that crossing-over does not have genetic effects. The central segments, however, are protected from crossing-over because homologous parts are prevented from establishing synaptic contact. The genetic differences among the various chromosome complexes in the genus are concentrated in these segments, and so the sets of genes characterizing the various complexes are highly permanent. Rarely, crossing-over does occur in the central segments; and when it does, radically different character combinations occur. In early genetic literature these were called *half mutants* (because they affect half of the progeny). Thus the distinctive genetic features of the genus *Oenothera* are simple results of the fact that these plants are permanent translocation heterozygotes. The demonstration of these facts was one of the early triumphs of evolutionary genetics.

Chromosomal Mutants of *Datura*

In the genus *Datura,* the jimsonweeds (Figure 12–9), A.F. Blakeslee and his collaborators have found translocation phenomena that are, in some respects, even more complicated than those found in *Oenothera*. The normal chromosome number is twenty-four, so that there should be twelve pairs of chromosomes on the metaphase plate of the first maturation division. This is generally the case, but in interracial and interspecific crosses, rings of four and of six are commonly found. This fact shows that the various species and the races within a single species differ by a few translocations for which the several races and species are themselves homozygous. Each translocation complex appears to be associated with a different phenotype.

Trisomics are also common in this genus; that is, one chromosome of the set may be present in triplicate, making a total of twenty-five chromosomes ($2n + 1$) rather than the usual twenty-four in the zygote. Any one of the twelve chromosome pairs may be so affected, and the phenotype of the plant depends upon the particular chromosome that is present in triplicate. Each is called a *prime type*. Although any trisomic causes phenotypic changes throughout the plant, trisomic phenotypes have been named on the basis of changes in morphology of the seed capsules. At meiosis, the extra chromosome should be distributed to half of the gametes, but it tends to lag behind the other chromosomes and to be destroyed in the cytoplasm, and so less than half of the gametes formed by a trisomic plant will transmit the trisomic condition. As the $n + 1$ condition is a pollen lethal, it can be transmitted only by the ovule parent. Such trisomics, in which a completely normal chromosome is present in triplicate, are called *primary* trisomics. All of the twelve possible primary trisomics in *Datura* are known. At pachytene, two of the three chromosomes form a small ring to which the third is appended like a tail.

Also known are *secondary* trisomics in which the extra chromosome represents only half of a normal chromosome—the half, however, being duplicated. Thus, if the ends of the chromosomes are numbered so that the normal first chromosome is $1 \cdot 2$, its secondary trisomics would be $1 \cdot 1$ and $2 \cdot 2$. At pachytene, the three members of a secondary trisomic form a ring. Again, each secondary trisomic is associated with a distinctive phenotype. Twenty-four secondary trisomics are possible, but not all of these have been found.

Finally, there are *tertiary* trisomics, in which the extra chromosome is a translocation product made up of halves of two different chromosomes. These again are each characterized by a distinctive phenotype. At meiosis, the extra chromosome binds together the two tetrads with which it shares homology, thereby making a ring of *five* on the metaphase plate.

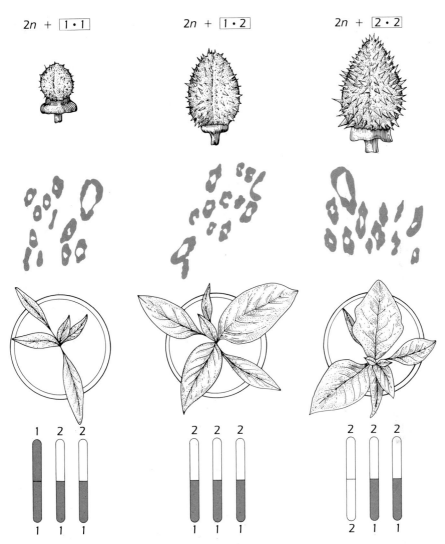

$2n + \boxed{1 \cdot 1}$ $2n + \boxed{1 \cdot 2}$ $2n + \boxed{2 \cdot 2}$

Figure 12–9. *Trisomics in* Datura. *The center column shows seed capsule, meiotic chromosomes, plant, and chromosome diagram for the primary trisomic* $2n + 1 \cdot 2$, *called "rolled." The left-hand column shows one of its secondary trisomics,* $2n + 1 \cdot 1$, *or "polycarpic." The right-hand column shows its other secondary,* $2n + 2 \cdot 2$, *or "sugarloaf."* Data from A.F. Blakeslee, *Journal of Heredity* 25(1934):19.

Evaluation of the Data

Thus in some well-demonstrated instances, homozygous inversions and translocations are among the characters that differentiate races and species. Less-detailed evidence is available with respect to duplications and deficiencies

(Chapter 8), yet such rearrangements are known to be involved in the differentiation of *Sciara* species; and differences in the metaphase chromosomes of many insects for which the salivary gland chromosome technique is not available are most easily understood in terms of duplications and deficiencies.

This brings us back to the question with which this discussion began—that is, to what extent do chromosomal rearrangements function as genetic isolating mechanisms? And are they also responsible for the phenotypic differentiation of related species, or is this a matter of independently accumulated gene mutations?

With respect to the latter question, all geneticists agree that position effects are a common result of chromosomal rearrangements. However, many rearrangements have been investigated without any evidence of corresponding position effects coming to light, particularly in maize (*Zea mays*), perhaps the most thoroughly known of all plants, genetically and cytologically. Hence the majority of geneticists doubt that chromosomal rearrangements play a major role in the phenotypic differentiation of species. W.R. Singleton has found it useful to disregard the distinction between chromosomal and gene mutations, and Goldschmidt regarded the former as the basis for systemic mutations of fundamental importance for speciation.

With regard to the first question, there is less difference of opinion. Undoubtedly, chromosomal rearrangements do serve an important function in isolating related populations from one another. B.T. Kozhevnikov performed a particularly interesting series of experiments on this problem. By combining, in a single stock of *Drosophila melanogaster,* two translocations between the second and third chromosomes, he obtained a strain that he regarded as a synthetic species, *Drosophila artificialis* (Figure 12–10). *D. artificialis* forms four types of gametes, which combine to form sixteen types of zygotes. However, only four of these survive, as the other twelve contain large deficiencies and duplications. *D. artificialis* is completely isolated from the parent species, *D. melanogaster,* for all zygotes formed by the "interspecific" cross are inviable because of large deficiencies and duplications. A "species" with only 25 percent viability would not be likely to fare well in nature. If, however, an additional rearrangement should stabilize this genotype so that it became viable in the homozygous condition, a good species would for all practical purposes have been synthesized in the laboratory, by means of a fortunate combination of a few chromosomal rearrangements.

Although much progress has been made in identifying and classifying isolating mechanisms and although they have been successfully fitted into the neo-Darwinian theory as a fundamental part of the mechanism of evolution, less progress has been made in analyzing the genetic basis of isolating mechanisms. The chromosomal rearrangements are an outstanding exception; even though their exact role in evolution is not clear, everyone can agree with Dobzhansky that "there can be little doubt that chromosomal changes are one of the mainsprings of evolution." There is, however, a general feeling among geneticists

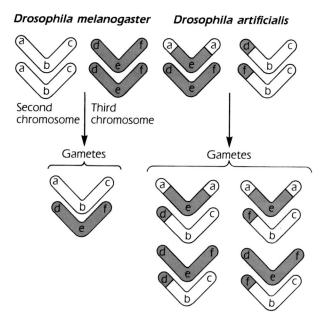

Figure 12–10. Drosophila artificialis. *Normal second and third chromosomes of* D. melanogaster. *Sections of these chromosomes are identified by letters (*abc, *second chromosome; and* def, *third chromosome). All gametes of these flies include one member of each pair. A series of translocations resulted in the chromosome set at the right:* aea, def, dbc, *and* fbc. *This results in the four types of gametes shown in the lower right. Crosses of these translocated flies to normal* D. melanogaster *are completely sterile because all combinations include large deficiencies and duplications. Because of this, Kozhevnikov named his translocated stock as an experimentally produced species,* Drosophila artificialis. *When crossed among themselves,* D. artificialis *shows only 25 percent fertility, for only four of the sixteen possible zygotic combinations are viable.*

that both isolating mechanisms and phenotypic differentiation are based upon quantitative characters that are influenced in a small degree by many different pairs of genes.

If Goldschmidt were right in his opinion that systemic mutation, in contrast to gradual accumulation of small mutations, is the basis for speciation, then only genetic isolation would be important for speciation. Other types of isolating mechanisms, although they would then be important for the biology of a species, could not affect the species' evolution above the subspecific level. The decisive chromosomal changes, which would at once be the basis both for genetic isolation and for phenotypic differentiation, could arise in a continuous population as well as in a population that was already partially isolated. Moreover, even with very long isolation by other means, the decisive chromosomal isolation might fail to occur.

FAILURE OF ISOLATING MECHANISMS

Regardless of the manner in which we evaluate the role of isolating mechanisms, one of their important properties is their occasional failure. Subspecies and species that are ordinarily separated by one or several isolating mechanisms may occasionally produce hybrids. Such cases are most instructive. If the barrier is geographical, natural or human-engineered events must overcome this barrier before hybridization can occur. Typical of these are the many cases in which populations were separated during the glacial ages, during which time they diverged, to a specific degree or perhaps only to a subspecific degree. With the recession of the glaciers, the diverging populations have again migrated into their former range, and now hybridization may occur. Mayr has assembled many such cases from the fauna—particularly the avifauna— of central Europe. During the Pleistocene glaciation, Scandinavian and Alpine ice caps approached within about 300 miles of each other in central Europe, with the result that the temperate flora and fauna, which had formerly inhabited this zone, were forced to take refuge either in southern France and Spain or in the Balkans. Thus segments of formerly continuous populations were isolated from each other at opposite ends of the Mediterranean Sea. While thus isolated, the two populations of the various species diverged. With the recession of the ice, the eastern and western populations again moved into their original, common territory. In North America, glaciation had a similar effect, so that pairs of subspecies or species, one eastern and one western, are common.

The behavior of these once-isolated, but now sympatric, populations varies greatly in different cases. In some, no interbreeding occurs, indicating that the divergent populations have reached the status of distinct species. In others, such as the hedgehogs *Erinaceus europaeus* (western) and *E. roumanicus,* hybrids are rare, but they do occur. These are treated as good (distinct) species, but their status may be debatable. In still other cases, for example the crows *Corvus corone* and *C. cornix,* a rather stable hybrid population occurs where the eastern and western groups meet. These are now treated as a *Rassenkreis,* under the former name.

In the case of the several species of plane trees (*Platanus*) the great geographic barriers have been spanned by human intervention. This genus was once widely distributed throughout the Holarctic region. Sometime during the Tertiary period, however, its distribution became discontinuous. The submersion of the Bering Straits area separated the Old World and New World populations. In the Old World, the elevation of the great mountain ranges restricted its range to Asia Minor and the eastern Mediterranean region, where it has been described under the name *P. orientalis.* Meanwhile, the elevation of the western mountain ranges and formation of deserts caused the plane trees in the United States to break up into three discontinuous populations. One of these, which inhabits most of the United States east of the Rocky Mountains, is called *P. occidentalis.* The southwestern representative is called *P. wrightii,* and the California representative is called *P. racemosa.* There is also a Mexican species.

In Western Europe, the London plane tree, *P. acerifolia,* is a common cultivated shade tree. It is generally regarded as hybrid between *P. occidentalis* and *P. orientalis,* although these species cannot survive in Western Europe today. All of the *Platanus* species are fully interfertile when crossed artificially. Whether ecological or other barriers would limit the crossability of these "species" if the hybrids were not cultivated is problematical. The possibility that hybridization might lead to the formation of new species has often been discussed. For those who are able to accept *P. acerifolia* as a good species, this is a clear-cut example. But the least that can be said is that the *Platanus* species are less distinct than are typical good species, and little support for the idea of speciation via hybridization is available elsewhere. We will discuss an important exception, allopolyploid plants, in Chapter 13.

INTROGRESSIVE HYBRIDIZATION

A more important result of the hybridization of species is what E. Anderson has called *introgressive hybridization.* This imposing name covers a simple phenomenon: if a natural hybrid is formed, it is very probable that it will be mated not to another hybrid but to one of the pure parental species. As a result of this backcross, some of the genes of each parental species will *introgress* into the genotype of the other. C.B. Heiser's study of the sunflowers *Helianthus annuus* and *H. Bolanderi,* exemplifies this phenomenon well (Figure 12–11). *H. Bolanderi* is restricted to the west coast of the United States. *H. annuus* appears to have been originally an eastern species, but it has been introduced into the coastal states by humans, and it has become well established and widely distributed there. There is some ecological separation of the two species, but they are found together in areas disturbed by civilization so that habitats intermediate between those usually occupied by these species result. In such areas, several natural hybrids were found, and one large *hybrid swarm* was found along a roadside. A hybrid swarm is a population in which F_1, F_2, and later generations of hybrid segregation are intermingled with backcross progeny of various degrees. Naturally, such a population shows extreme variability. The hybrids of recognizable degree all showed some reduction in fertility, sometimes a drastic reduction, with only about 3 percent of the gametes being viable. On backcrossing to either pure species, the fertility increased. Thus genes from each of these species can be transferred to the other. Heiser has found that each of these species of sunflower varies *in the direction of the other.* Although this fact could be accounted for by parallel mutation in the two species, he believes that introgression is a much more likely explanation.

Although introgressive hybridization has been studied less in animals than in plants, it has been found in every major group in which it has been sought, including all classes of vertebrates, insects, and especially butterflies and moths.

As we noted, hybridization of species is particularly likely to occur in areas of human disturbance. Anderson, who has made extensive studies on

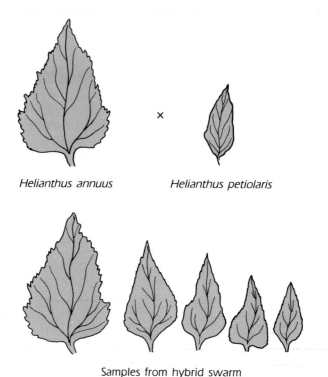

Helianthus annuus   *Helianthus petiolaris*

Samples from hybrid swarm

Figure 12–11.   Introgressive hybridization in sunflowers, Helianthus. *(Above) typical leaves of* H. annuus *and* H. petiolaris. *Like the cross of* H. annuus *and* H. Bolanderi *cited in the text, the hybrids produced by these species have very low fertility, but if the hybrids are repeatedly backcrossed to either parental type, successive generations approach normal fertility. The leaves from such a hybrid swarm (below) are highly variable, and the extremes are quite similar to the parental species.*

introgression in plants, believes that this interference may be necessary for the survival of hybrid swarms because the hybrids are likely to require habitats intermediate between those to which the parent species are adapted. He therefore speaks of the "hybridization of the habitat." As people disturb nature and their manifold activities, such "hybrid habitats" are likely to be formed, thus facilitating the exchange of genetic variability between related species and increasing the range of types from which selection can choose the most advantageous.

SUMMARY

In order for new species to form, gene flow among populations of a parental species must be reduced. Subspecies represent series of populations for which relative isolation has permitted modest or moderate genetic differentiation to occur, but members of one subspecies may breed with members of another subspecies without difficulty if opportunity presents. If a barrier between two subspecies is removed, the two may merge into genetic identity. However, if the barrier imposes genetic isolation for long enough, *allopatric speciation* may occur. Allopatric species reintroduced into each other's ranges are said to be *sympatric*, and such species will encounter genetic barriers to reproduction with each other. These barriers, called *isolating mechanisms*, may operate to prevent mating (*premating isolating mechanisms*) or to reduce the effectiveness of such matings that do occur (*postmating isolating mechanisms*). Fertility barriers among closely related species are often not absolute. Isolating mechanisms may either be the direct result of genetic differentiation during allopatry or may be selected for in sympatry due to the reduced success of the F_1 or F_2.

Examples of premating isolating mechanisms are geographical isolation, ecological isolation, temporal isolation, ethological isolation, and mechanical isolation. Major geographic features such as great rivers, lakes, and mountains serve today to separate species, and geological features of the past have accomplished the same. For instance, the Pleistocene glaciers bisected both North America and Europe, and many east-west pairs of species of both plants and animals are found today. The erection of the Panama land bridge between North and South America in the Pliocene formed a highway for land animals but imposed an absolute barrier between the Caribbean and the eastern Pacific for marine organisms. *Ecological isolation* results when organisms living in the same geographical range develop habitat preferences that keep them apart. *Temporal isolation* because of seasonal differences in breeding time forms a very effective barrier to breeding. Species of trees and of amphibians provide excellent examples of this phenomenon. *Ethological isolation* occurs when organisms (e.g., fruit flies, crabs, frogs, birds) employ elaborate species-specific mating rituals. In such cases, a male of one species is unlikely to court a female of a related species successfully. *Mechanical barriers* to mating due to the incompatibility of male and female genitalia in related species may sometimes occur. *Character displacement* is an important phenomenon in which species that are broadly allopatric accentuate premating isolating mechanisms in a zone of overlap, with the result that the probability of unproductive matings is reduced.

Postmating isolating mechanisms are of two major types, interspecific sterility, in which crosses fail to produce a viable F_1, and hybrid sterility, in which the F_1 is sterile. Commonly, however, reduced fertility rather than absolute sterility characterizes crosses between closely related (i.e., recently differentiated) species, particularly among plants. Chromosomal rearrangements as well as accumulated gene mutations are frequently detected in analysis of cases of hybrid sterility. These are well known among species of *Drosophila*. Chromosomal translocation complexes in evening primroses (*Oenothera*) and jimsonweeds (*Datura*) are particularly striking. A complicated series of trisomics (in which a single chromosome is represented by an extra copy) make the latter especially interesting.

When recently formed species come into sympatry, rare hybrids may form despite the existence of isolating mechanisms. In this case, hybrids are more likely to breed with one or the other of the parental species. This *introgressive hybridization* results in the introduction of genes from one species into the pool of the other. Hybrids tend to occur in areas of human disturbance where "hybrid habitats" exist unlike those encountered by either parental species in nature.

REFERENCES

Anderson, E. 1968. *Introgressive Hybridization.* Hafner, New York.
  A reprint of a book first published in 1949, it is still the major work on this subject, especially if supplemented by his review of 1953 in Biological Reviews *28:280–307.*
Dobzhansky, T. 1970. *Genetics of the Evolutionary Process.* Columbia University Press, New York.
  Contains an excellent chapter on isolating mechanisms.
Goldschmidt, R.B. 1955. *Theoretical Genetics.* University of California Press, Berkeley and Los Angeles.
  The last major contribution of this evolutionary dissenter, including important material on genetic isolation.
Grant, P.R. 1975. The classical case of character displacement. In *Evolutionary Biology,* vol. 8, edited by T. Dobzhansky, M.K. Hecht, and W.C. Steere, 237–238. Plenum Publishing, New York.
  A valuable review and updating of this concept.
Grant, V. 1981. *Plant Speciation,* 2nd ed. Columbia University Press, New York.
  Includes an excellent review of introgressive hybridization and of other topics germane to this chapter.
Mayr, E. 1970. *Populations, Species, and Evolution.* Harvard University Press, Cambridge, Mass.
  Includes a good chapter on isolating mechanisms; but they are more thoroughly (but less currently) covered in his book of 1963, Animal Species and Evolution, *of which this is a condensation.*
White, M.J.D. 1978. *Modes of Speciation.* Freeman, San Francisco.
  Includes valuable material on genetic isolation.

CHAPTER
13

Polyploidy

In 1917, O. Winge published a study of the numbers of chromosomes in plants. Although he noted diploid numbers ranging from four to well over two hundred, the frequencies of these numbers were by no means random. Twelve was the most frequent number, and eight was the second most frequent. About 50 percent of all plants have numbers below twelve. Among those plants with higher chromosome numbers, the most frequent numbers are multiples of the lower ones. Within a single genus, it commonly happens that there is a series of species in which the chromosome numbers of some are multiples of that of another species. For example, there are species of wheat with fourteen, twenty-eight, and forty-two chromosomes. Seven chromosomes appears to be the basic haploid number in this genus. If we were to plot the haploid numbers of many plants on a frequency histogram, none of the maxima would fall on prime numbers.

POLYPLOIDY: A MAJOR PHENOMENON IN PLANT EVOLUTION

Winge concluded that the most probable explanation of these facts lay in the assumption that over half of the higher plants were *polyploids*—that is, their gametic chromosome set consisted of two or more basic sets of chromosomes side by side in the same nucleus. This phenomenon could occur in one of two ways: either a single haploid set of chromosomes might be present more than twice (autopolyploidy) or two different sets of chromosomes might be present, making a total of more than two genomes (allopolyploidy). Winge ventured the guess that the second type would prove to be the more frequent, basing his opinion on the assumptions that lack of homology would prevent pairing of the chromosomes in species hybrids and that the necessity of pairing would, therefore, stimulate doubling of the whole chromosome complement in such hybrids. Winge's conclusions have been substantiated. The development of polyploid series, one of the major phenomena of plant evolution, is now one of the most thoroughly understood.

Colchicine Induction of Polyploids

The study of polyploidy has been greatly facilitated by the development of experimental techniques for its artificial production. Many moderately effective methods have been introduced. These include selecting bud variants with typical polyploid characters, treating seed with temperature shocks, radiating seed, and cutting the stem of a plant, then selecting tetraploids from among the callus shoots (shoots that develop just below the wound). The last method gives a yield of about 15 percent tetraploids in the tomato. In other plants, treating of the wound with growth hormones may be necessary in order to get good results.

All of these methods have been made obsolete by the colchicine method. Colchicine is an alkaloid drug derived from the root of *Colchicum autumnale,* the autumn crocus. That this drug interferes with the metabolism of nucleic acid (a major constituent of the chromosomes as well as of some other parts of the cell) has been known for nearly a century; but in 1937 scientists learned that this drug is a mitotic poison. The prophase of colchicine-influenced mitosis is apparently normal, and the doubling of the chromosomes proceeds as usual. However, the spindle is either defective or absent entirely, and so the chromosomes, which have already been duplicated, are all included in a single restitution nucleus. Thus a tetraploid condition is established. Yields of 50 to 100 percent tetraploid plants are not uncommon when colchicine is used.

Gigas Habitus

The most common polyploids are tetraploids, with four haploid genomes ($4n$) in the somatic chromosome complement. The gametes are therefore $2n$. Autotetraploids are known both in nature and in experimental materials. As a matter of fact, one of the bases of the mutation theory of De Vries was a mutant strain of *Oenothera lamarckiana* (Figure 13–1) which has turned out to be a spontaneously produced tetraploid, because it has twenty-eight chro-

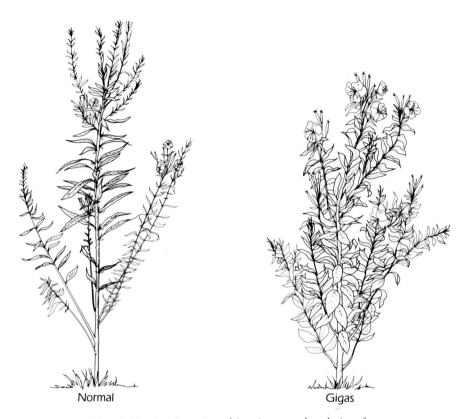

Figure *13–1.*   Oenothera biennis, *normal and gigas forms.*

mosomes instead of the usual fourteen. This plant illustrates well a complex of characters that are generally, though by no means universally, found in tetraploid plants. First, it is considerably larger than diploid *O. lamarckiana.* Because of this, De Vries named it *O. gigas,* regarding it as a new species. The stems are thicker, and the leaves are shorter, broader, and thicker than those of the diploid plants. The most obvious physiological difference is a slower growth rate, but increased vitamin content has been reported for tetraploid tomatoes, and the physiology of the tetraploid plant is probably modified as much as is its morphology. These traits collectively are referred to as the *"gigas* habitus," because they so commonly characterize tetraploids. Yet there are exceptions to all of them. There are tetraploids, for example, that are dwarfed as compared to the diploid.

We could regard the tetraploid *Oenothera lamarckiana* as either an autopolyploid or an allopolyploid. Because the *O. gigas* variety is derived from a single diploid parent species, we could regard it as an autotetraploid. But because the diploid *O. lamarckiana* itself is a permanent structural hybrid, we could regard the *O. gigas* variety as an allotetraploid. Generally no such confusion exists, yet the distinction between these two types of polyploidy is perhaps always one of degree rather than of absolute distinction. By definition, the several haploid sets that constitute the chromosome complement of an autopolyploid do not differ from one another in any greater degree than do the two haploid sets of the corresponding diploid. But the first requirement for the formation of an allopolyploid is that the two parent species must be able to form a viable (though not necessarily fertile) hybrid. As discussed in Chapter 12, forming of a viable hybrid becomes less and less probable as the relationship of the potential parent species becomes more remote. The most usual situation is that both parental species belong to the same genus, although we know of many cases in which allopolyploids have been formed between different genera of a single family. The relationship is usually close enough so no doubt exists that there is a significant degree of homology between the chromosomes of the parental species. Frequently, this homology is shown by a limited amount of synapsis in the F_1 hybrid. Some degree of homology between the several genomes of a polyploid is probably necessary for its formation. In autopolyploids, this homology is substantially complete; in allopolyploids it is markedly incomplete.

AUTOTETRAPLOIDY IN PLANT EVOLUTION

By the use of colchicine, a large number of experimental autotetraploids have been formed and investigated, but autotetraploids are also known from nature. Autopolyploidy does not seem to lead to the formation of new species but only to well-marked varieties. Yet there is considerable reproductive isolation between a diploid and its autotetraploid, because the hybrid between them is a triploid (three haploid genomes in each somatic cell). Triploids are highly sterile because the distribution of the chromosomes at meiosis is irregular.

Good examples of naturally occurring autopolyploids may be found in such diverse genera as the grasses *Phleum* and *Festuca* and the garden flowers *Viola, Dianthus, Chrysanthemum,* and *Nasturtium.* A particularly instructive example is that of the diploid and tetraploid races of *Tradescantia canaliculata,* the spiderwort, in southern United States. The chromosome numbers are twelve and twenty-four. This species is widely distributed over the great plains from the Rocky Mountains eastward to the Mississippi River. Over most of this area, the plants are tetraploid; but in a small area in northern Texas, the diploid race occurs. Geologically, this area is the oldest part of the total range. Only the tetraploid race appears to have been able to invade those territories more recently opened up to flora colonization. This greater aggressiveness of polyploid plants is characteristic. Another species of the same genus, *T. occidentalis,* overlaps the eastern part of the range of *T. canaliculata.* Interestingly enough, this species, which extends eastward, is also tetraploid over most of its range, but the diploid form occurs in the same refuge in northern Texas. In both of these species, the tetraploid races are much more vigorous than are the diploid races. Crosses between the races can be made, but the triploid offspring are sterile.

Experimental autotetraploids have been thoroughly studied genetically and cytologically. Although the plants are frequently hardier than their diploid relatives, they have not been outstandingly successful because they show markedly reduced fertility and tend to revert to diploidy. A study of meiosis in the autotetraploids revealed the cause of these characteristics. For normal meiosis, the chromosomes should form only pairwise associations. However, in auto-tetraploids *four* homologues of each kind are present. Frequently, only pairs are formed, and normal gametes result; but sometimes three chromosomes of a kind synapse, with the fourth one behaving independently. In such a case two of the three synapsed chromosomes pass to one pole, and one passes to the opposite pole. The independent chromosome may now balance this situation, or it may make it even more unbalanced. Again, all four chromosomes might synapse to form a tetravalent. Ordinarily, a tetravalent divides so that two members go to each pole, thus yielding normal gametes; but occasionally it splits into three and one. Now if only one or a few chromosomes are missing or are in excess, an otherwise tetraploid zygote may be successful. Such trisomic and monosomic strains are well known to plant breeders. If more than a small portion of the chromosome pairs are so unbalanced, lethality results. Lethality as a result of chromosomal imbalance is believed to be the cause of the reduced fertility of the autotetraploids. Reversion to the diploid condition is probably based upon development of unfertilized ovules.

The study of meiosis in naturally occurring autotetraploids reveals ab-normalities comparable in kind and in degree to those of the experimental autotetraploids. In the face of such a disadvantage, we may well ask how these have ever become established in nature, let alone become more widespread than the diploid parent, as is so commonly the case. However, the tetraploids are commonly more vigorous and adapted to more severe environments, and their selective value probably more than compensates for their reproductive liability. Yet most of the naturally occurring polyploids that have been analyzed have been of the allopolyploid type, and this reproductive liability may well have restricted the role of autopolyploidy.

ALLOTETRAPLOIDY IN EXPERIMENT AND IN NATURE

Allotetraploidy has also been produced experimentally. Many methods are available, of which much the simplest is the treatment with colchicine of the F_1 hybrids between two species. Other methods, however, give more insight into the means by which allotetraploids may occur naturally. One method is to cross two different autotetraploids. For example, if *A* and *B* each represent different haploid chromosome sets, then *AAAA* and *BBBB* would be the corresponding autotetraploids. With normal reduction, these will form gametes with the formulae *AA* and *BB*. Upon cross-fertilization, the allotetraploid, *AABB*, will form. Because an allotetraploid actually consists of two different diploid groups, existing side by side in the same nucleus, the terms *amphidiploid* and *double diploid* are often used as synonyms for allotetraploid. But because allotetraploids are more common in nature than are autotetraploids and because to cross two tetraploids is generally more difficult than to cross the corresponding diploids, this method has probably not had general importance in nature.

Two methods are based upon the occasional failure of reduction divisions, which is especially frequent in plants with chromosome complements that do not synapse readily. Thus, in the cross *AA* × *BB* (using the terminology introduced earlier), the F_1 should be *AB*. If there is insufficient homology between the chromosomes of the *A* and *B* genomes to permit synapsis, the probability of nonreduction becomes considerable. Thus a significant percentage of *AB* gametes may be produced. In a self-fertilized plant, some *AB* ovules would probably be fertilized by *AB* pollen, thus producing the allotetraploid, *AABB*, at once. Each chromosome now has a homologous mate; and, since there is little tendency of *A* chromosomes to synapse with *B* chromosomes, there is no tendency to form complex synaptic associations. The meiotic divisions, therefore, proceed perfectly normally, with gametes of constitution *AB* resulting and with no impairment of fertility.

Another method of production of allotetraploids involves a two-step utilization of nonreduction. If the hybrid *AB* is cross-fertilized, backcrossing to one of the parental species, let us say *AA*, is especially likely. If *AB* has undergone nonreduction, the backcross progeny will then be *AAB*. These plants may again undergo nonreduction, producing gametes of formula *AAB*. If backcrossed to parent *BB*, the resulting progeny would again be the allotetraploid, *AABB*. The allotetraploid might also be formed by doubling of the chromosomes in the zygote of the original hybrid, *AB*, in a fashion comparable to that produced by treatment with colchicine. These allotetraploids show characters of both parental species, together with some typical tetraploid characters in new and distinctive combinations. They breed true and are effectively isolated from the parent species by the sterility of the hybrid between the allotetraploid and either of the parental species. This sterility results from the behavior of the chromosomes in meiosis. In the hybrid between *AABB* and *AA*, the chromosomal formula will be *AAB*. The chromosomes of the two *A* genomes will synapse normally and will be distributed to the daughter cells normally. But the chromosomes of the *B* genome have no synaptic mates, and so they are distributed at random. They might all go to one pole, so that gametes of formulas *A* and *AB* would be produced in equal numbers, or they might pass in equal numbers (but not equal genetic endowment) to each pole, or any intermediate result might occur. The last three combinations would be inviable; and, since they comprise most of the gametes of an *AAB* plant, these triploids are highly sterile. Some viable gametes would also be produced by nonreduction.

Thus allotetraploids may properly be regarded as good species, produced in one or a few steps. It has even been suggested that new genera, families, and higher categories might be produced in this way. Available cases indicate that some species and genera are formed through allopolyploidy, whereas higher groups are not. Stebbins believes that this is due to the fact that no really new genetic material (mutations) is involved, but only new combinations of old and related genomes.

So far we have discussed only tetraploidy, the simplest and the most common type of polyploidy. Many higher degrees are known both in nature and in experimentally produced plants, but no new principles are involved in the higher polyploids, and so the whole discussion applies to all polyploids. Any polyploid having an even number of genomes in the somatic cells (tetraploid, hexaploid, octaploid, decaploid, and so on) should be fully fertile; whereas those with an odd number of genomes (triploid, pentaploid, septaploid, and so on) should be highly sterile because of abnormal meiosis and so should survive in nature only if they have efficient means of asexual reproduction.

Species Synthesis

In 1924 G.D. Karpechenko produced a very interesting allotetraploid while experimenting on crosses between the radish, *Raphanus sativus,* and the cabbage, *Brassica oleraeca* (Figure 13–2). In each of these species, the haploid chromosome number is nine. In the hybrid, there were eighteen chromosomes (9R + 9B), but they behaved as eighteen univalents at meiosis, with little or no tendency to synapse. As a result, most gametes were inviable, but a few fertile hybrids were obtained. Cytological examination showed that these had eighteen perfectly synapsed *pairs* of chromosomes at meiosis. Thus allopolyploidy had evidently arisen spontaneously in a few of the hybrid plants. These hybrids showed a combination of characters of the two genera that was completely different from anything previously encountered. The allotetraploid bred true and was reproductively isolated from both parents, so Karpechenko felt justified in describing this new plant as a new, synthetically produced genus, under the name *Raphanobrassica.*

R.E. Clausen and T.H. Goodspeed performed a similar synthesis of a new species of tobacco. They crossed commercial tobacco, *Nicotiana tabacum,* with a wild species, *N. glutinosa.* The hybrid was generally sterile, but a single plant was fertile. This bred true, had distinctive morphological traits, including the *gigas* habitus, and was reproductively isolated from the parent species. Hence the fertile hybrid was described as a new, synthetically produced species, *Nicotiana digluta* (Figure 13–3). Its synthesis was later repeated with much greater facility by the colchicine method. *Nicotiana tabacum* has twenty-four

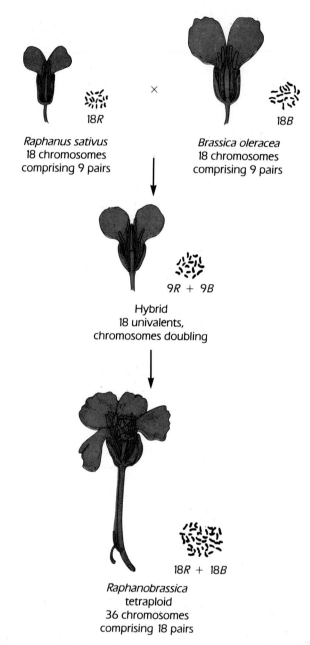

Raphanus sativus
18 chromosomes
comprising 9 pairs

18*R*

×

Brassica oleracea
18 chromosomes
comprising 9 pairs

18*B*

9*R* + 9*B*

Hybrid
18 univalents,
chromosomes doubling

18*R* + 18*B*

Raphanobrassica
tetraploid
36 chromosomes
comprising 18 pairs

Figure 13–2. *Experimental species synthesis by polyploidy. The radish,* Raphanus sativus, *has small flowers and 9 pairs of chromosomes. The cabbage,* Brassica oleracea, *has larger flowers and also has 9 pairs of chromosomes. The intergeneric cross results in hybrids (center) with flowers of intermediate size, with 9 chromosomes from each parent (9R + 9B). These hybrids are sterile because of failure of synapsis unless the reduction divisions fail altogether, in which case unreduced gametes may combine and give tetraploid progeny with 18R + 18B. Because normal pairing can now occur, the tetraploid plants, called* Raphanobrassica, *are fully fertile, but they are sterile in crosses with both parental species.*

286

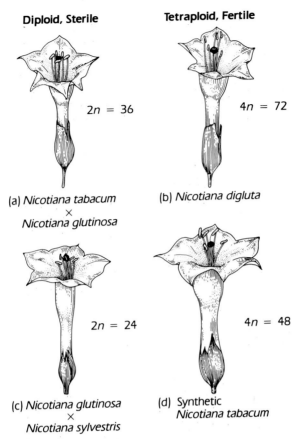

Diploid, Sterile

Tetraploid, Fertile

$2n = 36$

$4n = 72$

(a) *Nicotiana tabacum*
×
Nicotiana glutinosa

(b) *Nicotiana digluta*

$2n = 24$

$4n = 48$

(c) *Nicotiana glutinosa*
×
Nicotiana sylvestris

(d) Synthetic
Nicotiana tabacum

Figure 13–3. Nicotiana *hybrids. (a)* N. tabacum ×
N. glutinosa *(36 chromosomes) is sterile, while (b), its
colchicine-induced tetraploid with 72 chromosomes (= N.
digluta), is fertile. (c), N. glutinosa × N. sylvestris
has 24 chromosomes and is sterile, while its colchicine-induced
tetraploid, (d), with 48 chromosomes, is fertile. Note that
all parts of the tetraploid flowers are larger than the corresponding
parts of the diploid flowers.*

pairs of chromosomes; N. *glutinosa* has only twelve pairs. As expected, the
sterile hybrids had thirty-six chromosomes (not pairs), but the fertile hybrids
had thirty-six pairs of chromosomes and so were allopolyploids. Actually, these
were allohexaploids, for N. *tabacum* itself is an allotetraploid species, as will
be shown.

Genome Analysis

There is much evidence that allopolyploids are widespread in nature. For the most part, the evidence consists of series of chromosome numbers in a single genus or family that are multiples of a low number found in the same group. In addition, the species with higher chromosome numbers frequently show combinations of the characters of more than one of the basic species. However, the most-convincing evidence is what Clausen called *genome analysis*. A series of experimental crosses is made in order to establish the actual source of the different genomes of a suspected allopolyploid.

The analysis of *Nicotiana tabacum* is an example. The first step in genome analysis is to select probable parent species on the basis of morphological characters held in common by the allopolyploid and the suspected parents. Any species so chosen must of course have a chromosome number allowing it to contribute one or more genomes to the species being analyzed. On morphological grounds, Clausen and Goodspeed regarded *N. sylvestris* and some member of the *N. tomentosa* group as the most probable parent species of *N. tabacum*. Examination of the chromosomes, the second step in the analysis, showed that *N. tabacum* had twenty-four pairs and *N. sylvestris* and *N. tomentosa* each had twelve pairs. Thus, the numbers corresponded to the requirements of the problem. Further, the chromosomes of each of the basic species resembled some of those of the polyploid species in size and shape. Finally, a series of crosses was made to test the actual relationship of the chromosomes.

The genomes of *N. sylvestris* and *N. tomentosa* may be designated as S and T respectively, so that the normal diploid plants will have the formulas SS and TT. On the assumption that these genomes have been differentiated by mutation and by chromosomal rearrangements since the original formation of *N. tabacum*, this species is designated by the formula $S^1 S^1 T^1 T^1$. In the hybrid between *N. sylvestris* and *N. tabacum*, $S^1 S T^1$, *twelve pairs and twelve univalents are formed at meiosis. This fact indicates that all of the chromosomes of N. sylvestris* had a close enough homology with those of *N. tabacum* to synapse normally. This hybrid is, of course, sterile because of the irregular distribution of the unsynapsed T^1 chromosomes to the gametes. Similarly, if *N. tomentosa* is crossed to *N. tabacum*, the primary gametocytes show twelve pairs and twelve univalents. Thus the twelve chromosomes of *N. tomentosa* also have their homologues among the twenty-four chromosomes of *N. tabacum*. The next step was to cross *N. sylvestris* and *N. tomentosa*. This cross produced a viable but sterile plant ST, similar to the haploid *N. tabacum*, which can be produced experimentally. A small amount of pairing of chromosomes did occur in this hybrid—2.5 pairs per meiotic cell was the average. This fact demonstrated that there is some, but not much, homology between the chromosomes of *N. sylvestris* and *N. tomentosa*, since pairing frequently was completely absent.

This hybrid, *ST,* was then treated with colchicine to produce the allotetraploid, *SSTT.*

The resulting plant resembled *N. tabacum* much more closely than it did either of the parent species. Although it does not morphologically duplicate any of the many known varieties of *N. tabacum,* it does not differ from them more than they do from one another. The possibility remains that, if the right varieties of *N. sylvestris* and *N. tomentosa* were selected, a more exact duplication of naturally occurring *N. tabacum* might be achieved. Therefore, we may regard this procedure as a fairly successful duplication of the natural origin of one species, *N. tabacum.* Yet there is one disappointing feature. Although the artificially produced *N. tabacum, SSTT,* is fully fertile when used as a pollen parent, it is female sterile. For this as well as other reasons researchers suspect that another member of the *N. tomentosa* group may be the actual parent of *N. tabacum,* probably *N. otophora.*

A more frequently quoted example of genome analysis is Müntzing's study of *Galeopsis tetrahit,* the hemp nettle. This species has a haploid number of sixteen, whereas most species of the genus have haploid numbers of eight. This fact led Müntzing to suspect polyploidy. After comparing the morphology of *G. tetrahit* with that of the eight-chromosome species, he selected *G. pubescens* and *G. speciosa* as the most probable parent species. The hybrid between *G. pubescens* and *G. speciosa* was highly sterile, but some viable gametes were formed. A single F_2 plant was obtained, and this proved to be a triploid, apparently the product of an unreduced gamete and a gamete with the *G. speciosa* genome. This triploid was backcrossed to a pure *G. pubescens* plant, and again a single viable plant was obtained. As this plant was tetraploid, it must have been formed by the union of an unreduced gamete from the triploid parent with a normal gamete from the *G. pubescens* parent. This "artificial" *G. tetrahit* was fully fertile, both to itself and to natural *G. tetrahit,* which it resembles. It breeds true and is sterile when crosses are attempted with either of the parent species. Thus there can be no doubt that Müntzing has duplicated the natural "synthesis" of *Galeopsis tetrahit.*

The Wheats

A polyploid series of great economic importance is that of the genus *Triticum,* comprising wheat and some related grasses. This is a complex of nearly thirty species, of which only four are cultivated. Wheat falls into three natural groups, based on chromosome number. One array of species has seven pairs of chromosomes. Of these, only *T. monococcum* (Einkorn wheat) is cultivated, and its economic importance is minor. A second group of species has fourteen pairs of chromosomes. Two of these, *T. timopheevii* and *T. turgidum,* are cultivated, but only the latter (Emmer wheat) is important. Finally, several species have twenty-one pairs of chromosomes. Only one of these, *T. aestivum*

(Vulgare wheat), is cultivated, but that one is possibly the most important of all crop plants.

This series of gametic chromsome numbers—7-14-21—strongly suggests a polyploid series, with diploid, tetraploid, and hexaploid members. If *T. monococcum* is crossed with *T. turgidum,* seven bivalents (pairs) and seven univalents appear in meiosis; hence the monococcum genome, designated as *A,* also occurs in *T. turgidum.* The second genome of *T. turgidum* is designated as *B.* Similarly, in the cross *T. monococcum* × *T. aestivum,* seven bivalents and fourteen univalents appear, hence genome *A* is found in all of these species. Finally, the cross *T. turgidum* × *T. aestivum* yields fourteen bivalents and seven univalents, so that all of the chromosomes of *T. turgidum* are also found in *T. aestivum.* The third genome of *T. aestivum* is called *D.* The genomic formulae are thus *AA* for *T. monococcum, AABB* for *T. turgidum,* and *AABBDD* for *T. aestivum.* Wild members of the genus have several other genomes, including *C.*

How distinct are these genomes and from what sources have they come? Producing haploid plants of *T. turgidum* and *T. aestivum* is possible. In these haploids (*AB* and *ABD*), almost no pairing occurs, and so researchers formerly thought that the three genomes were only distantly related. Accordingly, they considered the wheats to be an ideal allopolyploid series. A specific gene in the fifth chromosome of the *B* genome, which suppresses pairing (the *5B* effect), has, however, been identified. If this chromosome is eliminated (which is possible because the other two genomes will cover its deficiency), then extensive pairing occurs—as many as nine bivalents among the twenty-one chromosomes of haploid *T. aestivum* (*ABD*). The three genomes, then, are rather closely related, and the data support the principle that the difference between autopolyploidy and allopolyploidy is not absolute, but only a matter of degree.

Genome *A,* derived from *T. monococcum,* is found in all of the cultivated wheats. Genome *B* is found in both of the polyploid species. Its derivation is less clear. It is similar to the genomes of several diploid species, such as *T. speltoides* and *T. searsii,* yet it is not identical to any of these. It is probable that either the *B* genome was derived from a diploid species that is now extinct or that the *B* chromosomes have been sufficiently rearranged (e.g., by inversions), thus making it difficult to identify the *B* chromosomes with those of its source species. Genome *D* is almost certainly derived from *T. tauschii,* a diploid species formerly classed as *Aegilops squarrosa.*

Let us synthesize these data with the aid of Figure 13–4. In the Middle East, cultivated wheat has for millenia been grown in fields infested with other members of the genus *Triticum* that grow as weeds. These weeds include various diploid and tetraploid species that were formerly classed in a separate genus, *Aegilops.* Because they combine with *T. monococcum* to form polyploid species of *Triticum,* they are now included in the latter genus. Hybridization

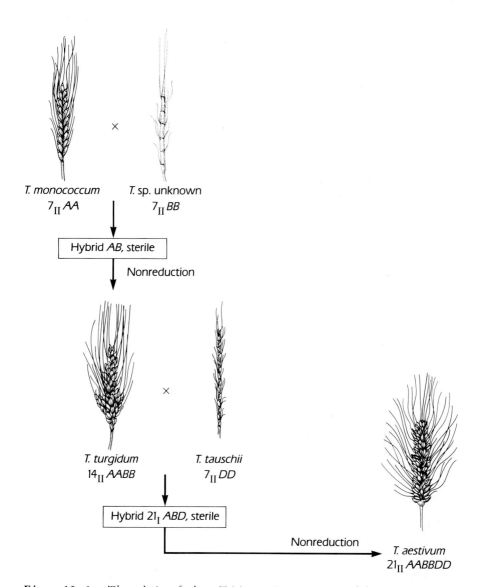

T. monococcum
$7_{II} AA$

T. sp. unknown
$7_{II} BB$

Hybrid *AB*, sterile

Nonreduction

T. turgidum
$14_{II} AABB$

T. tauschii
$7_{II} DD$

Hybrid 21_I *ABD*, sterile

Nonreduction

T. aestivum
$21_{II} AABBDD$

Figure 13–4.   *The evolution of wheat,* Triticum. *Primitive species of this genus have seven pairs of chromosomes, with similar genomes designated by the same letter:* T. monococcum, AA; T. searsii *and* T. speltoides, BB; *and* T. tauschii, DD. T. monococcum *hybridized with an unknown member of the* searsii-speltoides *group to produce a sterile hybrid,* AB. *Fertilization of nonreduced gametes produced a fertile tetraploid,* AABB, T. turgidum, *which then hybridized with* T. tauschii. *Again, the hybrid,* ABD, *was sterile, but fertilization of nonreduced gametes resulted in a fertile hexaploid,* AABBDD, T. aestivum, *the most widely planted of all wheats and possibly the most important of all plants economically.*

among the members of this complex occurred spontaneously. The cross between *T. monococcum* and the appropriate wild parent yielded the sterile hybrid *AB*. Spontaneous doubling of the chromosomes, probably by nonreduction, resulted in the fertile allotetraploid *T. turgidum* (*AABB*). Next, natural hybridization occurred between *T. turgidum* and *T. tauschii* (*DD*). Again, the hybrid (*ABD*) was sterile, but spontaneous doubling gave fully fertile *T. aestivum* (*AABBDD*), a species of which literally thousands of varieties are grown throughout the world.

Polyploidy in *Bromus:* A Complex Case

Stebbins and his collaborators analyzed an even more complicated case in the genus *Bromus,* a widespread complex of range grasses. Stebbins believes that seven is the basic number of chromosomes in this group, but most of the American species, of which *B. carinatus* is typical, have twenty-eight pairs of chromosomes, or fifty-six chromosomes in the somatic tissues. Hence they are octoploids. These chromosomes include twenty-one medium-sized and seven large pairs. A South American species, *B. catharticus,* however, has only twenty-one pairs of chromosomes, and these are all medium-sized. Hence this species is hexaploid. A single American species, *B. arizonicus* has been found to have eighty-four chromosomes, all medium-sized. Hence this species is a duodecaploid (twelve-ploid)! Crosses between these species have been made in order to study the behavior of the chromosomes at meiosis in the hybrid. When *B. carinatus* and *B. catharticus* are crossed, twenty-one pairs are formed by the medium-sized chromosomes; whereas the seven large chromosomes from *B. carinatus* behave as univalents. The three sets of medium-sized chromosomes are called *A, B,* and *C;* the set of large chromosomes is called *L.* On this basis, the hexaploid species, *B. catharticus,* has the formula *AABBCC,* and the octoploid species, *B. carinatus,* has the formula *AABBCCLL.* When *B. carinatus* is crossed to *B. arizonicus,* a complex meiotic pattern results (Figure 13–5). Such a hybrid will receive forty-two medium-sized chromosomes from *B. arizonicus,* and twenty-one medium-sized together with seven large chromosomes from *B. carinatus.* At meiosis, the seven *L* chromosomes of *B. carinatus* and fourteen of the chromosomes of *B. arizonicus* behave as univalents. Fourteen bivalents (normal tetrads) are present, indicating that two sets of chromosomes, arbitrarily designated as the *A* and *B* sets, are held in common by the two species. In addition, as many as seven trivalents (complex associations of three pairs of chromosomes) may be formed. Thus one of the *B. carinatus* sets of chromosomes appears to have considerable homology with *two* sets in *B. arizonicus.* This *B. carinatus* set and one of the corresponding genomes in *B. arizonicus* are called C_1, and the other set in *B. arizonicus* is called C_2. The two sets of chromosomes in *B. arizonicus* that are not represented at all in *B. carinatus* are called *D* and *E.* Thus the hexaploid species, *B. catharticus,* has

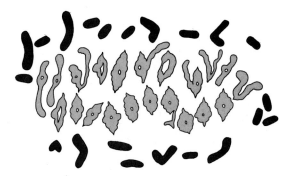

Figure 13–5.   *Complex meiosis in the cross of* Bromus carinatus *(octoploid)* × B. catharticus *(hexaploid). The chromosomes in the center (light gray) have formed bivalents and trivalents; univalents (dark gray) are distributed peripherally.*

the formula $AABBC_1C_1$; the octoploid species, *B. carinatus*, has the formula $AABBC_1C_1LL$; and the duodecaploid species, *B. arizonicus*, has the formula $AABBC_1C_1C_2C_2DDEE$.

SOME GENERALIZATIONS ON POLYPLOIDY

Allopolyploidy, then, is a widespread phenomenon. In 1942 T.H. Goodspeed and M.V. Bradley published a review in which they listed 124 well-authenticated cases of allopolyploidy, including both natural and experimental examples. No doubt the list would be much longer now if it were brought up to date. Since natural polyploids seem to be generally allopolyploids rather than autopolyploids, this type evidently has especial interest for evolution. As we may note from inspecting tables of chromosome numbers, over half of the higher plants are probably polyploids of one degree or another. Although the formation of polyploid series does not entail any new genic material, it does produce new combinations upon which selection can act, combinations that may be very different from anything formed in any other way. Therefore, that polyploids often invade territories not occupied by their diploid parents is not surprising.

Polyploids seem to be much more aggressive invaders of new territory than are their diploid relatives. For example, Anderson showed that *Iris versicolor* is almost certainly an allopolyploid derived from *I. virginica* and *I. setosa*. *I. virginica* is widely distributed in the United States; *I. setosa* is found as two widely separated races, one on the coast of Alaska, the other in Labrador, Nova Scotia, and Newfoundland. These species are regarded as remnants of the preglacial floras of their respective areas, and they have not extended their ranges into the glaciated parts of North America. But *I. versicolor,* their allopolyploid offspring, is distributed from Labrador and northeastern United States westward through the Great Lakes region to Wisconsin and Manitoba, throughout the glaciated part of North America.

An interesting, and largely unsolved, problem is the effect of polyploidy on the occurrence of new genetic variability due to mutations. Some geneticists have expressed the opinion that polyploidy should increase the total genetic variability rapidly, because random mutations might occur in any of the genomes and becomes homozygous. Others have pointed out that a new recessive mutant occurring in one genome would now be covered by its dominant allele in *three* other genomes (or more in higher polyploids), so that its phenotypic expression would be much less probable than in a diploid. These viewpoints may not be completely irreconcilable. From a short-range viewpoint, the latter is probably correct; but when time is provided on a geological scale, differentiation of the genomes by random mutation should finally result in greater total variability than could be achieved in a diploid.

Generally, diploids occur in the older part of the total range of a group; polyploids invade the geologically more recent parts. Thus, in a group in which polyploidy is common, the diploids tend to become relics and the more diversified polyploids fill the available niches. Stebbins pointed out one consequence of this phenomenon: when conditions become unfavorable for such a group, the diploids are the first to become extinct. When merely a few species remain of a once-prominent group, they are likely to be polyploid species without near relatives. For example, the entire order Psilotales, once a moderately important group, is now represented by only two genera, which are regarded by many botanists as monotypic. These have over a hundred pairs of chromosomes, and so they are almost certainly the last remnants of a once-great polyploid complex.

POLYPLOIDY IN THE ANIMAL KINGDOM

Although polyploidy has been a major phenomenon in plant evolution, its role in animal evolution has not been adequately assessed, and it is generally considered to be of minor importance. The reason for this difference between the kingdoms is not known with certainty, but Muller has suggested that it may be based upon the fact that the sexes are usually separate in animals, while plants are usually hermaphroditic (monoecious). Random segregation of the several pairs of sex chromosomes in a polyploid organism would result in sterile combinations. This explanation has been widely accepted. In 1937 A. Vandel reviewed all of the known cases of polyploidy in animals, and the data he assembled lend support to Muller's theory. If the list were brought up to date, it would be only a little longer, and the conclusions would be unchanged. Vandel found only eleven cases among plants in which polyploid plants are also dioecious (having separate sexes). One case is *Fragaria elatior,* a hexaploid species of strawberry that has been proven to have only one pair of sex chromosomes. Whether the other two pairs have lost their original sex-differentiating function or whether polyploidy first developed in a monoecious ancestor, with the separation of the sexes occurring later, cannot be ascertained. Vandel favors the former hypothesis. In any event, polyploidy, so common among plants in general, appears to be rare among those plants that are dioecious.

The majority of the animals that have been reported to be polyploids are parthenogenetic. The common water flea, *Daphnia pulex,* occurs in a diploid, bisexually reproducing form and in a hexaploid, parthenogenetically reproducing form. *Artemia salina,* the brine shrimp, occurs in tetraploid and octoploid parthenogenetic forms, but a tetraploid form that reproduces bisexually also occurs. The European sow bug, *Trichoniscus elisabethae,* is a triploid parthenogenetic species. Whether the parthenogenetic ostracod *Cypris fuscata* is triploid or tetraploid is uncertain. The walking stick insects *Carausius morosus* and *C. furcillatus* are, respectively, triploid and tetraploid, as well as parthenogenetic. The psychid moths *Solenobia triquetrella* and *S. lichenella* are also parthenogenetic tetraploids; but a bisexual, diploid race of the former is known. Finally, the parthenogenetic beetle, *Trachyphlaeus,* also appears to be triploid. Polyploidy is also common in hermaphroditic animals, such as flatworms and earthworms.

Although most polyploid animals are invertebrates, interesting examples occur among the vertebrates, particularly among the lizards. *Cnemidophorus* is a genus of lizards of southwestern United States and northern Mexico. Diploid and triploid species are common, and tetraploids have been produced experimentally. Diploid species, such as *C. tigris* and *C. inornatus,* have twenty-three pairs of chromosomes; and they reproduce bisexually. Although diploid, *C. neomexicanus* reproduces parthenogenetically; it appears to be a stabilized hybrid of *C. tigris* and *C. inornatus,* with one set of chromosomes from each parent species. Triploid species are always parthenogenetic. For example, *C. perplexus* has one set of chromosomes from *C. tigris* and two from *C. inornatus.* Other triploid karyotypes are known in *Cnemidophorus* and in other genera, and these lizards are always parthenogenetic.

A very few tetraploid animals reproduce bisexually. *Artemia* has already been mentioned. *Parascaris equorum* (*Ascaris megalocephala* of older literature), an important nematode parasite of horses, is known in diploid, tetraploid, and hexaploid forms, all of which reproduce bisexually. The starfishes *Asterias forbesii* and *A. glacialis* and the sea urchin *Echinus microtuberculatus bivalens* are tetraploid forms that reproduce bisexually. The golden hamster, *Cricetus auratus,* may belong here, for it has twenty-two pairs of chromosomes, whereas its nearest relatives have eleven pairs. This species has been claimed, but not proven, to be an allotetraploid of *Cricetus cricetus* and *Cricetulus griseus.* Thus it appears that any sexual imbalance that may be caused by random segregation of the sex chromosomes in polyploids can be overcome. But that it is not frequently overcome is indicated by the rarity of polyploids among bisexually reproducing plants and animals and by their overall rarity in the animal kingdom, in which separate sexes are the rule.

A related phenomenon is fragmentation of the chromosomes, with the result that multiples of a basic number appear in a series of chromosome counts. Vandel listed examples in most of the major groups of animals in which species show chromosome numbers that are twice the number of some more primitive species. Like polyploidy among plants, increase in the number of chromosomes by fragmentation appears to go hand in hand with increasing specialization among animals. Thus, among the lower mammals, the diploid number is often twenty-four; among the Eutheria, it is commonly forty-eight; among the highly specialized eutherians, such as the ungulates, it is often sixty; and among such highly specialized mammals as the rodents, it may be as high as eighty-four. Why fragmentation of the chromosomes should be related to evolutionary specialization is unknown, but it may be based upon position effects or comparable phenomena.

SUMMARY

Polyploidy is the occurrence of more than two basic sets of chromosomes in somatic sets, and hence more than one set in the gametes. Much the most common type is *tetraploidy,* in which there are four basic sets of chromosomes in the somatic cells and two sets in the gametes. Polyploidy is one of the major phenomena of plant evolution, yet it is an incidental factor in animal evolution. Because no new genetic material is involved, polyploidy commonly results in the multiplication of species without giving rise to higher categories. Like many other generalizations, this one probably has exceptions. *Raphanobrassica,* a synthetically produced genus, is not unique. Stebbins believes that genera have occasionally arisen by polyploidy, and he suggests *Platanus, Aesculus,* and *Tilia* as probable examples. Further, he believes that some subfamilies and even families, Salicaceae (willows) for example, may have arisen by way of polyploidy.

Polyploids often show the *gigas* habitus. In contrast to the corresponding diploids, the plants grow more slowly but to a greater size; the stems are thicker; and the leaves are shorter, broader, and thicker. There are also physiological differences from diploids.

There are two major types of polyploidy: *autopolyploidy* and *allopolyploidy.* In autopolyploids, all sets of chromosomes are derived from the same species, so that only allelic differences occur among them. Consequently, there is a tendency to form multivalents at meiosis. The multivalents divide irregularly, with the production of inviable gametes. Because of this instability, autopolyploidy has been of minor importance in evolution. In allopolyploids, the chromosome sets are derived from two or more species. The cross of two species, AA ×

BB, yields the hybrid *AB.* Because the chromosomes of the two sets cannot synapse properly, reduction divisions may fail altogether, so that unreduced gametes result. Fertilization of an *AB* ovule by an *AB* pollen then establishes the allotetraploid, *AABB.*

The early studies on polyploidy were made by exploiting those polyploids that occurred spontaneously, but scientists always try to produce experimentally the phenomena that they study. The use of the drug colchicine, which disrupts the mitotic spindle, proved to be a very efficient method for the production of polyploids. Experimentation with colchicine has made polyploidy one of the most thoroughly understood of evolutionary processes. Researchers using the technique of genome analysis have identified the parent species of many polyploids, and they have successfully duplicated the formation of many natural species in such diverse genera as *Nicotiana, Triticum, Spartina,* and *Galeopsis.* In these and in a great array of polyploid species, they have demonstrated with certainty the exact sequence of events in the origin of species. For this reason, polyploidy has enormous theoretical importance.

Some polyploid series may also have great practical importance. Wheat (*Triticum*) is of fundamental importance for Western civilization. The genus comprises a polyploid series with diploid, tetraploid, and hexaploid members. Researchers have proven that this is an allopolyploid series in which genome *A* is derived from *T. monococcum,* genome *B* from an unknown species that was closely related to *T. speltoides* and *T. searsii,* and genome *D* from *T. tauschii.* The principles of polyploidy are basic to the science of wheat breeding, and thus it may be said that evolutionary biology keeps bread upon the tables of the world.

REFERENCES

Darlington, C.D. 1963. *Chromosome Botany and the Origins of Cultivated Plants,* 2nd ed. Hafner, New York.
   Much material on polyploidy by a master of cytogenetics.
Feldman, M., and E.R. Sears. 1981. The wild gene resources of wheat. *Scientific American* 244(1):102–112.
   The wheat story brought up to date by two leading researchers.
Goodspeed, T.H. 1954. *The Genus* Nicotiana. Chronica Botanica, Waltham, Mass.
   Full details on polyploidy in tobacco.
Grant, V. 1971. *Plant Speciation,* Columbia University Press, New York.
   Includes two excellent chapters on polyploidy.
Stebbins, G.L. 1950. *Variation and Evolution in Plants.* Columbia University Press, New York.
   After more than thirty years, this book is still the essential starting point for studies in plant evolution. Polyploidy is especially well covered; includes all of the older literature.
———. 1970. Variation and evolution in plants: Progress during the past twenty years. In *Essays in Evolution and Genetics in Honor of Theodosius Dobzhansky,* edited

by M.K. Hecht and W.C. Steere, pp. 173–208. Appleton-Century-Crofts, New York.

Valuable updating.

———. 1971. *Chromosomal Variation in Higher Plants.* Arnold, London.

Treats polyploidy, as well as all other types of chromosomal variation.

———. 1976. Chromosome, DNA and plant evolution. In *Evolutionary Biology,* vol. 9, edited by M.K. Hecht, W.C. Steere, and B. Wallace, pp. 1–34. Plenum Publishing, New York.

Includes a review of polyploidy, as well as of other topics.

White, M.J.D. 1973. *Animal Cytology and Evolution,* 3rd ed. Cambridge University Press, London and New York.

Includes a very complete summary of what was known about polyploidy in animals up to the date of publication.

14

Quantitative Aspects of Macroevolution

In Chapter 10, we were concerned largely with processes by which populations are differentiated within a species. These processes are clearly basic to the formation of subspecies, which Darwin called incipient species. In most cases crosses among the species of a genus are not possible, and much less more distant crosses, so that direct genetic analysis of relationships above the specific level would seem to be out of reach. Whether the same processes that account for evolution at the level of genetic analysis—*microevolution*—also apply to the evolution of higher groups—*macroevolution*—continues to be very controversial. Nonetheless, much quantitative information available bears upon the processes of macroevolution, and we now turn to some examples.

ALLOMETRY

As an organism grows, the various parts grow at different rates, so that its proportions change. Thus a baby's head is relatively large, but its growth does not keep pace with that of the rest of the body, and hence its proportionate size in the adult is more moderate. The head is said to show negative *allometry*

(Greek for "differential measurement") or *heterogony* (Greek for "unlike development"). On the other hand, the adult teeth are larger in proportion to the head than are the milk teeth, and so tooth development is positively allometric. D'Arcy Thompson first showed that trends of allometric growth can be analyzed mathematically. They correspond to the equation $y = bx^k$, where y is the size of the organ studied, x is the size of the animal as a whole (or of some part used for comparison), b is a constant determined by the value

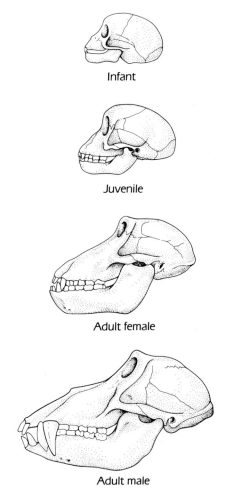

Infant

Juvenile

Adult female

Adult male

Figure 14–1. Allometry in baboon skulls, showing the great increase of facial length with respect to cranial length as total size increases. Here, k = 4.25, a very high figure.

of y when x is one, and k is the exponent of allometric growth. If k is less than one, allometry will be negative; if k is more than one, allometry will be positive.

The most obvious application of allometry is to ontogeny. For example, young baboons are only moderately prognathous, but their jaws protrude ever more strongly as the animals approach full size (Figure 14–1). In this series, $k = 4.25$, a very high value. In deer, antlers show positive allometry. In the red deer, *Cervus elaphus,* k is about 3 in young bucks, but it declines to about 1.6 as they grow fatter. Similarly, the "tails" of swallowtail butterflies show positive allometry (Figure 14–2).

Thompson and Huxley showed that phylogenetic changes in relative size can be analyzed in the same way as are ontogenetic changes—that is, if selection favors increase in total body size, as long as the genes for allometry remain unchanged, it will also entail increase in the proportionate size of specific parts. Thus the Irish elk, *Megaloceros* (Figure 14–3), shared with other deer the positive allometry of the antlers. The species grew steadily larger throughout the Late Pleistocene; and the antlers, continuing their allometric growth, became truly immense. It is often said that the Irish elk became extinct because its antlers grew so large that it could not hold up its head. There is no evidence for this improbable assertion. Had selection against increased antler size been so severe, it would no doubt have favored genes for smaller body size, or for a smaller k. S.J. Gould believes that selection favored large antlers for their value in the ritualized combat of males and in attracting females. He also believes that climatic change caused their extinction. Similarly, comparison of *Australopithecus africanus* with *A. boisei* has shown that their major features and even the fine details of facial anatomy are consistent with a single set of genes for allometric growth acting upon difference of total size.

Particularly in highly variable groups, general body outline can be analyzed by Cartesian transformation, a special aspect of allometry. The outline of a primitive member of the group is drawn upon a rectangular grid, then the grid is distorted by stretching particular parts in one direction or another. The results simulate the outlines of related species with different factors for allometric growth. Figure 14–4 shows Cartesian transformations for a series of crabs. Diagrams like this one are a strong indication that such evolutionary changes are based upon mutations of the genes influencing allometry. This conclusion is also supported by Kurtén's study on bear teeth.

Allometric changes often have adaptive significance in terms of maintaining adequate function at increased body size. For instance, respiratory requirements of an organism are determined by its mass, which varies as the cube of body dimensions, whereas respiratory exchange is mediated by surfaces that vary as the square. Thus, to maintain adequate respiratory function at increased body size, the area of the respiratory surface (alveoli of the lungs in tetrapods) must increase at a greater rate than does body length, for example. A similar

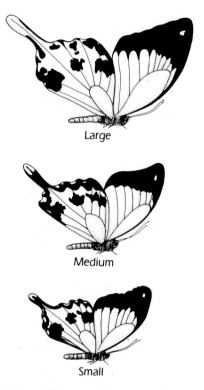

Large

Medium

Small

Figure *14–2. Allometry of the "tail" of the swallowtail butterfly,* Papilio dardanus. *Growth of the tail is somewhat more rapid than that of the wing as a whole, so that the tail is disproportionately larger in large than in small butterflies.*

argument holds for the width of limbs; it must increase at a greater rate than does body size, which explains why the legs of elephants are proportionately so much stouter than those of mice.

Allometric coefficients are also subject to selection in other ways. For instance, the antlers of moose are much smaller relative to body size than would be predicted for the allometric curve from red deer to Irish elk. However, moose live in dense forests, in which extremely broad antlers would hinder movement; whereas the Irish elk apparently lived on open ground, where they were not subject to selection for more modest allometry of the antlers. As a general rule, measurements of the central nervous system, such as the size of the eye or the brain, are negatively allometric, for they must be well developed at birth or soon afterward. Sexual features such as horns, crests, and frills are positively allometric; rudimentary at birth, when they have not yet any adaptive value, they develop rapidly as the time of maturity approaches.

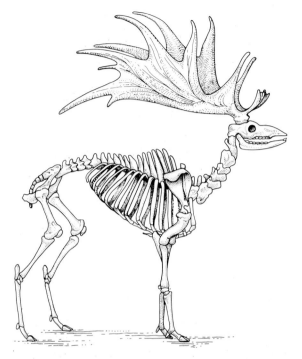

Figure 14–3.   Megaloceras, *the Irish elk, showing extreme allometry of the antlers.*

BIOCHEMISTRY AND MACROEVOLUTION

A long controversial tenet of the modern synthesis was that the processes of microevolution that produce subspecies also produce species and higher groups when continued over long reaches of time. As long as only Mendelian methods were available, the controversy was largely unresolvable, because crosses above the specific level were usually not successful, and crosses between distantly related species were almost never successful. Studies in biochemical genetics, however, have possibly bridged that gap, and meaningful comparisons of genetic constitutions of distantly related species are now possible.

A gene is a specific sequence of nucleotides in the DNA. It is first transcribed into mRNA, then translated into protein. Analyses of the base sequences of DNA and RNA are technically feasible, and good sequences are available for both. Amino acid sequences of many proteins have also been determined, and from these we can infer the probable sequence of nucleotides in the genes. The principal problem with the method relates to the degeneracy of the code, that is, to the fact that most of the amino acids are represented by more than one codon. This limitation has not proven serious.

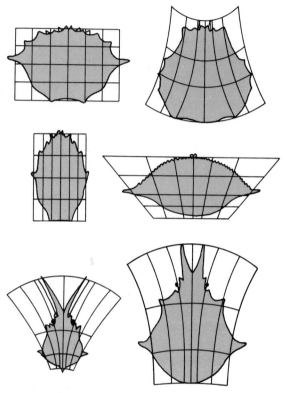

Figure 14–4. *Derivation of crab carapaces of very different proportions by Cartesian transformation of a single original type (upper left). Mutations for changes of allometric growth of the various regions of the body control these transformations.*

Insulin among the Vertebrates

Because a gene can be defined by the amino acid sequence of the protein it specifies, we may ask how a specific gene differs from species to species if the amino acid sequence is known for a given protein in a series of related species. Let us consider the gene that specifies insulin. The hormone consists of two polypeptide chains, A having twenty-one amino acids and B having thirty. It is therefore a rather small protein. The two chains are bound together by sulfur-to-sulfur bonds of their cysteines. Comparison of the insulins of cattle and sheep, members of the same order, shows that they differ only in one amino acid, the ninth in the A chain. The ninth is serine in cattle and glycine in sheep. Both are represented by several codons; but, to take a specific case, the codon AGU for serine could be changed to the codon GGU for glycine by a single nucleotide replacement—one point mutation. Cattle and horses, however, members of different orders, differ in positions 8, 9, and

10 (cattle have alanine-serine-valine; horses have threonine-glycine-isoleucine). Each of these, again, represents a single nucleotide substitution. Figure 14–5 compares the insulins of three species.

The common ancestor of horses and cattle was not too remote, perhaps only 60 million years in the past. The common ancestor of *H. sapiens* and the elephant was well over 60 million years in the past, yet they differ in only two amino acids in the *A* chain and none in the *B* chain. But the common ancestor of these mammals and of fish was much more remote, for they diverged nearly 400 million years ago. Nine of the twenty-one amino acids of the *A* chain of the codfish differ from those of the mammals, while seven of the thirty in the *B* chain differ. Not all of these differences could be achieved by a single mutation, so that the total amount of mutational difference is much greater. Further, there has also been more radical change, for the codfish has one more amino acid at the beginning of the *B* chain and one less at the end. Evidently, the genes diverge with time, just as we might expect.

Even fish and humans, however, share some sequences of amino acids in their insulins. It is likely that mutations have occurred in the codons for these apparently invariant amino acids, but such mutations have resulted in inactive insulin, and so the mutations were rejected by natural selection. Adaptively neutral molecules may diverge without the restraint of natural selection, and this, too, is illustrated by insulin. Although the insulin molecule consists of two protein chains, it is specified by a single gene that leads to a chain of eighty-four amino acids. A middle section of thirty-three amino acids is then removed, leaving the *A* and *B* chains, which are then linked by sulfur bonds. The middle section has no known function, and it varies much more widely than do the *A* and *B* chains, presumably because natural selection has no effect upon it.

Cytochrome c and Phylogeny

An even better example is afforded by cytochrome *c,* a respiratory enzyme that occurs in all eucaryotic cells and in some bacteria. The amino acid sequences of more than thirty species, which are widely distributed taxonomically, have been determined. The length of the chain varies from one hundred three to one hundred twelve amino acids. Of these, thirty-five are the same in all species studied; and seventy-seven differ from species to species. As expected, species that are taxonomically close have few amino acid differences; those that are taxonomically distant have many substitutions with respect to one another. Thus *H. sapiens* and chimpanzee have identical cytochrome *c,* whereas that of the rhesus monkey differs only at position 66, where it has threonine instead of isoleucine. The horse differs from *H. sapiens* at only eleven positions; the gray kangaroo, at ten. The snapping turtle has sixteen amino acid substitutions; and the bullfrog, eighteen. The tuna has twenty substitutions and one amino acid missing; the dogfish (a small shark) has twenty-four substitutions.

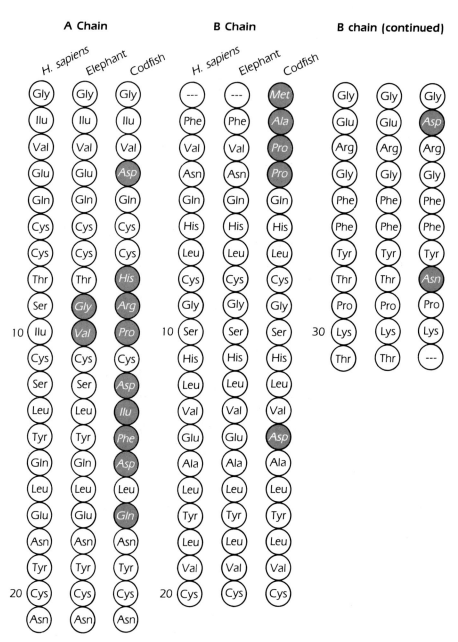

Figure 14–5. *The amino acid sequences of the A and B chains of insulin from three vertebrates:* H. sapiens, *elephant, and codfish. Notice that the A chains of the two mammals differ in only two of the twenty-one amino acids, whereas the codfish differs from both of the mammals in nine. In the B chain, the two mammals are identical. The codfish, however, has one more amino acid at the beginning of the chain and one less at the end. In between, there are six differences from the mammals. The differences are indicated by gray circles.*

By estimating the number of mutations needed for each substitution and then analyzing the data by computer, researchers have constructed a phylogeny for the thirty-three species whose cytochrome *c* sequences have been fully determined (partially illustrated in Figure 14–6). The parallel between the cytochrome *c* phylogeny and that based upon older methods is close but not exact, a result that should be expected when only a single character is used to construct a phylogeny.

Myoglobin, Hemoglobin, and Duplication of Genes

The examples are based on proteins that have diverged in amino acid composition while retaining the same function. Even more interesting from an evolutionary point of view is the duplication of genes with the acquisition of new functions, as is well illustrated by the evolution of myoglobin and hemoglobin.

Vertebrate hemoglobins are derived from a similar but simpler compound, myoglobin. It consists of a single chain of about one hundred fifty amino acids complexed with a heme group. Heme (Figure 14–7) is an iron-containing porphyrin group. When complexed with an appropriate protein molecule, heme will bind oxygen reversibly. Myoglobin functions to store oxygen in muscles so that it will be quickly available for muscular activity. Hemoglobin of the lamprey is closely similar to myoglobin of the same animal. It appears that the gene for myoglobin duplicated; one of the products of duplication continued to specify myoglobin, but the other diverged by a series of mutations and produced a closely similar protein-heme complex, which was then concentrated in blood cells, where it functioned in the transport of oxygen from the gills to the tissues of the body. At the relatively high oxygen tension of the gills, hemoglobin takes up oxygen; whereas at the low oxygen tension of the tissues, it releases the oxygen.

Early in the history of the fishes, the gene for hemoglobin duplicated, and the two resulting genes diverged by mutations for replacement of some amino acids and by some deletions. The proteins specified by these genes are called the *alpha* and *beta chains*. They now acquired the property of aggregating in fours—two alpha chains and two beta chains—always complexed with the heme nucleus. These tetramers are more efficient oxygen carriers because dissociation of oxygen from the protein-heme complex is more complete at the low oxygen tension of the tissues.

During its life, a mammal lives in two contrasting environments. It begins life in the uterus of the mother and must get oxygen from the maternal circulation by diffusion across the placental barrier. Its oxygen supply is, therefore, at a lower partial pressure than is normally found in the lungs of adults. It requires hemoglobin capable of exploiting an oxygen supply at that low pressure. The mammalian beta gene has again duplicated to produce two diverging genes: one still specifying the beta chain, the other the gamma chain. In the fetus, the beta gene is inactive, but the gamma gene produces

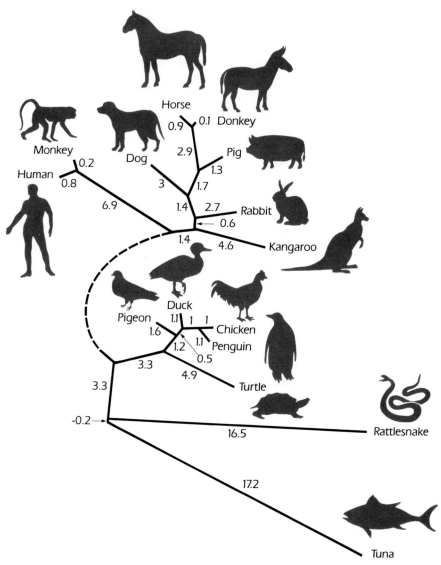

Figure 14–6. *Phylogeny of an array of vertebrates, based on the structure of the cytochrome c molecule. Data were analyzed by computer. Numbers beside the lines indicate mutational distances, and lengths of lines are proportional to these distances. The results are similar to conventional phylogenies based on morphology. There are some anomalies, but these will probably disappear when several series of molecular data are considered together.*

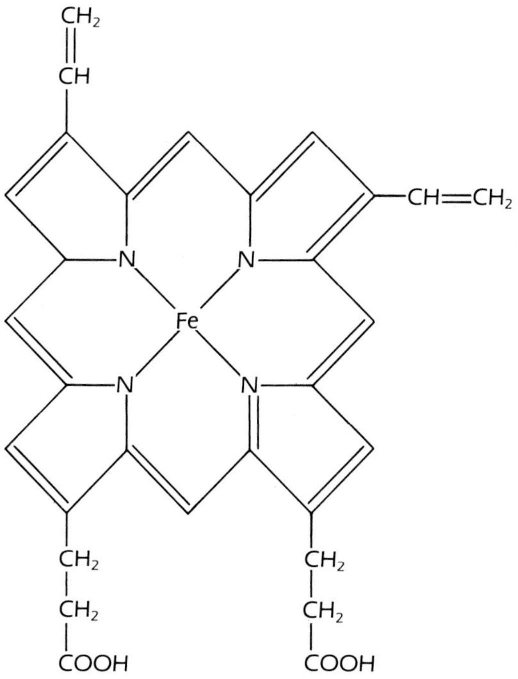

Figure 14–7.   The heme nucleus, an important biochemical unit in the evolution of myoglobin and hemoglobin.

gamma chains that aggregate with alpha chains to form a tetramer called *fetal hemoglobin.* The latter is capable of taking up oxygen at the low partial pressure of the placenta. After birth, when the mammal moves into an oxygen-rich environment, the gamma gene is repressed, the beta gene is activated, and the typical alpha-beta tetramers are produced.

Finally, the beta gene has again duplicated to produce a gene called *delta.* This produces a delta chain that is similar to the beta chain but that has a number of amino acid replacements. The delta chain forms a minor part of normal hemoglobin. So far as is known, it has no special function. This somewhat simplified history of the myoglobin-hemoglobin series (Figure 14–8) shows that the possibilities of mutation within such a family are great and that the changes can be followed across great taxonomic distances. Many phylogenies have been worked out on the basis of proteins. In general, they agree well with phylogenies grounded on morphology or other traditional bases. Some anomalies occur but these will probably disappear when the data from several different proteins are considered together.

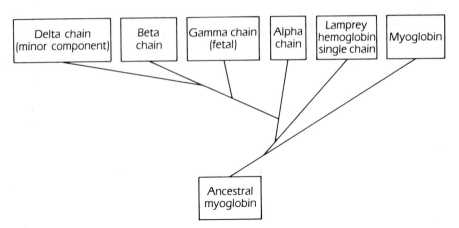

Figure 14–8. *A simplified diagram of the evolution of the hemoglobin chains, with special reference to human hemoglobins. The delta chain is found only in the higher hominoids (humans and great apes).*

Mutational Distance

The basic datum in the construction of a phylogeny from protein sequences is the number of mutational events necessary to convert one sequence into another. Thus, cytochrome *c* of rhesus monkeys differs from that of humans only by the substitution of threonine for isoleucine at position 66, a substitution that could be accomplished at one step (for example, the codon AUA for isoleucine could become ACA for threonine in one step). If, however, we were concerned with the replacement of histidine (codons CAU and CAC) by phenylalanine (codons UUU and UUC), two mutational steps would be necessary. Thus, the minimal number of base changes may be larger than the number of amino acid substitutions, and the former gives the better measure of the amount of evolution that has occurred. In practice, we tabulate the minimal *mutational distances* between species (the number of mutations required to change one polypeptide to another). Using hypothetical species A, B, and C, the following is a typical table:

	A	B	C
A	0	24	28
B	24	0	32
C	28	32	0

If we assume that the degree of relationship is inversely proportional to the mutational distances, then we can diagram the relationships by connecting the species via branching lines of lengths proportional to the mutational distances. There are twenty-four differences between A and B, twenty-eight between A and C, and thirty-two between B and C. These are the minimal numbers of mutations separating these species with regard to the protein under study. As A and B are most similar, they are joined with lines diverging from a hypothetical common ancestor, x, and both are joined to C through a common ancestor, y, in such a way that the distances A–B, A–C, and B–C add up to the values on the table. The following diagram of relationships results:

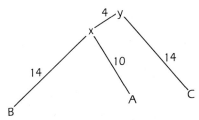

The data do not actually define the distances of the ancestral sequences x and y. We may be able to estimate these distances if we assume that A, B, and C have diverged at equal rates, an assumption that may be insecure. The calculation is further complicated by the fact that back mutation (return to a previous state) may occur and that some codons may mutate more frequently than others. When extensive data are available, as in the case of cytochrome c sequences, the data are generally analyzed by computer.

The Neutralist-Selectionist Controversy

In Chapter 9, it was pointed out that about one-fourth of all possible single base substitutions result in synonymous codons and hence in no change in the proteins. Further, many amino acid substitutions do not change the functions of the proteins. Cytochrome c occurs in all eucaryotes and in some procaryotes. Across this immense taxonomic range, most of the amino acids have been substituted. Only thirty-five of the one hundred twelve are invariable throughout the world of life. M.C. King and T.H. Jukes, therefore, suggested that much molecular evolution was neutral rather than controlled by natural selection and that these neutral biochemical changes were regulated by genetic drift. Although genetic drift may result in rapid fixation or elimination of alleles in small populations, it is an extremely slow process in very large populations.

Because the time of separation of the various groups in the cytochrome c phylogeny is known from paleontological data, we can use the sequence data to estimate the rate of change in the cytochrome c molecule or in other protein phylogenies. The data suggest that it requires a minimum of 4 million years for each allelic replacement. The rate of substitution may be different for different proteins, and estimates range up to 7 million years. M. Kimura has been an enthusiastic advocate of the neutralist viewpoint. Because such neutral mutations occur at a more or less constant rate and are gradually fixed or eliminated by genetic drift, these processes constitute a molecular clock that supplements geological estimates of time.

What the relative roles of neutral and selective evolution are is an extremely difficult question, partly because the two viewpoints are best able to explain different arrays of data. The adaptive value of morphological variants may often be intuitively evident (and perhaps sometimes intuitively deceptive, too!), and hence we customarily explain morphological differences in terms of natural selection. An enzyme like cytochrome *c* may exist in many alternative forms (*allozymes*), and they seem to function equally well in cell respiration. Any mutation that seriously reduces the efficiency of the enzyme will be quickly eliminated by natural selection. The thirty-five apparently invariable amino acids of cytochrome *c* probably consist of sequences that are essential for the functioning of the enzyme, so that mutational changes are quickly eliminated. Electrophoretic data show that allozymes are extremely common, and the inference that the members of a given family of allozymes do not differ significantly in selective value is justified. Both natural selection and neutral evolution, then, play significant roles in evolution, but their relative roles are difficult to estimate.

How Much DNA?

Many characteristics of organisms, then, vary quantitatively in evolution, and they are all controlled by the DNA. Does the DNA itself undergo a similar quantitative evolution? The quantity of DNA per cell can be measured photometrically. The usual unit of measure is the picogram (pg $= 10^{-12}$ g). One picogram of DNA corresponds to about 2×10^9 base pairs, and DNA quantity is sometimes expressed as so many billions of base pairs.

In Figure 14–9, the haploid values for a variety of animals are arranged on a phylogenetic diagram. The range of values is great, from as little as 0.0017 pg in some bacteria (average 0.007 pg) to as much as 100 pg in some urodele amphibians. Although the lowest values are at the base of the phylogenetic tree, there is no simple progression up the phylogenetic scale. In relationship to DNA content, all organisms seem to fall into four groups. If we disregard the viruses, which have a maximum of about 10^4 nucleotides—a mere trace of a picogram—the first group comprises only bacteria, and the quantity of DNA in these ranges from 0.0017 to 0.01 pg. The second group includes only fungi, which, with an average of about 0.09 pg, have significantly more DNA than do the bacteria yet less than do other eucaryotes. The third group includes most eucaryotes, whether plants or animals, with a range from about 0.1 to 10 pg. Most species in group three have less than 6 pg of DNA. Finally, group four includes both plants and animals with more than 10 pg of DNA, and often very much more. Examples include urodele amphibians

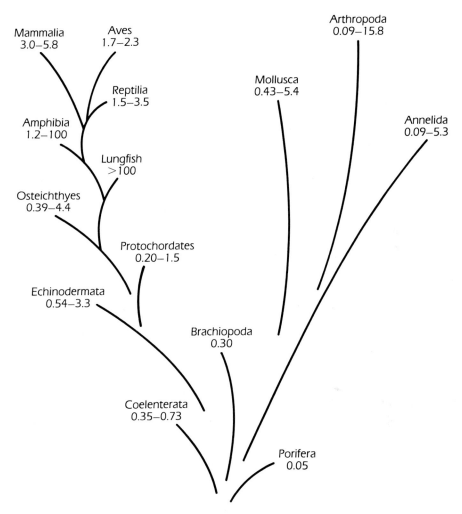

Figure 14–9.   *Haploid amounts of* DNA *in picograms in various phyla and classes, arranged approximately phylogenetically. In most groups,* DNA *content has been determined for only a few species.*

and ferns, both of which are reported to range up to 100 pg, and psilopsidans (primitive plants; see Chapter 18), which are reported to range up to more than 300 pg per cell!

We might expect that within a variable group of organisms the variations would be distributed on the normal probability curve, but in fact the curve is usually skewed to the left. Thus, teleost fishes range from about 0.4 to 4.4 pg, but the peak of the curve is near 1.0 pg. Again, we might expect close correlation between chromosome number and DNA content; yet in a given group, variation of DNA content may be much greater than that of chromosome number. Thus, all species of the salamander genus *Plethodon* have $n = 14$, yet the DNA content of that genome varies from 18 to 69 pg in different species. On the other hand, in many groups of plants (and more rarely in animals), whole sets of chromosomes may be multiplied (polyploidy; see Chapter 13); and in such series, chromosome number and DNA content are proportional. In fact, group four plants are generally highly polyploid.

Polyploidy is clearly a major factor in changing the quantity of DNA in group four plants. In other groups, duplications and deletions appear to be the major factors, with repeated small changes adding up to large effects. The interpretation of the quantity of DNA is greatly complicated by the existence of highly repetitive DNA as well as unique sequence DNA (see Chapter 7).

SUMMARY

In Chapter 10, we saw that quantitative methods were highly productive in the study of *microevolution,* that is, evolution at the level of genetic analysis. Are such quantitative methods also applicable to the study of *macroevolution,* the evolution of higher categories? The problems of macroevolution are more difficult than those of microevolution, but a gratifying degree of success has been obtained.

Allometric, or differential, *growth* is a common problem of ontogeny, as shown by the changing proportions of the face of the baboon or the "tail" of the swallowtail butterfly during development. Similar changes of proportion occur in phylogeny and have been successfully analyzed, with the aid of the equation for allometric growth, in the deer family, which culminated in the Irish elk, and in the facial proportions among the australopithecines that were fossil hominids.

Some of the best applications of quantitative methods to macroevolution derive from comparative biochemistry. Particular enzymes, hormones, and other proteins may occur over a wide taxonomic range. Thus, insulin occurs in all vertebrates, and cytochrome c occurs in all eucaryotes and in many procaryotes. When amino acid sequences are determined for a given protein in such a series, we find few amino acid substitutions between taxonomically close species and numerous ones between taxonomically distant species. Because each amino acid substitution corresponds to one or more mutations, we can calculate minimal mutational distances between species.

In some cases, the structure of proteins demonstrates that genes have duplicated in the course of evolution. Thus, myoglobin, consisting of a single chain of about 150 amino acids complexed with a heme group, functions to store oxygen in muscle tissue. Hemoglobin of lampreys is closely similar to myoglobin, but it is contained in red blood cells and functions in oxygen transport. In early fishes, the gene for this molecule duplicated, so that two-chain hemoglobin was produced. Mutations then occurred independently in the two genes, with distinct alpha and beta chains resulting. In higher vertebrates, these chains unite to form tetramers, which are extremely efficient oxygen carriers.

About one-fourth of all single base substitutions in the DNA results in synonymous codons. Also, in a typical enzyme, many amino acids can be substituted without impairment of function. Some suggest that such changes constitute neutral evolution, uninfluenced by natural selection. Such changes occur by genetic drift. Although genetic drift may accomplish rapid changes in small populations, changes by drift in large populations are extremely slow. By comparing paleontological and molecular data, we can estimate that the minimal time for an allelic substitution by drift in the large populations of typical species is four million years or more; and we can use this estimate as the basis for a molecular clock. The relative importance of neutral and selective evolution is difficult to estimate.

Because the data of molecular phylogeny span great taxonomic gaps and long reaches of time, they make possible answers to some questions that were formerly debated in the absence of decisive data. It is clear, for example, that mutations consisting of simple replacements of nucleotides have characterized the gene for cytochrome *c* throughout the history of the eucaryotes and even further back in time to their procaryotic ancestors. Mutations of these genes have been accumulated over long reaches of time, and they span great taxonomic distances. This strengthens greatly the probability of the assumption that the processes of mutation and selection, which produce subspecies, are also the ones that, carried out over long periods of time, produce species, genera, and even the highest groups. Of course, this does not preclude an important supplementary role for duplication, inversion, polyploidy, and other changes at the chromosomal level. It has also often been debated whether life arose only once, or whether it arose several or many times. Again, the universality of cytochrome *c* and the persistence in it of certain sequences throughout the eucaryotes and in certain bacteria strongly support the proposition that all of the organisms of this vast realm are part of one grand phylogeny. Other molecular data, such as the identity of the histones of cattle and of peas, strongly suggest that plants and animals evolved from a common ancestor among the primitive eucaryotes.

REFERENCES

Ayala, F.J., ed. 1977. *Molecular Evolution.* Sinauer Associates, Sunderland, Mass.
    There is a wealth of valuable material in this book, but see especially W.M. Fitch's evaluation of the molecular clock.
Barker, W.C., and M.O. Dayhoff. 1980. Evolutionary and functional relationships of homologous physiological mechanisms. *BioScience* 30:593–600.
    Many examples of biochemical phylogenies.

Dayhoff, M.O. 1969. Computer analysis of protein evolution. *Scientific American* 221(1):86–95 (July, 1969).
A clear and simple presentation of the cytochrome c story, told by a leading protein chemist.

Doolittle, R.F. 1981. Similar amino acid sequences: Chance or common ancestry? *Science* 214:149–159.
A comprehensive review.

Ferguson, A. 1979. *Biochemical Systematics and Evolution.* Blackie, Glascow, Scotland.
Extensive coverage of electrophoretic research and its results.

Gould, S.J. 1966. Allometry and size in ontogeny and phylogeny. *Biological Reviews* 41:587–640.
A comprehensive account of the functional and evolutionary significance of relative growth.

Hinegardner, R. 1976. Evolution of genome size. In *Molecular Evolution,* edited by F.J. Ayala, pp. 179–199. Sinauer Associates, Sunderland, Mass.

Huxley, J.S. 1932. *Problems of Relative Growth.* Methuen, London. Reprinted in 1972 by Dover, New York.
A classic study of allometry.

Simpson, G.G., A. Roe, and R.C. Lewontin. 1960. *Quantitative Zoology,* rev. ed. Harcourt Brace Jovanovich, New York.
An unusually lucid introduction to biometry, based upon zoological examples.

Thompson, D. 1952. *On Growth and Form,* 2nd ed. Cambridge University Press, London.
A beautifully written classic, and the foundation of the concept of allometry.

Wright, S. 1968–1978. *Evolution and Genetics of Populations,* vols. 1–4. University of Chicago Press, Chicago.
An authoritative and encyclopedic treatment of the subject, but rather difficult reading.

CHAPTER

15

Distribution of Species

Biogeography is the study of the patterns and causes of the distribution of plants and animals. Modern studies have developed in two directions: ecological biogeography and historical biogeography. The former focuses on patterns of diversity and processes of dispersal that can be observed in action and that are susceptible to experimentation and quantification. Historical biogeography relates patterns of distribution to major events in earth history. The time frame is very much longer, and there is no experimentation (or we might say that natural experiments have already been performed). Biogeography of any kind depends on the prior existence of good taxonomy that expresses degrees of relatedness among various taxa.

In one sense, problems of distribution overlap those of isolation. The minimum biogeographical requirement for relationship between two species is that their ancestors must at some time have lived in the same area. In other words, permanent isolation and relationship are mutually exclusive. Thus, on the one hand, if we learn that the magnolias of southeast China are related to those of the southeastern United States, we know that their present isolation did not characterize all past ages. As a matter of fact, that their distribution was continuous in the Tertiary is well established. On the other hand, once we know this fact, it becomes important to study the development of geographic isolation between the related populations. Such study is a significant aspect of descriptive biogeography.

From the time of Linnaeus until well into the post-Darwinian era, biogeography was a very active field, in which many of the most distinguished biologists worked. New discoveries of major importance could be expected as the reward of competent work in this field. New phyla were described frequently; new classes and orders, with regularity. But by the end of the nineteenth century, all of the major biogeographical realms were fairly well known, their floras and faunas were cataloged, and only problems of detail and revision confronted young biogeographers. Meanwhile, the rise of experimental biology made biogeography and taxonomy rather passé, and their study was often regarded as hack work. The modern revival of evolutionary studies brought with it a renewed interest in taxonomy, and more recently analytical biogeography has shared in this renaissance. The *Journal of Biogeography* was established in 1974, and practically every issue of *Systematic Zoology* contains at least one paper on biogeography. There are problems in evolution that can be profitably attacked only if detailed biogeographical information, both present and past, be studied. We will consider some of these problems in this chapter.

REVITALIZATION OF BIOGEOGRAPHIC THOUGHT

Three ideas have brought a major shift in biogeographic thought during the past decade. These are the acceptance of *continental drift* and the concepts of *vicariance biogeography* and the *equilibrium theory of island biogeography.* The first of these was the outcome of a revolution in geological thinking that began in the 1960s and was largely complete by the early 1970s. (Drift had been tentatively suggested much earlier, but had not been widely accepted.) The concept of *continental drift,* or *plate tectonics,* holds that the continents were not fixed in their present positions throughout geological time. During the Permian, there was a great assembly of continents called Pangaea. In the Triassic, the northern continents (Laurasia) began to separate from the southern continents (Gondwana) and from each other. During the second half of the Mesozoic and the early Cenozoic, the southern continents separated from each other and began to drift apart (Figure 15–1). The continents are carried on the backs of crustal plates. New crustal material is formed along midoceanic ridges (Iceland and the Falklands are volcanic islands at opposite ends of the mid-Atlantic ridge). Newly formed crust flows away from the ridge in either direction (east and west) at the rate of around 1 cm per year, a rate that is more than adequate to account for the hypothesized drift in 200 million years. Old crust is subducted and consumed in oceanic trenches.

It is readily apparent that a world with wandering continents requires interpretations that differ radically from those based on stable continents, the

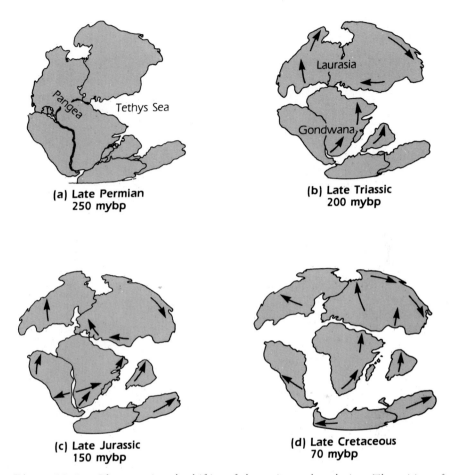

(a) **Late Permian**
250 mybp

(b) **Late Triassic**
200 mybp

(c) **Late Jurassic**
150 mybp

(d) **Late Cretaceous**
70 mybp

Figure 15–1.   *Plate tectonics, the drifting of the continents through time. The positions of the continents are illustrated (a) at the time of greatest assembly 250 million years before the present (mybp); (b) at the beginning of the breakup 200 mybp; and (c) and (d) at successive times since then. Note the slow formation of the Atlantic Ocean, the persistence of the connection between eastern North America and Europe, and the relatively recent contact between South America and North America and of western North America and Asia.*

paradigm that had long prevailed. Formerly, biogeographers were greatly concerned with centers of origin and means of dispersal. *Ad hoc* land bridges across ocean gaps had even been invoked to account for biotic similarities over great distances, irrespective of the lack of geological evidence for these. Various and often-conflicting rules were used to determine centers of origin of major groups. Some felt that current centers of diversity corresponded to ancient centers of origin. Others felt that areas inhabited by the most primitive members of a group represented the place of origin or—a variant of this belief—that the site of the oldest-known fossil marked the spot. Still others believed that primitive living members were situated farthest from the point of origin.

Vicariance biogeography is predicated on the effects of plate movements and other major events of earth history on the distribution or organisms, especially on whole biotas. It is an analytical method of historical biogeography. The origin of this concept is traced to the Venezuelan L. Croizat, who in 1964 privately published a book entitled *Space, Time, Form.* The book itself had very little impact, but, beginning in 1974, G. Nelson, D.E. Rosen, and C. Patterson have championed his ideas with vigor and have wedded this concept with cladistic taxonomy. In a nutshell, vicariance biogeographers believe that the most important process in biogeographic history is the severing of widely dispersed biotas by geological processes, with sister species forming allopatrically on either side of the barrier. Dispersal is downplayed in the vicariance model, whereas it was paramount in earlier schemes.

The third great idea, this one in ecological biogeography, is the *equilibrium theory of island biogeography,* formulated by R.H. MacArthur and E.O. Wilson in 1967. This quantitative analytical technique relates biotic diversity on islands (regardless of which particular species of a given group are present) to such factors as surface area of the island, distance from the source of new immigrants (thus, immigration rate), and local extinction rate. For an island in equilibrium, establishment of a new immigrant species will tend to cause the extinction of an established species, the total number of species remaining constant. Faunal equilibrium is thus dynamic, with turnover being a constant process. The equilibrium theory has been very useful in the study of the ecology of islands, and it has been extrapolated into evolutionary time as well. We begin our detailed consideration of biogeography by considering islands.

ISLAND LIFE

In many cases islands represent natural genetic or ecological laboratories in which we may observe the effects of diversifying evolution over a geologically brief span of time. Island ecosystems may be simplified as compared with mainland systems, with fewer species of a lower number of families or orders than on the adjacent mainland. Disharmony of the ecosystem (in the sense that some important ecological niches may be unfilled) is characteristic, with certain kinds of animals absent (e.g., amphibians, carnivores) and other kinds disproportionately abundant (e.g., rodents, insects). A fine example of floral disharmony occurs on Saint Helena in the South Atlantic. No trees succeeded in colonizing this remote island; in default of trees, weedy plants of the family Compositae have evolved into large, full-canopied, arborescent forms—merely trees to the casual observer, but whose flowers reveal their origins.

The study of the flora and fauna of the Galápagos Islands first caused Darwin to consider the possibility that species might be mutable, and the island life of the Malay archipelago played a major role in bringing Wallace to the same conclusion. Since their time, the study of oceanic island life has always been an important aspect of the study of evolution. Many features inherent in the island locale combine to make it so. The fact that the inhabitants of any island generally resemble those of the nearest mainland leaves little doubt that the island dwellers have migrated from the mainland. However, unless an island is very close to the mainland, the difficulties of crossing the water barrier will keep out a large portion of the potential migrants. Thus competition is less rigorous, and the selection pressure is lower than on the mainland. In the absence of the check of normal interspecific competition and often in the absence of predators, populations may reach prodigious numbers relative to the space available. Yet the total population is small compared to continental populations, and it is likely to be broken up by ecological factors into much more restricted breeding colonies. In such a situation, genetic drift may be stronger than selection and may produce island forms that are less well adapted than their mainland relatives. Apparently (but not actually) contradictory to this is the fact that closely related island dwellers may become adapted to situations so different that only remotely related organisms of the mainland could be compared to them.

Genetic Drift on Oceanic Islands

A celebrated example of genetic drift is that of land snails of the genus *Partula* on the islands of Tahiti and Moorea, neighbors in the Society Islands, about 2,400 miles south of Hawaii. These are typical volcanic islands, characterized by a central volcano from which deep valleys and narrow separating ridges radiate to the sea. The snails feed on the plants of the valleys, and the intervening ridges are almost impassible barriers to them. As many as eleven species have been described, but the most recent study, by B. Clarke and J. Murray, proposes just two species complexes, *P. suturalis* and *P. taeniata,* with many subspecies. Many of the subspecies (formerly described as species) interbreed freely in the laboratory, and there is some evidence that they also interbreed in nature. Two members of a species group that behave as distinct species (no interbreeding) at one locality may intergrade freely in another locality only a short distance away. The inhabitants of every valley constitute a distinct race—distinct in characters such as size, direction of coiling, details of shape, and color. The variation of the several races of a species definitely is not clinal. Races of neighboring valleys may be strongly divergent, whereas races at opposite ends of an island may be quite similar. Different species in a single valley may have identical ecological requirements and may live on the same food plants, and yet they may show no tendency toward parallel variation.

All attempts to interpret this situation in terms of different selective forces in the different valleys have failed. Probably, then, all of the races of *Partula* are subject to substantially identical selective forces, and the differences among them result from genetic drift. But genetic drift on so grand a scale is possible only because the geographic features of the islands enforce almost complete isolation upon all of the local breeding populations. These snails have been studied intensively four times, at wide intervals, in the past century, and significant changes in many of the races appear to have occurred even in that short time.

Rapid Selective Differentiation on Oceanic Islands

One of Darwin's studies in the Galápagos Islands was made upon a subfamily of finches, the Geospizinae, that had become adapted to an amazing range of ecological niches in the islands, so that their superficial differentiation was much greater than usual within the confines of a single family. Another family of birds, the Drepanididae, or Hawaiian honeycreepers, has undergone a similar adaptive radiation within a geologically brief time. The modern Hawaiian Islands appear to have originated no earlier than the Pliocene (but the ancestral Hawaiian-Emperor chain, represented by submerged seamounts to the northwest, appears to date back at least 50 million years). As the honeycreepers are forest-dwellers, their original ancestor could not have reached Hawaii (by migration from Central America) before the forests themselves were established. This might have occurred by Mid-Pliocene or later, thus giving a maximum of five million years for the differentiation of the family in Hawaii.

When the remote ancestor of the Drepanididae first migrated from the mainland to the Hawaiian forests, it found a rich terrain for which it had no competitors. The result was a rapidly expanding population that soon provided its own competition—that is, the population began to outstrip the food supply available *by the original method of feeding.* This drepanidid progenitor may well have been similar to the living *Loxops virens chloris,* which has a moderate-sized, slightly curved bill, adapted to feeding upon insects in foliage (Figure 15–2). *L. v. chloris* also occasionally digs for insects in loose bark or probes flowers for nectar and insects. Now the development of races or species with different feeding habits would permit the survival of a much larger total drepanidid population. How this might have occurred is perhaps indicated by living members of the genus *Loxops. L. v. chloris* is widely distributed through the islands, but on Kauai, one of the most-isolated islands, this species is represented by another subspecies, *L. v. stejnegeri.* Another species, *L. parva,* is found only on Kauai, and the characteristics of the bills of these two *Loxops* representatives are most suggestive. The bill of *L. v. stejnegeri* is somewhat larger, heavier, and more strongly recurved than that of *L. v. chloris.* Although *L. v. stejnegeri* still has feeding habits similar to those of *L. v. chloris,* it

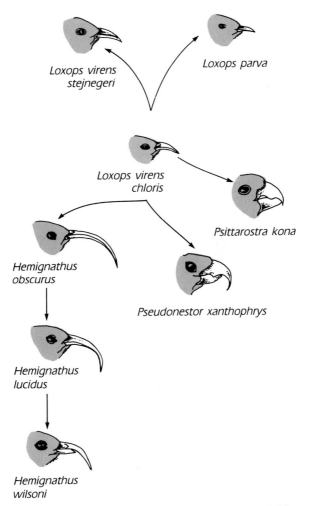

*Loxops virens
stejnegeri*

Loxops parva

*Loxops virens
chloris*

Psittarostra kona

*Hemignathus
obscurus*

Pseudonestor xanthophrys

*Hemignathus
lucidus*

*Hemignathus
wilsoni*

Figure 15–2.   *Evolution of bills among the Drepanididae
of Hawaii.*

depends more upon digging for insects in loose bark. It also visits flowers for
nectar and small insects. The bill of *L. parva* has deviated from that of *L. v.
chloris* in just the opposite way. It has become shorter and straighter. *L. parva*
depends primarily on insects on the surfaces of branches and leaves. Its bill
is not well adapted to digging, which it rarely attempts. Although it is not
well adapted to visiting flowers, it does frequently visit acacia flowers. Probably,
when these two species of *Loxops* were brought into competition, any variations
that tended to adapt them to different sources of food were strongly favored
by natural selection. The result is that these two species, so strongly divergent
in bill structure, differ little in other respects.

The genus *Loxops* only begins to indicate the range of variations of bills in the Drepanididae. The genus *Hemignathus,* most members of which have become extinct in recent times, showed a much greater range of variation than usually characterizes whole families. *Hemignathus obscurus,* which was probably closely related to *Loxops virens,* had a very long, slender, and strongly curved bill. It was adapted for probing fine crevices in the bark of trees and for visiting flowers. In *H. lucidus,* the upper mandible was much like that of *H. obscurus,* and it was similarly used for sweeping insects out of crevices in the bark of trees. But the lower mandible was much shorter and thicker and was used in woodpecker-fashion to pry or chisel bits of bark to expose the bird's prey. This species rarely visited flowers because its bill was not adapted to feeding on nectar. *H. wilsoni,* which is still extant, differs from the others in that its lower mandible is very heavy and chisel-like. It uses its lower mandible to break away wood and bark, then sweeps its upper mandible though the crevices so exposed. It never visits flowers. A closely related genus, *Pseudonestor,* has a heavy bill that is adapted for crushing dead twigs to expose beetle larvae. In a fourth genus, *Psittarostra,* which appears to be closely related to *Loxops,* the bills are like those of finches—that is, they are short, heavy structures for cracking hard seeds.

Thus within this family a very wide range of bill structure and feeding habits has been achieved within a short span of time. Differentiation of other structures has been much slower. Competition for a limited food supply by closely related birds has given a strong adaptive value to variations that tend to open up new food sources to them, stimulating rapid adaptive radiation. This case is an excellent example of character displacement (Chapter 12).

ENDEMISM

Closely related to the phenomena of island life is endemism. *Endemic* is defined as restricted to or prevalent in a particular district. Thus defined, all species are endemic, for all are confined to a definite area, even though that area may be great. For practical purposes, a species is regarded as endemic if its distribution is very much more restricted than that of typical species. Thus, the northern white pine, *Pinus strobus,* widely distributed over northeastern United States and Canada, is not regarded as an endemic; but the redwood, *Sequoia sempervirens,* which is confined to the coastal valleys of California, is an endemic. A still more restricted endemic, discovered only in 1946, is the living member of the genus *Metasequoia,* most of the members of which are extinct. This species is confined to a single valley in central China.

Two different types of endemic species are recognized. A species may have a very restricted distribution because it is a *young* species and has not had time to expand its range. During the first half of this century, J.C. Willis defended this type of endemism as typical. Introduced species of plants and animals have, however, spread over whole continents in less than a century in historic times, hence this type of endemism cannot have general importance. Or a species may have a restricted range because it is the last remnant of an old group nearing extinction. Most species of restricted distribution fall into the latter class. They have few living close relatives, and they are often well represented in paleontologic series, as for example, *Sequoia*.

Island life, particularly that of oceanic islands, is replete with endemics. It has been estimated that more than 90 percent of the flora of the Hawaiian Islands is endemic; probably the figure for animals would be comparable if calculated. As already mentioned, the high proportion of endemics in the Galápagos Islands was among the factors that directed Darwin's thoughts toward the possibility of the transformation of species. Mayr has pointed out that many of the birds of South Pacific islands that he has studied are endemics. Why the flora and fauna of an oceanic island should be endemic is easy to understand. The great water barrier prevents all but pelagic species from extending their ranges.

Endemism is by no means confined to islands. Plant geographers have given much attention to this problem. The cases of *Sequoia sempervirens* and *Metasequoia* were cited earlier. *Sequoia gigantea* is similarly restricted to the Sierra Nevada. S.A. Cain lists thirteen localities in which no fewer than twenty-five species of plants are endemic. These examples of endemic plants are sufficient to indicate the prevalence of the phenomenon. Why are continental endemics restricted to so limited an area? In the case of young endemics, the answer is obvious: they simply have not had time to achieve the range extension that may be expected in the future. As noted above, however, such cases are uncommon. More usually, endemics appear to have little genetic variability, with the result that they are adapted to only a narrow range of environments. This limitation must be based in part upon a low mutation rate. Also, such species may have lost much variability by genetic drift as the population contracted.

EQUILIBRIUM THEORY OF ISLAND BIOGEOGRAPHY

Precisely because islands are of circumscribed size, their study can be quantified very readily. In their 1967 publication, MacArthur and Wilson quantified a series of empirical observations concerning the diversity of island biotas. Surface area, for instance, is an excellent predictor of diversity: a tenfold increase in area of an island approximately doubles the number of species of a given group, say of insects or of birds. Extinction rates are higher on smaller islands because population sizes of individual species are smaller and more subject to fatal random fluctuations. Similarly, equilibrium numbers are lower on more remote islands than they are on islands of similar size that are closer to a mainland and thus have a higher rate of influx of new immigrants. The prediction of species turnover, the balanced dynamic of immigration and local extinction, is one of the most important heuristic consequences of the equilibrium theory.

The equilibrium theory has been subjected to empirical and experimental testing and it has fared well. D. Simberloff carried out a series of experiments on tiny mangrove islands along the coast of Florida. He counted species of insects and other arthropods before exterminating them by fumigation. He then monitored the buildup of species at regular intervals thereafter. He found that the more distant islands repopulated more slowly than did the closer islands; but after a year, each island had returned to approximately its pre-fumigation level of diversity. However, species composition was different from that which existed before the artificial disaster and even from that which obtained during midrecolonization. This fact confirms the dynamic nature of the process, one in which species are continuously immigrating and going extinct. Natural experiments on a grander scale show the same thing. For instance, when the volcanic island of Krakatoa in Indonesia exploded in 1883, it was effectively sterilized. Thirty years later, birds there had returned to pre-explosion levels of diversity, but there were differences in species composition. Since then species have continued to turn over slowly.

An interesting natural experiment on colonization of an island is now in progress on Surtsey (Figure 15–3), an island about 35 kilometers south of Iceland, which rose from the ocean floor (120 meters deep) as a result of

Figure 15–3. Surtsey in eruption, November 1963.
Courtesy of Professor Sturla Fridriksson, Agricultural Research Institute, Reykjavík, Iceland.

volcanic activity that began in November 1963 and ended in spring 1967, by which time the island had attained an area of nearly 3 square kilometers. Colonization of the island began almost at the time of its origin, for a gull was observed on it in November 1963. Bacteria and molds were established on the island while the volcano was still active, probably from spores brought by birds and by winds from the mainland. Blue-green bacteria are among the colonists, and they are actively fixing small amounts of nitrogen. Algae soon followed, and over 100 species have been identified. In 1970, species of lichens were found on the island. Mosses were growing on Surtsey as early as 1967, and by 1972 the number of species had reached 63. They appear to have been established by wind-borne spores, originating either on the mainland or on the Westman Islands, a small archipelago extending between Surtsey and the mainland at distances of 5 to 25 kilometers.

The first vascular plant on Surtsey, the sea rocket *Cakile maritima,* appeared during the summer of 1965. Three years later a second species appeared. By 1982 (19 years after the creation of the island), there were 19 species, including a horsetail, ferns, grasses, sedges, and other small plants. A sand dune community is beginning to emerge.

Animals were not long in following. The first dipteran was observed in 1964; 5 more species were collected in 1965. By 1970, 112 species of insects and 24 of arachnids inhabited the island, and other arthropods were also well represented. Protozoans, too, were early arrivals, with representatives of several orders being collected during the 1960s. Nematode worms were found as early as 1971.

Gulls were the earliest-known visitors to Surtsey. Three species of marine birds—black-backed gulls, kittiwakes, and arctic terns—breed on Surtsey, and others may be expected to follow. In addition, 60 species of birds have been observed as occasional visitors, mostly during periods of migration. Finally, mammals are represented by seals, which frequently come ashore on Surtsey.

Thus, Surtsey is populated by a wide variety of procaryotes, plants, and animals; yet these populations do not constitute a balanced ecological community, and equilibrium has not been reached. The biota is made up of pioneer species that are capable of being transported from neighboring islands by air, by water, or by hitchhiking on birds and that are capable of exploiting an extremely harsh and poor environment.

The essence of islands is their isolation. The concept of habitat islands is a useful one in ecology. Thus, an oasis is an island in the desert; a mountain surrounded by lowland jungle is an island in the tropics. There is no minimum size for an island nor is there a maximum one. That the continent of Australia is an island is self-evident; but to a certain degree each continent can be treated as an island in a historical sense, with periods of isolation alternating with periods of biotic interchange.

S.D. Webb applied the equilibrium theory to late Cenozoic mammals of

North America with considerable success. He substituted the evolution of new species *in situ* for the immigration of species in the MacArthur and Wilson formulation. He demonstrated the existence of a general constancy (equilibrium) in the numbers of genera of mammals through the late Cenozoic. During the Ice Age, however, despite the heavy rate of extinctions, there was a net increase in the equilibrium number as a greater number of smaller animals replaced a smaller number of larger animals (rodents flourished while mastodons, mammoths, glyptodonts, and ground sloths became extinct). Later in this chapter we discuss the application of the equilibrium theory to the "great American interchange," the Plio-Pleistocene exchange of mammals between South America and North America.

CONTINENTAL DISTRIBUTION

Islands are the laboratories in which we may observe the processes of biogeography in action, under somewhat simplified conditions. The continents are the stages on which the evolutionary processes molding the biotic regions of the globe are played out through time. In historical biogeography the time scale shifts, and extensive reference to the fossil record becomes extremely useful. Who would make a guess, based only on the distribution of living equids (horses) in Africa and Asia today, that North America was the center of the evolution of horses? Yet this is exactly what the fossil record suggests.

A species originates in a definite place at a definite time. It may then expand its range to the extent that its ecological requirements and its ability to disperse permit, until it is prevented from expanding further, either by ecological barriers (e.g., competition) or by physical or climatic barriers. Many species—house sparrows and dandelions, for instance—achieve transcontinental distribution. Distribution may be halted by the continental margin, although geological conditions may change and permit the species to cross from one continent to the next. Conditions thoughout the range of the organism, however, may change beyond the organism's capacity to respond, and it may become rare over large parts of its range. Finally, only an isolated endemic population remains. With the extinction of this remnant, the history of the species comes to an end.

G.G. Simpson has related the history of mastodonts, extinct members of the order Proboscidea, in these terms. Mastodonts first appear in the fossil record of North Africa in the Oligocene. Because other Oligocene faunas, well known from all parts of the world, contain no mastodont fossils, North Africa was quite probably the place of origin of the mastodonts, and they probably did not occur elsewhere at that time. They subsequently expanded their range through active migration, and, by the beginning of the Miocene, they had reached midcontinental Africa to the south, the Baltic area to the north, and India to the east. By Mid-Miocene, they occupied about half of Africa and most of Europe and Asia except the most northerly parts. Late in the Miocene, they crossed the Bering Strait to North America and spread over that area during the Pliocene. Only toward the end of the Pliocene did the mastodonts reach South America, and during the Pleistocene they spread over much of that area. At about this time the mastodonts became extinct in the Old World; by the Late Pleistocene the Americas were their last refuge. Finally, the American mastodonts also disappeared.

The facts of the mastodont case seem simple enough. Yet vicariance biogeographers strongly contest the generality of dispersal; instead they attribute patterns of distribution to ancient geological events. Dispersalists ask how an organism crossed a particular barrier; vicarists ask when the barrier arose to separate populations and permit allopatric speciation. To vicarists, the fact that birds and earthworms may share a common pattern of distribution, although they contrast almost completely in dispersal ability, suggests the existence of *generalized tracks*—that is, avenues along which whole biotas (creepers, crawlers, swimmers, flyers, hoppers, seed-setters, and so on) passed in the absence of barriers. Vicarists feel that dispersalists pay too much attention to chance events affecting single species. A creative tension between these two schools of thought is evident in biogeographic writings today.

DISCONTINUOUS DISTRIBUTION AND BRIDGES

It is in connection with this type of history that the problem of widely discontinuous distributions may be understood. In any particular case, discontinuity may be brought about by extinction in the intermediate parts of a wide distribution, by the bridging of a barrier between two distinct areas, or by a combination of the two processes.

How then do organisms pass from one continent to another or from a continent to an island? G.G. Simpson designated three types of bridges: corridors, filter bridges, and sweepstakes routes. Drawing on plate tectonics, M. McKenna has added two further means: Noah's arks and beached Viking funeral ships. A *corridor* is a broadly continuous connection, existing over a long period of time, that permits an extensive interchange of the floras and faunas of the connected regions. Such a connection now exists between Europe and Asia, and these two accordingly constitute a single biogeographic region. A *filter bridge* is more temporary in duration and more restricted in extent. Conditions are more uniform upon it, with the result that it "filters" the flora and fauna that might use it; only those with appropriate characteristics can pass. The Bering Strait acted as a filter bridge for mammals during the Pliocene. Only those mammals that were capable of making a rapid crossing and of withstanding cold weather could cross. A *sweepstakes route* does not involve migration across a land connection; rather it depends upon accidental transportation in the absence of any real connection. Corridors and filter bridges operate equally well in either direction and cause the exchange of numerous forms; a sweepstakes bridge operates in only one direction, and only a few forms succeed in crossing it. It is a one-chance-in-a-million phenomenon, but such odds allow many chances indeed if time be available on a geological scale.

A *Noah's ark* occurs when a plate breaks away from a continental margin and drifts off, carrying with it the resident biota. Such an interpretation has been advanced for the endemic vertebrate fauna of the Greater Antilles. Sweepstakes interpretations invoking hurricane-propelled chance crossings of the Caribbean from Yucatán had been favored. Recently D.E. Rosen and B.J. MacFadden have suggested that early in the Cenozoic the "proto-Antilles" split off from what is now Central America and drifted eastward to the present positions of these islands. This vicariant explanation accounts for the endemic nature of the Antillean fauna with both South American and North American representation, without having to resort to chance dispersal over water.

The *beached Viking funeral ship* concept represents a cautionary note in the plate tectonic system: lateral translation of continental plates can carry fossils to areas where they did not exist in life. We shall consider some important examples of these phenomena in more detail.

Corridors

Corridors are most striking after they no longer exist—that is, after geological events have separated land masses that once were continuous. Simpson has pointed out that New Mexico and Florida can be regarded as being connected by a corridor at present. As expected, they share most of the major groups and a large number of genera, but species are likely to be different because the climatic conditions of the two states require different adaptations. We seldom think of such a situation as being a "bridge," simply because it seems natural that the areas concerned should have a substantially similar flora and fauna. But it is perfectly possible that they might be separated in future ages by, for example, an inland sea occupying the present Mississippi drainage basin. Then paleontologists might use the fossils of our time or common genera of their own time to prove that a corridor once existed between New Mexico and Florida.

Such is the situation with respect to eastern Asia and eastern North America. By the Upper Cretaceous, North America had drifted into a position such that biotas could be exchanged across the North Pacific between the two continents. Thus, broadly continuous flora and fauna were established and flourished for many millions of years. Early in the Tertiary, most of this connection was submerged, leaving only the islands of the Bering Strait. Until sometime in the Eocene, an extensive exchange of floras and faunas occurred across this North Pacific bridge. How complete this exchange was is indicated by the fact that 156 living genera of plants are known to be common to the two regions at present. In some cases, such as skunk cabbage, *Symplocarpus foetidus*, the species appear to be identical and even the races are closely similar. The exchanged genera are by no means confined to such rapid migrants as herbs; many genera of trees are common to the two areas, such as *Acer* (maple), *Catalpa*, and *Magnolia*. Many plants known as fossils in both areas are now living in only one. Thus *Castanea*, the chestnut tree, still survives in eastern North America; and *Ginkgo*, the primitive maidenhair tree, still survives in Asia.

Exchange across the North Pacific has been studied less in animals, but parallel examples are known. For instance, alligators are known only from the United States and China, and the salamanders *Triturus* and *Cryptobranchus* are also found in these remote places. Geologic and climatic changes brought about the present isolation of these forms. Geologically, the North Pacific corridor was largely submerged, and the mountains of western North America were elevated, making climate and topography unfavorable for the former inhabitants of this area. Further, the climate of the entire northern part of the world became colder, with the result that these temperate and subtropical organisms became extinct over much of their former range, leaving the distributions as they are found today. In vicariant terms, the generalized track across the North Pacific was disrupted by the climatic and geological events so described.

Filter Bridges

The main characteristic of a filter bridge is that it filters out many of the organisms of the connected regions, while permitting the passage of others. Also, a filter bridge is typically of brief duration, whereas a corridor lasts for periods that are long even on the geological time scale. Although genuine corridors between continents appear to have been rare—perhaps only those described above having existed—filter bridges have been fairly common.

As the great continental ice sheets formed during the Pleistocene, huge volumes of water were withdrawn from the seas and were locked into the ice. Sea level dropped as much as 100 meters worldwide, thus extending the margins of the continents outward. Such was the case along the Bering Strait, across which Asia and North America were temporarily joined, or nearly so. The Bering Strait had the character of a filter bridge repeatedly in the Pleistocene (Figure 15–4). Its filtering action was probably due in part to the fact that the land connection may not have been completely continuous. Any organisms for which small expanses of salt water formed an impassable barrier would have been unable to cross. Also, because of its short duration, it was crossable only by plants and animals capable of migrating fairly rapidly. A more important factor was the location of this bridge and the prevailing climatic conditions: the location was just below the Arctic Circle, and the climate was cold during the Pleistocene. Thus, only those Palearctic and Nearctic animals that were

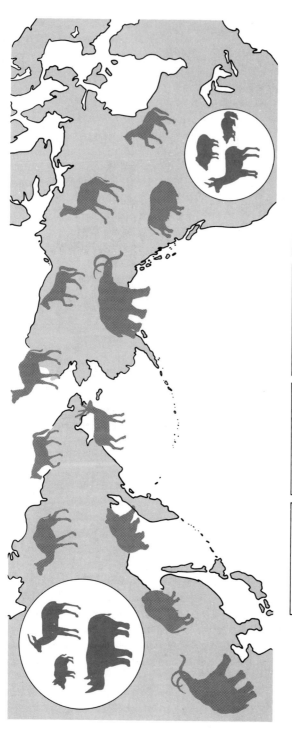

"Stay-at-Homes" Asia	Immigrants to North America		"Stay-at-Homes" North America	Emigrants from North America
Pig Rhinoceros Gazelle	Bear Cat Elk Caribou	Moose Mammoth Bison	Peccary Pronghorn Raccoon	Horse Camel

Figure 15–4. Pleistocene filter bridge across the Bering Strait.

adapted to cold could make the crossing. Temperate and tropical animals were excluded simply because they were unable to approach the bridge. Mainly mammals seem to have made the crossing, perhaps because of their superior powers of locomotion. Animals such as bears, cats, bison, deer, and mammoths crossed from Asia to North America; and dogs, horses, and camels crossed in the other direction.

Another well-known filter bridge is the Panamanian bridge that mediated the "great American interchange" between North and South America in the Plio-Pleistocene. South America enjoyed a long period of what Simpson called "splendid isolation." It was sundered from Africa 100 million or more years ago. The two continents lumbered away from each other at the rate of approximately 2 centimeters per year as the nascent South Atlantic filled the slowly widening gulf. Although some sweepstakes exchange between North and South America seems to have occurred around the beginning of the Cenozoic, a definitive land connection did not join the two continents until the Panamanian land bridge was established 3 million years ago. This vicariant event separated a common marine fauna of the eastern Pacific and the western Caribbean and permitted allopatric speciation to begin. It also made possible one of the great terrestrial faunal exchanges of all time.

Throughout the Cenozoic, South America had an endemic fauna consisting of a variety of marsupial types (including wolflike and saber-toothed catlike predators), edentates, ungulatelike herbivores, rodents, and primates. The fauna shows no particular affinity with that of North America; African origins for ceboid monkeys and hystricomorph rodents have been postulated (but not proven). With the establishment of land connections in the Pliocene, North American mammals moved southward and South American mammals moved northward. Immigrants to South America included raccoons, cricetine rodents, skunks, peccaries, canids, felids, bears, camels, deer, horses, tapir, mastodonts, squirrels, shrews, and rabbits. South American immigrants to North America included armadillos, capybaras, porcupines, opossums, tree sloths, ground sloths, New World monkeys, anteaters, agoutis, and spiny rats (Figure 15–5).

L.G. Marshall and his colleagues have recently studied quantitative aspects of the exchange in detail. They found that predictions from the equilibrium theory held quite well on both continents. Each continent had 34 families of mammals before the exchange. South America has 35 today; North America, 33. In each continent, the number of families swelled immediately following the exchange, but then extinctions occurred. Today North American mammals contribute 40 percent to the family-level diversity of South American mammals, and the corresponding figure for South American families in North America is 36 percent. At the family level, the exchange was remarkably balanced and symmetrical.

At the generic level, however, major differences appear. In North America, pre- and post-interchange numbers of genera are about equal (131 genera

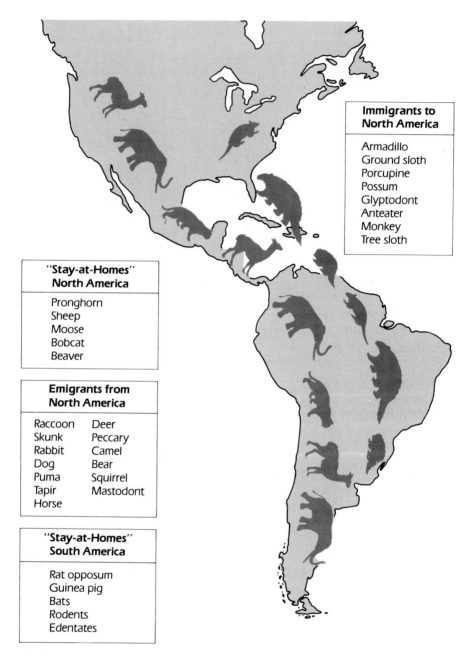

Immigrants to North America

Armadillo
Ground sloth
Porcupine
Possum
Glyptodont
Anteater
Monkey
Tree sloth

"Stay-at-Homes" North America

Pronghorn
Sheep
Moose
Bobcat
Beaver

Emigrants from North America

Raccoon	Deer
Skunk	Peccary
Rabbit	Camel
Dog	Bear
Puma	Squirrel
Tapir	Mastodont
Horse	

"Stay-at-Homes" South America

Rat opposum
Guinea pig
Bats
Rodents
Edentates

Figure 15–5. The Great American Interchange, the Plio-Pleistocene filter bridge exchange of fauna between North and South America that ended millions of years of splendid isolation on the southern continent.

before, 141 genera today), in keeping with the equilibrium theory. Post-interchange decline in native genera is 13 percent in South America, 11 percent in North America. The most-striking difference is that during the 6 million years prior to the interchange, the equilibrium number in South America was 72 genera. During the interchange, the numbers rose steadily to the present level of 170 genera, 50 percent of which are of North American origin (21 percent of North American mammals today are of South American origin). What the equilibrium theory failed to predict was the great diversification of the northern animals on the southern continent: 12 South American immigrant genera gave rise to 3 new genera in North America, and 21 North American genera gave rise to a staggering 49 new genera in South America! Carnivores, elephants, perissodactyls, artiodactyls, and especially cricetine rodents participated in this explosive diversification. Because South America has had such a long history of splendid isolation and because the Panamanian land bridge is so recent in time and so well defined in space, the "great American interchange" is among the best understood, most informative, and least controversial of biogeographical problems.

Sweepstakes Routes

Sweepstakes routes are much less tangible because no actual land connection is present. Natural rafts—driftwood or uprooted trees, for example—might carry plants, particularly as seeds, and animals from one place to another. This is the type of transportation we most commonly envisage in connection with sweepstakes routes. But the concept need not be restricted to water barriers. It can apply to any type of barrier if the crossing is improbable but not impossible. Simpson has summarized the characteristics of a sweepstakes route as follows: Generally, only small animals, and particularly arboreal types, can cross. The chances are much greater for some of these than for others, but at any particular time, the probability of a successful crossing by any organism is small. Chance is the major factor in determining that a crossing is made. Less likely immigrant species may succeed and more probable ones may fail. This situation is comparable to a lottery (whence the term sweepstakes) in which a person who holds only one ticket may win, and a person who holds many tickets may lose. Finally, a sweepstakes is likely to be a one-way route, in contrast to corridors and filter bridges. Island life is commonly established by the sweepstakes method, and because of this it is likely to be unbalanced.

Simpson has summarized the results of the sweepstakes colonization of Madagascar in Figure 15–6. Lions, elephants, apes, antelopes, and zebras are selected to typify animals that cannot cross the Mozambique Channel to Madagascar either because they are too large for the natural rafts or because they do not approach the seashore for ecological reasons. They do not hold

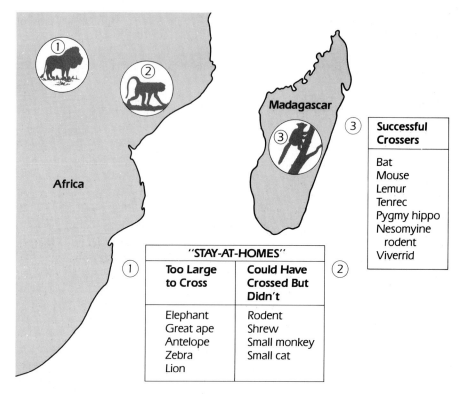

Figure 15–6. *The African-Malagasy Sweepstakes. According to Simpson, as few as five chance crossings from Africa to Madagascar between the Paleocene and the Pleistocene are sufficient to account for the mammalian fauna of Madagascar.*

sweepstakes tickets. There is no apparent reason why many other animals could not have crossed the channel as easily as those that have done so, yet these "disappointed ticket-holders" have not crossed. Out of the multitudes of ticket-holders in the African fauna, only a few have "won"; and these winners have been determined by chance, not by the characteristics of the animals. The winners are typified by endemic nesomyine rodents; by certain catlike carnivores of the family Viverridae; by lemurs; by tenrecs, an endemic group of insectivores; and by a pigmy hippopotamus that may have crossed the channel by swimming. These are all types represented in Africa by widely diversified groups. Had a filter bridge, let alone a corridor, existed, a much broader representation of each order should have entered. For example, the carnivores of Africa include a wide variety of cats, but only the related viverrids reached Madagascar. The primates of Africa include many monkeys and apes, but none of these reached Madagascar. Only two ungulates, out of a large African ungulate fauna, reached Madagascar. These are a bushpig and a pigmy hippopotamus, which is now extinct.

A vicariant explanation simply notes that the fauna of Madagascar is of the evolutionary level of the Eocene of the rest of the world. A common fauna was shared with Eastern Africa until a water barrier separated Madagascar from Africa in the Eocene, so that the island did not participate in the rest of Cenozoic mammal evolution. Even within this framework, sweepstakes dispersal is necessary to account for some Malagasy residents—pygmy hippos, for instance.

Probably no aspect of classical biogeography makes vicariance biogeographers more unhappy than sweepstakes dispersal, with its reliance on chance events— for instance, hurricanes accounting for the colonization of the Greater Antilles. However, G.K. Pregill views the geological evidence of the Caribbean as incompatible with the Noah's ark explanation and has concluded that sweepstakes dispersal remains the best explanation.

A continuing process of over-water dispersal is an observable feature on every oceanic island today; dispersal is a real phenomenon. Island biotas formed by vicariance ought to be harmonic or ecologically balanced; those resulting from sweepstakes dispersal ought to be disharmonic or ecologically unbalanced— that is, certain ecological niches ought either to be vacant or to be filled by an adaptive radiation based on only a few types (e.g., the composite "trees" of Saint Helena). Thus criteria exist for choosing one explanation over another.

PROBLEMATICAL DISTRIBUTIONS

At first inspection discontinuous distributions of many species may appear difficult to understand. As geological data are assembled, these discontinuous distributions frequently become readily understandable on the basis of corridors or filter bridges that are no longer extant. Of course, if the bridge is still extant, there is no difficulty at all. Some distributions, however, seem to require either sweepstake dispersal or plate tectonic transport. These cases are likely to remain in dispute even after thorough study, because many biologists are unable to accept the reality of so indeterminate a route as a sweepstakes. They prefer to look for land connections and then to search for some factor that would prevent a broad sampling of the floras and faunas from using them. The great difficulties to which this quest has at times led may indicate the artificiality of the goal.

Cain lists seventeen paris of areas between which major discontinuities exist. He does not regard his list as exhaustive but suggests that other authors may wish to extend it.

Eastern Asia–Eastern North America Case

When the biotic similarities of eastern North America and eastern Asia were first discovered, an explanation was by no means easily given. Here were two closely similar biotic regions separated by nearly half of the circumference of the globe and a great ocean, yet they had many genera and even some species in common! Then geological and paleontological studies revealed that the Old World and the New World had exchanged biotas many times during the Mesozoic and Cenozoic, sometimes via Greenland and Europe to the northeast, at other times via Alaska and Siberia to the northwest. The climate was much warmer, and thus these temperate and subtropical plants and animals were able to inhabit the great space now separating them.

Bipolar Mirrorism

G.E. Du Reitz has studied what he calls *bipolar mirrorism* of floras of the northern and southern hemispheres, in which the same or closely related species may be present in the temperate and boreal zones of both hemispheres. Thus, he finds a botanical correspondence between the Mexican plateau and Peru, between Texas and Argentina, between Chile and California, and between the Straits of Magellan and Arctic North America. The plants he studied are principally mosses, lichens, sedges, grasses, and other plant species generally unfamiliar to laymen. Some examples are the crowberry shrub, *Empetrum,* found both in North America and in the Andes; the alder tree, *Alnus,* found in the mountains of Mexico and Central America and also in the Andes of Peru; the beech tree, *Fagus,* widely distributed in the northern hemisphere; and the closely related southern beech tree, *Nothofagus,* found in both South American and Australia. Many mosses, grasses, and sedges show comparable distributions. Neither a corridor nor a filter bridge explanation seems to satisfy the details of bipolar mirrorism, and so the sweepstakes route seems to be the most probable explanation. Transport of seeds by migrating birds could be a factor.

Australia–South America Case

Yet another difficult problem of distribution is the similarity between the inhabitants of South America and of the Australian region, including New Zealand. The zoological evidence centers around the marsupials. The fact that the mammalian fauna of Australia is predominantly marsupial is better known than any other fact of Australian natural history. In the Tertiary period, South America had a large marsupial fauna, predominantly carnivorous and closely paralleling the Australian marsupial carnivores. Although most of these became extinct when the placental mammals from North America arrived in the Pleistocene, there are still more species of marsupials in South America than on any other continent except Australia. Likewise, many genera of plants are common to the two continents. *Nothofagus,* the southern beech, is noteworthy. Other examples include the sedge *Carex* and the moss *Sphagnum.*

Marsupials have a very good Upper Cretaceous fossil record in western North America, at a time when they are barely known from South America and completely unknown everywhere else and when placental mammals are very rare in North America. Under a model of fixed continents, it was easy to presume that North America was the place of origin and that marsupials dispersed to Australia by crossing the North Pacific and spreading southward through Asia. A serious difficulty with this theory, however, is that marsupials do not at present exist in Asia, nor is there any evidence that they ever did. Today, biogeographers consider South America to be a more probable center of origin of marsupials (although the fossils neither affirm nor disprove this theory, since the Cretaceous fossil record of South America is still too poorly known); and they postulate a sweepstakes entry of marsupials into North America in the Upper Cretaceous—the reverse of what was previously believed. Dispersal from South America to Australia would have occurred via Antarctica, elements of Gondwanaland that were still in close proximity at the end of the Cretaceous 63 million years ago. The prediction of marsupials in Antarctica was borne out in 1982 by the discovery of a 40-million-year-old (Eocene) specimen from Seymour Island, just off the Antarctic coast. As satisfying as the discovery was in one respect, it shed no light on the mystery of the absence of placental mammals from Australia. Marsupials and placentals were contemporaries in South America at this time. The most likely theory is that exchanges occurred by a sweepstakes route via Antarctica and intervening oceanic islands during warmer ages. Interestingly, the only extant vascular plant of Antarctica, the grass *Deschampsia antarctica,* is represented by other species of the same genus in both South America and New Zealand.

CONCLUSION

The great impetus that Darwin and especially Wallace gave to the science of biogeography did not continue far into the post-Darwinian years, and interest and research in this field soon lagged. In the 1920s and later, J.C. Willis attempted a biogeographic synthesis based upon the supposed correspondence between ages of species or higher groups and areas of their distribution, but his synthesis failed. His contemporary, W.D. Matthew, made profound studies of the distribution of mammals, both fossil and living. Although Matthew was much more successful than Willis, perhaps his most important contribution was his influence on G.G. Simpson. The work of the latter on the general principles of distribution and especially on the types of bridges between disjunct populations has been the solid basis upon which current theories of biogeography are being constructed. With the advent of plate tectonics, vicariance, and the equilibrium theory, we may expect continued active development of biogeography in the remaining years of this century.

SUMMARY

Biogeography is the study of the distribution of plants and animals. During the past decade has come the general acceptance of plate tectonics, the notion that the continents have drifted about the globe on the backs of crustal plates.

These plates are formed on the midoceanic ridges and are subducted in the deep ocean trenches. The fact that the continents were assembled in a vast supercontinent called Pangaea ("whole earth") as recently as 250 million years ago and have been drifting apart from each other since has profound implications for our understanding of plant and animal distributions.

Islands form natural laboratories for the study of biogeographic phenomena. Islands may show biotic disharmony, in which adaptive types common on the mainland are absent (e.g., amphibians, carnivores) and in which familiar niches are filled by improbable organisms (e.g., the composite trees of Saint Helena). Genetic drift is an important phenomenon on islands. Spectacular endemic radiations may occur in which a single colonizing species gives rise to a swarm of species spanning a wide range of adaptive types; the geospizine finches (Darwin's finches) of the Galápagos Islands and the honeycreepers (Drepanididae) of Hawaii are fine examples. R.H. MacArthur and E.O. Wilson in 1967 presented the *equilibrium theory of island biogeography,* which quantifies, for series of islands, relationships among such variables as surface area, distance from mainland, immigration rate, and extinction rate as determinants of diversity. Their theory emphasizes that although diversity (i.e., numbers of species) on an island at equilibrium may be constant over time, the actual species composition may change continuously. The essence of islands is isolation, irrespective of size. Australia obviously is a large island, but all continents have experienced isolation to varying degrees at times in their histories. Both North and South America have been successfully studied by means of equilibrium theory.

Classical biogeographers (dispersalists) emphasize the abilities of organisms to disperse across geographic barriers, and they have recognized several types of bridges. *Corridors* are broad connections between regions that allow free interchange of biota in both directions, as between Europe and Asia or between eastern North America and western North America today. *Filter bridges* are narrow connections between land masses that may be of relatively short duration and may permit only certain kinds of organisms to pass in each direction. The Pleistocene Bering Straits land bridge between Siberia and Alaska and the Plio-Pleistocene bridge between Central America and South America are excellent examples. *Sweepstakes dispersal* represents improbable events that occur rarely; for instance, overwater colonization of islands. G.G. Simpson advanced this explanation for the endemic mammalian fauna of Madagascar; he invoked single overwater crossings in the Paleocene, Eocene, Oligocene, Miocene, and Pleistocene to account for the fauna today.

Vicariance biogeographers feel that too much emphasis is placed on the dispersal abilities of individual organisms. They emphasize *generalized tracks,* common patterns of distribution of many members of the biota, including crawlers and plants, as well as walkers, runners, and fliers. Vicarists emphasize the fractionation of widespread biotas by the action of geological events (e.g., the erection of mountains, the separation of continents), with subsequent allopatric speciation. M. McKenna named a new dispersal mechanism that follows from plate tectonics: *Noah's ark*. The classical explanation for the origin of the fauna of the Greater Antilles is by sweepstakes dispersal from Yucatán. A vicariant explanation of the same is that a "proto-Antilles" plate split off from Central America and drifted to its present position, carrying its biota with it (Noah's ark).

The "great American interchange," the joining of North America to South America in the Plio-Pleistocene, is the best known of all biogeographic events. A filter bridge to dispersalists and a vicariant event to vicarists, it allowed a reciprocal exchange of faunas between both continents. The exchange followed the predictions of the equilibrium theory at the family level on both continents, and at the genus level in North America. In South America, however, there was an unpredicted increase in the number of genera by a factor of more than two. South America remains an evolutionary laboratory of great interest.

REFERENCES

Cain, S.A. 1944. *Foundations of Plant Geography*. Harper & Brothers, New York; reprinted 1971 by Hafner Publishing Co., New York.
Sound botanical geography. (DuReitz, Fernald)

Carlquist, S. 1965. *Island Life*. Natural History Press, Garden City, N.Y.
A nicely illustrated source of information on life on islands, written at an elementary level.

Cox, C.B., and P.D. Moore. 1980. *Biogeography: An Ecological and Evolutionary Approach*. Halsted Press, Wiley, New York.
Biogeography as seen by a paleontologist and an ecologist.

Darlington, P.J. 1957. *Zoogeography*. Wiley, New York.
This very important book was the first thorough reassessment of its subject since Wallace. (Amadon, Brown)

Darlington, P.J. 1965. *Biogeography of the Southern End of the World*. Harvard University Press, Cambridge, Mass.
A valuable supplement to the author's earlier book.

MacArthur, R.H. 1972. *Geographical Ecology*. Harper & Row, New York.
Tragically, the final summary of this brilliant young man on the geographical determinants of evolution.

MacArthur, R.H., and E.O. Wilson. 1967. *The Theory of Island Biogeography*. Princeton University Press, Princeton, N.J.
Difficult reading, but a work of fundamental importance for the understanding of the dynamics of evolution on islands.

Matthew, W.D. 1939. *Climate and Evolution*. New York Academy of Sciences, New York.
A reprint of a classic.

Nelson, G., and N. Platnick. 1981. *Systematics and Biogeography: Cladistics and Vicariance*. Columbia University Press, New York.
A defense of the two disciplines of the subtitle by a pair of enthusiasts.

Nelson, G., and D.E. Rosen. 1981. *Vicariance Biogeography—A Critique*. Columbia University Press, New York.
The proceedings of an important symposium.
Pielou, E.C. 1979. *Biogeography*. Wiley-Interscience, New York.
An excellent, exciting introduction to biogeography against an ecological background.
Scientific American. 1972. *Continents Adrift*. Freeman, San Francisco.
A highly readable source on this subject.
Simpson, G.G. 1967. *The Geography of Evolution*. Chilton Book, Radnor, Pa.
Very important, readily understandable essays on biogeography.
Simpson, G.G. 1980. *Splendid Isolation: The Curious History of South American Mammals*. Yale University Press, New Haven, Conn.
A report on a lifetime of research on this subject by a master biogeographer and one of the architects of the modern evolutionary synthesis.
Tarling, D.H., and S.K. Runcorn, eds. 1973. *The Implications of Continental Drift to the Earth Sciences*. Academic Press, London and New York.
A major symposium summarizing the current understanding of continental drift, including its zoogeographical implications. See especially the paper by M. McKenna, "Sweepstakes, filters, corridors, Noah's arks, and beached Viking funeral ships in paleogeography."
Wallace, A.R. 1876. *The Geographical Distribution of Animals*. Macmillan, London.
Wallace, A.R. 1911. *Island Life*, 3rd ed. Macmillan, London.
These two books are still sound and deserve careful study by every serious student of biogeography.

Phylogeny: Evolution above the Species Level

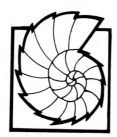

The earliest studies in evolution were largely devoted to the history of life, the derivation of groups, and, most especially, the ancestry of our own species. This complex of problems is still important, and progress toward its solution is published every year.

Parts 2 and 3 dealt for the most part with experimentally verified data and the reasonably secure conclusions drawn from them. In Part 4, the situation is radically different. The late Libbie H. Hyman, one of the most brilliant and learned of zoologists, cautioned that "the exact steps in the evolution of the various grades of invertebrate structure are not and presumably never can be known. Statements about them are inferred from anatomical and embryological evidence and in no case should be regarded as established facts."* She thereby indicated not only the tentative character of phylogenies but also her belief that they must be based primarily on comparative anatomy and embryology. Elsewhere she stated that biogeography and paleontology were also helpful, but she specifically disavowed the value of physiological and behavioral data. In contrast, G.G. Simpson proposed that everything known about animals ought to contribute

* L.H. Hyman, The Invertebrates, vol. I (McGraw-Hill, New York: 1940), p. 255.

to taxonomic and phylogenetic decisions. Hyman's viewpoint describes traditional practice; Simpson's describes the ideal to be sought.

In practice, we look for homologies, that is, characters shared through common inheritance. Shared primitive characters indicate remote common ancestry, and shared derived characters indicate closer relationship. At the level of relationships among phyla, the concept of shared characters is very difficult because by definition each phylum is based on a distinctive body plan that is hard to derive from any other such plan. As a result, conjecture plays a larger role in diagnosing relationships among phyla than is usual in scientific decision making. This is the reason for Hyman's judgment that the phylogenies "in no case should be regarded as established facts." When a well-graded series of fossils connects ancestral and descendant groups, relationships are clear; but such a series is far rarer than might be hoped.

Techniques more modern than comparative anatomy and comparative embryology may also be applicable to problems of phylogeny. For example, pollen analysis has been very useful in attacking the problem of the origin of the flowering plants (Chapter 18), cytochrome *c* and other protein analyses (Chapter 14) have confirmed the broad outlines of phylogeny as worked out by the older methods, and radioactive dating (Chapter 5) has been of great help.

Subject to these cautions, then, we present in Chapters 16 through 21 a provisional history of the world of life, and in Chapter 22 we attempt to derive evolutionary generalizations from the record.

16

Origin of Life

The expression "from amoeba to man" is commonly used to encompass the grand extent of evolution. Yet this expression is misleading, for there is a vast realm of life that is much more primitive in organization than the amoeba. Within the Protozoa, the Mastigophora (flagellates) are now universally recognized as more primitive than the Rhizopodea (the class to which amoeba belongs), and are probably ancestral to it. Indeed, many flagellates have chlorophyll (for example, the well-known *Euglena*) and other typical plant characters and thus form a link between the plant and animal kingdoms.

But the flagellates are highly complex organisms, hardly a starting point on the scale of life. The Cyanobacteria, or blue-green algae, are more primitive. In the latter there is typically no morphological separation of nucleus and cytoplasm; rather, the chromatin is distributed throughout the cell. Nonetheless these organisms are still significantly advanced beyond the bacteria, for the blue-green algae do, by virtue of the catalytic properties of chlorophyll, synthesize sugar from carbon dioxide and water, in the presence of sunlight.

Even the bacteria can hardly be called simple. The chemical analysis of their protoplasm shows a composition not too different from that obtained by the analysis of the protoplasm of higher plants and animals. Their morphology and colonial characteristics are sufficiently distinct to serve as guides to identification even where the smallest bacteria are concerned. But there are pathogenic agents so small that they pass the finest filters and are invisible with the best light microscopes. They multiply within the protoplasm of an appropriate host and cause the production of disease symptoms in the host organism. These are the viruses. Viruses have been crystalized, and the crystals are nucleoproteins—very simple when compared to typical protoplasm but very complex when compared to inorganic or to most organic molecules. Whether the viruses are living is debatable, but they are the simplest things with respect to which any such debate is possible, and so they bring us squarely before the problem of the origin of life.

THEORIES OF THE ORIGIN OF LIFE

Spontaneous Generation

Of the many theories of the origin of life, perhaps the oldest is the theory of spontaneous generation, according to which even complicated forms of life might arise spontaneously from nonliving matter. Aristotle believed that mosquitos and fleas arose from putrefying matter. Tadpoles, worms, and many other small organisms were supposed to arise from mud. Flies were supposed to be formed from putrefying flesh. Before the motor age, every child heard that a horsehair, left in water, would transform into a horsehair worm. Meal worms were often supposed to arise from flour spontaneously. Even such large and complex animals as rats were supposed to have arisen spontaneously from nonliving matter.

F. Redi, an Italian physician of the seventeenth century, attacked the theory of spontaneous generation experimentally, and he left it badly damaged. He exposed meat in containers that were covered over with fine mesh cloth. No maggots appeared on the putrefying meat, but flies laid their eggs on the cloth covers, and maggots developed there. It was evident, then, that the maggots that ordinarily appeared in spoiling meat were not spontaneously produced but were developed from eggs laid in the meat by adult flies. A century later, similar experiments were performed by the Italian priest L. Spallanzani, who showed that if meat were boiled in a sealed container, no organisms developed in it, even if it had been previously infected. This fact was soon applied to the practical problem of food preservation by means of canning.

After the work of Redi and Spallanzani, the theory of spontaneous generation no longer commanded the respect of biologists until the discovery of bacteria. Here were organisms simpler than any previously imagined. Their occurrence was practically universal, and it was difficult to exclude them from any medium suitable for their growth. The possibility that they might be produced spontaneously within every sort of organic medium seemed most plausible, and it had many adherents. The famous experiments of L. Pasteur disproved this theory completely. Pasteur kept boiled broth in a closed container, with air entering by a capillary tube that was bent to form a trap for solid particles. Thus the broth was freely exposed to oxidation; yet no bacteria appeared in it. Hence it was evident that air-borne bacteria ordinarily infected exposed broth (or other suitable media) and that the bacteria themselves arose only from preexisting bacteria. This discovery dealt the death blow to the theory of spontaneous generation of complex organisms.

Cosmozoic Theory

A second theory of the origin of life, the cosmozoic theory, is that the original spores of life reached the earth accidentally from some other part of the universe. This theory is completely unsatisfactory for two reasons. First, because of the intense cold, extreme dryness, and the intense radiation of interstellar space, the probability that even the most resistant living spores could withstand exposure to interstellar space is vanishingly small. Second, the theory does not explain the origin of life at all but merely changes the scene of origin from the earth to some remote and undefined part of the universe. In spite of these problems, the theory has been resurrected recently by the astronomers F. Hoyle and C. Wickramasinghe.

Viruses and the Origin of Life

The discovery of viruses (Figure 16–1) has placed the problem of the origin of life in a new light. The higher forms of life cannot be descended from viruses of modern type, for they are all parasitic, and parasites must always be descended from free-living ancestors. But the viruses show a unique combination of characters of living and nonliving systems. Viruses, lacking ribosomes, cannot metabolize, but they commandeer the metabolic machinery of the host and cause it to synthesize viral rather than host products. Chemically, a virus consists of a core of nucleic acid surrounded by a protein sheath. Partly because of this nucleoprotein structure, researchers suggested, in the 1940s, that viruses might be "escaped" genes. Like genes, viruses are ordinarily reproduced without change, but they can mutate as genes do, that is, they can undergo an inheritable change that does not interfere with their capacity for self-reproduction. Such mutations are detectable by a change in disease symptoms produced by the virus or by a change in the degree of its toxicity. However, investigation of viral genetics has shown that a virus contains a whole system of genes, so that, if the "escaped" gene theory be held, it must be modified to include related systems of genes.

In contrast, unlike undoubted organisms, viruses do not respire. The most striking property that viruses share with nonliving systems is the fact that

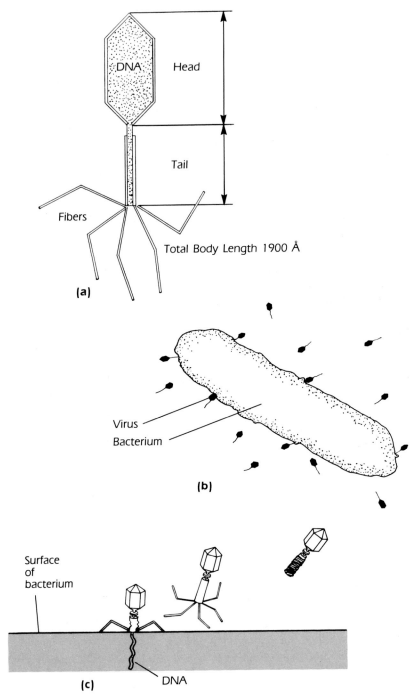

Figure 16–1. *Structure of a virus, based on T4. (a) A virus showing the DNA core surrounded by a protein sheath, the tail, and the attachment fibers. (b) Many virus particles attacking a bacterial cell. (c) Attachment of the virus to the bacterial cell and penetration of viral DNA into the cell.*

they can be crystallized and stored indefinitely without loss of infective powers. The chemist W.M. Stanley first demonstrated this fact in 1935 by succeeding in crystallizing the virus that causes tobacco mosaic disease. The crystals turned out to be a nucleoprotein. Chemical purity is indicated not only by crystallization but by a sharp sedimentation boundary when a virus suspension is ultracentrifuged. Finally, viruses can be broken down to a protein and a nucleic acid, both of which are inactive; then they can be recombined to form infective virus again. Therefore, the viruses appear to be homogeneous, or nearly so, in contrast to all undoubted organisms.

Thus viruses are on the borderline between the living and the nonliving, even though their parasitism cannot be primitive. The existence of these bodies, which are intermediate between the living and the nonliving and which have fairly simple chemical properties, suggests the possibility that something like a free-living virus may have been produced by chemical evolution under the influence of the unique conditions that prevailed when the earth was a young planet just cooling down toward a temperature range that could support life. Such a free-living, self-reproducing unit might be regarded as a simple group of genes. Mutation could then lead to formation of larger gene aggregates, with increasing differentiation within each aggregate. Accumulation of metabolites and ribosomes about such an elementary chromosome would approximate the structure of simple bacteria. With increasing amounts of protoplasm, without separation of cytoplasm and nucleus, larger bacteria would form, and with the addition of chlorophyll, the blue-green bacteria would emerge. Finally, separation into nucleus and cytoplasm by a membrane could then open up the potentialities of the cells of higher plants and animals.

A.I. Oparin (Figure 16–2) has worked out the possible details of such a chemical evolution of the most elementary forms of life particularly well, and we follow his account for the most part. Note that chemical evolution is actually a special case of spontaneous generation, although it is a plausible one because it does not involve direct origin of complicated organisms from nonliving matter.

OPARIN AND CHEMICAL EVOLUTION

The earth may have originated as a fragment from the sun, or the sun and the planets may all have originated as so many gravitational centers in a rotating mass of interstellar dust and gas. In either case, the probable chemical composition of the earth in the earliest geological periods can be surmised from spectrographic analysis of the sun and of stars in early stages of their physical evolution. All of the elements that enter into the composition of protoplasm were probably present as inorganic compounds. Free nitrogen, hydrogen, and oxygen, gases that now form so large a portion of the earth's atmosphere, were likely present at the very beginning, because they are present

Figure 16–2. A.I. Oparin (1894–).
Drawn by Peter Fortey.

as free elements in the sun. But they were probably soon lost to outer space, for it is unlikely that the gravitational pull of the earth is great enough to hold such light elements at the high temperatures that prevailed during the earliest ages of the earth's history. These elements were probably left only in compounds. Thus, there was a reducing atmosphere, in contrast to the familiar oxidizing atmosphere of today, and this difference is critical for some of the reactions that preceded the origin of life. A large quantity of the hydrogen and oxygen was probably united as water, but for long ages water was certainly present only as superheated steam. The hot vapors would rise toward the cold outer layers of the atmosphere, condense and fall as rain, only to be converted to steam again before striking the earth. Gradually the earth cooled sufficiently to permit the rainfall to strike the earth, then to begin to form pools and larger bodies of water. Optimum conditions for solubility and reaction existed, and the entire earth was a great crucible for random compound formation and re-formation.

Origin of Organic Compounds

In such a situation, in which the most important elements of organic compounds—carbon, nitrogen, oxygen, and hydrogen—are reacting at random with each other and with many other elements, forming countless compounds that in turn react in more combinations, it is highly probable that, sooner or later, organic compounds or their precursors would appear. Methane (CH_4), the simplest organic compound, is present in the atmospheres of some of the cooler stars. More complex hydrocarbons (compounds of carbon and hydrogen) have been found in meteorites, and so it is certain that they can originate without the intervention of living organisms. Compounds of carbon with metals or metallic carbides, give rise to hydrocarbons when treated with steam. So it is altogether probable that such compounds were formed in abundance while the cooling earth was still much too hot to permit the existence of life.

In a similar way, we would expect the formation of ammonia by the reaction of steam and metallic nitrides. Thus, $Na_3N + 3 H_2O \rightarrow 3 NaOH + NH_3$. Cyanide can be formed in many ways. Perhaps the simplest, at high temperatures, is the reaction of methane and ammonia: $CH_4 + NH_3 \rightarrow HCN + 3 H_2$. About as simple is the reaction of carbon monoxide and ammonia: $CO + NH_3 \rightarrow 2 HCN + H_2O$. Further, C.N. Matthews has shown that hydrogen cyanide is formed in abundance when mixtures of methane and ammonia are subjected to electrical discharges. Polymerization of the HCN then yields dimers, trimers, and tetramers, which in turn form larger complexes, as follows:

$$HCH \xrightarrow{HCN} (HCN)_2 \xrightarrow{HCN} \left[\begin{array}{c} H \\ | \\ -C-C{=}N \\ \| \\ NH \end{array} \right]_n$$

Other known pathways for the abiotic origin of organic compounds include the following: (1) Ultraviolet irradiation of formic acid, which yields large organic molecules, including amino acids; (2) Action of cosmic rays upon carbon dioxide and water vapor, which yields organic acids, for example:

$$CO_2 + H_2O \rightarrow \underset{\text{formaldehyde}}{H_2CO} + O_2; \text{ then } H_2CO \xrightarrow{O} \underset{\text{formic acid}}{HCOOH}$$

(3) Action of lightning upon an atmosphere of methane, hydrogen, ammonia, and water vapor, which yields a mixture of organic compounds, including amino acids, perhaps by such a reaction as:

$$2 \text{ H}_2\text{O} + 3 \text{ CH}_4 + \text{NH}_3 \rightarrow \text{CH}_3\text{CH(NH}_2)\text{COOH} + 6 \text{ H}_2$$

alanine

Researchers have duplicated all of these reactions in the laboratory. Simulating the last method by discharging electrical sparks into an artifical atmosphere in a sealed flask, S.L. Miller obtained a mixture of many amino acids and other organic compounds, some in fairly high yields.

Further reactions of the compounds just discussed have great potentialities. The hydrocarbons are not particularly reactive; however, one hydrogen atom in a hydrocarbon molecule could be readily replaced by chlorine or bromine. The new compound would be highly reactive. It might, for example, be hydrolyzed to form an alcohol and an inorganic acid. The alcohol could then be oxidized to form the corresponding aldehyde or ketone. These could in turn be further oxidized to form organic acids. Thus, if we start with ethyl chloride, we have:

$$\text{CH}_3\text{CH}_2\text{Cl} + \text{H}_2\text{O} \underset{\text{heat}}{\rightarrow} \text{CH}_3\text{CH}_2\text{OH} + \text{HCl}$$

In more general terms, if R represents any organic radical, then:

$$\text{RCH}_2\text{Cl} + \text{H}_2\text{O} \rightarrow \text{RCH}_2\text{OH} + \text{HCl}$$

heat   alcohol

Alcohols are readily oxidizable to yield aldehydes and organic acids, as:

$$\text{RCH}_2\text{OH} \overset{O}{\rightarrow} \text{RCHO} \overset{O}{\rightarrow} \text{RCOOH}$$

aldehyde   organic
acid

The simplest of the aldehydes, formaldehyde, will polymerize to form glucose ($\text{C}_6\text{H}_{12}\text{O}_6$) in the presence of sunlight (a pathway different from biosynthesis of glucose). Although the reaction proceeds extremely slowly, it is conceivable, given time on a geological scale, that large quantities of sugar might accumulate. If a double alcohol, or glycol, were formed, then one alcohol group could be oxidized to form an acid, and the other alcohol group could react with ammonia (NH_3) to give water and an amine. The result would be an amino acid, one of the building blocks of the proteins. The simplest possible amino acid is glycine, $\text{CH}_2 \cdot \text{NH}_2 \cdot \text{COOH}$.

These compounds can enter into reactions that lead to compounds intimately associated with protoplasm. Sugars can be polymerized to form starches, glycogen, and cellulose. The essential precursors of fats are long-chain hydrocarbons, in which an end member has been oxidized to form an acid, and glycerine. Glycerine is simply a three-carbon chain in which one hydrogen on each carbon has been replaced by an hydroxyl group ($\text{CH}_2\text{OH} \cdot \text{CHOH} \cdot \text{CH}_2\text{OH}$). Now if each hydroxyl group reacts with a long-chain organic acid, the resulting compound is a fat molecule. But the most significant possibility lies in the amino acids, which can react with one another to form aggregates of great molecular weight—the proteins. The acid group of one amino acid reacts with the amino group of another in what amounts to a salt-forming reaction, called the *peptide linkage* (see Figure 7–8). As it always leaves an acid radical exposed on one of the reacting amino acids and an amino group on the other, the reaction is subject to indefinite repetition, and thus it allows the great molecular weights of the proteins to build up.

This reaction may underlie Miller's findings. Nonetheless, peptide linkages form only with great difficulty in the absence of enzymes. Matthews has reported experiments on a pathway of abiotic synthesis of proteins that evades this difficulty. The large polymers of hydrogen cyanide react with water spontaneously to form polypeptides directly, without the preliminary formation of amino acids:

$$
\left[\begin{array}{c} H \\ | \\ -C-C=N- \\ \| \\ NH \end{array} \right]_n \xrightarrow{HCN} \left[\begin{array}{cc} H \\ | \\ -C-C-NH- \\ \| \quad | \\ NH \quad R \end{array} \right]_n \xrightarrow{H_2O} \left[\begin{array}{cc} H \quad H \\ | \quad | \\ -C-C-N- \\ \| \quad | \\ O \quad R \end{array} \right]_n \xrightarrow[-CO_2]{H_2O}
$$

$$
\begin{array}{ccccccc} & R & O & & R & O & & R & O \\ & | & \| & & | & \| & & | & \| \\ ---N- & CH- & C- & N- & CH- & C- & N- & CH- & C--- \\ & | & & & | & & & | \\ & H & & & H & & & H \end{array}
$$

amino acid polymer = polypeptide

In the Miller experiment, HCN is formed very early. Maximal concentration of HCN is reached within twenty-four hours, then it gradually diminishes. Matthews believes that this is due to polymerization of HCN to form polypeptides by the pathway just described, with free amino acids resulting from the hydrolysis of polypeptides.

Colloids, Coacervates, and Individuality

By the time the earth had cooled sufficiently to permit the formation of permanent bodies of water, a large amount of organic material of great variety was probably present, and the variety was continually increasing through reaction between whatever substances could react and happened to be brought together. Many of these large organic compounds would tend to form colloidal solutions in water. Wherever such colloidal particles included electrically active groups, as all proteins do in abundance, water molecules would tend to become oriented around the surface of such a particle. If colloidal particles of opposite electrical charges are mixed, then they mutually precipitate to form droplets of a complex mixture called a coacervate. Such coacervates would absorb water on their surfaces to form a sort of membrane, which would thus establish the beginnings of individuality.

Researchers have investigated synthetic coacervates of many types (Figure 16–3). A mixture of gelatin and gum arabic is a simple example. Tiny droplets of colloid, each separated from the medium by a sphere of water molecules that simulates a membrane, are dispersed throughout the solution. The dispersed phase need not be limited to two components. Mayonnaise, a beaten mixture of egg yolk, butter or olive oil, vinegar, water, and a variety of condiments, is a complex coacervate mixture of everyday experience. The hydrated surfaces of the droplets function as membranes that concentrate substances from the medium. They may also be selectively permeable. Biological reactions are commonly facilitated by adsorption of reactants on membranes. When glucose is added to the gelatin–gum arabic system, it is concentrated in the droplets and polymerized to form starch. No wonder Oparin proposes coacervate formation as an essential step in the origin of life!

Autocatalytic Systems, Genes, and Viruses

These large, complex, colloidal aggregates, the coacervates, continue to undergo random chemical reactions in the course of which still larger aggregates may be formed, or the existing ones may be broken down. It is most probable that some of the systems thus formed would be enzymatic in character, that is, they would tend to increase, or catalyze, the rates of specific reactions. Many enzymes catalyze the reactions of substances unrelated to the enzyme itself. Thus lipase will break down fats to glycerine and long-chain acids, or it will catalyze the synthesis of fats under appropriate conditions. Other

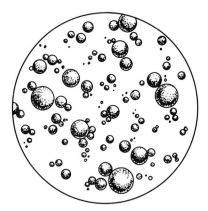

Figure 16–3. Structure of a coacervate.

catalysts, given a suitable substrate, tend to cause the production of more of the catalyst itself. The replication of DNA is an example par excellence. Such enzymes are said to be autocatalytic. If an autocatalytic substance were formed in the primordial seas, it would increase in quantity at the expense of organic compounds that lack this property. Thus the characteristic of self-reproduction would arise. Compounds possessing this property might be regarded as free genes; indeed, DNA is the most important autocatalytic compound. We need only add the characteristic of mutability, that is, the capacity for undergoing changes that are reproduced without interfering with the autocatalytic properties of the molecule, and we have the most fundamental characterisics of the hereditary units, the genes. Such a structure would be comparable to a free-living virus; and, like the viruses of today, these primitive beginnings of life were probably nucleoproteins. These nucleoproteins would utilize the complex organic compounds of their environment for the autocatalytic synthesis of more nucleoprotein, that is, for reproduction. M. Calvin has shown that even in the preorganismic stages, a sort of natural selection, based upon thermodynamic principles, favors the events described.

Origin of Bacteria

As larger self-reproducing units formed, mutation might cause parts of each aggregate to differentiate. Or the same result might be obtained by the coalescence of originally different "genes." At this stage, the structure would resemble some of the smallest known bacteria and might properly be called an *organism*. The organisms dealt with up to this point would have to be completely heterotrophic, that is, dependent upon complex food materials present in their environment and "feeding" only by absorption. The fact that viruses can function and reproduce only within the protoplasm of higher organisms indicates that they are also heterotrophic. They can utilize proteins, starches, fats, and vitamins already present, but they cannot synthesize them from amino acids, sugars, and other simpler organic precursors. In contrast, all of the undoubted organisms are able to synthesize at least some of their required foods from simpler compounds. Green plants synthesize all their foods from the elements. Animals cannot do so, but they can synthesize proteins from amino acids, complex carbohydrates from sugars, and some of the vitamins from simpler precursors. These syntheses are possible because the organisms possess enzymes that are specific for the necessary reactions. Such reactions are frequently complex, requiring a long series of steps with a different enzyme controlling each step. For instance, the synthesis of arginine, an amino acid, by the mold *Neurospora* has been shown to require at least seven different genically controlled enzymes. Each of these is useless without all of the others. That such "useless" genes should be preserved by natural selection is incomprehensible, but the probability that all of them should appear simultaneously and so become useful is equally so.

N.H. Horowitz devised a clever solution to this dilemma: the enzymes might have been originally acquired in a sequence opposite to that in which they are used by existing organisms. Suppose a primitive organism requires a substance *A,* which is abundantly present in its environment, as are also substances *B* and *C,* from which *A* could be synthesized in the presence of an appropriate enzyme. As long as *A* remains abundantly present in the environment, presence or absence of the enzyme can have no bearing on survival. Eventually the food requirements of the growing population of primitive organisms would outstrip the stockpile of organic compounds that was built up during the abiotic ages. Thus when *A* becomes rare in the environment, those organisms that have the enzyme for the production of *A* from *B* and *C* will have a selective advantage and will replace the original type. Such *preadaptation* is a common phenomenon in evolution. The enzyme has now become a permanent part of the organism's biological equipment. Likewise, *B* might be synthesizable from *D* and *E* in the presence of the necessary enzyme. Then when *B* becomes scarce, the possession of this enzyme, acquired by mutation, will also have selective advantage. There is no theoretical reason why such a process should not continue until an organism has acquired the ability to synthesize all of its requirements from the elements. This has happened in the case of the green plants. Such an organism is said to be autotrophic.

It seems probable, then, that the primitive free-living viruses and bacteria gradually used up the available supply of proteins and other complex compounds in the environment and that they simultaneously, step by step, developed the systems of enzymes necessary for the biosynthesis of the same compounds. Or it may be that only the primitive bacteria were sufficiently complex to permit this development, the viruses remaining completely dependent upon preformed compounds. If so, this would account for the viruses' obligate parasitism, for only within the protoplasm of undoubted organisms could an environment be found to contain all necessary food substances in the completely elaborated form. The evolution of the bacteria made parasitism possible; the impoverishment of the earth's supply of complex organic compounds made parasitism necessary for organisms that could not synthesize their food requirements, at least in part.

Respiratory Systems and Photosynthesis

The most primitive organisms, living in an environment that included an abundance of the most complex substances that the organisms might require, must have had a metabolism that was mainly catabolic—that is, it consisted of the breaking down of complex compounds to release the energy stored in them. As the environmental supply of high-energy compounds was reduced, a new type of metabolism became advantageous—*anabolic metabolism*—whereby complex compounds are built up from simpler ones, thus storing energy.

Anabolic metabolism requires a source of energy, that is, a system of respiration. Selection favored mutations by which several types of respiratory reactions and correlated nutritive systems have been exploited. These may be classified as heterotrophic and autotrophic systems. The heterotrophic bacteria are generally parasitic, deriving their energy by oxidation of the carbohydrates or other organic compounds of their hosts. Or they may be *saprophytic,* living by absorption of dissolved organic matter from their environment, much like the primitive ones in the scheme of Oparin. On the other hand, the autotrophic bacteria derive energy from chemical reactions involving simple inorganic compounds. They are thus independent of external sources of high-energy compounds.

There are three principal groups of such *chemotrophic* bacteria: the sulfur bacteria, the nitrifying bacteria, and the iron bacteria. Within each group, many different species and several different respiratory reactions occur. Here are a few examples, all in molar quantities:

$$2\ H_2S + O_2 \rightarrow 2\ H_2O + 2\ S + 81{,}600 \text{ calories}$$

$$S + 2\ H_2O + 3\ O_2 \rightarrow 2\ H_2SO_4 + 214{,}000 \text{ calories}$$

$$Na_2S_2O_3 + 2\ O_2 \rightarrow Na_2SO_4 + S + 75{,}000 \text{ calories}$$

$$2\ NO_2 + O_2 \rightarrow 2\ NO_3 + 46{,}000 \text{ calories}$$

$$2\ NH_3 + 3\ O_2 \rightarrow 2\ NO_2 + 2\ H_2O + 156{,}000 \text{ calories}$$

Iron bacteria oxidize ferrous compounds to ferric compounds with a much lower energy yield than is indicated for the above reactions. Even the best of these is vastly inferior to photosynthesis, the autotrophic mechanism of all green plants and perhaps the real basis for the evolution of the higher plants and animals, for it has made it possible for the world of life to tap the great reservoir of radiant energy from the sun. The photosynthetic reaction may be summarized as:

$$6\ CO_2 + 6\ H_2O + 677{,}000 \text{ calories} \rightarrow C_6H_{12}O_6 + 6\ O_2 \text{ per mole}$$

This large energy reserve is subsequently released by the sugar metabolism of the organism. Photosynthesis, then, for which the chlorophyll of green plants is the catalyst, not only vastly increases the energy potentially available to organisms, but it also releases oxygen from its compounds, thus making possible the oxygen respiratory systems of animals and of some bacteria. Photosynthesis begins among the bacteria, and some of the sulfur bacteria contain a green pigment that is capable of absorbing sunlight, thus making some contribution to metabolism. Blue-green bacteria, however, have true chlorophyll, and they synthesize sugar with the release of oxygen.

Oparin was of the opinion that this chemical evolution of life could have occurred only once because it would require a sterile environment—that is, it could not occur in a world already inhabited by organisms ready to seize upon any beginnings of organic compounds and utilize them as food. But, as C.R. Plunkett pointed out ten years earlier, this is an assumption for the proof of which no data are available, and it is entirely possible that life is in process of origin on earth at the present time. Plunkett acknowledged, however, that there is also no evidence for this latter possibility.

Prions

Kuru is a degenerative disease of the human nervous system (fortunately of very limited occurrence), and scrapie is a similar disease of sheep and goats. They develop only long after infection (months or even years), and hence they have been called *slow* diseases. Efforts to identify the causative organisms were fruitless until quite recently, when S.B. Prusiner found that protein particles in the infected nervous tissue seemed to meet the requirements. These particles are smaller than viruses, and all efforts to implicate nucleic acids in their reproduction have failed. It may be that they are autocatalytic proteins. Prusiner calls them *prions*. Their relationship to the problem of the origin of life is far from clear. W.S. Fox has long argued that life may have originated with autocatalytic protein particles, and that the association of those particles with nucleic acids came only much later. If the prions do indeed replicate without the intervention of nucleic acids, then they may support Fox's argument.

SUMMARY

The origin of life was long regarded as a purely speculative problem, beyond the possibility of scientific investigation. The *theory of spontaneous generation* had been accepted, but it was demolished by the experiments of L. Spallanzani, F. Redi, and L. Pasteur. The *cosmozoic theory,* that the germs of life came from outer space, has recently been revived, but few scientists are interested in it. Viruses are a focal point of interest in discussions of the origin of life, for they share properties with both living and nonliving systems. Like the latter, they do not respire, they are crystallizable, and their solutions give a sharp boundary when ultracentrifuged. Like the former, they are constructed of nucleic acid and proteins, and mutations occur in the nucleic acid (which may be either DNA or RNA, according to the species). Thus, viruses appear to be on the borderline between the living and the nonliving. Because systems of genes have been identified in viruses, they may now be treated as true organisms, if extraordinarily simple ones.

When A.I. Oparin published his study of the origin of life in 1938, he convinced biologists that this subject was indeed amenable to scientific study. He showed that, under the conditions of the primitive earth, a vast array of organic substances, including amino acids and polypeptides, could have formed spontaneously. The experiments of S.L. Miller demonstrated that the Oparin reactions do in fact occur in the laboratory. A plausible pathway from complex organic systems to viruslike particles and then to the simplest bacterial cells was introduced, with enzyme systems being evolved by the pathway described by N.H. Horowitz. S.B. Prusiner then identified *prions,* infective organisms that appear to be extremely small protein particles, with neither DNA nor RNA. The relationship of prions to the question of the origin of life is unknown and is a serious problem for future research.

Thus, where only a few decades ago the problem of the origin of life appeared to be beyond the possibility of scientific investigation, a plausible theory of the biochemical evolution of life now exists, and a considerable amount of experimental evidence lends credence to that theory.

REFERENCES

Note: Names in parentheses indicate important scientists whose works are covered in the references more extensively than in this chapter.

Blum, H.F. 1969. *Time's Arrow and Evolution,* rev. ed. Princeton University Press. Princeton, N.J.
> *Thermodynamic considerations relevant to this chapter. A new edition is needed. (Horowitz, van Niel.)*

Calvin, M. 1969. *Chemical Evolution. Evolution towards the Origin of Living Systems on the Earth and Elsewhere.* Oxford University Press, New York.
> *An excellent and fairly simple account of the subject.*

Chaisson, E. 1981. *Cosmic Dawn: The Origins of Matter and Life.* Atlantic Monthly Press/Little, Brown, Boston.
> *The viewpoint of a highly literate astrophysicist.*

Dickerson, R.E. 1978. Chemical evolution and the origin of life. *Scientific American* 239(3):70–86.
> *A readily understandable presentation of the subject matter of the present chapter and more.*

Farley, J. 1977. *The Spontaneous Generation Controversy from Descartes to Oparin.* Johns Hopkins University Press, Baltimore, Md.
> *A comprehensive and interesting report on much of the material in this chapter.*

Fraenkel-Conrat, H., and R.C. Williams, 1955. Reconstitution of active tobacco mosaic virus from its inactive protein and nucleic acid components, *Proceedings of the National Academy of Science* (Washington) 41:690–698.
> *The technical report on a most important experiment.*

Hoyle, F., and N.C. Wickramasinghe. 1979. *Lifecloud: The Origin of Life in the Universe.* Dent, London; and Harper & Row, New York.
> *In this book and in the following one, the authors, a distinguished astronomer and a mathematician, respectively, revive the cosmozoic theory of the origin of life.*

Hoyle, F., and N.C. Wickramasinghe. 1981. *Evolution from Space.* Dent, London.

Kerr, R.A. 1980. Origin of life: New ingredients suggested. *Science* 210:42–43.
> *This brief* Science *research news note reports work that suggests that the primitive atmosphere was only mildly reducing; it contained much CO_2, but little if any, NH_3.*

Matthews, C.N. 1974. The origin of proteins: Heteropolypeptides from hydrogen cyanide and water. *Origins of Life,* 5.
This interesting and controversial paper presents the theory that polypeptides were formed directly, not by polymerization of amino acids.

Miller, S.L., and H.C. Urey. 1959. Organic compound synthesis on the primitive earth, *Science* 130:245–251.
A classic report.

Nisbet, E.G. 1980. Archean stromatolites and the search for the earliest life. *Nature* 284:395–396.
The earliest possible date is 3.7×10^9 years before the present; the earliest probable date is 3.0×10^9 before the present; and the earliest definite date is 2.7×10^9 years before the present.

Oparin, A.I. 1968. *Genesis and Evolutionary Development of Life.* Academic Press, New York.
The most recent revision of the theory of the origin of life by the man who started the modern studies of the problem.

17

Separation of the Kingdoms

As soon as organisms arose (Chapter 16), they began to diversify. This process, still going on, has so far produced at least three phyla of procaryotes, two divisions* of fungi, six divisions of eucaryote algae (a term meaning seaweeds, based upon the larger members of this diverse array), two divisions of land plants, the Protozoa, and some thirty phyla of metazoans. This chapter will discuss how the broadest groups—the kingdoms—were differentiated. Most of us are probably already familiar with many of the groups that are critical for this discussion, but for orientation and review, Table 17–1 summarizes the organization of the world of life. We begin with the most fundamental problem, the separation of procaryotes from eucaryotes.

---

* The most inclusive taxonomic units within a kingdom have traditionally been called *divisions* by botanists, *phyla* (singular, *phylum*) by zoologists. Because *phylum* means a line of descent, some botanists used the term until 1972, when a new *International Code of Botanical Nomenclature* was adopted. The Code requires that descriptions of new taxa be written in Latin. As *phylum* is also the Latin word for *stem*, it would obviously be a source of confusion if it were also used as a synonym for *division*. Hence, the Code requires the use of *division* rather than *phylum*.

Table 17-1. Broad classification of organisms

Category	Comments
PROCARYOTES*	Unicellular organisms lacking nuclear-cytoplasmic differentiation; mitosis and meiosis unknown; no cytoplasmic organelles other than ribosomes.
Kingdom I? Archaebacteria	Bacteria living in atypical habitats and with unusual biochemical traits; status uncertain.
Kingdom II. Monera Phylum 1. Virulenta	Viruses; lacking ribosomes, and so dependent upon the metabolism of other organisms.
Phylum 2. Schizophyta	Bacteria; ribosomes present.
Phylum 3. Cyanobacteria	Blue-green bacteria, sometimes called blue-green algae; cytoplasm more abundant than in bacteria; true chlorophyll present.
EUCARYOTES*	Nuclear membrane separates nucleus from cytoplasm; cytoplasmic organelles present; mitosis, and often meiosis, occurs.
Kingdom III. Animalia Phylum 1. Protozoa	Predominantly unicellular or acellular animals.
30-odd metazoan phyla	Exact number depends upon the particular taxonomy followed.
Kingdom IV. Plantae Subkingdom Thallophyta	Plants with minimal specialization of cells; not divided into roots, shoots, and leaves.
Division 1. Eumycophyta	True molds and yeasts; all saprophytic or parasitic.
Division 2. Myxomycophyta	Slime molds.
Division 3. Chrysophyta	Yellow-green algae, golden-brown algae, and diatoms.
Division 4. Pyrrophyta	Cryptomonads and dinoflagellates.
Division 5. Euglenophyta	*Euglena* and its allies. Divisions 3, 4, and 5 are all unicellular, but may form moderate-sized colonies.
Division 6. Phaeophyta	Brown algae.
Division 7. Rhodophyta	Red algae. Divisions 6 and 7 are predominantly multicellular; may form large seaweeds.

Table 17–1. (Continued)

Category	Comments
Division 8. Chlorophyta	Extremely varied; may be unicellular, colonial, or highly organized multicellular organisms.
Subkingdom Embryophyta	Plants in which embryo is retained within maternal tissues; predominantly land plants.
Division 9. Bryophyta	Mosses and their allies.
Division 10. Tracheophyta	Vascular plants, including all land plants of common experience.

* The division of the world of life into procaryotes and eucaryotes is fundamental, and it could be recognized in formal taxonomy by treating each as an *empire* including one or more kingdoms. However, the Linnaean taxon of *empire* was dropped long ago, and taxonomists are reluctant to revive it; hence these great divisions are recognized only implicitly.

PROCARYOTE-EUCARYOTE DISCONTINUITY

In Chapter 16 we traced evolution to simple bacterial cells (Phylum Schizophyta). Such cells generally consisted of a double helix of DNA in a small amount of protoplasm, in which there were ribosomes to synthesize proteins. Further evolution certainly involved an increase in the quantity of protoplasm and probably an increase in the quantity and variety of genes.

Bacterial Cytology

Study of the mechanics of cell division in bacteria is difficult because of their small size; and it was long believed that simple fission, without equal division of the genes, was the rule. If this were true, it would mean that bacteria have a genetic system with rather few equivalent genes scattered throughout the cell, so that division need not result in genetically unlike daughter cells. Thus, occasional mutation and the action of natural selection on the mutants would be the only means of evolution. Directly opposing this view, a series of researches during the 1940s and the early 1950s sought to demonstrate typical mitosis in bacteria obscured simply by their minute size. After some encouraging preliminary results, the effort failed.

It has since become clear that both viewpoints are wrong. Bacteria are essentially haploid, and the genes are parts of a "chromosome" that is a double helix of DNA in the form of a circle, with little or no associated protein. This chromosome is free in a small mass of protoplasm that contains ribosomes but no other organelles. The genes are arranged in serial order on the bacterial chromosome; for some bacteria, such as the colon bacillus *Escherichia coli,* they have been mapped in great detail. Before division occurs, this simple chromosome is replicated, and the daughter chromosomes then separate. This nuclear replication may not correlate closely with division of the cell body. Although sexual reproduction does not occur, J. Lederberg and others have discovered various "parasexual" processes, by which recombinations of genes between different mutant strains of a species can occur.

Because both "nuclear" and "cytoplasmic" aspects of these cells contrast so strongly with the cells of higher organisms, they are called *procaryotes,* as opposed to the *eucaryotes,* which have a nucleus separated from the cytoplasm by a nuclear membrane, chromosomes in which the DNA is complexed with large amounts of protein, mitotic division and often meiosis and sexual reproduction as well, and a complex array of cytoplasmic organelles.

Phylum Cyanobacteria

The Cyanobacteria, often called blue-green algae, are also at the procaryotic level of organization. They are always unicellular, but cells may form colonies of moderate size. No differentiation, however, occurs within such colonies. Although the protoplasm is typically more abundant than that of other bacteria, there is still no separation of nucleus and cytoplasm and, of course, no nuclear membrane. No evidence of sexual processes has ever been observed in blue-green bacteria, reproduction apparently occurring exclusively by simple fission. In eucaryote plants, the chlorophyll is confined to cell organelles, the plastids, or chloroplasts, but in the blue-green bacteria it is spread throughout the outer part of the cell. Their bluish color is caused by a blue pigment, phycocyanin, and a red pigment similar to that of the red algae may also be present. Unlike most of the eucaryote algae, they lack flagella and are generally nonmotile. Blue-green bacteria are among the oldest fossils, found in rocks nearly three billion years old, and their descendants of today are regarded as the little-changed survivors from that remote past. In fact, blue-green bacteria living today in nitrogen-rich soil at the foot of Harlech Castle, Wales, are morphologically indistinguishable from *Kakebekia,* a genus found in northern Ontario in rocks about two billion years old.

Because the blue-green bacteria contain chlorophyll, they release oxygen as a by-product of photosynthesis. When they first arose, a reducing atmosphere still enveloped the earth. Their photosynthetic activity slowly added oxygen to the atmosphere, but only after long ages did the oxygen concentration reach a level at which it could support organisms with aerobic respiratory systems.

Kingdom Monera

Thus, all the other bacteria and the blue-green bacteria share the procaryotic characters: the chromosome is a simple strand of DNA that is not isolated by a nuclear membrane; neither mitosis, nor meiosis, nor sexual reproduction occurs; and the protoplasm does not contain organelles such as mitochondria, centrioles, or chloroplasts. In all of these respects, they contrast with the more complex organisms, the eucaryotes. (Not all eucaryotes reproduce sexually, but sexual reproduction does occur in all major groups of eucaryotes.) Biochemical studies have reenforced rather than weakened this contrast, and it now appears that the discontinuity between the procaryotes and the eucaryotes is the most decisive discontinuity in the entire world of life. We must, then, recognize the procaryotes as a distinct kingdom of life—the Kingdom Monera*—for they contrast with all other organisms far more sharply than animals contrast with plants.

Viruses as Monerans

In Chapter 16 we left unresolved the question of whether or not viruses are living organisms. Studies on the genetics of viruses, however, have revealed extensive *systems* of genes, a previously unsuspected degree of complexity. The most important difference from small bacteria lies in the lack of ribosomes, a deficiency for which they compensate by commandeering those of the host organism. For these reasons, many biologists now believe that viruses are best considered as living organisms and should be included in the taxonomic system. If so, then they belong in the Kingdom Monera, for they share the procaryotic characters, except ribosomes. Thus, the Kingdom Monera includes viruses, bacteria, and blue-green bacteria, all very simple organisms that arose during the earliest ages of life.

Kingdom Archaebacteria?

Until very recently, all procaryotes were assigned to one kingdom, the Monera. A few groups of bacteria, however, live in unusual habitats and have unusual characters. These are the methanogens (producing methane), the extreme halophiles (living in strong brine), and the thermoacidophiles (living in hot springs). Their biochemical characters include transfer and ribosomal RNAs that differ from those of all other organisms; cell walls of protein rather than the peptidoglycan cell walls of other bacteria; and ether linked lipids. C.R. Woese and his collaborators at the University of Illinois have argued that these differences separate them so widely from all other bacteria that they must comprise a distinct kingdom, the Archaebacteria, the relationship of which to the other kingdoms is not clear. Woese envisions a tripartite organization of the world of life, with Archaebacteria, Monera, and Eucaryotes as the three

---

* The name *Monera* was proposed in 1866 by E. Haeckel for hypothetical organisms lacking a nucleus. His type specimen was an amoeboid organism that later proved to be a fragment of cytoplasm from a large amoeba. Because the type specimen belongs to a different kingdom, the name should be invalid, and so G. Enderlein in 1926 proposed the name *Mychota* for this assemblage. However, there are no rules of nomenclature at this level, and *Monera* is currently in vogue whereas *Mychota* is not.

primary kingdoms. (Are plants and animals, then, secondary kingdoms or subkingdoms?). The concept of the Archaebacteria as a separate kingdom has been well received, although not without dissent. How they will be integrated into concepts of broad taxonomy remains to be seen.

While we have emphasized the gulf that separates the Monera from the higher kingdoms, nonetheless it must have been crossed, for the eucaryote kingdoms could only have arisen from Moneran ancestors. We now turn to the question of how the eucaryotes originated.

BRIDGING THE GAP: THE ORIGIN OF THE EUCARYOTES

Before the gulf separating the procaryotes and the eucaryotes had been clearly demonstrated, various relationships between them had been suggested. The sulfur bacteria contain a green pigment, *bacteriochlorophyll* or *bacteriochlorin*, which catalyzes a type of photosynthesis, although with a much lower energy yield than that of green plants and with the release of sulfur compounds rather than oxygen. Researchers suggested these bacteria as a possible ultimate source of the green plants, with the blue-green bacteria as an intermediate. Spirochaetes have a type of flagellar motility which suggests that of some flagellate Protozoa, but otherwise they are typical procaryotes. They were classified as bacteria by some biologists and as protozoans by others. Finally, the slime bacteria were compared to slime molds, and phylogenetic relationship was suggested, although the slime of the former is a nonliving secretion of the bacteria, whereas that of the latter is the actual protoplasm. A more probable solution to the problem of relationship of procaryotes and eucaryotes has developed along lines radically different from any of the above.

Intracellular Symbiosis

Over a period of a century, many biologists have suggested, on evidence of various degrees of adequacy, that mitochondria and sometimes flagella or cilia share characteristics with bacteria, and that these organelles may have arisen as intracellular bacterial symbionts. Similarly, chloroplasts of green plants may have originated as symbiotic blue-green bacteria. Perhaps the majority of biologists dismissed such ideas as the products of a little coincidence of morphology and a lot of unrestrained imagination. Since about 1960, however, researchers have demonstrated that all of these organelles have their own DNA and that their DNA is closer to that of procaryotes than it is to the nuclear DNA of eucaryotes in both its morphological and its biochemical characteristics. As a result, the idea that the organelles of eucaryotic cells originated as symbiotic procaryotes has gained new prestige and widespread acceptance.

The work of L. Margulis has been especially useful in synthesizing the applicable data into a coherent and plausible theory of the origin of the eucaryotes by means of a series of intracellular symbioses. Her starting point is a moderately large, amoeboid, heterotrophic, anaerobic procaryote. This organism fed by ingesting smaller procaryotes. Some of the ingested organisms were aerobic, and they became established as symbionts, thus conferring the advantages of aerobic respiration upon the host cell. In effect, they became mitochondria in the host cell. Next, this neo-aerobic procaryote ingested flagellate procaryotes with the 9 + 2 structure (Figure 17–1), and once again they were integrated into the host cell as obligate symbionts, which then functioned as cilia or flagella, with the nuclear region of the symbiont serving as a basal granule. The new symbionts thus became locomotor organelles. In some instances, however, only the nuclear region remained, and the rest of the symbiont was lost. These nuclei became centrioles or centromeres. The evolution of mitosis and meiosis thus became possible. Once they were realized, the eucaryotic cell was complete.

Margulis has emphasized that every such eucaryotic cell contains at least three genomes (sets of genetic material) of quite different origin: the nuclear genome, which is investigated in conventional Mendelian genetics, and the genomes of the mitochondria and of the 9 + 2 organelles, which were brought into the cell by intracellular symbiosis. But this is not the end of the story, for some of these neo-eucaryotes also ingested blue-green bacteria, which became established symbiotically as chloroplasts; these cells became green plants, with *four* genomes in each cell, three of them living within the eucaryotic cell as obligate symbionts.

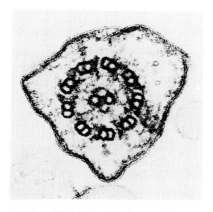

Figure *17–1. Flagellum of* Polytomella, *a green alga; electron micrograph of a cross section. Notice the circlet of nine double microtubules, with two more microtubules at the center. This "9 + 2 structure" is characteristic of a whole series of organelles—flagella, cilia, basal granules, centrioles, and the centromeres of the chromosomes.*
Courtesy of Dr. David L. Brown.

On first reading, this theory may seem far-fetched; yet it is the result of the synthesis of an enormous amount of information—morphological, physiological, genetic, and biochemical—which has been accumulating for a century (since 1883). It deserves serious consideration, and there is good probability that it is correct, at least in its broad outlines, yet it remains controversial.

Alternative Hypothesis

Although Margulis's theory has evoked great interest among biologists, decisive proof is lacking. Uzzell and Spolsky offer an interesting alternative that does not involve symbiotic relationships. They propose that the basic change in the ancestral procaryote that led to the formation of eucaryotic cells was the invagination of the cell membrane to form double membrane structures within the protoplasm, such as the endoplasmic reticulum that characterizes all eucaryotic cells. In procaryotes, characteristically lacking close correlation between replication of the chromosome and division of the cell, a single cell might already contain several identical chromosomes. Thus, if the invaginating membrane should surround *two* chromosomes, a eucaryote-type diploid nucleus would result, whereas if the invaginating membrane should enclose only a single chromosome, it might form a haploid organelle, such as a mitochondrion or a chloroplast. Uzzell and Spolsky point out that this would at once account for the fact that these organelles are always surrounded by a double membrane and that the inner membrane is predominantly under the control of the DNA of the organelle and the outer membrane is predominantly under the control of the diploid cell nucleus.

Also interesting is the *plasmid* theory of R.A. Raff and H.R. Mahler. Procaryotic cells often contain plasmids, small accessory chromosomes that carry a few nonessential genes. Although they are best known for genes such as those for resistance to antibiotics, plasmids do exchange DNA (crossover) with the main chromosome of the cell, and so in principle they might carry any genes. If a plasmid with the genes for synthesis of the respiratory enzymes were to be included in a vesicle detached from the cell membrane, it would effectively be a mitochondrion. Raff and Mahler propose such an origin for the mitochondria. In the present state of knowledge, there are data favorable to all of these theories. None is decisively proven, although at present the evidence for the multiple symbiosis theory is perhaps the strongest. All theories should be considered with an open mind.

KINGDOM PLANTAE, SUBKINGDOM THALLOPHYTA

The simplest divisions of plants were formerly grouped together as a single division, Thallophyta. The Thallophyta are now treated as a subkingdom and include all plants that reproduce without the formation of an embryo within the ovary of the maternal plant. This subkingdom includes eight divisions, ranging from simple unicellular algae to large algae nearly as complex as the simpler vascular plants.

Two of these divisions, the Myxomycophyta, or slime molds (450 species), and the Eumycophyta, or true molds and yeasts (47,000 species), together with the bacteria, which clearly no longer belong in one group with eucaryotes, constituted the old group Fungi. The relationships of the two divisions of molds are not at all clear, but they appear to be terminal groups, that is, they have not given rise to others. The slime molds move by amoeboid locomotion. They reproduce by means of flagellate, amoeboid swarm cells. These swarm spores fuse in pairs, and thus they have a simple type of sexual reproduction. Many mycologists feel that the Myxomycophyta are more similar to the Protozoa than to any other group. They have even been classified as Protozoa under the name Mycetozoia. Those who support this viewpoint regard them as at least transitional between the kingdoms.

The Eumycophyta are of similarly obscure origin. In structure and function, they show many parallels to the green algae, and so they have been thought to be derived from this group. Their zoospores, however, resemble those of the slime molds and show amoeboid movement. Hence it is possible that the Myxomycophyta have given rise to the Eumycophyta. These flagellated zoospores closely resemble some Protozoa, and the viewpoint that the Eumycophyta may be of protozoan origin has much support. The multiple symbiosis theory of the origin of the eucaryotes could account for the origin of the two divisions of molds in three ways: (1) they may have arisen from the protoeucaryote before the earliest symbiosis with blue-green bacteria; (2) they may have arisen from amoeboid protozoans, perhaps at a later time; or (3) they may have arisen from more-advanced algae by loss of chlorophyll. In the third case, the amoeboid phase of the life cycle might represent recapitulation of the protoeucaryote stage.

The remainder of the old division Thallophyta consists of six divisions of algae (Figure 17–2), ranging in complexity from the extremely simple, unicellular algae to the green algae of large size and complexity only a little less than that of the vascular plants. Most of these divisions are of uncertain origin and have given rise to no further groups, hence they need not be discussed here in detail, even though some of them have attained a considerable degree of specialization. These are the division Chrysophyta, including yellow-green algae, golden brown algae, and diatoms (some 11,000 living species); the division Pyrrophyta, including cryptomonads and dinoflagellates (1,000 species); the division Phaeophyta or brown algae (1,500 species), which L.S. Dillon and S.H. Hutner believe to be of especial importance for the origin of animals; and the division Rhodophyta or red algae (4,000 species). Two other divisions of algae are of more especial interest for the present discussion: these are the Euglenophyta, including *Euglena* and its allies (450 species), and the Chlorophyta, or green algae (7,000 species).

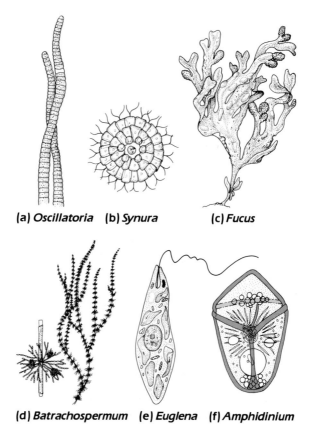

(a) *Oscillatoria*   **(b)** *Synura*   **(c)** *Fucus*

(d) *Batrachospermum*   **(e)** *Euglena*   **(f)** *Amphidinium*

Figure 17–2.   *Representative algae of several phyla. (a)*
Oscillatoria, *a blue-green alga; (b)* Synura, *a golden brown
alga; (c)* Fucus, *a brown alga; (d)* Batrachospermum, *a
red alga, showing growth habit and details of one whorl;
(d)* Euglena; *(f)* Amphidinium, *a dinoflagellate.*

EUGLENOPHYTA AND SEPARATION OF
THE EUCARYOTE KINGDOMS

The Euglenophyta, typified by the common *Euglena* of elementary biology
laboratories, show many advances over the blue-green bacteria. They have a
definitely organized nucleus, which is separated from the cytoplasm by a
membrane. The chlorophyll is no longer free in the cytoplasm, but rather it
is concentrated in numerous ovoid bodies, the chloroplasts. Further, the color
is grass green rather than blue-green. Unlike the green algae, the Euglenophyta
are not provided with a cellulose cell wall. The cells are provided with one
or two flagella, and the euglenas are active swimmers. There is a gullet at

the anterior end, yet it appears that *Euglena* is autotrophic. Near the gullet there is a red-pigmented eyespot, which seems to be sensitive to light. Reproduction is always by simple mitotic division, although sexual reproduction has been reported for one genus (*Scytomonas*). In short, Euglenophyta are typical eucaryotes.

The Euglenophyta are also described in zoological works as the order Euglenida of the class Phytomastigophorea and the phylum Protozoa. This class includes most of the unicellular algal divisions, and the class Zoomastigophorea includes the more animal-like flagellates. Both are included in the subphylum Mastigophora. The group as a whole shows a curious mixture of plant and animal characteristics.

Plant or Animal?

Consider the typical differences between plants and animals. Generally speaking, the mode of life of animals is active, and that of plants is passive. Animals are heterotrophic, eating other organisms to obtain the complex organic compounds that they require as foods. Plants, in contrast, are generally autotrophic, synthesizing all of their food requirements from the elements. But there are exceptions in both kingdoms. Some advanced plants, such as the sundew (*Drosera*), have developed mechanisms for the capture and digestion of insects; and many animals are saprozoic, that is, they absorb decaying organic matter from their environment. In plants differentiation of organs is predominantly external; in animals it is predominantly internal. In plants, growing tissue, the meristem, is present at all stages of the life cycle; whereas most animals have a definitely limited growth. Finally, plants are generally sedentary, and the individual plant cells are surrounded by a rigid cellulose wall; animal cells generally lack such rigid walls, and the animal moves about freely in its environment.

Against such a group of criteria, *Euglena* and its allies are difficult to place. *Euglena* itself is supplied with an abundance of chlorophyll, yet it can be raised on completely inorganic media only with difficulty. Traces of amino acids or peptones facilitate culture. Nonetheless, there is no evidence that *Euglena* ever ingests other organisms; it seems more probable that its normal nutrition is predominantly holophytic (by photosynthesis), with a supplement obtained saprozoically. Some other flagellates, however, are completely holophytic. Others lack chlorophyll, and these of course cannot be autotrophic. Some of them are entirely saprozoic, but some ingest other organisms in typical protozoan fashion and so may be said to be holozoic. Thus the whole range of nutritional possibilities occurs within a single group, and opposite extremes may occur within a single genus. Plants ordinarily store food as starch; animals store it either as glycogen (similar to starch) or as fat. Euglenoids store it as paramylum, a carbohydrate different from both starch and glycogen. The embryological criteria of external or internal organ formation and presence or absence of a continuously growing meristem obviously have no applicability to unicellular organisms. Like an animal, *Euglena* moves freely in its environment, but its near relatives include sedentary forms. Finally, although euglenoids lack a cellulose cell wall, they have a pellicle that in some species is rigid.

One result of this mixture of plant and animal characteristics within the unicellular, flagellate organisms is confusion in taxonomy. Some biologists have treated the whole array as plants, a procedure that makes it necessary to treat as plants such organisms as the trypanosomes, blood parasites of vertebrates that do not show any plantlike characteristics. Others have tried to designate some forms as plants and others as animals on the basis of the above criteria or similar ones. This division creates the absurd situation of the assignment of different members of the same genus to different kingdoms, in some instances. Often the whole series of algae and Protozoa have been lumped together as a single kingdom Protista, with only the Metazoa left in the animal kingdom and only the vascular plants and the bryophytes left in the plant kingdom. This system too is unsatisfactory because the higher algae are obviously much more closely related to the vascular plants than to many of the Protozoa, and conversely the animal nature of many of the Protozoa, such as the ciliates, is not open to doubt.

How Many Kingdoms?

H.F. Copeland, who devoted his long career to this problem, marshalled much good evidence in favor of a system of broad taxonomy based upon four kingdoms. Kingdom I, the Mychota, would include the bacteria and the blue-green bacteria, that is, all organisms in which the nuclear-cytoplasmic differentiation is not complete. (He regarded viruses as nonliving.) Kingdom II, the Protoctista, would include most of the algae, the fungi, and Protozoa. Thus the Kingdom Protoctista would include all of those primitive organisms from which more complex plants and animals may have arisen, as well as their relatives that have given rise to no further groups. The Kingdom Plantae is thus restricted to green algae, vascular plants, and bryophytes. The Kingdom Animalia is similarly restricted to the Metazoa.

Although there is much to justify this classification, it presents some serious difficulties. First, the Kingdom Protoctista is subject to the same criticism as is the Protista: the extreme members are just as clear-cut plants and animals as are the vascular plants and Metazoa, respectively. Copeland's system also substitutes two areas of confusion for one, for we would have to decide whether an organism were a protoctistan, a plant, or an animal. The Mychota have been much more clearly defined since Copeland's work, thereby eliminating confusion at the mychotan-protoctistan border.

Two broad taxonomies have emerged from the synthesis of many lines of evidence that bear upon the separation of the kingdoms. One of these has been developed by R.H. Whittaker, who starts by giving kingdom status to the procaryotes (under the name Monera), then emphasizes the evolution of three nutritional modes among the eucaryotes. He recognizes a Kingdom Fungi for organisms that are saprophytes; a Kingdom Plantae for the autotrophs; a Kingdom Animalia for the Metazoa, all of which are heterotrophs; and a Kingdom Protista for those microorganisms and their relatives that are difficult to sort out in this way.

One of the present authors (E.O.D.) has proposed an alternative system, which also begins by recognizing the kingdom status of the procaryotes (under the name Mychota), then continues by emphasizing interrelationships among the various groups of flagellates ("the crossroads of the kingdoms," according to Copeland), among fungi and protozoans, among the various groups of algae, especially between the green algae and the higher plants, and finally between protozoans and metazoans. This complex nexus of interrelationships suggests that these organisms are fairly close to the protoeucaryote ancestor of both plants and animals, which need not be thought of as either plant or animal, but as the progenitor of both. The living algae and protozoans would then represent various stages in the evolution of the characteristic differences between the two higher kingdoms, together with specific adaptations to the acellular grade of construction. There is, theoretically, no reason why a sharp line of separation should exist between those developing along plantlike lines and those developing along animal-like lines; indeed, it would conflict with the idea of origin by evolution. Perhaps the most important feature of *Euglena* and its allies is *intermediacy between the kingdoms,* suggesting as it does the probability that the euglenoids may be primitive organisms, fairly close to the stem group from which both plants and animals have come.

We recommend a broad taxonomy of three kingdoms (Monera, Animalia, and Plantae), then, based on probable interrelationships among those primitive groups that are critical for the problem of separation of the kingdoms. Whittaker, however, drawing upon much the same array of data, has emphasized the nutritional differences and so has arrived at a five-kingdom system (adding Fungi and Protista). In the present state of knowledge, both systems are defensible, and both promote understanding of the broad taxonomy of the world of life. Final decision between these two systems must be deferred until further knowledge is brought to bear upon the problem. Whittaker has suggested that systems of broad taxonomy function like different optical systems to bring out different features of the array of organisms studied. This apt metaphor suggests that it may be desirable to use both systems in parallel, just as it is desirable to use more than one optical system in studying an organism.

CHLOROPHYTA

The Chlorophyta or green algae are an extraordinarily varied group. The simplest members are unicellular, but there is a definite separation of nucleus and cytoplasm, and the chlorophyll is contained in a single plastid. Multicellular species may show considerable specialization of different cells, and the higher green algae may attain large size. Although the more primitive species reproduce by simple fission, sexual reproduction and alternation of generations are well developed in the division. The green algae appear to be on or near the main line of evolution leading to the higher plants; hence great interest attaches to this division.

Chlamydomonas and the Origin of Sex

Sex probably originated in an unknown green alga resembling the living *Chlamydomonas* (Figure 17–3). Each plant consists of a single cell. It has a well-defined nucleus and a single large chloroplast. It swims by means of two flagella, which are located at the anterior end of the cell. The cell is protected by a heavy cellulose wall. The plant may reproduce by simply dividing to form two, four, or eight *zoospores* (so called because they swim actively, like an animal) within the cellulose wall. These zoospores are then released by the dissolution of the cell wall, and each swims away, an independent plant, like the parent in every respect except size. This difference is soon bridged by growth. However, sexual reproduction may also occur, for the parent plant may divide to form eight, sixteen, or thirty-two gametes, cells that resemble the zoospores and the adults, except that they are much smaller. Like the zoospores, these gametes are released into the water, where those from different parent cells unite in pairs to form zygotes. The zygote forms a thick wall about itself and remains quiescent for a time. It is in this highly resistant encysted condition that the plant survives unfavorable conditions such as the drying of ponds. In time the zygote again becomes active. It then undergoes

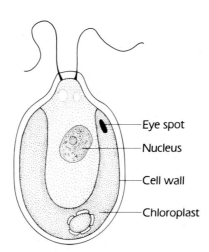

Figure 17–3. Chlamydomonas, a green alga.

two divisions—the meiotic, or reduction, divisions—producing four zoospores, which are released to form adult algae.

Reproduction in *Chlamydomonas* may represent an early natural experiment in sexuality and thus may afford some insight into the origin of sex. Most of its life cycle is passed with only the haploid number of chromosomes, for the reduction divisions occur as soon as the zygote becomes active. Haploidy, that is, single representation of each type of chromosome or genetic factor, was undoubtedly the normal situation for all organisms before the origin of sex, and it is still normal for organisms that do not reproduce sexually. Diploidy is a necessary consequence of sexual reproduction, for the union of two gametes can have no other result. Gametes could not be reduced below the haploid condition without qualitative loss of genetic material. In these organisms, so close to exclusively haploid ancestors, meiosis appears to serve primarily to restore the ordinary, physiological, haploid chromosome number; whereas diploidy is introduced as a temporary concomitant of a "new" method of reproduction.

Why sexual reproduction should ever have developed under such circumstances is not clear. A "hunger theory of sex" has been proposed, according to which the gametes are simply undersized spores, individually lacking the food and energy necessary to complete development. They therefore pool their resources by means of two-by-two fusions, and thus each zygote obtains a sufficient supply of the materials necessary for development. This theory would be more satisfying were it not for the fact that the four haploid zoospores produced by the zygote are also small cells, yet they complete their development satisfactorily without fusions. Whatever the original stimulus to sexual reproduction may have been, the great selective advantages that have made it so nearly universal among higher plants and animals are clear. Sexual reproduction causes a relatively rapid reshuffling of the various characters among the progeny, so that the most favorable combinations may be formed and tested by natural selection. Further, diploidy, which results from sexual reproduction, makes it possible to accumulate a store of genetic variability in the heterozygous state. Diploidy may also be physiologically advantageous because of the increased production of nuclear enzymes or in other less obvious ways.

The degree of sexual differentiation varies greatly in different species of *Chlamydomonas*. The adults are morphologically identical and cannot be sexed by inspection. In most species (forty out of fifty-four on which good data are available), the gametes are also morphologically identical. They are said to be *isogamous*. Yet they are physiologically differentiated, for fertilization is possible between some pairs of clones (asexually produced descendants of a single cell) but not between others. Because they are not morphologically identifiable, they are referred to as + and − mating types rather than as sexes. In some species, like *C. reinhardi,* these are determined by a single pair of genes, so that meiosis always results in two cells of + type and two of − type.

Researchers have identified as many as eight mating types (sexes?) in some species, so the male-female alternative of higher organisms is clearly not applicable. In other species, both *macrogametes* (egglike) and *microgametes* (spermlike) are produced, both of which are flagellate and actively motile. These *anisogamous* species suggest a male-female alternative, yet there are species like *C. elongata* and *C. intermedia* in which fertilization may occur in the combinations large-large and small-small as well as large-small. Finally, there are three species of *Chlamydomonas—coccifera, pseudogigantea,* and *suboogama*—in which the macrogamete is nonflagellate and must be sought by the flagellate microgametes, approximating the condition of egg and sperm in higher animals or of ovule and pollen in higher plants (*oogamy*). Thus, the early experiments in sexual reproduction range from isogamy through various levels of anisogamy to oogamy (Figure 17–4).

Closely related to *Chlamydomonas* is the family Volvocaceae, known to all students of elementary biology as the standard example of an evolutionary trend toward increasing complexity of colonies. The simplest members of this family (*Gonium*) consist of four to thirty-two identical cells, any of which can form gametes. They are advanced beyond *Chlamydomonas* in that gametes are always morphologically different from the vegetative cells. The most-specialized member of the family, *Volvox,* is made up of as many as 40,000 cells arranged as a hollow sphere (Figure 17–5). Most of these cells are purely vegetative and substantially identical. The sex cells are localized in antheridia (containing sperm) and oogonia (containing eggs). The eggs of *Volvox* are always large, nonmotile cells, and the sperm are always actively swimming flagellate cells, which seek out the egg. As both of these are produced by a single colony, the colony may be regarded as hermaphroditic, a condition that is much more common in plants and in lower organisms generally than it is among higher animals. It may be, therefore, that hermaphroditism is primitive.

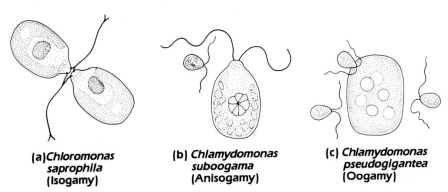

(a)Chloromonas saprophila (Isogamy)   **(b) Chlamydomonas suboogama (Anisogamy)**   **(c) Chlamydomonas pseudogigantea (Oogamy)**

Figure 17–4. *Members of the family Chlamydomonadaceae show a wide range of types of fertilization. (a)* Chloromonas saprophila *is isogametous, (b)* Chlamydomonas suboogama *is anisogametous, and (c)* Chlamydomonas pseudogigantea *is oogamous.*

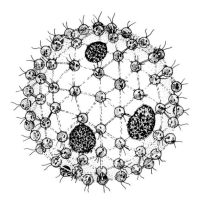

Figure 17–5.   Volvox.

Multicellular Individuals

In another line of descent, typified by *Ulothrix, Draparnaldia,* and *Ulva,* the green algae have developed not merely colonies of substantially independent organisms but multicellular individuals, the various cells of which are inter-dependent (Figure 17–6). These are the filamentous algae, the typical "seaweeds" of laymen. From this group of algae the bryophytes and vascular plants appear to have developed. *Ulothrix* is a simple, unbranched, multicellular filament.

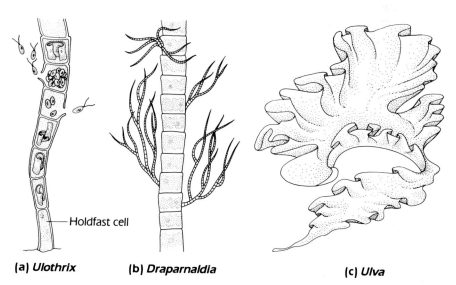

Holdfast cell

(a) *Ulothrix*   **(b)** *Draparnaldia*   **(c)** *Ulva*

Figure 17–6.   *(a)* Ulothrix, *(b)* Draparnaldia, *and (c)* Ulva.

The basal cell is specalized as a holdfast to anchor the plant to a rock or other substrate. Cell divisions occur; but, whereas this results in asexual reproduction in the unicellular algae, in *Ulothrix* and its allies mitosis results in growth without reproduction.

Reproduction may occur either sexually or asexually. Asexual reproduction occurs by fragmentation of the plants with subsequent regeneration by each fragment, or it occurs by the formation of zoospores from the vegetative cells of the plant. These zoospores are not unlike those of *Chlamydomonas,* except that they have four flagella instead of two. Each zoospore develops into a plant like the parent plant. Any vegetative cell of the plant may also give rise to gametes. These are smaller than the zoospores, and they have only two flagella. All of the gametes are identical morphologically, but physiologically there must be a sexual differentiation, for zygotes are always formed by the union of two gametes from different plants. After a quiescent period, the zygote undergoes two meiotic divisions, forming four zoospores, which then develop into adult plants. Thus, most of the life cycle of *Ulothrix* is passed with only the haploid chromosome complement; the zygote alone is diploid.

Draparnaldia is advanced over *Ulothrix* principally by the more complex development of the vegetative body. Whereas the latter is composed only of simple, unbranched filaments, the former has a major, basal filament from which many branches arise, and from these come secondary branches. Reproduction again is accomplished either by zoospores or by isogametes.

Ulva, the common sea lettuce, is advanced over the previously discussed algae in several ways. The vegetative body forms large, leaflike sheets that, for the first time, are more than one cell thick. Reproduction is again either by zoospores or by isogametes, but a significant development occurs in the life cycle. In all of the algae discussed, the reduction divisions occur in the zygote, with the result that in the whole life cycle only the zygote is diploid. In *Ulva,* however, the divisions of the zygote are ordinary mitotic divisions, with the result that a diploid plant is formed. Some of the cells of the adult plant then undergo the meiotic divisions, which result in the formation of haploid zoospores. This diploid plant, which reproduces by the formation of haploid zoospores, is called a *sporophyte.* The spores develop into haploid plants, much as do the spores of other algae. These haploid plants then reproduce by isogamy. This haploid, gamete-forming plant is called a *gametophyte.* The sporophytic and gametophytic generations of *Ulva* are morphologically indistinguishable. This alternation of a diploid, sporophytic generation with a haploid, gametophytic generation is one of the most fundamental features of plant biology. Alternation of generations is first introduced with the origin of sex, and it is present in all sexually reproducing algae; but in more primitive algae, the sporophytic generation is represented only by the zygote. In *Ulva,* and in many other green algae, however, the sporophyte is as highly developed as is the gametophyte.

Toward Vascular Plants

Thus, within the algae at large and more especially within the green algae, advances have been made that approach the condition of the simpler vascular plants. The nucleus has been delimited from the cytoplasm by a membrane, and the mitotic mechanism has been perfected. The grass green chlorophyll, no longer masked by the blue phycocyanin, is contained in chloroplasts like that of vascular plants, not dissolved in the cytoplasm as it is in the blue-green bacteria. The most primitive algae are unicellular, but colonies of increasing complexity have been formed, leading finally to true multicellular individuals. The simplest of these are unbranched filaments, but these have given rise to branched and rebranched plants, and finally to large, fleshy plants with considerable differentiation of tissues. These may include rootlike, stemlike, and leaflike structures. Cellulose walls are present in some of the algae. Also, a great range of reproductive mechanisms has been developed among the algae. The most primitive algae reproduce only asexually, either by simple fission or by the formation of clusters of zoospores. Isogamy, the most primitive form of sexual reproduction, probably began with the pairwise fusion of undersized zoospores. Later, these gametes became differentiated into small microgametes and large macrogametes, both of which were motile (heterogamy). Finally, the macrogamete became a large, nonmotile cell, which was sought by the microgamete. Meanwhile, the alternation of generations developed, with the diploid generation, at first a minor incident in the life cycle, becoming increasingly prominent.

The most-advanced characters of the algae are all carried over into the vascular plants. No single alga exhibits all of these characters, yet the trend of development in the green algae especially is clearly toward the type of organization characterizing the vascular plants, and it is highly probable that the most primitive vascular plants were derived from green algae. The evolution of the vascular plants and the bryophytes will be taken up in Chapter 18.

The events discussed in Chapters 16 and 17 are undoubtedly among the most important in the whole history of evolution. In point of time, they must have occupied most of the history of life. Yet all of these events must have occurred long before the earliest known useful fossils were formed. Thus it is altogether probable that decisive fossil evidence on the problems discussed in these chapters will never be obtained, and that these subjects must always remain speculative, even though some inferences may be made with a fair degree of probability on the basis of primitive or archaic organisms now living, and of experiments performed under presumably primitive conditions.

SUMMARY

The greatest discontinuity in the world of life is that which separates the *procaryotes* (prenuclear organisms, including archaebacteria, viruses, bacteria, and blue-green bacteria) from the *eucaryotes* (nuclear organisms, including algae, fungi, land plants, protozoans, and metazoans). How this gap was crossed may never be conclusively solved. In the past, biologists generally believed that the simple structures of the procaryotes were gradually transformed into the complex structures of the eucaryotic cell, although no transitional stages were known. An alternative theory is that a series of intracellular symbioses may have established the major organelles of the eucaryotic cell. This theory was not taken seriously until recently, but it has gained much support because of the demonstration that organelles of eucaryotes have their own DNA and that organellar DNA has more in common with that of procaryotes than it does with the nuclear DNA of its own cell.

Traditionally, only two kingdoms, Plantae and Animalia, have been recognized. The kingdom Plantae was divided into the subkingdom Thallophyta (comprising bacteria and an array of algae) and the subkingdom Embryophyta (comprising the two divisions of land plants). The kingdom Animalia comprises some thirty-odd phyla.

Microorganisms have made it impossible to maintain this simple, Linnaean taxonomy. The demonstration of the great gulf between procaryotes and eucaryotes requires the recognition of the procaryotes as a kingdom apart (or even as two separate kingdoms). And among eucaryotes, the lines are blurred between different groups of microorganisms by the flagellates and the amoeboid organisms. One response to this blurring of the lines is to use the two eucaryotic kingdoms and accept the blurred areas as the inevitable consequence of their evolutionary origin. Another response is to recognize four eucaryotic kingdoms based on nutrition: Fungi, saprophytic; Protista, including all types of nutrition (most algae and protozoans); Plantae, holophytic (green algae and land plants); and Animalia, holozoic (metazoans). Both systems are useful.

Sex probably originated in a green alga similar to *Chlamydomonas*. Microorganisms include isogamous, anisogamous, and oogamous types, even in the same genus. In primitive algae, the reduction divisions occur in the zygote, so that the zygote is the only diploid cell in the life cycle. In more-advanced algae, the zygote may divide mitotically to produce a diploid sporophyte that alternates with a haploid gametophyte. The relative importance of these two generations in the life cycle has changed during the evolution of the algae: initially, the haploid gametophyte was dominant almost to the exclusion of the sporophyte, but the importance of the latter gradually increased. Increasing importance of the sporophyte, multicellularity, use of cellulose to the exclusion of other compounds in cell walls, and differentiation of parts suggesting roots, stems, and leaves all portend the transition to land plants.

REFERENCES

Dodson, E.O. 1971. The kingdoms of organisms, *Systematic Zoology* 20:265–281.
    A fuller exposition of the three-kingdom system.

Dodson, E.O. 1979. Crossing the procaryote-eucaryote border: Endosymbiosis or continuous development? *Canadian Journal of Microbiology* 25:651–674.
A comprehensive review of the problem indicated.

Margulis, L. 1981. *Symbiosis in Cell Evolution.* Freeman, San Francisco.
This prime mover of the endosymbiotic theory presents her case.

McAlester, A.L. 1977. *The History of Life,* 2nd ed. Prentice-Hall, Englewood Cliffs, N.J.
This brief book presents much valuable information. It develops plant taxonomy along much the same lines as here, following the principles established many years earlier by O. Tippo.

Uzzell, T., and C. Spolsky. 1974. Mitochondria and plastids as endosymbionts: A revival of special creation? *American Scientist* 62:334–343.
This paper presents an alternative theory of the origin of the eucaryotes.

Whittaker, R.H. 1969. New concepts of the kingdoms of organisms. *Science* 163:150–159.
The most important statement of the five-kingdom system.

Woese, C.R., L.J. Magrum, and G.E. Fox. 1978. Archaebacteria. *Journal of Molecular Evolution* 11:245–252.
The principal advocates of this new kingdom present their case.

18

Evolution among Land Plants

It is almost a point of definition for the algae that they are aquatic plants, although some of them have invaded moist habitats on land. The fossil record becomes rich and varied only in the Cambrian period of the Paleozoic era, some 600 million years ago; and the Cambrian record, so far as plants are concerned, consists entirely of a wide variety of algae and bacteria. In fact, the early Paleozoic is often referred to as the age of algae and invertebrates. The first fossil land plants appear in Upper Silurian rocks dated at 405 million years before present. When these first colonists left the waters to invade the more difficult but potentially more-varied habitats on land, the algae remained the dominant members of the earth's flora. Soon (geologically speaking) the land dwellers surpassed their aquatic progenitors. One of the crucial problems that had to be solved before plants could invade the land was the protection of the zygote against drying. All land plants solve it, with important differences in the details, by the retention of the zygote and the developing embryo within the sex organs of the maternal plant. For this reason, the land plants are known collectively as the subkingdom Embryophyta. This subkingdom includes only two divisions, the Bryophyta and the Tracheophyta, but the latter is greatly varied. T.N. Taylor has recently suggested elevating the classes of land plants to the level of divisions, a suggestion that would seem to have merit.

BRYOPHYTES

The division Bryophyta comprises some 20,000 species of mosses, liverworts, and hornworts, all small organisms. These are the amphibians of the plant world, for they have met only minimum requirements of adaptation to the terrestrial environment. They are restricted to wet habitats, and all of them require water for reproduction, at least as a film over the surface of the plant in which sperm can swim. Because they lack cells specialized for support or for conduction of nutrients, they are said to be nonvascular land plants.

The bryophytes share with other land plants certain adaptations that permit them to utilize terrestrial habitats. The embryos, which are always multicellular, are retained within the female sex organs and are thus protected from drying. The plants are always oogamous, that is, the egg is a large, nonmotile cell that must be sought by the sperm. The sex organs of both sexes include a jacket layer of sterile, protective cells. All of the aerial parts of the plant are covered by a waxy cuticle that protects the plant against drying. Finally, alternation of generations is well developed; the diploid sporophyte is substantially a parasite upon the haploid gametophyte in the bryophytes (Figure 18–1), for the former is always attached to the latter, and it contains inconsequential amounts of chlorophyll, if any.

Most of these characters, other than retention of the embryo in the maternal sex organs, already occur in the algae, although no alga has all of them in combination. Other bryophyte characters tie these plants in with the algae more closely. Like the algae, the bryophyte body is a *thallus,* a simple cell mass that is not differentiated into roots, stems, and leaves. Structures that resemble these parts are present, but the same may be said of many of the algae. The gametophyte is larger than the sporophyte, as in most algae, and photosynthesis is largely restricted to the gametophyte.

PHYLOGENY OF BRYOPHYTES

In view of the simplicity of the bryophytes and of the many characters they share with green algae, they evidently derived from an ancestor among the green algae. The hornworts show such progressive characters that they were once considered likely ancestors of vascular plants, but this is not possible, for vascular plants precede bryophytes in the fossil record.

Another possibility is that bryophytes arose from primitive vascular plants by *loss* of vascular tissue. Stomata on the epidermis, like those on the leaves of vascular plants, are the basis of the attempts to join them to vascular

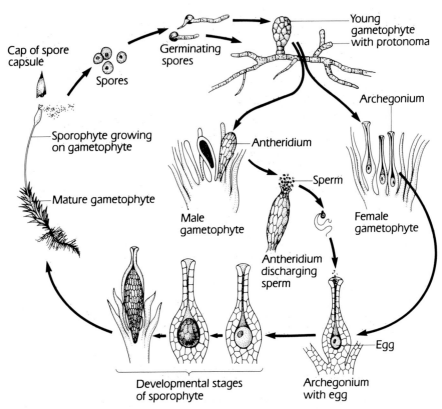

Cap of spore capsule

Spores

Germinating spores

Young gametophyte with protonoma

Archegonium

Sporophyte growing on gametophyte

Antheridium

Sperm

Mature gametophyte

Male gametophyte

Antheridium discharging sperm

Female gametophyte

Developmental stages of sporophyte

Archegonium with egg

Egg

Figure 18–1.   *Life cycle of a moss (clockwise from the left): Mature gametophyte with a sporophyte growing on it and discharging its spores. The cap of the spore capsule. Ripe spores. Germinating spores. Young gametophyte with protonema. Portions of male and female gametophytes showing antheridia and archegonia. Antheridium discharging sperm. Archegonium with egg. Developmental stages of the sporophyte.*

plants. The current view is that bryophytes represent a sterile group that did not give rise to anything else. Lower Devonian *Sporogonites* from Norway is perhaps the oldest convincing fossil bryophyte, but the verification of Silurian forms would not be surprising. They are a rather minor component of the fossil record.

ORIGIN OF VASCULAR PLANTS

The division Tracheophyta* is an ancient and highly varied group that includes the dominant plants of today. Typically, the division Tracheophyta is divided into five subdivisions:

* The Botanical Code specifies that the ending *-phyta* be used to denote divisions; *-phytina* for subdivisions; *-opsida* for classes; *-ales* for orders; and *-aceae* for families.

subdivision Psilophytina: primitive vascular plants

subdivision Lycophytina: club mosses

subdivision Sphenophytina: horsetails

subdivision Pterophytina: ferns

subdivision Spermophytina: seed plants including conifers
and flowering plants

In intense study of Silurian and Lower Devonian vascular plants, H.P. Banks and others have shown that the psilophytes are a polyphyletic assemblage of three distinct groups of early vascular plants. Accordingly, the subdivision Psilophytina may now be replaced by:

subdivision Rhyniophytina

subdivision Zosterophyllophytina

subdivision Trimerophytina

This further classification raises the number of subdivisions to seven. Taylor elevates the subdivisions to divisions, splits the polyphyletic gymnosperms into five divisions, adds a transitional fossil group, and elevates the class Angiospermae of most taxonomists to the level of division (which he calls the Anthophyta), thereby achieving thirteen divisions whereas others usually recognize five or seven. Given the tremendous size of the class Angiospermae, whose nearly 300,000 species outnumber all other tracheophytes combined by a factor of about 10, it is difficult to quarrel with an elevation in rank for them. This scheme was proposed in 1981, and it remains to be seen what degree of acceptance it will achieve.

We do not know how the vascular plants arose from their algal ancestors, and the fossil record throws little light upon this important question. Early in this century the French botanist M.O. Lignier published a highly speculative theory on this subject, and in the meantime such fossils as have been discovered have been consistent with his theory, even if they haven't proved it. According to Lignier's scenario, the ancestor of the vascular plants must have been a green alga characterized by branching filaments; and this plant must have been a tide-flat dweller. As the land was elevated, tide pools became isolated and then dried up. Much of the flora became extinct, but if one or more of the branches of such an alga had penetrated the ground, it might have become transformed into a root system capable of supplying the plant with water and minerals. Some of the branchings might then have straightened out, leaving a main stem or trunk with branches. Because the entire plant was no longer immersed in water, a conducting system was now necessary, and only those plants that evolved one could survive. Thus the stems became thickened, and the ends flattened out to form organs specialized for photosynthesis—the leaves. Now the conducting system had to operate in both directions, carrying water and salts up from the roots and carrying organic compounds down from the leaves. Finally, only those plants that had developed a cuticle over the aerial parts could have escaped extinction through drying. Although critical proof of this theory is still lacking, evidence accumulated in the intervening years is consistent with it, and no more probable theory has yet been proposed.

DIFFERENTIATION OF EARLY LAND PLANTS

Rhyniophytina

The study of early vascular plants began with the discovery by J.W. Dawson in 1859 of some extremely simple, leafless, dichotomously branching, rhizome-bearing plants, which he named *Psilophyton princeps*. These fossils, from Devonian rocks of the Gaspé in Quebec, did not fit into the then-prevailing botanical classification, so Dawson's discovery was disparaged or ignored. But in 1917, R. Kidston and W. Lang discovered three genera of similar plants in a silicified bog of Devonian age in Rhynie, Scotland. The preservation of the Rhynie fossils is so good that the cell walls can be seen in thin sections. The fossils are complete and abundant, and a number of different kinds belonging to several divisions have since been described. Since then, additional discoveries of early vascular plants have been made in Europe, North America, and Australia.

Rhynia (Figure 18–2) and *Horneophyton* are characteristic fossils of the

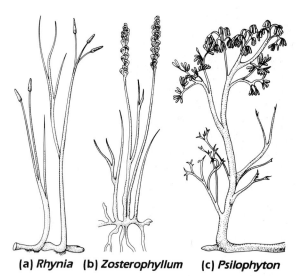

(a) Rhynia   (b) Zosterophyllum   (c) Psilophyton

Figure 18–2.   Early Devonian land plants. (a) Rhynia, representative of the very primitive, dichotomous rhyniophytes. (b) Zosterophyllum, representative of the zosterophytes. (c) Psilophyton, representative of the progressive trimerophytes, in which a separation of sterile and fertile fronds appears.

Rhynie chert and characterize the Rhyniophytina, which are the simplest of all known vascular plants. They are erect, sparse, dichotomous-branching plants, with terminal sporangia. Various species ranged from 5 to 50 cm in height. Upper Silurian *Cooksonia* is the earliest known vascular land plant; and plants originating from green algae are not to be sought at a more remote time than that. Since *Cooksonia* lacked leaves, its green stems must have served in photosynthesis.

The living Psilotales (Figure 18–3), *Psilotum* and *Tmesipterus,* are similar to *Rhynia* and are presumed related, though a gap of 375 million years separates them. In nature they are confined to tropical and subtropical regions, but they can be grown in greenhouses anywhere. The major morphological difference between the living and extinct species is that the sporangia are located in the axils of the "leaves" of the living plants, whereas they were located at the tips of branches in the extinct species. Another point of considerable interest relates to the alternation of generations. Only the sporophytes of the psilophytes have been preserved in the fossil record, but the gametophytes of the modern genera are known. The gametophyte looks very much like a fragment of rhizome, bearing numerous archegonia and antheridia that are similar to those

Figure 18–3.   Psilotum *sporophyte. The nodules on the branches are sporangia.*

of the bryophytes. These may give an indication of the character of the gametophyte of the extinct psilophytes, but there is no assurance that they do. In any event, the Psilotales are regarded as the little-changed descendants of the ancient psilophytes.

Sphenophytina

The Sphenophytina, the horsetails, are a group of simple vascular plants that appeared in the Late Devonian, enjoyed some success in the Carboniferous coal swamps, then suffered major extinctions in the Triassic. They appear to have risen from rhyniophytes. Two growth forms were apparent during the Carboniferous: herbaceous species comparable to the horsetails today and arborescent (treelike) species 10 m tall that were significant contributors to coal beds. Modern horsetails comprise a single genus, *Equisetum* (Figure 18–4), with about 25 species.

Horsetails represent an advance over the most primitive vascular plants in the differentiation of the plant into definite root, stem, and leaves. The stems of horsetails have a pith region that is commonly hollow. The stem grows by means of concentrations of meristem at definite nodes, at which point the stem can be easily disjointed, to the delight of many a child. The leaves are arranged as whorls, and they are relatively small. Nonetheless, most are of the opinion that they originated through the flattening of the tips of branches rather than by emergence from the stem. The leaves are not adequate for the photosynthetic needs of the plant, and the entire stem has retained this function. Horsetails represent a line of descent from rhyniophytes that did not give rise to anything else.

Zosterophyllophytina

Zosterophyllophytes (Figure 18–2), the second of the three groups of very early vascular plants, appeared about 10 million years after the earliest rhyniophyte, that is, about 395 million years ago (Early Devonian). Growth habits of these plants varied and included the simple dichotomous pattern, but they differed from rhyniophytes in that the sporangia are borne on the stem, often in globose or kidney-shaped clusters. Dawson originally described *Sawdonia ornatum* as a new species of *Psilophyton,* and it took a hundred years to realize how different it was. *S. ornatum* was about 30 cm high, and its slender branches were covered by numerous, tapered spines. *Hicklingia* was a naked-stemmed representative of this group in the Rhynie flora, and its similarity to *Rhynia* underscores the probability of common ancestry of the zosterophyllophytes and the rhyniophytes from green algae. Numerous anatomical structures link zosterophyllophytes to the lycophytes.

Figure 18–4. Equisetum, *a horsetail. The inset shows fruiting bodies.* Inset courtesy of Ward's Natural Science Establishment.

Lycophytina

The Lycophytina, comprising the club mosses, derived from zosterophyllophytes in the Early Devonian. Living lycophytes, comprising 5 genera and 1,100 species, are all small plants, generally less than 30 cm high. But they were not always small; during the Carboniferous period (Mississippian and Pennsylvanian combined) there were 40-meter giant club mosses. These were the dominant plants of the time, and their fossil remains are a major part of the coal beds. These giant club mosses became extinct in the Permian period, perhaps as a result of inability to adapt to the severe Permian climate, for it was a time of extensive glaciation.

Figure 18–5. *A club moss,* Lycopodium.

The club mosses (Figure 18–5) are advanced over the zosterophyllophytes in many respects. The differentiation into root, stem, and leaves is complete. Their leaves, which are small and spirally arranged, are supplied with vascular tissue. The sporangia are enclosed in specialized leaves called *sporophylls,* clusters of which occur at the tips of branches. Such clusters are called *cones,* or *strobili.* The life cycle of *Lycopodium,* one of the living genera, is simple and not very different from that of the psilophytes. The spores are all alike and are scattered by the wind. Those that fall on favorable ground germinate to form small, subterranean, thalluslike gametophytes, which are monoecious. Sperm swim to the archegonia to fertilize the eggs. The zygote then produces a young sporophyte, which is at first a parasite upon the gametophyte, but a root is soon formed and the gametophyte decays. In another living genus, *Selaginella,* the life cycle is modified in a fashion suggestive of the seed plants. There are two types of spores: megaspores, which are contained in sporophylls in the lower part of the strobilus, and microspores, which are contained in the upper sporophylls. While still contained within the sporangia, these megaspores and microspores germinate to produce female and male gametophytes, respectively.

The gametophytes are thus parasites upon the sporophyte, although the female gametophyte may contain some chlorophyll. When the microspores are released, they fall down onto the lower sporophylls. When wetted by rain or dew the microspore wall splits and sperm are released. These swim to the archegonia of the female and fertilize the eggs while the female gametophyte is still contained in the spore wall in the parental sporophyte. The embryo thus develops in the female gametophyte while the latter is still contained in the sporophyte, a condition very similar to that of the seed plants.

Lower Devonian *Baragwanathia* from Australia is the oldest known lycophyte; a representative in the Rhynie flora, *Asteroxylon,* was originally lumped with the psilopsids. We might expect that a group with so many progressive characters would have produced further, more progressive descendants, but this has not been the case. The lycophytes are a terminal group. Soon after their origin, the group produced the dominant plants of the coal swamps; *Lepidodendron* towered 40 m and attained a diameter of 2 m. But the group became extinct before the end of the Paleozoic era, and only five genera of small, insignificant lycopods have survived to the present. These are often referred to as living fossils because their closest relatives are long-extinct plants.

Trimerophytina

The Trimerophytina (Figure 18–2) were the last of the three groups of primitive vascular land plants to appear in the Early Devonian, about 380 million years ago, and are believed to be more complex descendants of the rhyniophytes. Here resides Dawson's *Psilophyton,* by no means the most primitive of early plants but rather characterizing the group from which all higher plants are believed to have descended. A variety of growth habits occurred, including trichotomous branching; and although sporangia were terminal, specialization of branches into fertile, often pendulous fronds, quite distinct from vegetative ones, occurred. Although *Psilophyton* was usually in the range of 30 to 60 cm, *Pertica* attained a height of nearly 3 m. Middle Devonian *Ibyka* is difficult to classify because it seems to be transitional between trimerophytes and pterophytes (ferns); other types suggest affinities with progymnosperms and pteridosperms (seed ferns). Thus the trimerophytes seem to represent the ancestral stock among early vascular plants for the evolution of higher plants.

Pterophytina

The Pterophytina comprise some 300 genera and 10,000 species of living ferns that range in size from tiny, floating water ferns to tree ferns 20 m high with 4 m leaves. Like all tracheophytes beyond the level of the zosterophyllophytes and the trimerophytes, the plant is differentiated into true roots, stems, and leaves. The stem is often a rhizome, lying horizontally in or on the ground. In the Tropics, where epiphytic ferns are common, rhizomes may be aerial; in tree ferns the stem is an upright trunk. The leaves are large fronds; again except in tree ferns, these may be the only aerial parts of the plant. The sporangia occur in clusters called *sori* on the undersurfaces of the leaves. The spores, when released, germinate on moist ground, usually to form monoecious gametophytes, which are always small, thalluslike plants. Some are dioecious. The sperm of one plant swim to the archegonia of another, where fertilization takes place. A new sporophyte then develops from the zygote.

Although the ferns most familiar in temperate latitudes are all of moderate size, in the Tropics there are tree ferns in which the stem forms an erect trunk as tall as 23 m, and the leaves form a palmlike cluster at the top.

Ferns are abundantly represented in the fossil record from the time of their origin in the Devonian to the present. Fernlike plants, showing the pedigree of trimerophytic origin, appear as early as the Early Devonian and are encompassed in the class Cladoxylopsida. In the Carboniferous, forests of ferns of the extinct class Coenopteridopsida, along with great lycophytes and spenophytes, rained their debris down into extensive swamps, resulting in the formation of vast deposits of coal around the world. True ferns (class Filicopsida, order Filicales) appeared in the Carboniferous. The common cinnamon fern of eastern North America (*Osmunda cinnamonea*) is a member of the family Osmundaceae that traces back to the Permian. The family Polypodiaceae, which includes the majority of living ferns, did not appear until the Late Cretaceous. Thus, though ferns are of ancient pedigree, the modern fern flora did not take form until the Cenozoic.

PROGYMNOSPERMS AND EVOLUTION OF HIGHER VASCULAR PLANTS

In 1960 C.B. Beck recognized that an assemblage of Middle Devonian to Lower Carboniferous arborescent plants—among them, *Aneurophyton, Tetraxylopteris,* and *Archaeopteris*—shared properties that placed them intermediate between trimerophytes and gymnosperms. Progymnosperms (Figure 18–6), some of which were trees 30 m tall, combined the woody characteristics of gymnosperms, including abundant secondary xylem (and thus active cambium), with the reproductive characteristics of ferns. Within the group, progressive trends took place toward the development of leaves and of heterospory (the development of smaller numbers of larger megaspores that produce female gametophytes and larger numbers of smaller microspores that produce male gametophytes). Some Upper Devonian megaspores exceeded 2 mm in diameter. Continuation of this trend seemingly resulted in the formation of true seeds and pollen—the hallmarks of the higher vascular plants. A seed is a structure in which a single megaspore occupies a megasporangium, which is in turn invested with an integument. A *gymnosperm* ("naked seed") is a plant in which

Figure 18–6.   Archaeopteris, *a progymnosperm.*

the seed is borne on the surface of an appendage (e.g., scales of a cone); by contrast, an *angiosperm* ("vessel seed") is a plant in which the seed is enclosed within an ovule. The oldest known seed is *Archaeosperma* ("ancient seed") from the Upper Devonian of Pennsylvania; this gymnospermous seed is 4 mm long, borne on a structure 15 cm long.

The development of seeds was a major evolutionary event that represented the final step in the conquest of land, for it made reproduction independent of water possible. For instance, in conifers pollen is carried to the female cones by the wind (to the chagrin of hayfever sufferers).

Typically, the seed-bearing plants are united in the subdivision Spermophytina, with two major classes: the Gymnospermae and the Angiospermae. Recently, however, it has become clear that at least two separate lines of descent occur among the gymnosperms. Thus gymnosperms represent a grade or level of organization, not a natural monophyletic clade. Taylor has responded by creating five separate divisions in the place of a single class Gymnospermae.

Conifers

Conifers today include some 50 genera and 550 species of familiar, sometimes majestic, arborescent plants. Some are shrubs; most are evergreen. *Sequoia,* or California redwood, attain heights of nearly 100 m; bristlecone pines (*Pinus aristata*) reach ages of 5,000 years. Conifers differ most strikingly from lower tracheophytes in their reproductive cycles. At the outset the gymnosperms produce two types of spores, megaspores that develop into female gametophytes and microspores, or pollen, that develop into male gametophytes. In spite of their names, the two types may be of equal size, or the microspores may actually be the larger. The cones in which the spores are formed consist of specialized sporophylls, modified leaf clusters spirally arranged about a central axis. The microspores divide while still within the spore wall to form the male gametophytes. The pollen is shed in great quantity while in the four-cell stage and is carried by the wind, sometimes for great distances. Some of the pollen will reach female cones and become stuck in a sticky fluid exuded by the ovules, complex structures that include the female gametophyte. The pollen produces a tubelike growth that enters the ovule. Down this pollen tube pass the two sperm nuclei, one of which fertilizes the egg. The developing embryo (sporophyte) pushes into the mass of the female gametophyte, now called the *endosperm,* which serves as nutritive material for the embryo. This endosperm is in turn surrounded by a seed coat, which is actually a part of the parent sporophyte. The seed is shed; and if it falls on favorable ground, the seedling may develop into a mature sporophyte.

Some aspects of this reproductive cycle are especially noteworthy. For the first time in the phylogenetic series, reproduction is independent of water. The pollen is carried to the female cones by wind, and the sperm—mere nuclei rather than flagellate cells—are carried to the egg by the protoplasmic pollen tube. Both the male and female gametophytes are reduced to minute structures consisting of only a few cells; whereas the relative predominance of the sporophyte in the life cycle has become great. The gametophytes lack chlorophyll and are completely dependent upon the sporophyte. Finally, the seed, a new structure in the phylogenetic series, consists of an embryo (sporophyte, $2n$) contained within the endosperm (gametophyte, $1n$), which is in turn contained within a seed coat that is sporophytic tissue ($2n$) of the parental generation. Thus the appearance is very much as though the embryo sporophyte were produced directly by the parent sporophyte, with the gametophytes being simply organs of the parent sporophyte. Only comparison to ancestral plants reveals the true situation.

Conifers have an ancient pedigree. The order Coniferales dates from the Triassic, and indeed most living families date from then as well. Paleozoic antecedents of the conifers were the cordaites, some of which towered 40 m in Carboniferous swamps. These trees bore a whorl of branches in a crown near the top. Leaves were straplike, up to one meter long. Cordaites are believed to have arisen directly from progymnosperms in the Upper Devonian.

Other Gymnospermous Plants

There are two orders of nonconiferous living gymnospermous plants that deserve mention, ginkgos and cycads. Only a single species of ginkgo, *Ginkgo biloba*, the maidenhair tree, survives. It may be extinct in the wild but has been cultivated for centuries in temple gardens in China. The symmetrical, fan-shaped leaf is quite distinctive. Ginkgos have a good fossil record that takes them back through the Mesozoic and into the Permian, where they appear to have arisen from seed ferns. Cycads are an assemblage of ten genera and a hundred species of somewhat palmlike trees. The unbranched stem may be short or up to 10 meters long, with a terminal crown of long, leathery, compound leaves. As in ginkgos, sperm are mobile and flagellate, probably a retention from aquatic ancestors. The Mesozoic has been called the age of cycads, which flourished then. Cycads trace back to the Permian for certain and quite possibly into the Pennsylvanian. Their ancestry too appears to lie within the seed ferns (Figure 18–7).

The extinct pteridosperms, or seed ferns, hold the key to unraveling the relationships of gymnospermous plants. These arborescent plants resembled ferns in their foliage but had the wood, and, more importantly, the seeds of higher plants. They appeared in the Upper Devonian, possibly from trimerophyte ancestors (i.e., they were independent of the progymnosperms). Seed ferns showed great diversity in the later Paleozoic and in the first half of the Mesozoic. Then they dwindled but lasted into the Cretaceous. They in turn gave rise to cycads, to ginkgos, and possibly to angiosperms as well.

Flowering Plants

Angiosperms ("vessel seeds"), or flowering plants, dominate the world today, numbering something like 10,000 genera and perhaps nearly 300,000 species. Formerly considered the class Angiospermae, taxonomists now tend to recognize them as a separate division, for which they use the names Anthophyta or Magnoliaphyta. The familiar term angiosperm is unlikely to disappear, however. The variety of the angiosperms is enormous; they range from great trees to grass. Generally they are land plants, but they have become adapted to almost every available habitat, including marine; and while they are typically free-living green plants, not a few have become parasitic and some are saprophytic. Angiosperms include all of our food crops and many of our medicines.

The angiosperms share the major characters of the gymnosperms, being differentiated into true roots, stems, and leaves and reproducing by means of true seeds (but which are enclosed within the ovule—hence the name angiosperm meaning "vessel seed"). They have a highly developed vascular system. Fertilization is by means of pollen that are independent of water. The sporophyte is much the dominant generation; the gametophyte is minute and completely dependent. In addition, the sporangia of the angiosperms are within the flowers, which are modified cones surrounded by modified and often highly

Figure *18–7.*   *A reconstruction of* Medullosa, *a Carbon-iferous pteridosperm.*
Courtesy of Dr. H.W. Pfefferkorn, University of Heidelberg.

ornamental leaves (Figure 18–8). The microspores develop into pollen (male gametophytes) with only three nuclei: one tube nucleus and two sperm nuclei. The megaspores develop into female gametophytes with only eight nuclei, including the egg nucleus and two polar nuclei, that fuse to form the fusion nucleus. The macrospores are completely enveloped within modified sporophylls, the carpels, which become the fruit when mature. Pollination may be effected by wind as in the gymnosperms; but it is commonly effected by insects, occasionally by birds or bats, and rarely, by water (in the case of aquatic plants). Colorful and often fragrant flowers are the result of natural selection for attraction of specific pollinators. Coevolution of plants and pollinators is a major theme.

A unique characteristic of the angiosperms is the phenomenon of double fertilization. As usual, a single sperm nucleus unites with the egg nucleus to

Figure 18–8.   *Magnolia blossom, a primitive flower.*

form a diploid zygote from which the embryo develops. But in addition to this, the other sperm nucleus unites with the fusion nucleus to form a triploid (3n) cell from which the bulk of the endosperm is formed. In contrast, recall that the endosperm of gymnosperms remains haploid, typical gametophyte tissue. In both gymnosperms and angiosperms, the endosperm serves as a nutritional reserve for the embryo, but karyotypes contrast very strongly. Figure 18–9 illustrates the life cycle of a typical angiosperm.

Over a century ago, Darwin described the origin of the flowering plants as "an abominable mystery," and there the problem remained until recently. Currently, botanists are more optimistic. The applicable fossils are wood, leaves, fruits, and pollen. Of these, pollen have proven to be the most useful. We are now almost certain that the angiosperms arose during the Early Cretaceous. Supposed earlier genera have not fared well. Some, like *Palmoxylon*, proved to be genuine angiosperms, but of later date; others, like *Sanmiguelia*, were of early date (Triassic, in this case) but turned out to be gymnosperms that simply approached the angiosperms in some vegetative characters. The earliest Cretaceous deposits show no trace of angiosperms, although they have yielded some 1,500 species of gymnosperms and around 1,000 species of ferns and other lower tracheophytes. Among the gymnosperms, great variety existed, and many approached the angiosperms in some of their characteristics. In mid-Lower Cretaceous, a few types of apparently angiospermous pollen occur, and the same may be said for the simplest types of angiospermous leaves. Both become more abundant and more complex in mid-Cretaceous, and by Late Cretaceous some 20,000 species of angiosperms are known. More fossil evidence will be needed to identify the particular group that gave rise to the angiosperms, but current studies on fossil pollen are promising; odds would seem to favor an origin from seed ferns.

Whatever their origins, angiosperms rapidly became the dominant plants of the world and appear still to be on the increase. They are split into two

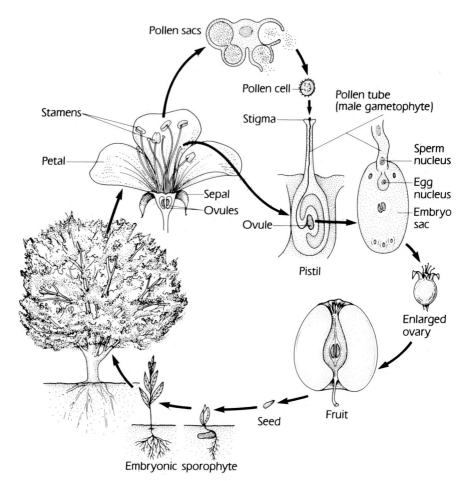

Figure 18–9. Life cycle of a flowering plant.

major groups (subclasses if angiosperms are held to be a class), the Dicotyledoneae and the Monocotyledoneae. The trait to which the names refer is the *cotyledon,* or seed leaf, two of which appear during embryogenesis in the Dicotyledoneae, and one in the Monocotyledoneae. A host of vegetative and reproductive characters correlate with this apparently simple difference. Another simple character is seen in the flowers: monocot flowers have three sets of parts, dicots have four or five. The 55,000 species of living monocots include the grasses (7,500 species), the lilies (4,200 species), the palms (2,600 species), and the vast assemblage of orchids (25,000 species). The 155,000 or more species of dicots include 2,500 species of potatoes; 3,000 species each of roses (a family that includes apples, peaches, strawberries, and many other fruits), carrots, and mustards; 4,500 species of mints; 8,000 species of euphorbs; 14,000 species of legumes; and 20,000 species of composites.

MAIN LINES OF PLANT EVOLUTION

We may now summarize the main lines of plant evolution with the aid of Figure 18–10. The six divisions of eucaryotic algae probably arose from the protoeucaryote by symbiosis of blue-green bacteria within the eucaryotic

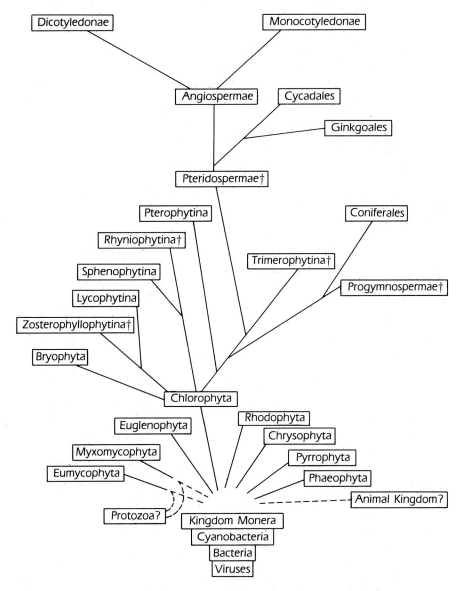

Figure 18–10. Summary of the probable lines of plant evolution. The dashed lines present highly problematical alternatives, and the daggers indicate extinct groups.

cell. Whether the different algal divisions arose independently or from a common stock is scarcely indicated by any available evidence. The origin of the two divisions of molds—Myxomycophyta and Eumycophyta—is also much in doubt. They may have arisen directly from the protoeucaryote, or from amoeboid protozoans, or from more-advanced green algae by loss of their chloroplasts. However much vexed the problem of the origins of these primitive groups may be, it appears to be probable that the two divisions of land plants—Bryophyta and Tracheophyta—arose from the green algae independently. The bryophytes have differentiated into three minor groups—the mosses, liverworts, and hornworts—but have not produced any more progressive types of plants. The primitive tracheophytes colonizing the land at the end of the Silurian quickly gave rise to varied types of spore-bearing lower tracheophytes that became important plants in the Carboniferous swamp habitats but were reduced in importance by the end of the Paleozoic. These were especially the lycophytes (club mosses), sphenophytes (horsetails), and pterophytes (ferns). The conquest of the land and the development of efficient vascular systems, roots, leaves, secondary xylem, and seeds all occurred in a span of only about 60 million years, from the Late Silurian to the Late Devonian, a remarkable period in the evolutionary history of plants. Structural innovations from that time until the Cretaceous, when angiosperms arose, were much more modest compared with this rapid outburst. Gymnospermous plants were separately derived from advanced spore bearers: conifers were derived from progymnosperms, with modern families appearing as early as the Triassic; and seed ferns were derived from trimerophytes. Seed ferns gave rise to cycads (which peaked in the Mesozoic), to ginkgos, and probably to angiosperms. Angiosperms underwent explosive adaptive radiation in the Late Cretaceous and in the Cenozoic; and these plants dominate the world flora today.

Evolution in the plant kingdom shows several general trends. Among the most primitive plants, the evolution of mitosis and meiosis has been a major trend. Associated with it were the evolution of sexual reproduction and the alternation of generations. Development of the colonial habit and multicellularity is a third major trend. Still another trend has been a transition from aquatic to terrestrial habitats, but little specific information is available regarding the steps in this process. Life on land required the development of vascular and strengthening tissues, among the most characteristic features of the land plants. These tissues, however, are found only in the sporophyte, so the sporophytic generation has increased in size and functional importance relative to the gametophytic generation until finally, in the angiosperms, the gametophytic generation is recognizable as distinct from the sporophyte only by comparison with more primitive plants. This trend is illustrated in Figure 18–11. Another important trend, especially in the angiosperms, has been coevolution with animals, particularly with the insect vectors of pollination and with herbivores (see Chapter 9 for details and examples).

Finally, in a general way, there has been a phylogenetic tendency for plants to increase in size. It is obvious among the algae, where the most

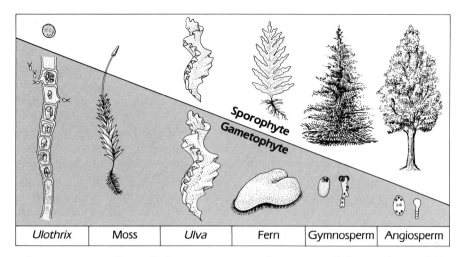

Figure 18–11. *The gradual increase in size and importance of the sporophyte and the corresponding decrease in size and importance of the gametophyte.*

primitive species are all unicellular and the more-advanced ones may be (but need not be) multicellular. While the most primitive land plants were smaller than large algae, all of the classes of tracheophytes have produced species much larger than the largest of the algae. Nonetheless, that evolution is necessarily accompanied by progressive size increase is far from axiomatic. The most successful members of many groups are characterized by small size. The club mosses and horsetails, for example, were once represented by great trees, but only the smaller members of the group have survived to the present. Among the angiosperms, the trees seem to have been primitive forms from which the shrubs and grasses have evolved. This reduction has been accompanied by the evolution of the annual habit, presumably as an adaptation to prevent extinction by winter-killing, because a dormant seed may easily survive severe weather that would kill a mature plant. Alternatively, annual plants may be best able rapidly to exploit resources that are only briefly abundant in climates with contrasting seasons. Thus acquisition of small size, rapid growth and maturity, and the annual habit constitute an adaptation to arctic, subarctic, and arid conditions; and such plants appear to be on the increase in temperate lands as well.

SUMMARY

In order for plants to invade the land, they had to confront the problem of desiccation. Thus all land plants retain the developing embryo within the sex organs of the maternal plant, a step that was accomplished during the

Silurian period, because land plants appear in Upper Silurian rocks more than 400 million years old. The most primitive of the living plants are the *bryophytes* (mosses, liverworts, and hornworts), which lack differentiated roots, stems, and leaves, as well as cells specialized for support and nutrient transport. Bryophytes show much in common with green algae and may have derived either directly from algae or from early vascular plants by secondary loss of vascular tissue.

Three groups of Upper Silurian–Lower Devonian land plants appear to have given rise to all vascular land plants (*tracheophytes*) that we know today. *Rhyniophytes* are the simplest of all known vascular plants. They lacked roots or leaves and stood erect, with sparse dichotomous branches bearing terminal sporangia. Living *Psilotales* appear to be little-changed descendants of rhyniophytes. Horsetails (*sphenophytes*) are rhyniophyte descendants with certain distinct advances: they had developed roots, stems, and leaves. Horsetails appeared in the Late Devonian and grew to heights of 10 m in the Carboniferous coal swamps.

Zosterophyllophytes, the second basal group of vascular plants, were slightly more advanced then rhyniophytes, carrying sporangia in globose clusters and sometimes carrying spines on the stem. The club mosses (*lycophytes*) are zosterophyllophyte derivatives. Club mosses have good roots, stems, and extensive leaves, all of which are vascularized. The sporangia are enclosed in specialized leaves, clusters of which, called *strobili,* occur at the tips of branches. Club mosses appeared during the Early Devonian and formed 40 m tall giants in the Carboniferous coal swamps.

Trimerophytes were the last and most complex of the basal groups of vascular plants to appear. These rhyniophyte derivatives showed a variety of growth habits, including trichotomous branching; and they showed a specialization of fertile fronds, which were distinctive from sterile, vegetative fronds. They probably gave rise to ferns (*pterophytes*), seed ferns (*pteridosperms*), and *progymnosperms.* Ferns appeared in the Early Devonian, were major contributors to Carboniferous coal swamps, and remain important today, particularly in the Tropics, where they form trees 20 m high. In ferns, the stems are often rhizomes, and the sporangia, clustered together as sori, occur on the undersurfaces of leaves.

Progymnosperms were a group of arborescent Devonian and Lower Carboniferous plants that were intermediate between trimerophytes and gymnosperms. They exhibited the woody characteristics of gymnosperms (with active cambium and secondary xylem) and the reproductive characteristics of ferns. A progressive trend in progymnosperms was the development of *heterospory,* in which large megaspores produced female gametophytes and abundant small microspores produced male gametophytes. Continuation of this trend resulted in the final step in the conquest of the land—the formation of seeds and pollen, a step achieved during the Late Devonian. The two great assemblages of seed-bearing plants are the *gymnosperms* ("naked seeds") and the *angiosperms* ("vessel seeds"). *Conifers* are dominant gymnosperms today and have existed since the Triassic; their antecedents, the *cordaites,* arose from the progymnosperms in the Late Devonian and were major plants in the Carboniferous coal swamps. The seed ferns (*pteridosperms*) arose from trimerophytes independently of progymnosperms and gave rise to the other naked seed groups—the *cycads* and the *ginkgos;* they may also have given rise to angiosperms in the Early Cretaceous.

The angiosperms, or flowering plants, underwent explosive diversification in the Late Cretaceous and became the dominant plants on land, numbering nearly 300,000 species today. The enclosure of seeds within the ovule correlates with a variety of other features that led to efficient reproduction, dispersal, and growth. Attractive flowers, fruits, and nectars are features that coevolved with animals to ensure the propagation of the plants. Angiosperms are enormously important to humans, both as food crops and as medicines.

REFERENCES

Allan, M. 1977. *Darwin and His Flowers: The Key to Natural Selection.* Taplinger Publishing, New York.
  A biography emphasizing Darwin's botanical research, which was more extensive than is often realized by zoologists.
Banks, H.P. 1968. The early history of the land plants. In *Environment and Evolution,* edited by E.T. Drake, pp. 73–107. Yale University Press, New Haven, Conn.
  A summary and interpretation of data on the origin of the higher plants.
Banks, H.P. 1970. *Evolution and Plants of the Past.* Wadsworth, Belmont, Calif.
  A well-illustrated short survey of fossil plants—reasonable starting place for further inquiry.
Beck, C.S., ed. 1976. *Origin and Early Evolution of Angiosperms.* Columbia University Press, New York.
  This subject is revitalized with important essays by major workers in the field.
Bold, H.C., C.J. Alexopoulos, and T. Delevoryas. 1980. *Morphology of the Plants and Fungi.* 4th ed. Harper & Row, New York.
  A thorough systematic treatment going far beyond the outline presented here.
Delevoryas, T. 1977. *Plant Diversification.* 2nd ed. Holt, Rinehart and Winston, New York.
  A concise introduction to paleobotany and plant phylogeny.
Doyle, J.A. 1977. Patterns of evolution in early angiosperms. In *Patterns of Evolution as Illustrated by the Fossil Record,* edited by A. Hallam, pp. 501–546. Elsevier, Amsterdam and New York.
  A good, concise account of pollen analysis as applied to the problem of the origin of angiosperms.
Jones, S.B., Jr., and A.E. Luchsinger. 1979. *Plant Systematics.* McGraw-Hill, New York.
  A very useful book explaining the botanical code and the application of the principles of plant taxonomy, with emphasis on examples from the angiosperms.
McAlester, A.L. 1977. *The History of Life,* 2nd ed. Prentice-Hall, Englewood Cliffs, N.J.
  This brief book presents much valuable information. It develops plant taxonomy along the same lines as we have, following principles established many years earlier by O. Tippo.

Stebbins, G.L. 1974. *Flowering Plants: Evolution above the Species Level*. Belknap Press, Harvard University Press, Cambridge, Mass.

The first complete reexamination of flowering plant evolution in many years, presented by one of its leading students.

Stewart, W.N. 1983. *Paleobotany and the Evolution of Plants*. Cambridge University Press, Cambridge, England.

A comprehensive, well-illustrated, and highly readable survey of current thinking on the history and phylogeny of plants.

Taylor, T.N. 1981. *Paleobotany*. McGraw-Hill, New York.

An excellent and up-to-date introduction to paleobotany.

Tippo, O., and W.L. Stern. 1977. *Humanistic Botany*. Norton, New York.

Sound botany with the emphasis on human applications.

19

Animal Evolution

As pointed out in Chapter 16, the Protozoa, the most primitive of animals, appear to have derived from primitive flagellate algae. Within single genera among the euglenoids there may be some species that show predominantly plant characters and others that show predominantly animal characters. Thus a high probability exists that the flagellates are close to the point of separation of the two kingdoms, if, indeed, the separation is complete here, for flagellates like *Trypanosoma* are undoubtedly animals; others like *Chlamydomonas* are undoubtedly plants; and the intermediate group, typified by *Euglena,* defies any indisputable assignment to the kingdoms.

DIVERSIFICATION OF PROTOZOA

Since 1845, when von Siebold separated unicellular animals into the phylum Protozoa, an immense literature has accumulated. By the end of the nineteenth century, the familiar four classes (Mastigophora, Sarcodina, Sporozoa, and Ciliata) had all been delineated. Exploring relationships among them, biologists debated whether the phylum corresponded to a natural unit or whether each class should be a phylum. As knowledge of protozoan taxonomy has become vast, however, protozoologists have become ever more conscious of the limitations of this nineteenth-century framework, which expresses neither the diversity of the group nor its possible polyphyletic origins.

The Society of Protozoologists in 1964 adopted a classification recognizing four subphyla and twelve classes within the phylum Protozoa. Soon, however, a consensus developed that they had not gone far enough. A new committee was appointed to review the problem, and this committee introduced a thorough revision in 1980, recognizing a subkingdom Protozoa with seven phyla and some two dozen classes. Clearly, analysis of relationships within this vast realm of more than 65,000 species must be left to specialized works on protozoology, but Table 19–1 will trace some of the broader lines. Because the 1980 revision has addressed the major concerns of protozoologists, we hope that this revision will remain serviceable well into the twenty-first century.

Phylum Sarcomastigophora

The phylum Sarcomastigophora is a great assemblage that includes all of the flagellate and amoeboid protozoans. It unites three subphyla: Mastigophora, Opalinata, and Sarcodina.

Subphylum Mastigophora.   The subphylum Mastigophora is the old class of the same name. It now comprises two classes: the Phytomastigophorea (treated in Chapter 17 as several divisions of algae), and the Zoomastigophorea. The latter is an extraordinarily varied array, which may share ancestry with other groups of protozoans and with the Metazoa.

Several groups of flagellates deserve special mention. The proteromonads include a wide variety of colorless flagellates, typically with two flagella. Reproduction is always asexual. Parasitic species may undergo multiple divisions, which are difficult to distinguish from spore formation. They probably share common ancestry with one or more of the spore-forming phyla. In the latter, however, sexual reproduction has also evolved. In closely related orders are *Trypanosoma* (Figure 19–1a), a genus parasitic in the blood of vertebrates (including that of *H. sapiens*), and the choanoflagellates. The latter possess a protoplasmic collar encircling the base of the flagellum, and the cell is called a choanocyte. The evolutionary importance of choanoflagellates lies in the fact that they share choanocytes with sponges and that colonial choanoflagellates, such as *Proterospongia* (Figure 19–1b), resemble simple sponges.

The order Hypermastigida includes the most complex of flagellates, with numerous flagella arranged in a specific pattern and with complex organelles. *Macrospironympha* (Figure 19–1c) is typical. All hypermastigids are parasites or commensals in the digestive tract of termites or cockroaches. They are essential to the nutrition of the host, for they digest the cellulose that the host eats. The host starves if the protozoans are removed experimentally.

Table 19–1. The subkingdom Protozoa, as classified by the Society of Protozoologists (1980)

Category	Comments
Phylum Sarcomastigophora	Includes former classes Mastigophora and Sarcodina as well as parts of former classes Sporozoa and Ciliata.
Subphylum Mastigophora	The flagellates.
Class Phytomastigophorea	Plantlike flagellates, the algae of Chapters 17 and 18; thirteen orders, some, like Euglenida, equivalent to divisions of the preceding chapters.
Class Zoomastigophorea	Animal-like flagellates; eight orders, including Choanoflagellida (*Proterospongia*), Rhizomastigida (*Mastigamoeba*), and Hypermastigida (*Trichonympha*).
Subphylum Opalinata	Leaf-shaped, ciliated, but with only one kind of nucleus; parasitic in cloaca of anurans and a few other vertebrates; may be transitional between flagellates and ciliates.
Subphylum Sarcodina	Protozoans with pseudopodia.
Superclass Rhizopoda	Pseudopodia for locomotion and feeding.
Class Lobosea	Several orders, including the best-known amoebae, such as *Amoeba* and *Entamoeba*, and shelled amoebae, like *Arcella*.
Class Eumycetozoea	The slime molds.
Class Granuloreticulosea	Includes the foraminiferans.
Five more classes	
Superclass Actinopoda	Four classes, all having slender, raylike pseudopodia; includes radiolarians and heliozoans.
Phylum Labyrinthomorpha	Small, recently described group, not well known at present.
Phylum Apicomplexa Phylum Microspora Phylum Ascetospora Phylum Myxozoa	Collectively, these comprise the old class Sporozoa; all are parasitic, no locomotor organelles, sexual reproduction usual, life cycles often complex.
Phylum Ciliophora	The ciliates, including the old classes Ciliata and Suctoria; micronucleus (genetic) and macronucleus (metabolic); sexual reproduction by conjugation; largest of the protozoan classes, its taxonomy has become very complex.

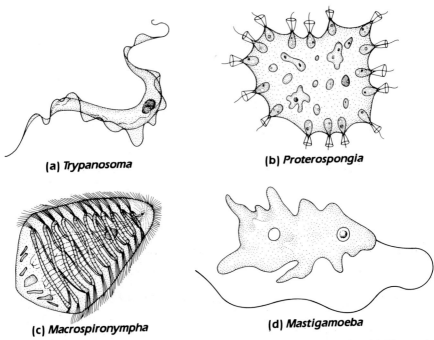

(a) *Trypanosoma*

(b) *Proterospongia*

(c) *Macrospironympha*

(d) *Mastigamoeba*

Figure 19–1. *Mastigophorans:* (*a*) Trypanosoma, (*b*) Proterospongia, (*c*) Macrospironympha, *and* (*d*) Mastigamoeba.

The order Rhizomastigida is of special interest because its members appear to be intermediate between the flagellates and the Sarcodina—amoeba and its allies. Many flagellates are capable of amoeboid movement, but members of the order Rhizomastigida possess a flagellum and are also permanently amoeboid, and so they make a nice connecting link between the subphyla. *Mastigamoeba* (Figure 19–1d) is a good example.

Subphylum Opalinata. The inclusion of the Opalinata in the phylum Sarcomastigophora is an interesting feature of the new classification. This small group, most of which are parasites in the cloaca of amphibians, had formerly been treated as aberrant ciliates. Although they have a complete coat of cilia, they lack the nuclear dualism of the ciliates. This placement of the opalinids was intended to suggest that the group may be transitional between flagellates and ciliates. If proven, this will have great theoretical importance.

Subphylum Sarcodina. The amoeboid organisms that make up the subphylum Sarcodina probably derived from ancient rhizomastigids. There are two superclasses: Rhizopoda, including the more typical amoebae, and Actinopoda, including radiolarians and heliozoans, which have slender, raylike

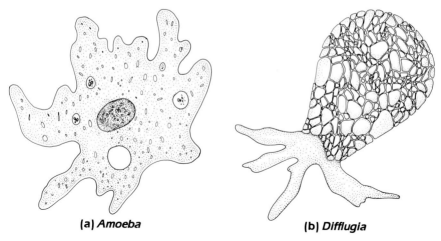

(a) *Amoeba*   **(b)** *Difflugia*

Figure 19–2. Sarcodinans: (a) Amoeba *and (b)* Difflugia.

pseudopodia. By far the best-known member of this or of any protozoan group is *Amoeba proteus* (Figure 19–2a), the familiar study material of every elmentary biology laboratory and the traditional example of a primitive animal. *Amoeba proteus* is typical of the class Lobosea, which also includes shelled organisms such as *Difflugia* (Figure 19–2b) and parasites like *Entamoeba histolytica,* which is found in the human digestive tract. The superclass Rhizopoda also includes the Foraminiferida, a marine order characterized by complex and often beautiful multichambered shells, through the pores of which extend many filamentous pseudopodia.

The rhizopods are generally regarded as terminal in evolution, yet they share amoeboid characteristics with the Myxomycophyta and the Eumycophyta, and it has been suggested that they may also share ancestry. The new classification treats the slime molds as the class Eumycetozoea, within the superclass Rhizopoda.

Included in the superclass Actinopoda are the radiolarians, floating marine protozoans with a central, chitinlike capsule and a siliceous skeleton, and the heliozoans, a somewhat similar group found mainly in fresh water. Both have many slender, semipermanent pseudopodia, which radiate from the cell body. Radiolarians are abundant in the fossil record.

Phyla Apicomplexa, Microspora, Ascetospora, and Myxozoa

The four phyla Apicomplexa, Microspora, Ascetospora, and Myxozoa make up the old class Sporozoa (Figure 19–3). All reproduce sexually, often with complex life cycles, and all are parasitic. They share characteristics with parasitic proteromonads, with which they may also share ancestry. They are all terminal groups, and we will, therefore, not follow them further.

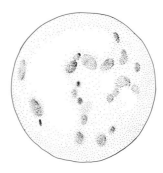

Figure 19–3. *A sporozoan,* Plasmodium, *in a red blood cell.*

Phylum Ciliophora

The phylum Ciliophora is rather isolated from other protozoans, although the opalinids may link them to the flagellates. Ciliates have a definite shape, which is maintained by a pellicle. There is a definite anteroposterior axis; and the animal is symmetrical, sometimes radially (more accurately, monaxially), sometimes bilaterally, or often it is an irregular deviant from one of these. Nutrition is holozoic, with minute organisms being ingested by mouth.

Protozoans are often said to be the simplest of animals. We may debate this statement, but there is little doubt that ciliates are the most complicated of cells. The cilia may be arranged in coordinated tracts that sweep a food-bearing current to the mouth, they may be fused to form undulating membranes, or they may be fused in tufts to form leglike cirri. Their movements are coordinated by a neuromotor system that exceeds in complexity the nervous systems of the simpler Metazoa. Alternating with the cilia are trichocysts, small bodies under the pellicle that can discharge filaments. Food vacuoles form at the mouth, pass through the cell by a regular route, and discharge at a definite point. In some species, it would scarcely be an exaggeration to say that there is a digestive tube and an anus. There are two nuclei, a micronucleus concerned with heredity and a macronucleus controlling metabolism. Finally, the ciliates have evolved a unique type of sexual reproduction called *conjugation.* The details are complicated, but, in brief, the maturation divisions result in a stationary and a wandering nucleus in each of a joined pair of conjugants, and the wandering nucleus of each fertilizes the stationary nucleus of the other.

The most primitive ciliates, exemplified by *Paramecium* (Figure 19–4a), are completely clothed in cilia and are strong swimmers. Specialization in ciliates has generally resulted in attachment to the substrate, restriction of cilia to limited areas related to feeding, and specialization of the ciliary tracts to form membranelles or undulating membranes. *Vorticella* (Figure 19–4b) is a good example of such specialization. One order, the Hypotrichida (Figure

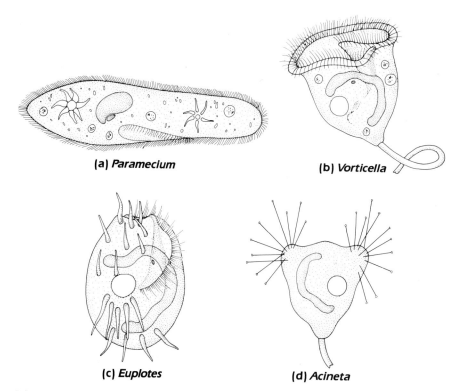

(a) *Paramecium*   **(b)** *Vorticella*

(c) *Euplotes*   **(d)** *Acineta*

Figure 19–4. *Ciliates: (a)* Paramecium, *(b)* Vorticella, *(c)* Euplotes, *and (d)* Acineta, *a suctorian.*

19–4c), has specialized for increased efficiency of locomotion. A band of typical cilia propels the feeding current; but locomotion is based upon tufts of fused cilia, the cirri, which function in a leglike fashion.

The suctorians (Figure 19–4d) are a small subclass of ciliates. The adults have neither cilia nor other locomotor organelles and they capture their food by means of tentacles. They are classed with the ciliates because they have the two kinds of nuclei and reproduce by means of conjugation. Further, the suctorian zygote develops into a free-swimming, ciliated organism, which only later settles down on a substrate and becomes a typical suctorian. This development suggests an embryological recapitulation of ancestral history.

ORIGIN OF METAZOA

As L.H. Hyman points out, no direct proof of the origin of the Metazoa (multicellular animals) from the Protozoa exists; yet the discussion of the origin of the Metazoa usually revolves around the question of which protozoan stock is the most probable progenitor of the Metazoa. Two broad possibilities exist by which the Metazoa could have evolved from the Protozoa. First, repeated nuclear division without cytoplasmic division might have led to formation of a plasmodium, like some of the Heliozoa. Formation of cell membranes would then result in multicellularity, and differentiation might then lead to the true multicellular individual. Second, the differentiation of cells within a colony of Protozoa, comparable to *Volvox,* might have led to interdependence and individuality.

A very different third possibility has been suggested. In protozoan colonies each cell ingests food, but even simple metazoans use a new method of feeding, one in which a digestive tract feeds for the whole organism. This transition may be difficult. A.C. Hardy suggests that simple plants, like *Volvox,* living in an environment deficient in nitrates and phosphates, may have satisfied the deficiency by capturing smaller organisms. Increasing utilization of this nutritive pathway, together with loss of photosynthesis, would lead to a simple metazoan. Insectivorous plants show the feasibility of such a nutritive mechanism, and the fact that unicellular plants seem to have given rise to protozoans more than once lends plausibility to the suggestion that multicellular plants may have achieved animalization at least once.

Paleontology is of no help in this problem, for the Metazoa were already well established at the beginning of the Cambrian. Probably, therefore, the origin of the Metazoa will always be speculative. But most zoologists favor the flagellates, through the differentiation and integration of the cells of a colony, as the most probable progenitors of the Metazoa. Their reasons are many. The flagellates are a highly variable group, which appear to have given rise to many groups of plants and to several, perhaps all, other groups of Protozoa. Further, some groups of flagellates show a tendency to form colonies of ever-increasing size and complexity. The evolution of sex occurred here, and the colonies are definitely divided into somatic and germinal "tissues." Oogamous reproduction is the rule for such colonial flagellates. The sperm are at once similar to some simpler, noncolonial flagellates and to the typical sperm of Metazoa. Such colonies may also show anteroposterior differentiation. These large, highly specialized colonies occur principally among the plant flagellates, yet colonies of a highly suggestive character also occur among the animal flagellates. None of these reasons is conclusive, yet collectively they carry considerable weight.

However much the origin of other Metazoa may be disputed, the Porifera (sponges) probably derived from choanoflagellates. It is a short step from the structure of the colonial choanoflagellate *Proterospongia* to that of the simplest sponges. *Proterospongia* (Figure 19–1b) consists of a small mass of gelatinous material in the surface of which are imbedded choanocytes (collared, flagellated cells) and in the interior of which are ameboid cells. The choanocytes can withdraw the collar and flagellum and move into the interior to become ameboid cells. In order to change to the structure of a simple sponge it would be necessary only to develop a system of channels through the gelatinous mass, to line these channels with choanocytes, and to cover the outer surface with a simple epithelium. No organ systems are present; the component cells individually carry on the functions of the sponge.

Evolution within the Porifera (Figure 19–5) has taken the form of elaboration of the canal system and of the supporting spicules or fibers. As relationships within the phylum are not at all clear, it may be more profitable to go directly to the problem of the relationships of the Porifera to other animals. The Porifera are a terminal group. Further, because they are so different from all other Metazoa in the absence of unified tissues, in the physiological independence of the individual cells, and in their embryology and adult anatomy, it is

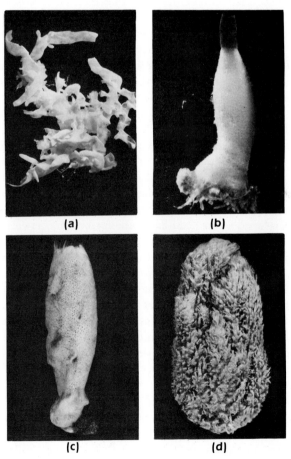

(a)   (b)

(c)   (d)

Figure 19–5. Sponge diversity. (a) Leucosolenia; (b) Grantia; (c) Euplectella, *a glass sponge;* (d) *a bath sponge.*

generally believed that their origin from the Protozoa must have been independent of that of the rest of the Metazoa. The Porifera are therefore assigned to a separate branch of the subkingdom Metazoa, the branch Parazoa, in contrast to the Eumetazoa. Some even claim that there should be a subkingdom Parazoa.

Gastrea Theory

Speculation on the origin of the Metazoa was long dominated by the Gastrea theory of E. Haeckel, a theory based on the literal application of the biogenetic law. In its original form Haeckel interpreted the egg as corresponding to an amoeboid ancestor, possibly to *Amoeba* itself. As evidence in favor of this, he pointed to the amoeboid eggs of sponges and of some coelenterates. Other types of eggs he assumed to be secondary specializations. The egg, of course, undergoes the cleavage divisions that result first in solid morula and then in a hollow ball of cells, the blastula. The morula was interpreted as corresponding to a simple hypothetical amoeboid colony, the *Synamoeba;* whereas the blastula was supposed to correspond to a colonial ancestor, the *Blastea,* which was more or less comparable to *Volvox,* but amoeboid rather than flagellate. The modern exponents of this theory assume flagellate rather than amoeboid ancestors for the reasons stated earlier. There being only a single layer of cells in the *Blastea,* all cell functions were at first shared by all cells, but then a division of labor occurred, with the posterior cells assuming the nutritive functions. These cells then invaginated, with the result that the organism became a two-layered gastrula, having an outer layer of flagellate cells (ectoderm) and an inner layer of digestive cells (entoderm). Haeckel called this hypothetical organism the *Gastrea,* and he believed it to be ancestral to all Eumetazoa. Some of the coelenterates he regarded as living gastreads. Next, the *Gastrea* developed a third cell layer, the mesoderm, between the first two. Haeckel thought that all of the structures of the higher phyla derived from these three layers. The development of the bottom-feeding habit led to elongation of the body and the formation of primitive worms, similar to the living Turbellaria. From these worms the more complex phyla developed.

The principal modern advocate of this theory, G. Jägerston, modifies it to account for gastrulation. He suggests that if the *Blastea* had settled on a substrate and adopted a crawling habit, it would have tended to elongate and become dorsoventrally differentiated. Ingestion of larger food particles would then be aided if the ventral cell layer were to invaginate. As the organism was already bilateral, he calls this the Bilaterogastrea theory.

The Gastrea theory is a beautiful simplification and synthesis of a vast amount of embryological, morphological, and taxonomic data, and it is almost without a serious competitor. Unfortunately, however, as Hyman pointed out, "it is probably one of those simplifications that are too beautiful to be true." Embryology is not a safe basis for construction of pedigrees, especially if comparison is made between embryos of advanced species and adults of their supposed ancestors, as Haeckel did (see Chapter 3). At most, embryology should be treated only as one of several corroborative lines of evidence. Moreover, even the embryological evidence does not give unequivocal support to the Gastrea theory. In the coelenterates, the group that is closest to the hypothetical *Gastrea,* gastrulation ordinarily occurs not by simple invagination of the posterior cells but by the wandering in of many cells from all parts of the blastula. This process at first produces not a typical gastrula but a ball of ectodermal cells filled by a solid core of entodermal cells. This type of larva is called a *planula.* Only later does this entodermal core hollow out and a mouth (blastopore) break through to form a typical gastrula. Although the type of gastrulation with which Haeckel dealt is known, for example, in the starfishes, it is not widespread in the animal kingdom, and it appears to be a secondary modification rather than a primitive character.

It is plausible, then, that the ancestor of the Eumetazoa was a blastulalike colonial flagellate, in which a differentiation occurred between somatic and reproductive cells, as in living *Volvox,* and then a differentiation between digestive cells and locomotor cells, with the former type moving into the interior of the organism to form either a gastrula or a planula. Yet decisive evidence on this basic question probably will never be obtained. That this primitive metazoan was not identical with any living type is almost certain.

The three most primitive living metazoan phyla are the Mesozoa, the Coelenterata (or Cnidaria), and the Ctenophora. These phyla are generally radially symmetrical (having one differentiated axis) or else biradially symmetrical (having two differentiated axes). Their general grade of organization is more advanced than that of the Porifera, because, although they have no organ systems, they have two well-defined tissues: the ectoderm and entoderm (or epidermis and gastrodermis). In most coelenterates and ctenophores, the mesoglea, a jellylike mass that may include some cells, is found between these layers. Thus it is not strictly true, as is often stated, that these phyla have only two cellular layers.

Phylum Mesozoa

The correct phylogenetic position of the Mesozoa (Figure 19–6) is a much vexed question. From a structural viewpoint, they are the simplest metazoans, consisting of an outer, generally ciliated, layer of cells enclosing a core of internal, reproductive cells. They thus resemble a planula, except that the internal cells are not digestive cells. Van Beneden gave the group its name in 1877, with the intention of indicating his judgment that the group was extremely primitive and intermediate between the Protozoa and the rest of the Metazoa. Some zoologists classify the Mesozoa as degenerate flatworms because all of them are parasitic and because of the supposed similarity of their life cycle to that of digenetic trematodes. E.A. Lapan and H.J. Morowitz, however, have shown that, unlike the Digenea, they have only a single host and are biochemically closer to protozoans.

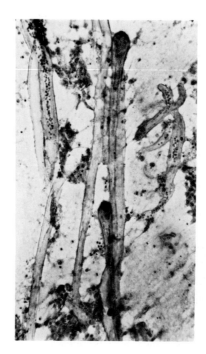

Figure 19–6. Dicyemmenea californica, *a typical meso-zoan. Total length is about 100 microns.*
From E. Dodson, *The Phenomenon of Man Revisited,* by permission of Columbia University Press (New York, 1984).

A decisive demonstration that the characters of the Mesozoa are primitive rather than degenerate would give great phylogenetic importance to the group, for we could then reasonably argue that the group must be little changed from the remote, Precambrian ancestor of the Metazoa. It would prove that the Metazoa were derived from a planula- rather than from a gastrea-type ancestor, and it would leave the Gastrea theory badly damaged. But the available evidence does not furnish a basis for a final decision on the taxonomic position of the Mesozoa. W.K. Brooks has said that "suspended judgment is the greatest triumph of intellectual discipline," and this appears to be an appropriate place for that achievement.

Phylum Coelenterata (or Cnidaria)

The Coelenterata have traditionally been treated as the most primitive of the Eumetazoa. Haeckel regarded them as the source of the flatworms and hence of all more complex phyla. The principal reason for this viewpoint is the obvious resemblance of a hydroid polyp to a gastrula. The hydroid consists of two simple cell layers, with no organ systems and with only a trace of noncellular mesoglea. The anatomical structure of the polyp could be derived from that of the gastrula simply by the elongation of the body and the drawing out of a circlet of tentacles around the mouth. Food is taken in and waste residues are expelled by the mouth, which is simply the blastopore of the blastula. Food is still digested by the protozoan method, that is, entodermal cells engulf food particles, and digestion is carried on intracellularly. But enzymes are also secreted into the gastrovascular cavity, and much digestion occurs there. Muscle tails (elongate processes, with contractile fibrils from the base of the cell) may form in connection with either the epidermis or the gastrodermis. A nerve net is formed from epidermal elements. Thus there is a high degree of tissue differentiation, but no organ systems. Outstanding specializations of the coelenterates include the development of nematocysts (organelles for food procurement and defense) and the alternation of a free-swimming, sexually reproducing medusa generation with a sessile, asexually reproducing polyp generation. This cycle is unrelated to the alternation of generations of plants, for both generations are diploid. Superficially so different, the polyp and the medusa are structurally similar; we can derive the medusa by simply inverting the polyp, greatly increasing the amount of mesoglea and its cellular contents, and drawing the circlet of tentacles away from the mouth, as illustrated in Figure 19–7.

Haeckel assumed that the ancestral coelenterate was a polyp, because of the ease with which a polyp can be derived (in theory) from the Gastrea. But study of what appears to be the most primitive order of hydroids, the Trachylina, has led to the conclusion that the medusa phase is primary and the polyp is

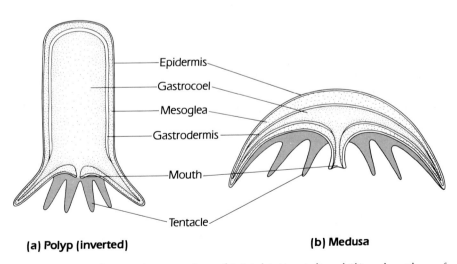

| Epidermis |
| Gastrocoel |
| Mesoglea |
| Gastrodermis |
| Mouth |
| Tentacle |

(a) Polyp (inverted)   **(b) Medusa**

Figure 19–7.   *Structural correspondence of (a) polyp (inverted) and (b) medusa phases of coelenterates.*

derived. Once formed, the Coelenterata (Figure 19–8) diverged along three major lines, each constituting a class of the phylum. In the most primitive class, the Hydrozoa, both generations are generally well developed. The class includes hydroids, such as the *Hydra* of elementary laboratories, and the much more typical marine colonial forms, such as *Obelia*. There are also some in which the polyp generation is reduced, as in the order Trachylina. In the class Scyphozoa, including the jellyfishes, the medusa is much the more prominent generation, with the polyp being reduced to a developmental stage. The final class, the Anthozoa, includes only the polyp phase, the medusa being suppressed entirely. The class comprises the sea anemones, the corals, and their allies. Many of these, including all of the corals, secrete a calcareous exoskeleton, because of which they have left an excellent fossil record going back to the Ordovician. All three classes are ancient, and it is probable that they diverged in Precambrian times from a primitive hydrozoan type not dissimilar to the Trachylina.

Closely allied to the coelenterates is the phylum Ctenophora, a small phylum of less than 100 species. All are small marine animals. They are commonly called "comb jellies" because they bear rows of comblike plates of

Figure 19–8. Coelenterate diversity. Obelia, *polyp (top left) and medusa (top right) phases; a fan coral (bottom left); and a large anemone,* Cerianthus *(bottom right).*

cilia. With the coelenterates, they share the tissue grade of construction, two cell layers separated by abundant mesoglia, and a gastrovascular cavity. They are biradially symmetrical. In contrast, ctenophores (except *Euchlora*), lack nematocysts, they are hermaphroditic, and they have no alternation of generations. There seems to be little doubt that the coelenterates gave rise to the Ctenophora, yet all attempts to relate them to any of the classes of living coelenterates have failed. Probably the Ctenophora derived from the same Precambrian, trachyline stock that gave rise to the three classes of the Coelenterata at about the same time.

PRIMITIVE BILATERAL PHYLA

Haeckel, basing his opinion as usual on the biogenetic law, believed that a primitive hydrozoan was the ancestor of the bilateral phyla. The evidence is insufficient; and it seems at least as probable that flatworms first derived from the same planulalike stock that gave rise to the coelenterates. In the flatworms, the middle cell layer (mesoderm) became more abundant and specialized, with organized muscle layers, a reproductive system, and an excretory system. Yet no coelom, or body cavity, developed within the mesoderm, as it does in most higher groups. The nervous system is formed from the ectoderm, although it is imbedded in the mesoderm. It is not a diffuse nerve net as in the coelenterates, but rather it is somewhat centralized, being organized about cerebral ganglia at the head end and longitudinal cords. There are organized sense organs, including eyes. Food is still distributed to the various parts of the body by the branches of the digestive tract, which is now called an intestine rather than a gastrovascular cavity. (As there is no anus, the digestive system is said to be incomplete.) Thus the flatworms are advanced much beyond the tissue grade of construction that characterizes the radiate phyla, for definite organ systems are present, principally in the mesoderm.

J. Hadzi has urged a radically different phylogeny. He believes that multinuclear ciliates became acoelous flatworms by the formation of cell membranes. In evidence, he points out that both are ciliated; hermaphroditism of flatworms he finds homologous with conjunction of ciliates; trichocysts of ciliates he believes are represented by rhabdites, rodlike inclusions in some of the epidermal cells of flatworms. He believes that the Anthozoa derived from flatworms by adoption of the sedentary life and that the other coelenterates then arose from anthozoans. He also believes that the more complex Metazoa derived from flatworms, and he groups them into only four phyla, a procedure that unites some highly diverse groups.

Although the phylum Platyhelminthes, and especially the most primitive class of this phylum, the Turbellaria, is commonly treated as the stem group from which the more complex phyla arise, this derivation is by no means established. Closely related to the Platyhelminthes, and on the same general level of organization, is the phylum Nemertinea. This group is less well known than the former because it is predominantly marine and because it is a difficult group to study. Nemerteans have some progressive characters that qualify them for consideration as forerunners of more complex invertebrates. For the first time, the digestive system is complete (an anus is present); they have a blood circulatory system with hemoglobin; and they have a nervous system that might afford a basis for the development of the nervous systems of higher groups. However, the fact that the nemerteans capture their food by means of an extensible proboscis, a mechanism found in no other group, suggests that they are a terminal group.

Yet another possibility is that a third group, an ancient unknown phylum of primitive flatworms derived from the primitive planula by the development of the mesoderm, may have been ancestral to the Platyhelminthes, the Nemertinea, and the higher phyla. Whatever their origin, once formed, the platyhelminths diverged along three main lines of descent. The first of these is represented by the class Turbellaria, comprising the free-living flatworms, of which *Planaria* (Figure 19–9) is the best-known example. The other two

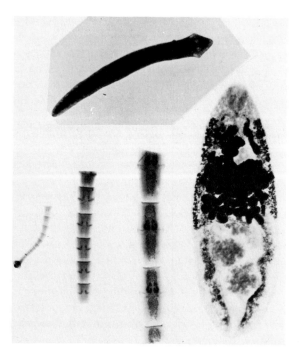

Figure 19–9.   *Platyhelminth diversity. (Above) planaria; (lower left)* Dipylidium caninum, *a tapeworm of dogs (three strips are shown); and (lower right)* Metorchis conjunctus, *a liver fluke of dogs.*
Photograph of planaria by courtesy of Carolina Biological Supply Company.

classes have become greatly modified for parasitism, and most of them have evolved complicated life cycles that permit their transfer from one host to another. These are the class Trematoda, the flukes, which have retained all of the organ systems of the Turbellaria, and the class Cestoda, the tapeworms, which have undergone an extreme degenerative evolution. Most of the Trematoda and all of the Cestoda are intestinal parasites of vertebrates.

On the same general level of organization as the acoelomate groups, there are eight minor phyla, of uncertain relationships, that are characterized by the possession of a pseudocoel. A coelom is, by definition, a body cavity formed within mesoderm. A pseudocoel, on the other hand, is a remnant of the cavity of the blastula, which may be partially filled by large, vacuolated cells. These pseudocoelomate organisms are among the most primitive bilaterally symmetrical animals; we do not have room here to explore their uncertain relationships. Of these, only the phylum Nematoda, the round worms, is important. It includes some well-known parasites of both plants and animals as well as a host of little-known free-living species. Hyman treated them as a class of the phylum Aschelminthes, along with five more groups of pseudocoels: the Nematomorpha, Rotifera, Gastrotricha, Kinorhyncha, and Priapulida. Most zoologists, however, treat each of these "classes" as an independent phylum. (The Priapulida have been reported as having a true coelom. If this is correct, they must be related to the annelids.) The remaining two phyla of pseudocoels are the Acanthocephala (spiny-headed worms) and the Entoprocta, which are polyplike animals. Texts on invertebrate zoology provide details on these groups.

THE PROTOSTOMOUS PHYLA

The remaining major phyla of the animal kingdom can be arranged in two diverging lines of descent, largely on the basis of embryological criteria. One line culminates in the Annelida, Arthropoda, and Mollusca; the other culminates in the Echinodermata and the Chordata. Certain minor phyla can be associated with one or the other of these main lines, with varying degrees of satisfaction. Haeckel believed that the echinoderm-chordate line derived from the Turbellaria, whereas C.A. Kofoid believed that both lines derived from the Nemertinea. It is at least as likely that both derived from the unknown, primitive, acoelous flatworm from which both the Platyhelminthes and the Nemertinea probably evolved.

The cleavage divisions of the annelid-arthropod-molluscan line are both spiral and determinate. In spiral cleavage, spindles are at right angles to those of the preceding division, so that the cells of each layer alternate with those of the next layer, like bricks in a wall. (Spiral cleavage does not occur in the large, yolky eggs of arthropods, yet their origin from annelids is clear from other evidence.) Determinate cleavage proceeds according to a set pattern, with the part of the body to be formed from each blastomere fixed from the start. Destruction of a blastomere results in a deficient larva. We can identify which blastomere will form ventral surface, which the gut, and so on. The mesoderm is formed from stem cells, which multiply to form a pair of ventral bands, growing forward from the posterior end of the larva. The coelom is formed from the splitting of these bands; hence, these phyla are said to be *schizocoelous*. The blastopore becomes the mouth of the adult; therefore, this whole series of phyla is called Protostomia.

Generally, though by no means always, protostomes develop a trochophore larva (Figure 19–10). This more or less spherical larva has an *apical tuft* of cilia dorsally, and a *prototroch,* or girdle of cilia, around the equator, by which it can swim weakly. There is a complete digestive system, consisting of a mouth, a short foregut, an enlarged stomach, a short hindgut, and an anus. Some mesoderm is present, and some mesodermal organs, such as a kidney, are formed. This type of larva is characteristic of the annelids and molluscs, but unique larval stages have evolved among the arthropods.

Collectively, these characters strongly suggest relationship among the protostomes. The precise relationships are, however, largely a matter of conjecture. Origin from some primitive, acoelous flatworm is probable, yet there is little basis for deciding which of the possible groups of flatworms was the ancestor, beyond the fact that spiral cleavage is also found in the Nemertinea and in the polyclad Turbellaria. Because of the widespread occurrence of the trochophore larva, an independent, trochophore-type animal was generally supposed to be intermediate between the flatworms and the present-day phyla of the annelid-arthropod-molluscan series. The only evidence for this is the interpretation of

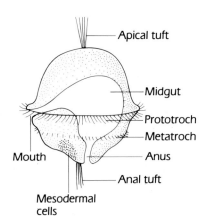

Figure 19–10.   A trochophore larva, possibly similar to the remote ancestor of all protostomes.
Reprinted from E. Dodson, *The Phenomenon of Man Revisited,* by permission of Columbia University Press (New York, 1984).

the embryological evidence according to the biogenetic law, a type of reasoning that is always insecure. Direct evidence is not likely to be obtained, for such an organism would be poor material for fossilization. It would have to be sought in Precambrian rocks, because the major invertebrate phyla were all present in the Cambrian.

Phylum Mollusca

Whatever the source, the phylum Mollusca was early established as a group divergent from the others. This great phylum, the second largest in the animal kingdom, has around 80,000 species, encompassing a great variety of forms and inhabiting almost every type of marine, freshwater, and terrestrial habitat. All of the extant classes of Mollusca were already abundantly represented in the Cambrian. The classes were just as distinct then as now, and so paleontology is of no help in deciding what the relationships within the phylum may be.

In addition to the general characters of all members of the protostome line, the phylum Mollusca is distinguished by many additional structural characters. In all, the body is divided into four regions: a muscular foot, a head, a visceral hump, and a mantle that generally secretes a calcareous shell. Basically, molluscs are bilaterally symmetrical, but this is obscured in the adults of some classes. The coelom is much reduced. Mollusca are traditionally described as unsegmented, yet in 1957 *Neopilina,* a new species found in the deep waters off the west coast of Mexico, was described. This animal has several segmentally arranged organ systems, and its affinities are with the class Monoplacophora, which had been thought to be extinct since the Devonian. It suggests that the primitive molluscs may have been segmented. On the basis of anatomical evidence, the class Amphineura is regarded as very primitive but not necessarily as the progenitor of the more specialized classes. This class comprises the chitons, or sea cradles, familiar parts of the fauna of rocky seacoasts; but there are no freshwater species. The chitons have a broad, flat, elongated foot, over which lies the visceral hump and at the anterior end of which is the head. The mantle covers the visceral hump and typically secretes a series of eight calcareous plates, the valves, which may be beautifully ornamented.

The class Gastropoda, which includes the snails, slugs, limpets, and their less well known allies, has evolved a much enlarged visceral hump. In most gastropods, this hump has grown asymmetrically, with the familiar coiling as a result. Correlated with this change has been a torsion of the visceral hump through 180°, so that structures that were originally posterior have moved to an anterior position, and vice versa. The class Scaphopoda, or tooth shells, is a small group that is very much specialized for digging. All species are marine, and there are only a few of them. The class Lamellibranchiata includes the bivalves, the familiar clams, mussels, and their allies (Figure

Figure 19–11.   Lyropecten, *a Miocene scallop from Chesapeake Bay. Note also the fragments of shells of encrusting barnacles.*

19–11). They are compressed bilaterally, and enclosed within a two-valved shell, hinged dorsally, and secreted by the two lobes of the mantle. The gills are immensely enlarged to form broad ciliated tracts that produce a current for feeding upon plankton and detritus. The head is very much reduced. The foot is generally wedge-shaped, unlike the broad, flat foot of the more primitive molluscs, but it can be protruded between the valves to serve a locomotor function. In some species, it is specialized for burrowing.

The class Cephalopoda includes the octopi, squids, chambered *Nautilus,* and the latter's extinct allies. They are the most complex molluscs, equal to the most-advanced insects and vertebrates in degree of complexity. The visceral hump, like that of the gastropods, is much enlarged, but it is still symmetrical. The mouth is surrounded by a ring of tentacles for seizing food. The head and foot are fused so completely that there is no agreement among specialists as to which these tentacles are derived from. There is a siphon by which water may be forcibly ejected from the mantle cavity. The coelom is better developed than in most molluscs. The nervous system is highly centralized, and efficient eyes, superficially quite similar to those of vertebrates, are present. Psychological experiments in problem solving show that they are among the most intelligent invertebrates. The more primitive members of the class have a chambered shell; the animal always lives in the most recently secreted chamber. Each chamber corresponds to a stage in the growth of the animal, like the successive moults of an arthropod.

The shelled cephalopods are today represented only by a single genus, *Nautilus* (Figure 19–12), the last remnant of a once-dominant group in the

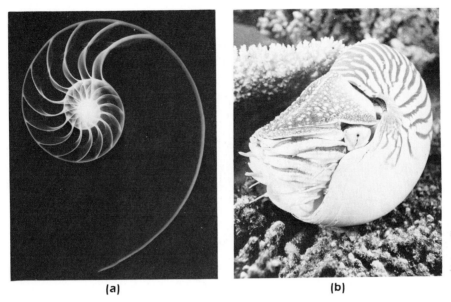

(a)   (b)

Figure 19–12.   Nautilus. *(a) A negative print of an X-ray photograph of the shell. Note the intricate design, with each chamber a little larger than its predecessor. They are formed to accommodate the continuing growth of the animal. (b) The living animal.*
Living *Nautilus* photographed on Palau, South Pacific by Dr. Bruce Saunders, Byrn Mawr College.

oceans of the world. The nautiloids appeared in the fossil record in the Late Cambrian, and rapidly assumed a dominant position in the marine fauna. They reached their peak diversity in the Silurian, and then gradually declined. During the Devonian, they gave rise to another suborder, the Ammonoidea, with which they competed unsuccessfully for a long period of the earth's history. The ammonites (Figure 19–13) were the dominant marine molluscs throughout most of the Mesozoic era; they dwindled and became extinct near the end of the Cretaceous. Meanwhile the nautiloids survived in small numbers and are known from a single genus at the beginning of the Cenozoic era. No longer in competition with the ammonites, the nautiloids underwent a rapid evolution at the beginning of the Cenozoic. Seven new genera appear during the Paleocene epoch; but only one of these, *Nautilus,* has survived to the present time. In the meantime forms with the shell much reduced and internal, the octopi and squids, have become the principal cephalopods.

Phylum Annelida

The second main branch of the protostome line is the annelid-arthropod branch. The phylum Annelida, best known for the common earthworms, includes a wide variety of worms arranged in four classes. All annelids are segmented; that is, the functional and structural units of the body are repeated

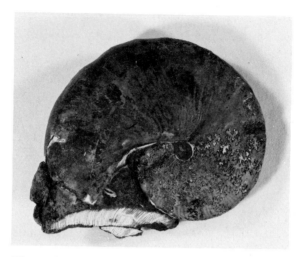

Figure 19–13.   *A Cretaceous ammonite. Note the distinctive, lacy suture lines, which are characteristic of ammonites.*

serially along the length of the body, and the segments may be marked externally. A well-developed coelom separates the digestive tract from the muscular body wall. There is an advanced nervous system, based upon a pair of cerebral ganglia, and a pair of ventral cords on which ganglia are located in each segment of the body. Whenever there is a larval stage, it is a trochophore. Most annelids have a thin cuticle, and the typical species have segmentally arranged chitinous bristles called chaetae.

The phylum Annelida is best typified by its most primitive class, the Polychaeta (Figure 19–14). These take their name from their numerous chaetae, which arise from limblike lobes, the parapodia, on each body segment. There is typically a distinct head, which may have appendages. The sexes are separate, and fertilization is external. There is a trochophore larva. The polychaetes are adapted to a wide range of habitats, including pelagic and bottom-dwelling, surface-crawling and burrowing, and permanent tube-dwelling, and they show striking morphological adaptations. They are important, if inconspicuous, members of shallow marine communities. In number of species, they undoubtedly exceed the remaining classes combined. Yet they are less well known because they are almost exclusively marine.

Better known, but far less typical of the phylum, is the class Oligochaeta, including the earthworms and many small freshwater annelids. The class appears to have evolved from the polychaetes by a process of reduction and simplification. The head region is much reduced and never includes appendages. The chaetae are reduced both in size and in numbers, and they are no longer set on parapodia. Oligochaetes are all hermaphroditic. Development is direct, without larval stages. The embryos develop in a cocoon, an adaptation to terrestrial life. The class is much more uniform than is the Polychaeta.

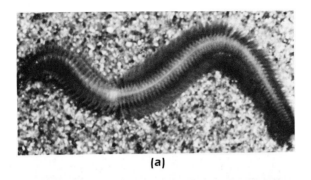

(a)

(b)

Figure 19–14.   *Polychaete annelids. (a)* Nereis, *a typical polychaete. (b)* Spirographis spallanzanii, *a more-specialized polychaete. That the range of variability of annelids includes such beauty may be surprising to many.*
Photograph of *Nereis* by courtesy of Carolina Biological Supply Company.

The class Archiannelida is a small group of extremely simplified marine annelids. These are all small worms that generally lack parapodia and chaetae. They are ciliated, like the young of polychaetes. Their nerve cords retain the primitive connection with the epidermis. The coelom is only slightly developed. The Archiannelida were named with the intention of implying that this class was the most primitive of the phylum and the probable progenitor of the other classes. But embryological and morphological studies have led to the conclusion that this class is not truly primitive; rather, it has derived from the Polychaeta by extreme simplification.

The class Hirudinea, or leeches, appear to have derived from oligochaete ancestors, for they share many characters with that class. However, all leeches are ectoparasites, and they are extremely modified for that mode of life.

Phylum Arthropoda

The protostome series is climaxed by the great phylum Arthropoda, the most successful of all groups of animals if number of species be an indication of success, for about three-fourths of all species of animals are arthropods. This phylum is also the most varied, for there are arthropods adapted to every imaginable habitat from the abyssal depths of the ocean (crabs and pycnogonids) to aerial heights (insects). Perhaps the key to evolution of the arthropods is their early development of a thick, chitinous cuticle. This thick cuticle minimizes water loss by evaporation, and it may be a major factor in their successful invasion of the land, for no other invertebrate phylum has so large a proportion of terrestrial species. Moreover, the rigidity of this cuticle made necessary the joints—thin places in the cuticle—from which the phylum takes its name. In order to move the hard pieces of the cuticle, the continuous muscular wall, inherited from polychaete ancestors, became broken up into specialized muscles, attached to ingrowths of the cuticle. The jointed limbs, far more adaptable than the parapodia of the polychaetes, specialized for many functions, including sensation, feeding, and locomotion. The concentration of the nervous system and the sensory organs in the head region (cephalization), begun in the polychaetes, is carried much further in the arthropods. The coelom is much reduced and has been largely replaced by a hemocoele. Segmentation is prominent externally, but less so internally.

As may be expected with so large a phylum as the Arthropoda, there is no general agreement on the taxonomic rank to be accorded to the major divisions of the phylum. For present purposes, five major subphyla will be recognized, and no attempt will be made to treat their component classes because of the vastness of the groups. These are the Trilobita, Crustacea, Myriapoda (centipedes and millipedes), Insecta, and Chelicerata (spiders and their many allies). These well-defined groups, with the exception of the Trilobita, are universally familiar, and so need not be defined here.

The trilobites are of especial interest because they are the most ancient arthropods known, and because their very generalized structure could conceivably have given rise to the other major groups of arthropods. The trilobites were dominant from the Cambrian through the Silurian. They then declined until the Permian, from which period only a single species is known. With this, the group became extinct. The trilobite body (Figure 19–15) was enclosed in a chitinous skeleton and was divided into three longitudinal regions or lobes by two longitudinal furrows; hence the name of the group. The body consisted of a head and a segmented trunk. The segments were generally movable, but a variable number of posterior segments were joined to form a rigid unit, the pygidium. There was a single pair of antennae. The remaining appendages were all simple, undifferentiated, biramous appendages. None was specialized as mouth parts, but all had gnathobases, basal processes that could be used for biting. All species were marine. X-ray photographs have even revealed some of their internal structures.

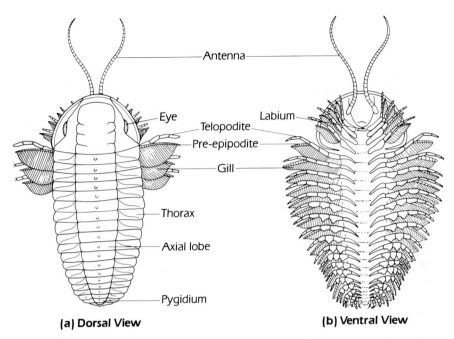

Figure 19–15. *A trilobite,* Triarthrus, *in (a) dorsal and (b) ventral views. Extraordinary preservation has made possible the observation of details of the structure of gills and antennae in the dorsal view.*

Fossil remains of transitional organisms from trilobites to the other major arthropod types are entirely lacking, and the several groups may have arisen independently. Indeed, this theory avoids some perplexing inconsistencies. Comparative morphology indicates that the crustaceans, myriapods, and insects form one line of descent, whereas the arachnids must have diverged very early. In the Crustacea (Figure 19–16) this transformation has involved the division of the body into a highly fused cephalothorax and an abdomen in which the original segmentation is retained. The appendages have become greatly diversified, while still retaining the biramous plan (Chapter 3). Almost all crustaceans are aquatic, but there are some terrestrial species, such as the sowbugs. Like the trilobites, the crustaceans are represented in early Cambrian rocks, and it is not improbable that they arose in Precambrian times.

The Myriapoda, including the millipedes, centipedes, and their allies, may have derived from a crustacean progenitor by the reduction of the exoskeleton and the loss of the gills. The gills were replaced by a system of tracheae, small tubes that carry air to the tissues of the body, thus permitting direct respiration. Although not well represented in the fossil record, myriapods are known from rocks dating from the Devonian.

Insects probably also arose in Devonian times or earlier, but good fossils

(a)

(b)

Figure 19–16.   Crustaceans. (a) A fiddler crab, with one chiloped greatly enlarged. (b) A hermit crab, adapted to live in the shells of snails.
Photographs by courtesy of Carolina Biological Supply Company.

are first seen in rocks of Pennsylvanian age. They may have derived from myriapods, or they may have come directly from crustacean progenitors. Whichever ancestry they had, the tracheal system is undoubtedly one of the important adaptations that permitted successful invasion of the land. The body became divided into three well-defined regions: head, thorax, and abdomen. The appendages of the head segments are all specialized, either for sensation or for eating. The thorax bears three pairs of walking legs. Dorsally, it usually bears two pairs of wings, but one or both pairs have been lost in some groups. The abdominal appendages are all suppressed, or greatly modified for other functions. An immense variety of insects has evolved; approximately 675,000 species are known, more than the combined numbers of described species of all other living groups. Moreover, the populations of many of these species are truly immense.

The subphylum Chelicerata includes a highly varied array of organisms that have developed along lines quite divergent from the series just discussed. This subphylum comprises the horseshoe crabs, scorpions, spiders, harvestmen, ticks, mites, and their allies. It is difficult to derive them from the trilobites, but it may be possible. The alternative is independent origin of the two major series of arthropods. There is a tendency toward suppression of segmentation, with only two distinct body regions, the anterior prosoma and the posterior opisthosoma. There are six pairs of appendages, of which four pairs are walking legs. Unlike all other arthropods except the trilobites, none of the appendages is specialized as jaws. Gnathobases of the anterior limbs serve this function.

The class Pycnogonida, the sea spiders, is a small and little-known group, which is usually placed with the Chelicerata simply because of their superficial resemblance to spiders. Yet their morphology is utterly different. J. Hedgpeth has studied the group thoroughly and has concluded that, although they are undoubtedly arthropods, they are so widely divergent that they cannot reasonably be grouped with any of the other arthropod types.

Even less justification can be found for the common practice of making the Tardigrada a class of chelicerates. This small and little-known group of minute, freshwater organisms shows some relationship to the Arthropoda and has usually been treated as a class of the Chelicerata. But specialists in the field feel that it should be regarded as an independent phylum of uncertain relationship to the Arthropoda.

The Onychophora: A Unique Evolutionary Link

As we have just said, the derivation of the Arthropoda from the Annelida is more certain than is the derivation of any other phylum. This conclusion depends upon our knowledge of the Onychophora, a group of about eighty species, all of which are assigned to a single genus, *Peripatus* (Figure 19–17). This group of animals shows a peculiar mixture of annelid and arthropod characters. Among the annelid characters is the general appearance of the organisms: they look much like polychaetes in which the parapodia do not bear chaetae. The cuticle is thin like that of annelids, and the muscles of the body wall are continuous. The excretory organs of both annelids and onychophorans are mesodermal tubules segmentally arranged (coelomoducts); those of the Arthropoda are usually entodermal or ectodermal in origin. The reproductive ducts of the Onychophora are ciliated, but cilia are unknown among the Arthropoda. The eyes of the annelids and onychophorans are simple, whereas those of the arthropods are compound. In contrast to the annelids, the Onychophora and the Arthropoda have jaws derived from appendages. The coelom in each is much reduced and largely replaced by a hemocoel; whereas the

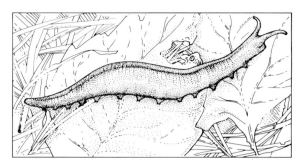

Figure 19–17.   Peripatus, *an animal that is intermediate between annelids and arthropods.*

coelom of the annelids is highly developed. The circulatory system of the Onychophora also resembles that of the Arthropoda rather than that of the Annelida. Finally, the respiratory system of the Onychophora consists of a set of tracheae, a characteristic known nowhere else but in the Arthropoda.

Because of this strange mixture of characters, the taxonomic position of the Onychophora has always been a much vexed question. They were originally treated as a class of the Annelida, but because of the presence of a hemocoel and especially because of the tracheal system, they are now generally treated as a class of the Arthropoda. These are not the only possibilities. They are sometimes treated as an independent phylum intermediate between the Annelida and the Arthropoda. Finally, S.F. Light has urged that, as they are not properly separable from either of the major phyla, the entire annelid-onychophoran-arthropod series ought to be recognized as one great phylum under the name Articulata, a phylum first proposed by G. Cuvier. With no other series of phyla are such considerations possible; hence the statement with which this discussion began, that the origin of the Arthropoda from the Annelida is more certain than the origin of any other phylum. Unfortunately, the Onychophora are scantily known in the fossil record. They are represented only by *Aysheia,* a mid-Cambrian genus from British Columbia, and by a doubtful Precambrian fossil.

Minor Protostomes

A few minor phyla also show the general characters of the protostome line. Two of these, the Sipunculoidea and the Echiuroidea, are wormlike burrowers of the tide flats. They are generally treated as minor annelids, an arrangement that is more convenient than accurate. Yet they probably are more closely related to the Annelida than to any other phylum. The three remaining phyla, the Bryozoa, Phoronida, and Brachiopoda, are more difficult to place. Like the Entoprocta, they feed by means of a lophophore; but unlike that phylum, they have a coelom. The Bryozoa are small, colonial animals that superficially resemble the Entoprocta but differ from them fundamentally by being coelomate. The Phoronida include only two genera. The animals are elongated, wormlike creatures that dwell in tubes on tide flats. When the tide is in, the lophophore projects into the water to feed on plankton and detritus. The phylum Brachiopoda (see Figure 5–6), or lampshells, bear a superficial resemblance to the molluscs because of their bivalve shells, but these are dorsal and ventral rather than right and left. Internally, they do not suggest the molluscs at all. They have a prominent lophophore. These marine animals are present in the earliest Cambrian deposits. The greatest interest in the group derives from the fact that a single genus, *Lingula*, has persisted from the Ordovician to the present time, a span of 500 million years. It may be the oldest genus in existence. These three phyla were formerly grouped together as a single phylum, the Molluscoidea. Yet they have little in common with one another except the lophophore, and there is little indication that any of them have a close relationship to the Mollusca. Hence it seems best to treat them as independent phyla of protostomes, of uncertain relationships to the larger phyla.

THE DEUTEROSTOMOUS PHYLA

The other major branch of coelomate animals, the Deuterostomia, comprises only five phyla: Chaetognatha, Pogonophora, Echinodermata, Hemichordata, and Chordata. This series contrasts with the Protostomia in embryological characters. The cleavage divisions are neither spiral nor determinate. The mesoderm is not formed from stem cells, but rather from outpocketings of the entoderm of the gut. This process simultaneously establishes the coelom, which is said to be *enterocoelous* (meaning simply that the cavity is established from the gut). Since the development of the coelom is radically different in the protostome and deuterostome series, these coeloms are probably not homologous but arose independently in the two series. The original blastopore becomes the anus, and a new mouth is formed in this series of phyla, hence the name Deuterostomia.

Development does not lead to a uniform larval type. The Chaetognatha have a unique larval type. The Echinodermata have several types of larvae, but great theoretical significance is attached to a hypothetical, average echinoderm larva, the *dipleurula* (Figure 19–18). A typical gastrula is formed. The entoderm and ectoderm first fuse at one end, then break through to form a mouth. The blastopore becomes the anus. The digestive tube buds off an anterior vesicle, which first divides into two lateral compartments and finally into three segments on each side. These are the coelomic pouches. The cilia that covered the blastula and gastrula evenly now become concentrated in a series of bands arranged around the concave ventral surface of the animal. It is this stage

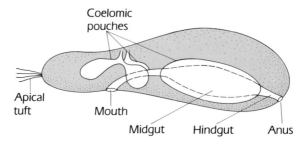

Figure 19–18. A dipleurula larva. This hypothetical larva is a sort of average of echinoderm larvae, similar to the tornaria larva of hemichordates and probably also to the remote, bilateral ancestor of the aberrant echinoderms, the more typical hemichordates, and the very progressive chordates.

upon which the hypothetical dipleurula is based. Because of the universality of the larval stage among the echinoderms, zoologists generally believe that the echinoderms must be descended from a dipleurulalike ancestor that was bilaterally symmetrical and free-swimming.

The Hemichordata have a larva, the tornaria, that is similar to the dipleurula and to the bipinnaria larva of starfishes. When first discovered, the tornaria was described as a larval starfish, and only much later was the error discovered. This resemblance of the larvae is one of the major arguments for a relationship of the Hemichordata to the Echinodermata. Larvae are rare among the Chordata, but the tunicates and amphibians have tadpole larvae, and the cyclostomes have a unique larva, the ammocoetes.

Minor Deuterostomes

The Chaetognatha are a small, uniform marine group, the arrowworms. Although they show the deuterostome characters, they suggest close relationship to no other phylum, so the phylum probably branched off the deuterostome line soon after its origin. Aside from a few doubtful specimens, the phylum is unknown in the fossil record. A superficial resemblance to amphioxus is misleading.

The Pogonophora is a phylum of deep-sea worms. The body consists of a short protosoma, with one or more tentacles, a short mesosoma, and a very elongate metasoma. The coelom is unpaired in the protosoma but paired in the other two regions; it is drained by paired nephridial ducts. Paired dorsal nerve cords arise from a nervous mass and ring in the protosoma. The circulatory system consists of two longitudinal vessels. The Pogonophora have no digestive system, and larvae are unknown. A.V. Ivanov, who described this phylum and studied it in detail, assigned the group to the deuterostomes, but some zoologists believe that they are protostomes because of annelidlike characters of the terminal segments of the body.

Phylum Echinodermata

The Echinodermata are best known by the sea stars (Figure 19–19), which typify the phylum well. Their unusual set of characters stimulated Hyman to "salute the echinoderms as a noble group especially designed to puzzle the zoologist." They are the only advanced group that is radially symmetrical, and they are typically pentamerous (having five radial segments). The dermis is usually armored with heavy, calcareous plates that may be fused to form a continuous test, or shell. Calcareous spines commonly protrude through the epidermis. These spines are the basis for the name of the phylum, which means "hedgehog-skinned." The blood circulatory system has been reduced to a minor remnant, but paralleling it is a water vascular system that is actually a derivative of the embryonic coelom. The tube feet, which extend in large numbers from the undersurface of the arms of a sea star, are the most conspicuous parts of the water vascular system. Most obviously they function in locomotion and food capture, but they also function in respiration and excretion. There are no respiratory or excretory systems, an astonishing fact for animals so large and complex. The muscular and nervous systems, too, are relatively underdeveloped. Indeed, part of the nervous system is still closely associated with the epidermis, with which it shares its embryological origin; another part may be of mesodermal origin, a situation unique in the animal kingdom.

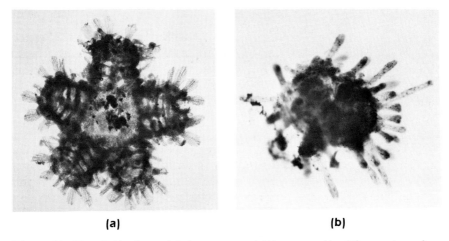

(a)   **(b)**

Figure 19–19. Echinoderms: (a) A sea star and (b) a sea urchin. These specimens have just metamorphosed and are less than two mm in diameter.

The sea stars constitute the class Asteroidea. They are allied with four more classes of living echinoderms and with seven extinct classes.

The class Ophiuroidea includes the brittle stars, so called because if a predator seizes a brittle star by an arm, the arm breaks off and the rest of the animal escapes. This process, called *autotomy,* takes place because of a specific plane of weakness in the arm and because of a pair of muscles that snap it on that plane. Whereas the conical arms of a sea star blend broadly into the central disc, those of the brittle stars are long, narrow, and sharply set off from the central disc. The class Echinoidea comprises the sea urchins and sand dollars. They have a globular test (flattened in sand dollars) armed with numerous long spines and with rows of tube feet extending along five meridians. They also have a beautifully elaborated, five-toothed jaw apparatus called Aristotle's lantern. The class Holothuroidea consists of the sea cucumbers, perhaps the most aberrant members of the phylum. The skin is leathery in texture, and there are no spines. The body is sausagelike in form, with the tube feet arranged in five longitudinal rows. Some of the tube feet are modified to form a circlet of tentacles around the mouth. Finally, the class Crinoidea comprises the sea lilies and feather stars. Their arms are much branched, and ciliary tracts on the oral surfaces of the arms sweep a food-bearing current toward the mouth. The sea lilies are attached to the substrate by a stalk that may be quite long; the feather stars swim free in the sea. The extinct classes are more similar to the crinoids than to the other extant classes, and so the crinoids are probably the most primitive living type. All echinoderms are marine.

The radial symmetry of echinoderms is clearly secondary, for all begin life as free-swimming, bilaterally symmetrical larvae. The development of radial symmetry may have been related to the change from a pelagic to a sessile mode of life by primitive echinoderms, for radial symmetry is a common characteristic of sessile organisms. The fossil record of the echinoderms, which goes back to early Cambrian times, is among the best; and excellent phylogenies can be constructed within each of the five extant and seven extinct classes. But the record throws little light upon the origin of the phylum or upon its possible relationships to other phyla. These questions depend at present entirely upon embryological evidence, with all of its limitations.

Phylum Hemichordata and Origin of the Chordata

The Hemichordata (Figure 19–20) are a small phylum of wormlike marine animals, which have been extensively studied because of their supposed relationship to the Chordata. In Haeckel's phylogeny of the vertebrates, the hemichordates were the next stage after the primitive flatworm. They were originally classed as a subphylum of the Chordata because they show the three basic characters of the Chordata: a dorsal nerve tube, a notochord, and a pharynx modified for respiration. The pharynx is pierced by numerous gill slits, which agree

Figure 19–20.   Balanoglossus, *a hemichordate. Unfortunately, the anterior end of the trunk of this specimen is rather distorted.*

in detail with those of amphioxus, an undoubted chordate. Although they function primarily as exit pores for the feeding current, these gills probably also handle some respiration; these facts may indicate that feeding was the primary function in the gills of the first true chordates. The other two chordate features are more equivocal. The dorsal nerve tube is confined to the collar region, and the main nervous system is a ventral nerve cord like that of many invertebrates. The notochord is a small structure, developed as an outgrowth from the digestive tract into the proboscis. It has also been considered to be homologous to the pituitary gland, casting its taxonomic value into doubt. Probably, then, these worms are related to the chordates, but it is not clear how. Their relationship to the echinoderms, indicated by a comparison of the tornaria and bipinnaria larvae, is discussed in Chapter 20.

OVERVIEW OF INVERTEBRATE EVOLUTION

The adaptive radiation of the protoeucaryotes resulted in a great array of algal, fungal, and protozoan organisms that exploited every conceivable ecological niche open to unicellular (or acellular) organisms. Among these was a great variety of colonial flagellates, and it is probably among these that the ancestry of the metazoans is to be sought. It is almost certain that the sponges share ancestry with colonial flagellates and probably with choanoflagellates.

The line leading to the Eumetazoa may have passed through a planula stage, which gave rise to the phyla at the tissue level of organization (Coelenterata, Ctenophora) and to the flatworms and other phyla at the acoelomate level. The common ancestor of the higher phyla may have been a turbellarian, but it is at least as likely that it belonged to some long-extinct, unknown group at the flatworm level of organization. The adaptive radiation of this group included several phyla at the same level as well as the somewhat more advanced pseudocoelomate phyla. More importantly, the common ancestor gave rise to two great series: the protostomes, including most of the invertebrates and culminating in the Mollusca, Annelida, and Arthropoda; and the deuterostomes, culminating in the Echinodermata, Hemichordata, and Chordata.

The phylum Chordata—the prochordates and vertebrates, including humans—is the subject of the next chapter. Obviously, this phylum is the last evolved and the most complex, not only of the obscure (from the viewpoint of relationships) Deuterostomia but also, perhaps, of the entire animal kingdom. Figure 19–21 summarizes one viewpoint on animal phylogeny.

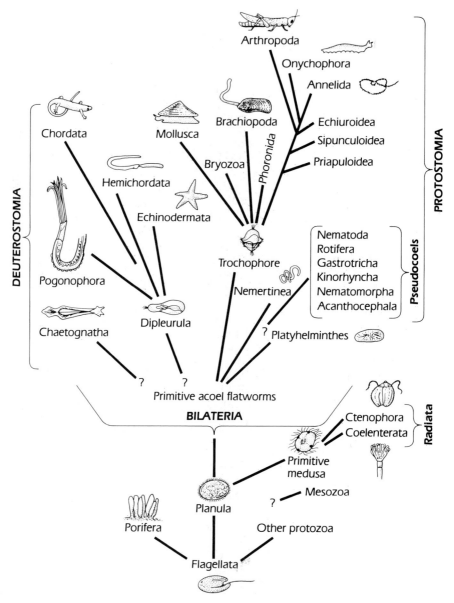

Figure 19–21.   Probable main lines of animal phylogeny.

REFERENCES

Anderson, D.T. 1973. *Embryology and Phylogeny in Annelids and Arthropods*. Pergamon Press, Oxford and New York.
Largely from embryological evidence, the author concludes that the present phylum Arthropoda should be divided into at least two and possibly three phyla: the phylum Uniramia, including the Onychophora, Myriapoda, and Hexapoda; the phylum Crustacea; and the phylum Chelicerata, including trilobites, spiders, and their allies.

Barnes, R.D. 1980. *Invertebrate Zoology*, 5th ed. Saunders, Philadelphia.
An authoritative and comprehensive text on the invertebrates.

Cisne, J.L. 1974. Trilobites and the origin of the arthropods. *Science* 186:13–18.
Relying mainly on paleontology, Cisne arrives at much the same conclusion as did Anderson.

Hadzi, J. 1963. *The Evolution of the Metazoa*. Macmillan, New York.
The major English language presentation of the author's rather radical views.

House, M.R., ed. 1979. *The Origin of the Major Invertebrate Groups*. The Systematics Association Special Volume no. 12. Academic Press, New York and London.
An authoritative treatment of the subjects of this chapter.

Hyman, L.H. 1940–1967. *The Invertebrates*, vols. 1–6. McGraw-Hill, New York.
Unfortunately, this series was never completed, but as far as it goes, it is the best, most comprehensive, and most profound treatment of the invertebrates in the English language. (Cleveland, Kofoid).

Ivanov, A.V. 1963. *Pogonophora*. Translated from the Russian by D.B. Carlisle. Consultants' Bureau, New York.
An extensive treatment by the man who first described this phylum.

Lapan, E.A., and H.J. Morowitz. 1972. The Mesozoa. *Scientific American* 227(6):94–101.
This interesting paper strengthens considerably the case for the Mesozoa as primitive rather than degenerate animals.

Lowenstam, H.A. 1978. Recovery, behaviour, and evolutionary implications of live Monoplacophora. *Nature* 273:231–232.
Current research on these living fossils.

Manton, S.M. 1977. *The Arthropoda: Habits, Functional Morphology, and Evolution*. Oxford University Press, London and New York.
This book is the outcome of the author's lifelong study of the arthropods.

Price, P.W. 1980. *Evolutionary Biology of Parasites*. Princeton University Press, Princeton, N.J.
An unusually successful evolutionary treatment of parasites.

Runnegar, B., and J. Pojeta, Jr. 1974. Molluscan phylogeny: The paleontological viewpoint. *Science* 186:311–317.
From a monoplacophoran ancestor, the authors derive two subphyla: the subphylum Cyrtosoma includes the Monoplacophora and culminates in the Gastropoda and the Cephalopoda; the subphylum Diasoma, passing from the monoplacophorans through the extinct rostroconchs, culminates in the Scaphopoda and the Pelecypoda.

20

Phylum Chordata

We now come to the chordates, perhaps a modest phylum from an arthropod's point of view, but a very important one from our perspective, and one that includes some of the most familiar forms in the living world. In Chapter 19 we presented the echinoderm theory as the most probable theory of the origin of the chordates. Evidence in its favor is not conclusive, and almost every invertebrate group has been suggested at one time or another as a possible ancestor of the chordates. We discuss a few of the more plausible suggestions here, then we follow evolution through the phylum Chordata.

THEORIES OF CHORDATE ORIGIN

Discarded Theories

C.A. Kofoid and A.A.W. Hubrecht argued for nemertean ancestors because of their nervous system of eight longitudinal cords. Development of the two dorsal cords at the expense of the ventral and lateral cords might explain the origin of the dorsal nerve tube. Other nemertean systems are sufficiently generalized that they could have given rise to chordate systems, but there is no evidence that they did. We can say much the same of the turbellarian theory of E. Haeckel. A stronger case has been made for annelid ancestry of the chordates, based on the resemblance of a primitive chordate, such as an ammocoetes larva, to an inverted annelid. In Figure 20–1 if we compare the upright (annelid) structure with the upside-down (chordate) structure, we can see the suggestion of a relationship. Serious problems, however, exist: a new mouth would be necessary, no annelid structure is even remotely like a notochord or like a gill slit, and annelids contrast with chordates in the whole array of embryological characters that define the two main groups of coelomates. Collectively, the objections to the annelid theory outweigh its merits.

Echinoderm Theory

We are left with the echinoderm theory, based upon conformity of both echinoderms and primitive chordates to that set of embryological characters described for Deuterostomia generally. The theory received its major impetus from the discovery that the tornaria larva, originally described as a starfish larva, was actually that of *Balanoglossus,* a hemichordate (Figure 20–2). At that time (the last quarter of the nineteenth century), hemichordates were generally regarded as the most primitive subphylum of the Chordata. This conclusion is reinforced by serological and other biochemical evidence. Many zoologists still accord them that position, but specialists on the group are

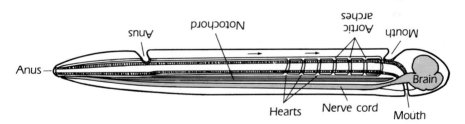

Figure 20–1. *Diagram to illustrate the supposed transformation of an annelid worm into a vertebrate. In normal position the figure represents the annelid with a "brain" at the front end and a nerve cord running along the underside of the body. The mouth is on the underside of the animal, the anus at the end of the tail; the bloodstream (indicated by arrows) flows forward on the upper side of the body, back on the underside. Turn the book upside down and now we have the vertebrate, with nerve cord and bloodstreams reversed. The supposed change is not as simple as it seems: it is necessary to build a new mouth and anus and close the old one; the worm really had no notochord.*

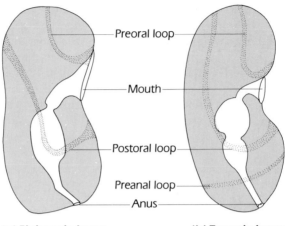

(a) Bipinnaria Larva   **(b) Tornaria Larva**

*Figure 20–2. (a) Bipinnaria larva of a sea star.
(b) Tornaria larva of* Balanoglossus. *Stippled areas represent
ciliary bands. Note preoral loop (of cilia), mouth, postoral
loop, preanal loop, and anus. The preanal loop is missing
from younger tornaria larvae, and it is present on the larvae
of many echinoderms.*

more inclined to feel that a separate phylum, Hemichordata, should be recognized.
Whether this taxonomy will decrease the prestige of the echinoderm theory
of descent remains to be seen. The phylum Hemichordata is generally agreed
to be closely related to the phylum Chordata, even if it be conceded that the
two are distinct. At any rate the echinoderm theory now has more support
than does any other theory of chordate origin, but few would care to claim
that it is securely established.

A variant on the echinoderm theory, formulated in exhaustive detail by
R.P.S. Jefferies, focuses on a group usually regarded as primitive echinoderms,
the carpoids, of Cambrian to Devonian age. Unlike typical echinoderms, they
ranged from asymmetrical to more or less bilaterally symmetrical. On the
basis of superbly preserved material, Jefferies claims to show such anatomical
details as a brain, cranial nerves resembling those of vertebrates, gills, and a
tail with a notochord. Yet carpoids have a skeleton of calcareous plates. Jefferies
regards them as primitive chordates, and he erected a subphylum Calcichordata
for them. The gap between calcichordates and vertebrates must be crossed by
replacing the skeleton of secreted plates by a bony one. While Jefferies has
done a fine job of elucidating the anatomy of an unusual group that may be
distinct from the echinoderms, his belief that they represent ancestral chordates
has won few adherents.

MAJOR DIVISIONS OF THE PHYLUM CHORDATA

The phylum Chordata (Table 20–1) includes three subphyla: the Urochordata, or tunicates; the Cephalochordata, including amphioxus, the favorite of most elementary zoology texts; and the Vertebrata, the most important subphylum, including the dominant animals of land, sea, and air. The first two, together with the Hemichordata, have been exhaustively studied for evidence of the origin of the vertebrates.

Table 20–1. Major divisions of the phylum Chordata

Taxon	*Description*
Phylum Chordata	Deuterostomes; with a dorsal nerve tube, notochord, and a pharynx perforated by gill slits at some time in the life cycle
Subphylum Urochordata	Tunicates; nerve cord and notochord in larva only
Subphylum Cephalochordata	Amphioxus and its allies; chordate characters prominent throughout life
Subphylum Vertebrata	Notochord supplemented or largely replaced by vertebral column; brain with ten pairs of cranial nerves protected by skull; epidermal sensory placodes; ventral heart; pronephric kidney or derivative; bone, enamel, or dentine usually present; neural crest
Infraphylum Protopisces	Suctorial mouth; neither jaws nor paired appendages
Class Agnatha	Ostracoderms, hagfishes, and lampreys
Infraphylum Gnathostomata	Mouth armed with jaws; paired appendages
Superclass Pisces	Includes three classes of fishes
Class Placodermi	Extinct fishes; jaws and paired appendages in considerable variety
Class Chondrichthyes	Sharks, rays, and their allies
Class Osteichthyes	Bony fishes; long dominant in the waters of the world
Superclass Tetrapoda	"Four-footed" vertebrates; fins have been replaced by legs or wings; generally terrestrial, although many are aquatic
Class Amphibia ⎫ Class Reptilia ⎬ Class Aves ⎪ Class Mammalia ⎭	These vertebrates are so well known that their characterization can be left entirely to the text

Morphologically, the tadpole larva of the tunicates is well described by the three fundamental characters of the chordates: (1) a dorsal nerve cord, (2) a notochord, and (3) a pharynx perforated by gill slits. For morphological reasons, therefore, N.J. Berrill believes that the vertebrates arose from ancient tunicates. In the adult, however, the nerve cord and the notochord degenerate. The gills, which are feeding mechanisms rather than respiratory structures, become highly developed. Adult tunicates are generally sessile, and three well-marked classes occur in the subphylum.

Amphioxus was long regarded as an ideal "ancestor" because of its beautiful simplicity and its organization around the basic chordate characters. However, later research showed that amphioxus is specialized in some characters, such as the inclusion of the gills within an atrium, the fact that the notochord runs the entire length of the body, and the inexplicable fact that the kidneys of amphioxus have more in common with those of annelids than they do with those of vertebrates. These primitive chordates (Figure 20–3) must have branched off from the main chordate stock very early, and their exact relationship to the vertebrates is no more clear today than it was when the question was first raised. The fossil record is of no help, for none of the protochordates are known as fossils at all (unless *Jamoytius* qualifies).

The vertebrates themselves are divided into two infraphyla and eight classes, of which one class is extinct. Four classes are entirely aquatic, the classes Agnatha, Placodermi, Chondrichthyes, and Osteichthyes. The last three are grouped together as the superclass Pisces, corresponding to the common term *fish,* yet they differ from one another more fundamentally than do the remaining four classes of land animals, which we recognize as distinct. These are the classes Amphibia, Reptilia, Aves, and Mammalia, which comprise the superclass Tetrapoda. There are roughly 44,000 species of living vertebrates, half of which are fishes, and the other half are tetrapods.

INFRAPHYLUM PROTOPISCES (CLASS AGNATHA)

The first undoubted vertebrates appear in marine rocks of Upper Cambrian age from Wyoming. They belong to the class Agnatha (Greek, "without jaws"), the sole class of the infraphylum Protopisces. All other vertebrates may be grouped in the infraphylum Gnathostomata (Greek, "jaw mouths"). The only surviving members, numbering about fifty species, are the cyclostomes (lampreys and hagfishes). Such naked, boneless animals are rarely fossilized, but the ancient agnaths had bony armor, which often became well fossilized. These ancient agnaths, best known from marine and freshwater sediments of Late Silurian and Early Devonian age, are known from areas as diverse as Utah, the Yukon, the Canadian arctic, Scotland, Spitzbergen, and China. They are called ostracoderms (Greek, "shell skin") because of the bony armor over the head. They comprise two subclasses: the Monorhina, which, like the

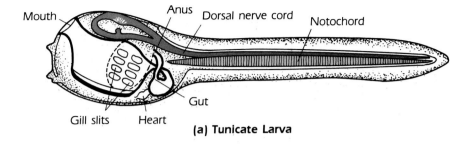

Mouth Anus Dorsal nerve cord Notochord

Gut

Gill slits Heart

(a) Tunicate Larva

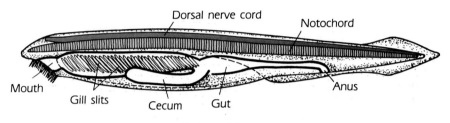

Dorsal nerve cord Notochord

Mouth Anus

Gill slits Cecum Gut

(b) Amphioxus

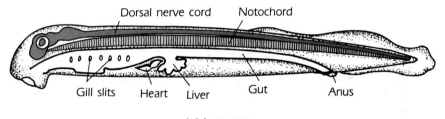

Dorsal nerve cord Notochord

Gill slits Heart Liver Gut Anus

(c) Lamprey

Figure 20–3. *Three primitive chordates: (a) Tunicate larva, (b) amphioxus, and (c) lamprey.*

living cyclostomes, had a single nostril and gill pouches that opened directly to the outside (Figure 20–4); and the Diplorhina (heterostracans of older literature), which had two nostrils (as is usual among vertebrates) and gills that opened into a common atrium with a single excurrent pore on either side. The head region of ostracoderms was covered by bony plates, either as a single shield or as a mosaic of small pieces, and the rest of the body was only lightly scaled (and so usually poorly preserved). Most were less than a foot long. They became extinct by the end of the Devonian.

Living cyclostomes (Figure 20–5) may have evolved from a naked monorhine (such as *Jamoytius?*) or from an ancestor of the monorhines; whereas the gnathostomes (fishes with jaws) may have evolved from diplorhine ancestors,

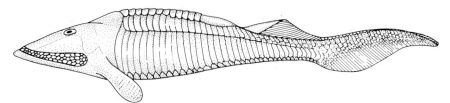

Figure 20-4. *A Lower Devonian ostracoderm,* Hemicyclaspis.

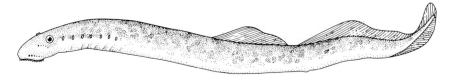

Figure 20-5. *A modern jawless fish, the lamprey,* Petromyzon.

but the details of both lines of descent and of the relationship between monorhines and diplorhines are unknown. Cyclostomes lead a highly specialized, semiparasitic life, but certain of their structures indicate their primitive nature: the vertebrae consist of only the dorsal arcualia, the notochord being the major part of the axial skeleton; there are only one (hagfishes) or two (lampreys) semicircular canals in the ear; the kidney is pronephric in hagfishes, although it is mesonephric in lampreys; and finally, the ammocoetes larva of the lamprey approximates the archetypal chordate more closely than does any other living form.

INFRAPHYLUM GNATHOSTOMATA: SUPERCLASS PISCES

Class Placodermi

During the Silurian and Devonian, ostracoderms shared the waters with another group of armored fishes, the placoderms (Greek, "armored skin"). The major features separating these and all later vertebrates from the more primitive ostracoderms were the presence of paired appendages and the acquisition of jaws. Jaws, a key evolutionary advance, made possible a predatory mode of life. Placoderm jaws were of very primitive structure, not readily homologized with those of other vertebrates, and most lacked teeth. As in ostracoderms, the armor was confined to the head and the anterior part of the thorax. Vertebrae were not ossified. Although they showed diverse sorts of paired fin structure, none approached the stable pattern of paired pectoral and pelvic fins, which characterizes higher fishes. Freshwater and early forms were generally

Figure 20–6. *A large, Upper Devonian placoderm,* Dinichthys.

small, comparable to ostracoderms; Late Devonian forms such as *Dinichthys* (Figure 20–6), found in black shale near Cleveland and also in Germany, were huge, attaining a length of 9 m. Because such enormous fishes were bottom dwellers, they could not gape their mouths by lowering their jaws; instead, they elevated their massive skulls by means of peculiar joints between head and trunk armor. At least one group of placoderms, the antiarchs, seemingly had lungs.

The placoderms were succeeded by sharks and by bony fishes, but their relationship to these more-advanced groups is not evident. It is possible that the chimaerids, which are usually treated as an aberrant offshoot of the sharks, are actually unarmored placoderms. Otherwise, placoderms became extinct at the end of the Devonian, largely because they could not compete successfully with the more-advanced fishes, which could swim better and obtain food more efficiently.

Class Chondrichthyes

Sharks, dating from the Middle Devonian, were the last class of fishlike vertebrates to appear. They completely lack bone and have a cartilaginous skeleton instead. (Chondrichthyes means "cartilage fish.") This character is evidently not primitive but a specialization in the course of their evolution from a bony, placodermlike ancestor. Where the placoderms generally had bony plates for biting, the sharks (like some placoderms) developed true teeth, with a core of dentine and an enamel surface. These teeth are identical in structure and in mode of development to the placoid scales that cover shark skin, and they may have derived from such scales by enlargement and invasion of the oral cavity. Their fossil record is based upon these teeth and upon calcified cartilage.

There are sufficient fossil data to recognize three major evolutionary stages. Small sharks were common in late Paleozoic* seas. They were characterized by long, terminal jaws; the upper jaw firmly attached to the brain case; teeth

* Throughout this chapter, the reader who is not versed in geology may be puzzled by what seems to be capricious use of capital letters. Here it is "late Paleozoic," earlier it was "Late Devonian." It is accepted geological practice to capitalize the adjective when it refers to a stratigraphically determined subdivision of a larger period and to use a lowercase letter when it is simply a rough estimate.

with tall, narrow cusps (from which these sharks are called cladodonts); broad, triangular pectoral fins of restricted mobility; a total lack of bone in the vertebral column; and a tail with a rather heavy, stiff skeleton. Sharks almost became extinct in the Permian, but they recovered during the Mesozoic, when a new type of shark, the hybodonts, arose. Hybodonts had low teeth appropriate for crushing, upper jaws more loosely attached to the brain case, pectoral fins with a narrower base and more flexible structure, and a tail in which the ventral lobe was reduced and more flexible, producing greater thrust and lift. Hybodonts were more mobile than the Paleozoic cladodonts, and the difference in the teeth suggests that they exploited a different food source. Modern sharks appeared in the Cretaceous. They are characterized by short, protrusible jaws, with the upper jaw movably suspended from the brain case; a strong rostrum that causes the mouth to assume a ventral, subterminal position; and calcified vertebral centra, which are common fossils in Tertiary deposits. Shark teeth are also common fossils in Tertiary deposits along the east coast of the United States, including teeth of the giant *Carcharodon,* whose 1.8 m jaws bore teeth 15 cm high. All told, there are about 500 species of cartilaginous fishes today. Modern sharks (Figure 20–7) are swift-moving, highly effective predators. Although they are less important now than they were in the Cretaceous seas, they are still an important part of marine faunas today, and a few sharks have invaded fresh water.

The adaptive radiation of the sharks in the Cretaceous included the formation of a new adaptive type, the rays (Figure 20–8). Rays are essentially flattened sharks, the teeth of which are modified to form plates for crushing the shells of the molluscs on which they feed. Chimaerids (ratfishes) seem widely divergent from sharks and rays, and they have an independent history that can be traced back into the Paleozoic. Nonetheless, these rare and unfamiliar fishes share so many characters with more typical chondrichthyans that their affinities seem beyond doubt. Although some modern sharks are as small as 15 cm, the great

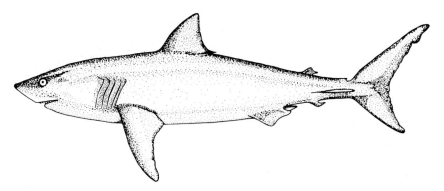

Figure 20–7. *The great white shark,* Carcharodon, *a formidable predator in modern seas.*

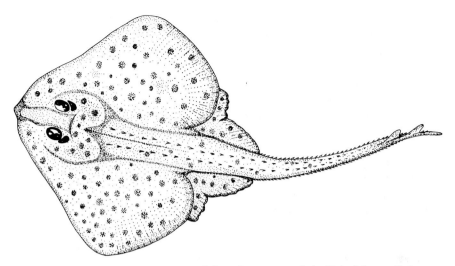

Figure 20–8. *A small skate of the Atlantic coast of the United States,* Raja.

white shark (*Carcharodon*), which reaches 12 m in length, is the largest predaceous shark; and the plankton-feeding whale sharks (*Rhincodon*) and basking sharks (*Cetorhinus*), which approach 15 m in length are the largest fishes in the sea.

Class Osteichthyes

The dominant fishes of the world are, and long have been, the Osteichthyes, or bony fishes. Following Haeckel, biologists long believed that these fishes took their origin from primitive sharks. This theory no longer seems likely, for the fossil evidence indicates that the Osteichthyes arose earlier than did the Chondrichthyes. Further, the bony fishes originated in fresh water and remained there for long ages, whereas the sharks originated in the sea. Thus it seems most probable that both groups arose independently. The name of the class is based upon its possession of a bony skeleton. Although this is a general character, it is variable; some "bony fishes" have largely cartilaginous skeletons. Because of the extensive bony skeletons of ostracoderms and placoderms, that of the Osteichthyes is now regarded as a primitive character, whereas the cartilaginous skeleton of the sharks and the partially cartilaginous skeletons of some "bony fishes" are regarded as based upon secondary retention of an essentially embryonic character.

Even at their first appearance in freshwater deposits of Early Devonian age, there were two major groups of bony fishes, the subclass Actinopterygii, or ray fins, and the subclass Sarcopterygii, or fleshy fins (formerly called Choanichthyes, or "nostril fishes"). As the names imply, they differed fundamentally in the structure of the fins. The former has fins of fanlike structure, with slender, bony rods for support (familiar to any fisherman); the latter has stout, fleshy fins, with a strong internal skeleton and much muscle. The evolutionary success of the two groups is measured by different standards. Ray-finned fishes today number several tens of thousands of species and are the most diverse group of vertebrates; lobe-finned fishes are almost extinct and number but four genera today. However, the main line of vertebrate evolution passed through the lobe-finned fishes, and from them all tetrapods have descended.

Three superorders of ray-finned fishes are recognized: Chondrostei, Holostei, and Teleostei. Their names, from the Greek for "cartilaginous bone," "complete bone," and "perfect bone," imply an evolutionary progression. The most primitive are the Chondrostei, whose name is misleading, for it implies that they all had a cartilaginous skeleton. This is largely true of such surviving members as the Holarctic sturgeons (Figure 20–9) and spoonbills, which are confined to the drainages of the Mississippi (*Polyodon*) and Yangtze (*Psephurus*) rivers. However Paleozoic chondrosteans, the paleoniscoids were invested in a sheath of heavy, enamel-coated ganoid scales and had well-ossified skulls. The jaws of paleoniscoids were long, and their jaw muscles were relatively small and weak. The vertebrae were unossified, and the tail had the upturned, heterocercal structure. The paleoniscoids had lungs to supplement their gills. Paleoniscoids were dominant until the Triassic, but only a few specialized chondrosteans have survived to the present.

Holosteans arose from chondrostean ancestors in the Permian, and they became dominant in the Triassic. Important changes in the feeding mechanism marked this evolutionary level. The jaws became shorter, the muscles of the jaws increased in size and in efficiency of orientation, and a coronoid process improved the leverage of the jaws. The tail became more symmetrical, although the vertebral column was still somewhat upturned toward the posterior end. In advanced holosteans, the scales lost their ganoine covering, leaving thin, flexible, bony scales, similar to those of teleosts. The vertebral centra were ossified. Both of these advances are found in *Amia,* the bowfin, one of the two surviving holosteans. The other, the garpike *Lepisosteus* (Figure 20–10), has thick, ganoid scales like those of the paleoniscoids. The lungs fused to form a swim bladder, which controls buoyancy but which is still important

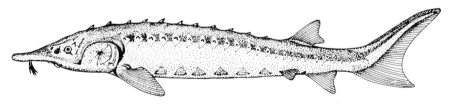

Figure 20–9. *A primitive ray-finned fish, the sturgeon,* Acipenser.

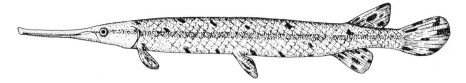

Figure 20–10. A holostean fish, the garpike, Lepisosteus.

for respiration. Holosteans frequently come to the surface to gulp air, and a garpike can be drowned by being held under water. Holosteans invaded the seas, where they underwent great diversification during the Mesozoic. Yet, curiously, the two genera that have survived to the present are both freshwater fishes. By the Cretaceous, holosteans had passed their peak and were declining toward their present low numbers.

Teleosts first appeared in the Jurassic. By the end of the Cretaceous, they dominated both fresh and salt waters. The transition to the teleost level occurred in a well-documented group of small, advanced holosteans, the pholidophorids. The set of characters that marks the teleosts includes a skeleton that is always entirely bony; scales that are thin, flexible chips of bone; a symmetrical, homocercal tail based primarily on the original ventral lobe and ideally adapted to developing forward thrust without unnecessary lift; and lungs that, if present, form a swim bladder. Holosteans are sluggish swimmers, but the thin scales and ossified vertebrae of teleosts are conducive to swift swimming. Primitive teleosts, such as herring, tarpon, and even salmon, have long jaws. A great adaptive radiation is that of the Ostariophysi, intermediate-level freshwater fishes including the characins of South America (notably the piranha) and Africa, as well as carps, minnows, and catfishes. These fishes "hear" with their swim bladders, or rather with vibrations transmitted from the swim bladder to the ear via a chain of special bones, the *Weberian ossicles.* Advanced teleosts are represented by another great radiation of fresh- and saltwater fishes that includes a large proportion of familiar aquarium, sport, and commercial fishes. Most numerous of these are in the order Perciformes (Figure 20–11). In perciforms, the first dorsal fin is very spiny and the pelvic fins have been repositioned under the pectoral fins, thus allowing fine control of swimming movements. The jaws are short, and the maxilla has been converted from a tooth-bearing bone to a mobile strut controlling the tooth-bearing premaxilla, which is protrusible. This highly versatile jaw mechanism has provided the basis for an explosive adaptive radiation into every niche possible for fishes. The largest teleost is probably the bluefin tuna (*Thunnus*). At 4.5 m it is considerably smaller than the more primitive sturgeons at 7 m maximum and the great sharks at 11 m to 15 m.

Today, teleosts comprise more than 95 percent of the fishes of the world. In the oceans, only the Chondrichthyes compete with them. In fresh water,

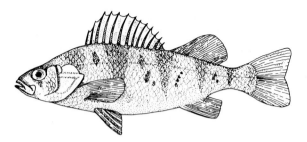

Figure 20–11.   *An advanced, spiny-rayed percaform fish, the yellow perch,* Perca.

only a few stragglers of the chondrosteans and holosteans and a very few unusual sharks are their competitors. They are one of the most successful and varied groups of vertebrates, and they are at the peak of their diversity today.

INFRAPHYLUM GNATHOSTOMATA: SUPERCLASS TETRAPODA

Subclass Sarcopterygii and Origin of the Tetrapods

Lobe-finned fishes originated in fresh water in Early Devonian times. They achieved a brief dominance, although they quickly yielded this position to the chondrosteans. From the very beginning, there were two types, the orders Crossopterygii and Dipnoi, which were as separate from each other as both were from the ray-finned fishes, although considerable similarities indicate a not too remote common ancestry for all three.

The most important distinguishing feature of the sarcopterygians was the structure of the paired fins. The pattern of skull bones was completely different from that in the ray-fins, and most had choanae, or internal nostrils. The scales were cosmoid rather than ganoid, emphasizing the dentinelike internal layer rather than the superficial enamel layer. The tail was initially heterocercal, as in other primitive fishes, but early in their history, they achieved symmetry by the extension of the vertebral column horizontally to the end of tail, where it formed a little tassel. Thus, in contrast to the teleosts, the external symmetry of this *diphycercal* tail reflects true symmetry of internal structure.

The Dipnoi, or lungfishes, are represented by three surviving genera: *Neoceratodus,* a little-changed descendant of the Triassic *Ceratodus,* in Australia; and two with filamentous pectoral and pelvic fins, *Protopterus* in Africa and *Lepidosiren* in South America. Dipnoans were once believed to be the stock from which the tetrapods derived (and this theory has been revived recently by D.E. Rosen and others). They show similarities to tetrapods in many aspects of their soft anatomy: in lungs, circulatory and nervous systems, and in the structure of the choanae. Most important, however, are their fins (Figure

20–12), which suggest an elm leaf, with a single row of basals running the length of the fin and a row of rays on either side. A muscular lobe extends along the row of basals. C. Gegenbaur, regarding this fin as the probable source of tetrapod appendages, called it an archipterygium ("primitive limb"). He conjectured that the tetrapod limb formed from it by suppression of all of the rays except the terminal five. However, it is difficult to derive a limb in which the second segment consists of two parallel bones from a limb with a single row of basals. Further, dipnoans have a mosaic of small skull bones that cannot be homologized with those of tetrapods, and they have paired upper and lower tooth plates (Figure 20–13) totally unlike those of other vertebrates. In spite of these difficulties, the dipnoan theory might still enjoy favor were it not that the Crossopterygii provide a more plausible solution to the problems of amphibian ancestry.

The Crossopterygii consist of two suborders, the rhipidistians and the coelacanths. The former are much more important for evolution: there is little doubt that they crawled out during the Late Devonian onto land that had formerly been colonized only by plants and by invertebrates. These freshwater fishes arose in the Devonian and became extinct in the Permian. They used

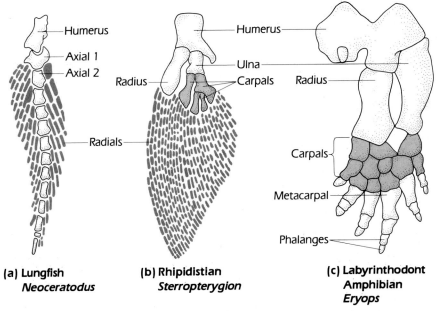

(a) Lungfish
Neoceratodus

(b) Rhipidistian
Sterropterygion

**(c) Labyrinthodont
Amphibian**
Eryops

Figure 20–12. Primitive forelimbs in the fleshy-finned fishes and early tetrapods. (a) Archipterygium of the living Australian lungfish, Neoceratodus. *(b) Dichotomous fin of the Devonian rhipidistian,* Sterropterygion. *(c) The dichotomous limb of the Permian labyrinthodont amphibian,* Eryops. *Notice the similarity of the last two and their contrast to the first.*

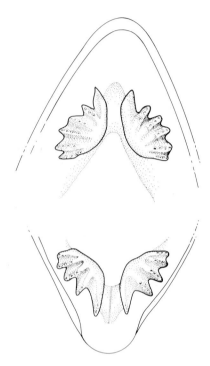

Figure 20–13. *Paired tooth plates from upper and lower jaws of the Australian lungfish,* Neoceratodus.

their lungs to breathe air and their stout fins to drag themselves about. The first forays on land may have been in search of new pools, when, in times of drought, the old ones dried. They left behind paleoniscoids that died and lungfish that burrowed in the mud to aestivate.

The skull of the rhipidistians has much in common with that of early amphibians (Figure 20–14). They differ mainly in that a few bones of the rhipidistians have been lost in the amphibians, and a few others have fused. Both have an opening for a pineal eye, internal choanae or nostrils, and lungs. In both, labyrinthodont teeth are present. These are peculiar teeth, known only in crossopterygians and in very primitive amphibians, in which the enamel forms deep ridges extending into the dentine. The skull of the rhipidistians could be flexed on a prominent joint between the anterior and posterior parts; even in modern amphibians there is a line of weakness in the skull at the same location.

The rhipidistian fin was of the dichotomous type, which, unlike the archipterygial fin, can be readily homologized with the limbs of primitive amphibians (Figure 20–12). Both are characterized by a single, heavy piece, the humerus, which articulates with the shoulder girdle, followed by parallel

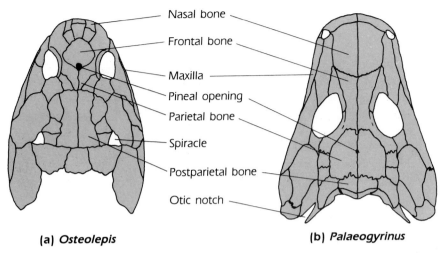

Nasal bone

Frontal bone

Maxilla

Pineal opening

Parietal bone

Spiracle

Postparietal bone

Otic notch

(a) *Osteolepis*   **(b)** *Palaeogyrinus*

Figure 20–14.   *Dorsal views of the skulls of rhipidistian: (a)* Osteolepis, *a crossopterygian, and (b)* Palaeogyrinus, *a primitive amphibian. The specialized gill opening, the spiracle, of many fishes was transformed in early amphibians into the otic notch for support of the ear drum to receive air-borne sounds.*
Data from Gregory, *Evolution Emerging* (New York: Macmillan, 1951).

radius and ulna, and then by a cluster of small, less exactly homologized bones at the end of the appendage. Finally, the vertebrae of both rhipidistians and primitive amphibians were diplospondylous—that is, in each body segment there were two centra, one developed from the pleurocentrum, the other from the intercentrum (Figure 3–5). Other parallels between the groups also reinforce the idea that these crossopterygians gave rise to the tetrapods.

The coelacanths were a specialized offshoot from the main line of the crossopterygians. Although they shared many characters with the rhipidistians, they lacked choanae, had short skulls and reduced jaws and teeth, and showed no tendency toward ossification of the vertebrae. From the Devonian to the Permian, they inhabited only fresh water, but during the Triassic they entered shallow seas, where they persisted until the Cretaceous. They had been believed to have been extinct for 75 million years until 1939, when, to the astonishment of the scientific community, a living coelacanth was caught by a fisherman and brought into a South African port. It was described under the name *Latimeria chalumnae* (Figure 20–15). Since 1952, many other specimens have been caught, mainly by native fishermen of the Comoro Islands, northwest of Madagascar. The natives lower deep lines from tiny boats at the edge of the reefs. Several specimens have been observed alive for a few hours before they died; and preserved specimens have been sent to France, England, the United States, Canada, and Russia for study.

Figure 20–15.   *The coelacanth,* Latimeria, *a living fossil.* Courtesy of the National Museum of Natural Sciences, National Museums of Canada.

Class Amphibia

The first amphibians that crawled out upon the mud banks of Late Devonian streams were little more than fishes with fins sufficiently modified to support the weight of the body. *Ichthyostega,* the earliest amphibian known, has been found in Upper Devonian rocks of Greenland. It was about .9 m long. Its limbs, though short, were not terminated by fin rays but by five digits of normal tetrapod pattern. The pelvis was attached to the vertebral column by a single sacral vertebra, unlike in its rhipidistian ancestors. The vertebrae were more completely ossified, and they were again diplospondylous. These Paleozoic amphibians are called labyrinthodonts because they had teeth with deeply infolded enamel (Figure 20–16), a direct carry-over from rhipidistians.

Labyrinthodonts became common, widespread, diverse, and successful in the faunas of the great coal swamps of the Pennsylvanian and in the red beds of the Permian. One line survived through the Triassic. Their skulls were broad and low, some attaining lengths of over a meter. The bones of the roof

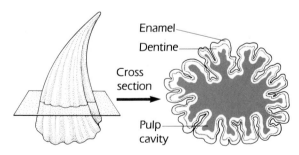

Figure 20–16.   *A labyrinthodont tooth and its cross section. This specialized type of tooth is found in both rhipidistian fishes and in labyrinthodont amphibians.*

of the skull were ornamented with pits and grooves, and there was a prominent posterior *otic notch* to support the eardrum. Some became adept at life on land, although they were, of course, still tied to the water for reproduction; others reverted to the water and became primarily aquatic. There were two major lines of labyrinthodonts: in the temnospondyls, the anterior intercentrum was emphasized at the expense of the pleurocentrum, which actually disappeared in advanced forms and left a vertebra with the familiar single centrum; whereas in anthracosaurs, the posterior pleurocentrum hypertrophied and the intercentrum atrophied. The true centrum of reptiles, birds, and mammals is a pleurocentrum, reflecting the fact that reptiles evolved from anthracosaurs (Figure 3–5).

All the while that large labyrinthodonts were the dominant tetrapods, there was a simultaneous radiation of small, rather inconspicuous amphibians, the lepospondyls. Lepospondyls had a single centrum much like that of living amphibians. One group was snakelike, a second looked very much like small reptiles (causing much confusion among paleontologists), and a third developed curious boomerang-shaped skulls and attained a length of over a meter. These were all important pond dwellers of Early Permian times.

Modern amphibians, including toads and frogs (anurans, 2,600 living species), salamanders and newts (urodeles, over 300 living species), and wormlike apodans (150 living species), are greatly modified animals that are utterly unlike their labyrinthodont ancestors. They have survived in a world dominated first by reptiles, then by mammals, because they specialized for very limited niches in which they could excel—that is, niches for small animals, in environments that are warm, at least seasonally, and wet (ponds or swampy areas, not usually large bodies of water). These living amphibians have naked bodies with thin, moist skin that serves as a respiratory membrane, in contrast to the skin of their ancestors, which was thick and keratinized and could not serve for respiration. The bones of the skull have been much reduced and simplified, obscuring their ancestry among the labyrinthodonts. The earliest known frogs are from the Triassic; the salamanders date from the Cretaceous.

Class Reptilia

The Amphibia play a minor role in the vertebrate fauna of today, and perhaps their greatest importance lies in their role as the source of the class Reptilia. The origin of the reptiles from primitive labyrinthodonts is unusually well attested, for there are many transitional genera. Reptiles first appear in the fossil record in the Pennsylvanian. By the Permian, they had already begun a great diversification, which led to the formation of six orders in the Permian and ten more in the Triassic. Throughout the Mesozoic, they were the dominant vertebrates, and hence the ordinary designation of this era as the "Age of Reptiles."

The original reptiles were substantially just amphibians adapted for permanent land life. As in modern reptiles, the skin was probably thickened and cornified to prevent drying of the animal. The early reptile had four short limbs set more or less at right angles to the body, so that it could lift its weight only clumsily. It had a large number of undifferentiated conical teeth.

Most important was a profound adaptive shift that separates the amniotes (reptiles and their descendants, birds and mammals) from other vertebrates— the development of the *amniote egg,* a momentous event in the history of life. Animals possessing this adaptation became reproductively free of water, making the final conquest of the land possible. Today, reptiles are highly successful even in deserts, for the fluid within the amnion provides the developing embryo with its own "pond." Fertilization in reptiles is internal, a more efficient method than the external fertilization in amphibians.

Such characters as amniote eggs and internal fertilization are not fossilized, so it is necessary to find skeletal traits to separate reptiles from amphibians. The earliest known reptiles were found in the trunks of fossil trees of Pennsylvanian age at Joggins, Nova Scotia. Called captorhinomorphs (an order of the superorder Cotylosauria), these small, lizard-sized animals were the "stem reptiles" from which all other reptilian types evolved. Their skulls were high and narrow, without an otic notch, and anapsid, that is, lacking openings in the temporal region, which characterize many reptiles. The teeth were not labyrinthodont, in contrast to those of their amphibian ancestors. Finally, they had a second sacral vertebra. Turtles are the only living reptiles that retain the anapsid condition. They appeared in the Triassic, and they remain even today an important part of the reptilian fauna, with over 200 living species.

Almost as old as the captorhinomorphs were the pelycosaurs, in which the temporal region was pierced by a single opening (for attachment of jaw muscles), the synapsid condition. Pelycosaurs, much larger than most captorhinomorphs, were dominant herbivores and carnivores in Lower Permian deposits of Texas and the southwestern United States. Some of the pelycosaurs bore large, sail-like structures composed of elongated vertebral spines interconnected by a web of skin. These may have been an early attempt at temperature regulation.

The immediate descendants of the pelycosaurs were the therapsids (Figure 20–17), mammal-like reptiles that are well known from Upper Permian to Mid Triassic sediments from South Africa and South America. These advanced synapsids were evolving some remarkable changes. In place of the usual isodont teeth, they developed incisors, canines, and even expanded cheek teeth, indicating that they chewed their food far more than does any living reptile. In typical reptiles, the dentary, the major tooth-bearing bone, is but one of many bones forming the jaw; in therapsids, the dentary grew at the expense of the other bones, which in advanced members were reduced to small, accessory ossicles in a shallow groove near the jaw joint. A strong, elevated coronoid process provided powerful leverage for the jaw muscles, which were evidently being reorganized into the mammalian configuration. The masseter muscle, unknown

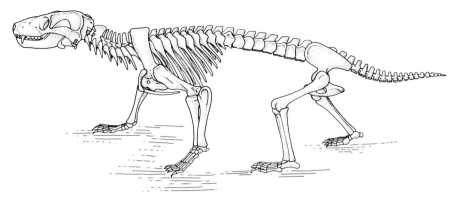

Figure 20–17. *Skeleton of a therapsid reptile,* Thrinaxodon, *from the Lower Triassic of South Africa.*

in other reptiles, now invested the outer surface of the coronoid process. The temporal opening enlarged enormously to accommodate the temporalis muscle mass. The occipital condyle, single in other reptiles, split into the double state as in mammals. There were apparently even turbinal bones in the nose, as in mammals. The sprawling posture common to most reptiles was giving way to a posture in which the knees and elbows were rotated underneath to lift the body. The toes became almost equal in length, with three phalanges in each toe except the first, as in unspecialized mammals (in reptiles, digits I to V have two, three, four, five, and three, respectively). The limb girdles also foreshadowed the mammalian condition, and the tail was shorter than in typical reptiles as the hindlimb retractor muscles shifted from the tail to the pelvis, as in mammals. Therapsids were dominant through the Mid Triassic, but they declined in the Late Triassic. Almost simultaneously, two highly important evolutionary events occurred: dinosaurs and mammals arrived on the scene. The therapsids lingered on to the Early Jurassic, then became extinct. We trace their descendants, the mammals, in a subsequent section.

Sauropsida

All reptiles other than the mammal-like groups have been collectively termed the Sauropsida. These reptiles dominated both land and sea in the Age of Reptiles, the Mesozoic era. They also took their origin among the captorhinomorphs, and their adaptive radiation has been extraordinarily complex (Figure 20–18). They included several kinds of aquatic reptiles, the euryapsids, with a single temporal opening high on the skull. During the Triassic large, superficially turtlelike placodonts, with pavements of blunt teeth for crushing molluscs, and nothosaurs, which had minimal adaptations for aquatic life, evolved. In the Jurassic and Cretaceous, plesiosaurs were highly successful and attained lengths of over 12 m. Their limbs were modified to long paddles, with many segments in each digit. Some had extremely long, flexible necks,

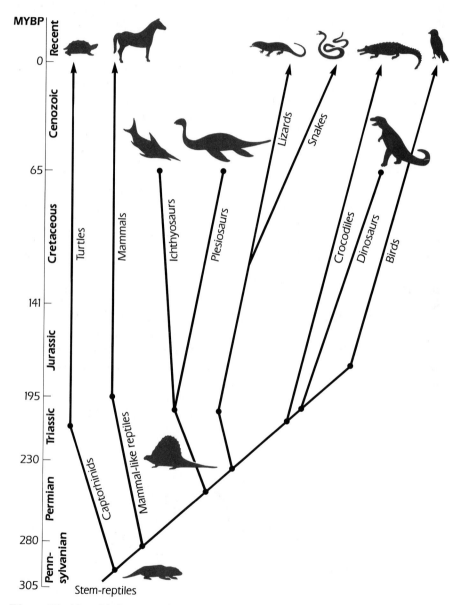

Figure 20–18. *Phylogenetic relationships and distribution in geological time of the major groups of living and fossil reptiles.*

with up to seventy-six cervical vertebrae. The paragons of aquatic adaptation among reptiles are the ichthyosaurs, "porpoises of the Mesozoic," which appeared in the Triassic and were common into the Early Cretaceous. They were of small or moderate size, mostly under 3 m long. With streamlined form, heterocercal tail (but with the vertebral column in the ventral rather than the dorsal lobe), and with pectoral paddles bearing extra digits as well as extra phalanges in each digit, they must have been superb swimmers. All euryapsids became extinct before the end of the Mesozoic, and they left no descendants.

Diapsida are "two-arched" reptiles, with two temporal openings on either side of the skull, one high and one low. There are two subclasses of diapsids, the Lepidosauria and the Archosauria, both of which have surviving representatives today. The relationship between the two is still quite unclear.

Lepidosauria

Lepidosaurs appeared in the Late Permian and achieved prominence in the Mid Triassic, when they included a group of moderate-sized herbivores of the order Rhynchocephalia, which was abundant in the southern hemisphere. The sole survivor is *Sphenodon punctatum* (Figure 20–19), a living fossil on small islands off the coast of New Zealand. Although it resembles a lizard externally, its skull is primitive, with two unreduced temporal arches (Figure 20–20). The next lepidosaurian stage is represented by the lizards (order Squamata, suborder Lacertilia). In lizards the bony arch of the lower margin of the lateral temporal opening has broken down (Figure 20–20), allowing the quadrate bone that supports the lower jaw to move back and forth. Their fossil record is not good until the Late Cretaceous, but by this time and the early Tertiary, modern families became well established and lizards assumed an important role in the fauna of small-bodied, cold-blooded animals. They are particularly successful in warm climates today, and they are the most diverse living reptiles, with nearly 3,000 species. In the Late Cretaceous one

Figure 20–19. Sphenodon, *the tuatara, a living fossil from New Zealand. Rare and protected, these animals are almost never seen in zoos.*

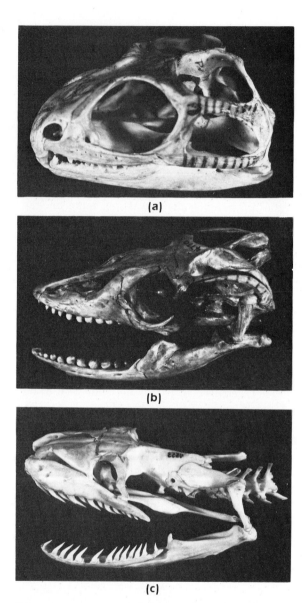

*Figure 20–20.   Structural levels of lepidosaurian skulls.
(a) Complete two-arched skull (both temporal arches striped)
of* Sphenodon. *(b) Skull of* Varanus, *showing loss of the
lower temporal arch in lizards, increasing the mobility of the
jaws (upper temporal arch striped). (c) Skull of* Python,
*showing loss of both temporal arches in snakes; although jaw
support is weak, mobility of the jaws is enormous.*

group of varanoid lizards, the mosasaurs, assumed aquatic habits. They became highly successful predators, attaining a length of 6.5 m and feeding at will on fishes, turtles, and ammonites. Their success may have been a factor in the decline of the ichthyosaurs. The largest living lizard is *Varanus komodoensis,* the Komodo dragon of Indonesia, which reaches a length of 3 m and can bring down a water buffalo!

Snakes (suborder Ophidia) are the most specialized lepidosaurs, and they have evolved a highly specialized jaw mechanism. The temporal arches have been lost completely (Figure 20–20), so that the jaw support is very flexible, thus allowing snakes to swallow objects of prodigious size. The structure is so weak, however, that they are incapable of breaking off small pieces of meat, and so they must swallow prey whole. They are, of course, legless, except for a vestigial pectoral girdle in the boas. Snakes first appear in the Cretaceous, but their fossil record is poor. Primitive snakes are constrictors (superfamily Booidea), which kill their prey by squeezing. In more-advanced snakes the maxilla is only loosely joined to the skull. The majority of the 2000 species of snakes of the world belong to the family Colubridae. For the most part they are not only harmless to humans but also beneficial because they eat insects, rodents, and other small vertebrates. Elapids are poisonous snakes (cobras, coral snakes, and sea snakes), with fixed, venomous fangs. Viperids, the most-advanced snakes, have erectile fangs and include rattlesnakes of North and South America and vipers of Europe.

Archosauria

The final subclass of reptiles is the Archosauria, the ruling reptiles that completely dominated terrestrial faunas of the Mesozoic, from the Late Triassic to the end of the Cretaceous, a period of some 140 million years. Archosaurs, like the lepidosaurs, were diapsid, but they showed no tendency to evolve a loose and mobile quadrate. A characteristic new bone, the laterosphenoid, was added to the braincase; and an opening evolved anterior to the orbit, perhaps related to the development of powerful pterygoid musculature. A characteristic pelvic structure evolved, with a strong ilium joining the limbs to the vertebral column, a pubis pointing ventrally and forward and an ischium pointing ventrally and backward. The limbs were slender, unlike the short, stocky limbs of typical, sprawling reptiles, and the metatarsals were elongated, another novel feature indicating adaptation for swift movement. The stem from which the ruling reptiles developed was a group called the thecodonts, with light bodies and shortened forelimbs. They were probably capable of bipedal locomotion. Their early radiation, however, also included some large and heavy quadrupeds, such as the phytosaurs, crocodilelike, aquatic reptiles with a long, conical, tooth-bearing snout. In some, the skulls were well over a meter long. The nostrils were on prominences immediately in front of the eyes, an interesting adaptation for breathing while mostly submerged. Thecodonts were common in the Triassic, but they disappeared abruptly at the end of that period.

The archosaurs that have survived to the present are the crocodiles. There are about twenty-three species, spread over all continents except Europe (where they persisted until the Pliocene) and Antarctica. They, too, date from the Triassic. Modern crocodilians are characterized by a complete secondary palate, which, like the posterior position of the nostrils in phytosaurs, enables the animals to breathe while mostly submerged. Crocodiles have been very successful in a limited niche—in near-shore, surface waters of warm climates. Although they have never been very diverse, neither have they been seriously challenged by comparable mammalian predators, and they remain important today in tropical and subtropical rivers and estuaries. *Phobosuchus* from the Cretaceous of Texas had a skull 2 m long and a total length estimated at 15 m.

Some thecodont descendants, the pterosaurs, took to the air on a flight membrane stretched from an elongated fourth finger. They ranged from the size of sparrows to four times the size of the largest birds. (*Pteranodon* from the Kansas chalk had a wingspan of 9 m; and gigantic *Quetzalcoatlus* from Texas spanned an estimated 15 m. Jurassic forms had long tails, which the more-advanced members lacked. The sternum was prominent, though unkeeled, and the stout bracing of the shoulder girdle to both the sternum and the vertebral column suggests powered flight, as opposed to simple gliding. Pterosaurs failed to survive to the end of the Cretaceous, perhaps because they could not compete successfully with the increasing numbers of true birds.

Dinosaurs

Dinosaurs (Figure 20–21) were the greatest triumph of the archosaurian radiation. Structural clues to the success of the dinosaurs are found in the characters associated with upright, bipedal posture, which was perfected in the earliest dinosaurs of Late Triassic age. The head of the femur was set at right angles to the shaft of the limb and inserted into a perforated acetabulum of the pelvis, an arrangement unknown in reptiles before or since. The knee joint of dinosaurs permitted movement only fore and aft, and the foot developed a new, birdlike ankle joint that also permitted only flexion and extension, unlike the complex joint of thecodonts, crocodiles, and mammals. The foot developed a birdlike symmetry, and the first and fifth digits were variably reduced.

The term *dinosaur* refers specifically to members of two orders of archosaurs, the Saurischia and Ornithischia, and not to any large reptile of the Mesozoic, of which there were many. The order Saurischia ("reptile-hipped") included two basic adaptive types, the herbivorous sauropods with long necks and tails and small heads, among which were the largest animals ever to walk the surface of the earth; and the carnivorous theropods (Figure 20–22), both small

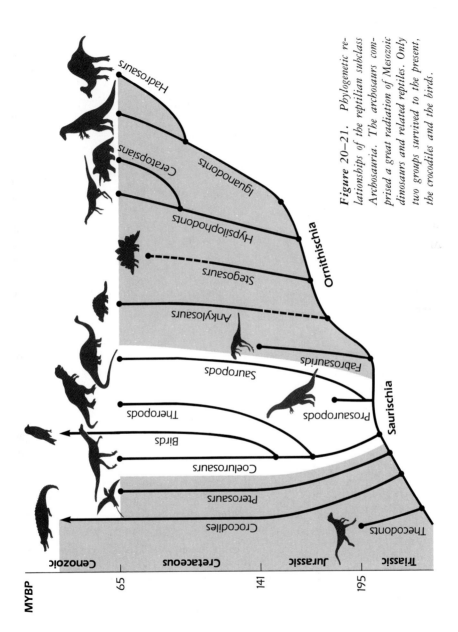

Figure 20–21. *Phylogenetic relationships of the reptilian subclass Archosauria. The archosaurs comprised a great radiation of Mesozoic dinosaurs and related reptiles. Only two groups survived to the present, the crocodiles and the birds.*

Hadrosaurs

Ceratopsians

Iguanodonts

Hypsilophodonts

Stegosaurs

Ankylosaurs

Fabrosaurids

Sauropods

Prosauropods

Theropods

Birds

Coelurosaurs

Pterosaurs

Crocodiles

Thecodonts

Ornithischia

Saurischia

Cenozoic

Cretaceous

Jurassic

Triassic

MYBP

65

141

195

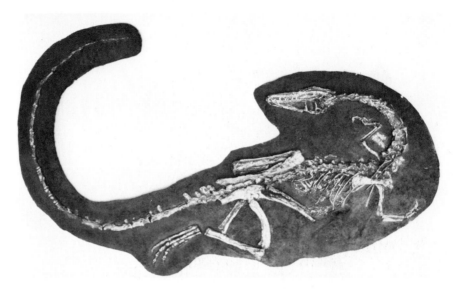

Figure 20–22. *Skeleton of* Coelophysis, *a small theropod dinosaur from the Upper Triassic of New Mexico. From an animal such as this, huge flesh-eaters like* Tyrannosaurus *must have arisen.*

(down to the size of a turkey) and large (*Tyrannosaurus rex,* the most awesome of predators). The saurischian pelvis was essentially like that of the thecodonts. Theropods and prosauropods were among the first dinosaurs of the Late Triassic. Theropods underwent gradual transition from advanced thecodonts, and they lasted to the end of the Mesozoic. Prosauropods gave way to sauropods in the Jurassic. Sauropods were the most important herbivores in that period, but they were much less common in the Cretaceous.

In ornithischians, the "bird-hipped" dinosaurs, the pubis had rotated backward to parallel the ischium, as in birds. These dinosaurs, too, date from the Late Triassic. Their ancestry is uncertain. All ornithischians were herbivorous. A crucial adaptive feature, unknown in any extant reptile, which helps to account for their success was the development of cheeks, which facilitate handling of food. There were four adaptive types, beginning with small, bipedal ornithopods originating in the Late Triassic. The Late Cretaceous duckbilled dinosaurs (Figure 20–23) from western North America and Asia were the culmination of ornithopod evolution in terms of size, diversity, complexity of dental apparatus, and elaboration of structures for species recognition and sexual recognition. Such structures included tall crests on the head, with long, elaborate narial passages that were quite possibly used for trumpeting and bellowing. They developed incredible batteries of teeth, up to 400 in a jaw, which were replaced continuously as older teeth wore out, making the jaws marvelously effective at reducing tough plant matter to fine particles.

Figure 20–23. Skull of a duck-billed dinosaur, Cory-thosaurus, *from the Cretaceous of Alberta. The crest, the top of which is broken off, carried air passages from the nose to the throat. This individual was a female; in males, the crest was much fuller.*

The stegosaurs were large, clumsy-looking quadrupeds of the Jurassic, with great triangular plates down the center of the back. The plates were probably defensive, but they must also have had display value; and they had a rich blood supply, so that they could have served for heat exchange. Stegosaurs were perhaps replaced ecologically by ankylosaurs, which were important in the Cretaceous. Ankylosaurs (Figure 20–24) were large, broad, low, and heavily armored, with bony scutes paving the back and stout spikes projecting laterally. The skull was triangular, broad posteriorly; and the teeth were small and rather weak.

Horned dinosaurs, the ceratopsians, were confined to the Late Cretaceous. The skulls were up to about 2.5 m long, with horns over nose or eyes, so that they invite comparison with rhinoceroses. A frill extending from the skull over the neck region provided some protection to the neck and seems to have served as the origin for powerful jaw muscles. The frill clearly had an important function in display. Sexual dimorphism has been recognized in the small, primitive *Protoceratops* (Figure 20–25) from Mongolia, and the elaborate pattern of horns and frills must have been functionally analogous to the horns of modern African bovids, used not primarily for defense but for

Figure 20–24. Ankylosaurus, *a large, armored, herbivorous dinosaur, and two specimens* of Thescelosaurus, *a smaller herbivore. These animals flourished in the Late Cretaceous in Alberta and Montana. While the form is well known from the fossilized skeletons, the pattern of spots on* Thescelosaurus *is entirely hypothetical.*
Drawn by Dr. R. T. Bakker, and reproduced by courtesy of the National Museum of Natural Sciences, National Museums of Canada.

species recognition and perhaps ritualized intraspecific combat. Ceratopsians seem to have been the swiftest of quadrupedal dinosaurs as well as the last dinosaurs at the end of the Cretaceous.

Warm-Blooded Dinosaurs

Dinosaurs were thus extremely successful animals. As we have seen, dinosaurs and mammals originated at almost the same time, and the early remains of both are found in the same place—South Africa. It is a curious fact that mammals, which are commonly considered "superior" to reptiles, played a subordinate role throughout the 140 million year reign of the dinosaurs. In explanation, it has been suggested that the dinosaurs were warm-blooded! As surprising as this theory may seem, it is not merely wild speculation. First, crocodiles today have almost complete four-chambered hearts, whereas birds, their closest living relatives, have complete four-chambered hearts and are, of course, warm-blooded. Dinosaurs are phyletically allied to both crocodiles and birds, and they may well have been closer to birds in their soft anatomy and physiology, as indeed they were in their skeletal structures. Second, the upright posture and long, slender limbs of dinosaurs appear to have been adaptations

***Figure** 20–25.   Skull of* Protoceratops, *a primitive ceratopsian dinosaur from Mongolia. Unlike most ceratopsians (horned dinosaurs), the horn is represented only by an arching of the nasal bones between the eyes and the nostrils. This individual was a very large female. A male of the same size would have had a more prominent horn.*

for continuous high levels of activity, which typically require endogenous heat control. Upright posture does not relate merely to supporting the bulk of behemoths; the small, early dinosaurs were completely upright, and similar types continued throughout the Mesozoic. Third, the bones of cold-blooded vertebrates are poorly vascularized, whereas those of birds and mammals are permeated by dense networks of Haversian canals. The histology of dinosaur bone is comparable to that of mammals in this regard. At the very least, this fact indicates that dinosaurs grew rapidly, and it suggests that they were warm-blooded. Finally, a mammalian carnivore requires about ten times as much food as does a cold-blooded carnivore of similar size. Therefore, in natural communities carnivores such as wolves and lions are only about 1 or 2 percent as numerous as their prey species, and the same is true of mammalian communities in the fossil record. Cold-blooded predators, as in the Permian of South Africa, by contrast seem to have numbered about 25 percent of their prey. The finding that carnivorous dinosaurs from the Late Cretaceous of Alberta numbered only a few percent of the herbivorous dinosaurs suggests that they may have been warm-blooded.

Although the concept of warm-blooded dinosaurs is still highly controversial, it certainly illuminates the spectacular success of the dinosaurs. If the dinosaurs were uninsulated, large body siz would have reduced heat loss, but cool temperatures at the end of the Cretaceous may have been a factor in their extinction. R.B. Cowles, however, has suggested that high temperatures may have caused their extinction by inhibition of spermatogenesis. In Chapter 7 we saw that sex ratios of living reptiles are determined by nest temperatures during incubation. If this were also true for dinosaurs, it could have played a role in their extinction at the end of the Cretaceous as temperatures cooled.

Class Aves

Birds, warm-blooded, feathered animals known by all and loved by many, have been called "glorified reptiles." Recent work has shown that "glorified dinosaur" would be even more valid, for birds almost certainly derived from small, carnivorous dinosaurs. Although *Archaeopteryx* (Figure 20–26) from the Upper Jurassic Solenhofen limestone of Germany is universally acclaimed as the first bird, it is decidedly more reptilian than avian, except for the presence of feathers. *Archaeopteryx* had a long tail, characteristic of theropods but lacking in all other birds. The pelvis was theropodan, except that the pubis had begun to rotate backward toward the avian position. There was no incorporation of presacral vertebrae into the rigid synsacrum of modern birds; instead, the presacral column was long and flexible. Fore and hind limbs and the pectoral girdle were all theropodan. The metatarsals were not fused as in modern birds. A characteristic avian feature, the wishbone, or furcula, representing fused

Figure 20–26.   Magnificent skull of the fifth specimen of Archaeopteryx, *found near Eichstätt, Germany. The skull is very reptilian in appearance, especially because of the numerous sharp teeth. The scale is in mm.*
Photo courtesy of Dr. John H. Ostrom, Yale University.

reptilian clavicles, was probably present. The jaws of *Archaeopteryx* bore sharp teeth, unlike those of modern birds. Sharp claws on the forelimbs (wings) leave no doubt of its predaceous habits. The structure of the feathers, whose impressions are faithfully preserved in the fine-grained sediment, is entirely comparable to that of modern birds. The feathers were not firmly anchored to the ulna as they are today, and the wing lacked complex locking joints that make the rigid wing such an effective flight organ in modern birds. J.H. Ostrom's studies of the anatomy of *Archaeopteryx* in relation to the origin of flight suggest that it evolved from a small, cursorial dinosaur.

The next glimpse of birds is not until the middle of the Late Cretaceous. *Hesperornis,* a large, aquatic, fully avian bird, was found in the Niobrara Chalk of Kansas and other sites in western North America. It still possessed teeth and had a primitive palate, but it was highly specialized in that it was flightless. It was a loonlike diver, but far more primitive than loons. A contemporary of *Hesperornis* was *Ichthyornis,* also toothed, but a good flier. It was tern-sized and ternlike; although possibly related to gulls and terns, it was much more primitive.

The fossil record of birds is not a good one because of their small size and the thinness of their bones. Much of the taxonomy of fossil forms is based on diagnostic fragments, such as distal ends of humeri and tarsometatarsals. The record is also strongly biased in favor of aquatic forms, for preservation is more likely in aquatic environments. By the end of the Cretaceous, loons, grebes, pelicans, herons, and gulls had appeared. Most of the twenty-four orders of modern birds had appeared by the Eocene.

One of the most-interesting groups of modern birds is that of the ratites of the southern continents, including ostriches of Africa, rheas and tinamous of South America, cassowaries and emus of the Australian region, and kiwis of New Zealand. These flightless birds are united by a common, primitive palate structure. Large members of the group resemble nothing so much as theropod dinosaurs of modest size (forgetting feathers for the moment!). Indeed cassowaries, although frugivorous, are armed with a long, sharp claw on the second digit of the foot, and they are quite capable of disembowelling a human with a single, slashing blow! Flightlessness evolved independently in each of these birds, and it has long been debated whether they are a natural group. Recent work on the electrophoresis of albumen proteins indicates that they are indeed a natural group.

During the Eocene in North America and Europe, and in the mid-Tertiary in South America, large, predaceous, ground birds that must have been awesome carnivores appeared. *Diatryma* of North America stood more than 2 m high, and *Phororhacos* of South America nearly 1.5 m high.

Birds have been highly successful and diverse in their ecological niches. There are some 9,000 living species, 5,000 of which belong to the order Passeriformes; these are the familiar perching songbirds. On the whole, they have exploited small size. Condors, pelicans, and swans approach the maximum weight for flight, about 15 kg, and ostriches weigh about 90 kg. Albatrosses have the greatest wingspan of extant birds, at around 4.25 m.

Class Mammalia

Mammals are easy to characterize today: they are warm and furry, and they suckle and care for their young. They have a four-chambered heart and systemic circulation developed from the left aortic arch. Once again, such characters cannot be found in fossils, and not all of them were necessarily present in the earliest mammals; in fact, we know that some were not. Several lineages of therapsids approached the mammalian condition closely in the Triassic; although apparently only one group, the advanced cynodonts, actually succeeded in crossing the mammalian boundary.

Paleontologists have had to define the mammalian condition arbitrarily in terms of useful fossil characters. The dentary bone in the jaw expanded at the expense of its neighbors until it contacted the squamosal bone of the skull directly, bypassing the old reptilian joint between the quadrate and the articular. These two bones were not lost but were transformed into the malleus and incus of the mammalian middle ear. The presence of a *dentary-squamosal jaw joint* is thus fundamental for defining mammals. Some early mammals retained accessory bones on the jaw. Continuous replacement of worn teeth is typical of reptiles, but even in the earliest mammals there were only *two generations of teeth,* the milk, or deciduous, dentition and the permanent dentition; this is a second defining character. A third character refers to the complex pattern of the crowns of the cheek teeth: in mammals, *three principal cusps* are arranged in a triangular pattern or in a pattern derivable from this one (Figure 20–27). The complex occlusal relationships of mammalian teeth result both from the fine control of jaw movements by the mammalian neuromuscular system and from the permanence of the teeth. Facets developed on the molar crowns as the teeth wear are important in guiding movements of the teeth and in controlling their effectiveness in chewing.

Mammals were present throughout most of the Mesozoic, from the Late Triassic on, but they attained no greater size than that of house cats. Most of the archaic lineages were extinct by the Cretaceous, but one group, the rodentlike multituberculates, were highly successful and abundant over the long period of time from Late Jurassic to earliest Oligocene. The monotremes of Australia and New Guinea, exemplified by the duckbilled platypus (*Ornithorhynchus*) and the spiny anteaters (*Tachyglossus, Zaglossus*), are often used as models for archaic mammals, but such an interpretation is misleading. These sluggish, sprawling mammals have skeletons more reptilian (therapsid) than mammalian. They have hair but lay eggs. Although they secrete milk for their young, they lack nipples. They maintain a constant body temperature, but at a lower level than that of other mammals. Because they lack means of disposal of excess heat, they are vulnerable to fatal overheating. Although

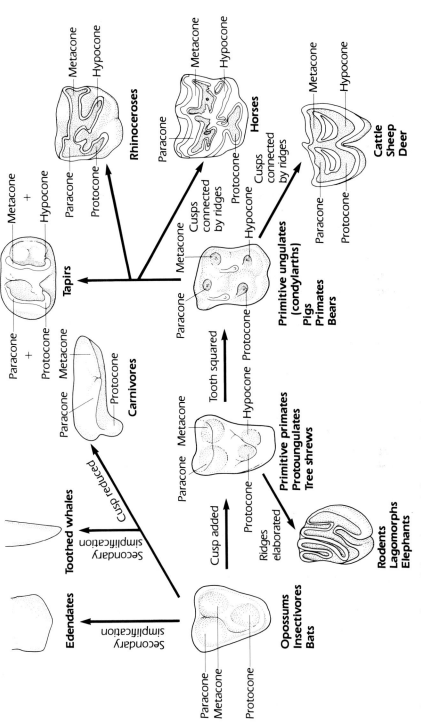

Figure 20–27. Schema of the evolution of the major kinds of placental mammals, based on the modification of the primitive tribosphenic (triangular) upper cheek tooth.

they show affinities to therapsid reptiles, they are highly specialized in snout and limb structure. Thus they do not typify archaic mammals.

Mammals of modern aspect, that is to say marsupials (metatherians) and placentals (eutherians), seem to have originated in the Cretaceous. In North America, marsupials and multituberculates were the principal mammals during most of the Late Cretaceous, but they remained small. Placentals, though present, were rare until the end of the period. Marsupials, living and fossil, are unknown in Asia, but placentals were much more common there, and they may have given rise by migration to North American placentals, although this is by no means certain. Marsupials almost became extinct at the end of the Cretaceous; however, they managed to reach South America in the Cretaceous. Not until the Miocene did they appear in Australia. Major radiations occurred on both continents in the absence of placental competitors. They produced a wide range of adaptive types, including a remarkable saber-toothed predator (*Thylacosmilus*) in South America, and the Tasmanian "wolf," *Thylacinus*. The North American marsupials were extinct by the mid-Tertiary. The marsupial fauna of Australia is still abundant and varied, comprising some 168 species that range in size from the 15 g honey possum (*Tarsipes*) to the 90 kg great gray kangaroo (*Macropus*). In South America major extinctions occurred late in the Tertiary when connections with North America were established and placentals invaded the southern continent. Only two families of marsupials now remain in South America, the didelphids (opossums) and the caenolestids (opossum rats), together about 73 species. The familiar Virginia opossum of North America, *Didelphis marsupialis*, is a Pleistocene immigrant from South America.

Reproduction in marsupials is curious (by placental standards). Marsupials are essentially *ovoviviparous*—the fertilized egg is retained in the uterus for a time, but no intimate vascular connection develops to enable the mother to contribute to the nourishment of the fetus. Intrauterine life, typically lasting from ten to thirty days, may be terminated by an immune rejection response by the mother. The extremely underdeveloped fetus, barely recognizable as a vertebrate, then spends from two to five times that length of time clinging to the mother's underside or housed in a pouch and suckling before it is sufficiently developed to forage on its own.

The major feature differentiating placentals from marsupials is that their embryos develop an efficient *placenta* for obtaining nourishment from the mother's bloodstream. This placenta also serves as an organ of respiration and excretion. That placental mammals are by far the dominant animals of the world today is in no small measure due to the prolonged period of embryonic development, which is made possible by the placenta, as well as by parental care and the enlargement of the cerebral hemispheres, diminutive in all other classes of vertebrates. These trends are already observable in the marsupials, but they are carried much further by the placental mammals.

The ancestral placentals were small animals, probably because only small animals could avoid competition with the reptiles of the Cretaceous. Their small size restricted them largely to a diet of insects, worms, and other small invertebrates, with perhaps some vegetable matter.

Because the extant and extinct orders appear so rapidly at the beginning of the Tertiary, it is difficult to trace probable relationships among them. On the basis of comparative anatomy, comparative serology, and paleontology, certain probable relationships have been drawn up, some much better supported by evidence than others. Living placentals number about 3,800 species, organized into 15 to 22 orders, depending on the authority. Mammalian taxonomists rely on such anatomical characters as teeth, carotid arteries, auditory bullae, and feet. Among living mammals, the order Insectivora (some 400 species of shrews, moles, hedgehogs, and their allies) are usually considered closest to the ancestral placentals, yet they are probably not very close. M. McKenna accords this distinction to the South American order Xenarthra, or Edentata (31 species of armadillos, anteaters, and sloths), and upon their sister group, the order Pholidota (8 species of pangolins, or spiny anteaters, of Asia and Africa). Insectivores first appear in the Paleocene, yet the placental radiation had already begun by the end of the Cretaceous. During the Cretaceous, true primates, condylarths (ungulate ancestors), creodonts (carnivore ancestors), and protoinsectivores (sometimes called the order Proteutheria) were present, but no true insectivores. Hence the Insectivora no longer hold the central position in mammalian phylogeny that they formerly did.

Nonetheless, several other orders were clearly derived from insectivores, most importantly the order Chiroptera (bats), which, with 850 species, constitute about 20 percent of all living mammalian species. They have been described as flying insectivores, because only adaptations for flight distinguish them from the parent order. They appear in the fossil record in the Eocene. The order Dermoptera (flying lemurs—which only glide and aren't lemurs!) includes only two living species, both in Southeast Asia. They appeared in the Eocene and are known from Wyoming and northern Canada. The order Primates is known from Late Cretaceous, and may have been derived from protoinsectivores. In common with the above orders, they retain such primitive characters as a clavicle and a generalized, five-fingered hand. The evolution of the primates is traced more fully in Chapter 21.

The order Rodentia is a great mammalian radiation. These enormously successful animals, dating from the end of the Paleocene, today number approximately 1,700 species, over 40 percent of all living mammalian species, and they are worldwide in distribution. They retain a clavicle, are generally small, and are often nocturnal. Specialized characters include ever-growing incisors for gnawing, great physiological hardiness (they flourish in deserts as few other groups of mammals can), and high reproductive rates. Ever-growing cheek teeth of some rodents represent a peak of dental evolution. Although most are small, beavers (*Castor*) of North America and Asia reach thirty kilograms, and the semiaquatic capybara (*Hydrochoerus*) of South America weighs up to sixty kilograms and is ecologically convergent upon the niche of the pigmy hippo of Africa. Other South American rodents converge upon the niches of small antelope, and the African springhaas (*Pedetes*) are like small kangaroos. Rodents, which were once incorrectly thought to be derived from insectivore ancestors, may have branched off from protoeutherian ancestors; the fossil evidence, however, is inadequate. They appear to be at an evolutionary peak now (or is there even greater diversity yet to come?) and are fertile ground for the study of such evolutionary processes as convergence, speciation, and biogeographic distribution.

The Lagomorpha (rabbits, hares, and pikas) are also gnawers with ever-growing incisors and may also date from the Late Paleocene. Superficially like rodents, they differ from them in an important group of characters, including an added pair of upper incisors and other changes of teeth, a facial opening in the skull in front of the orbit, and loss of ability to rotate the forearm. The fossil record shows no link between lagomorphs and rodents. In contrast to the rodents, lagomorphs have never diversified much, and today they number only 63 species.

Carnivore roots trace back to the Cretaceous, with an animal named *Cimolestes* in particular being implicated. Early placentals with carnivorous tendencies were called creodonts (Paleocene to Pliocene), and the earliest members of the order Carnivora were the miacids (Paleocene to Eocene). Miacids had the *carnassial teeth* of modern carnivores, which are flesh-shearing teeth formed by the laterally compressed fourth upper premolar and first lower molar working against each other. Their brains were larger and more complex than were those of most of their contemporaries. The first modern carnivores, the canids (dogs, wolves, foxes, jackals, coyotes), were in evidence by late Eocene, and they were followed in the early Oligocene by the felids (cats of all kinds). Today, carnivores number 284 species (253 if the aquatic Pinnipedia are recognized as a separate order). The most diverse (75 species) and poorly known living carnivores are the Old World viverrids (mongooses, civets, and their allies), although the cosmopolitan mustelids (otters, mink, badgers, skunks, and their allies) are almost as diverse with 71 species. Canids and felids are almost equal in numbers of species, but about two-thirds of the 37 species of cats are in the one genus *Felis,* and only 7 of 41 species of canids fall in the genus *Canis.* (The other canids are distributed among seven other genera.) Canids show more anatomical diversity, but felids show a greater range in body size: tigers (*Panthera tigris*) attain a weight of 275 kg; wolves (*Canis lupus*), a mere 75 kg.

Though carnivores are properly renowned for their flesh-eating proclivities, all carnivores probably consume plant material regularly—it may be a nutritional requirement. Coyotes, foxes, and jackals eat berries in season, and they may become nuisances around fruit farms. Domestic cats, particularly Siamese cats, are notorious grazers on houseplants. Larger cats may obtain plant material by consuming the stomach contents of their herbivore prey. Large carnivores, such as bears (Ursidae) and giant pandas (*Ailuropoda*), are either largely or strictly herbivorous.

The major large-bodied herbivores are the ungulates, so named because they walk on the ends of their toes, or unguals. Long limbs make them potentially swift runners (shorter-limbed carnivores keep pace with their stiff-backed prey by adding flexion and extension of their backs to their running strides). Both carnivores and ungulates have lost clavicles, which brings their shoulders closer together and is an aid to running. Ungulates trace their ancestry back to the condylarths, one of the first groups of placentals to differentiate at the end of the Cretaceous. At first appearance, condylarths were very similar to creodonts; thus the lion and the lamb trace back very nearly to a common ancestor. There are two orders of living ungulates: the Perissodactyla, or odd-toed ungulates, and the Artiodactyla, the even-toed, cloven-hoofed ungulates. Both appeared at the end of the Paleocene, and both derived directly from condylarths. Perissodactyls enjoyed considerable success early in the Tertiary and peaked in the Oligocene, when rhinoceroses and rhinolike titanotheres were abundant. Horses (family Equidae) and tapirs (family Tapiridae) were almost indistinguishable at first appearance. Tapirs remained in the forest and retained relatively simple teeth and spreading feet; in contrast, horses invaded the plains, where selection favored a compact, single-toed foot and elaborate, high-crowned, somewhat rodentlike teeth. Evolution of the horse is extremely well documented in the fossil record (see discussion in Chapter 5). Rhinoceroses had their peak in the Oligocene, when long-legged runners and barrel-chested hippolike forms complemented more generalized types and flourished in western North America. *Baluchitherium* from the Oligocene of Mongolia stood 5.5 m at the shoulders, the largest land mammal ever. There are only 16 species of perissodactyls today, 7 of equids (horses, zebras, and asses), 5 of rhinos, and 4 of tapirs. Horses disappeared from North America only about 40,000 years ago, and a scant 10,000 years ago tapirs strolled about as far north as Philadelphia.

The Artiodactyla have flourished at the same time that the perissodactyls have declined. Unlike the latter, the artiodactyls were slow to diversify and were uncommon in the Eocene. A defining character of the order, along with the even-toed symmetry of the foot (digits III and IV being equally long), is the existence of a specialized bone of the ankle, the so-called *double pulley astragalus*. From the indifferent success of early artiodactyls, we would not infer that such an arrangement of the foot conferred great selective advantage on its bearers. The most-generalized artiodactyls today are members of the suborder Suina, including 11 species of Old World pigs and New World peccaries. These animals have simple stomachs, separate toes, and unfused metacarpals and metatarsals. They have low-crowned teeth suitable for omnivorous diets. Recognizable suids appeared in the Oligocene. Hippos are large African members of the Suina; the resemblance to pigs is more than skin deep.

Advanced artiodactyls belong to the suborder Ruminantia. The development of *rumination*—a process of bacterial fermentation of cellulose in a complex, multichambered stomach—was a later development in artiodactyl history and probably held the key to their enormous success. The most primitive of the living ruminants are the camels, which have only three chambers in their stomachs. Camels appeared in late Eocene and were from the start characterized by long, slender limbs, fused metacarpal and metatarsal bones (cannon bones), and *selenodont* tooth structure (having multiple crescentic enamel ridges, a pattern common to all ruminants). Camels attained unguligrade stance, then later returned to the digitigrade stance of their ancestors. The "snowshoe" effect of the spreading digits is adaptive for movement over sandy ground. Camels flourished in North America for some 30 million years, then became extinct during the late Ice Age, only a few thousand years ago. South American camels (*Llama*) resemble those that were common in the Miocene of North America.

Advanced ruminants attained a four-chambered stomach. They split into two separate groups: the antler-bearing Cervidae (deer) and the horn-bearing Bovidae (cattle, antelope), both of which are very successful radiations of small, medium, and large herbivores. Cervids, which are primarily north temperate forest browsers, appear in the fossil record of Asia as early as the Oligocene. They entered South America in the Pliocene, and five genera are now endemic there. The 37 species of living deer range in size from the 6 kg *Pudu* to the 800 kg moose (*Alces*).

Bovids appeared in the Miocene of Europe and have enjoyed great success. There are 111 living species, more than half of them in Africa, and this number may be only half of the Pleistocene total. Typically, bovids are grazers on open lands and may live in vast herds; among cervids, only *Rangifer* (caribou and reindeer) fit this description. Because a number of different species of bovids may live sympatrically, horns may be elaborated into complex patterns to aid in species recognition. Unlike cervids, both sexes may bear horns. Bovids flourished in Eurasia but were driven southward during the Pleistocene glaciation. At that time, however, several *cold-adapted* bovids crossed the Bering land bridge from Siberia to Alaska and so entered North America. These immigrants are *Ovis* (bighorn sheep), *Bison, Oreamnos* (mountain goat, but really a goat-antelope), and *Ovibos* (musk ox, but literally the name means sheep-cow). None reached South America. Bovids, along with rodents, are the outstandingly successful mammals of the late Cenozoic.

Condylarths are the probable ancestors of a number of other orders too. Whales (84 living species of the order Cetacea) appeared full-blown in the Eocene. Molecular data suggest affinity between artiodactyls and whales. Condylarth ancestry is logical but unproven. Large whales have evolved the largest brains of all vertebrates; relative brain size (encephalization quotient) of small and medium-sized toothed whales (dolphins and killer whales) is second only to that of humans.

African aardvarks (order Tubulidentata), the most-isolated living ungulates, comprise a single genus and species (*Orycteropus afer*). Its known fossil record extends back to the Miocene, and no trends indicate relationship to artiodactyls. Again, condylarth ancestry is suspected. Although confined to Africa today, aardvarks have a fossil record in Europe and Asia as well. Their name means "earth pig" in Afrikaans, and they are rooting termite eaters.

Three additional orders originated in the Old World and are grouped under the heading *subungulates,* animals that bear neither hooves nor claws, but blunt, flattened, nail-like structures on their toes. At first glance, the grouping of elephants (order Proboscidea), hyraxes (order Hyracoidea), and dugongs and manatees (order Sirenia) seems surprising; collectively, they account for only 18 living species among them. However, fossil roots for each group extend back to the Oligocene or earlier; and various characters are shared among them, one of which is relatively poor temperature regulation. Elephants have dealt with this problem partially by achieving large size; sirenians by retreating to the thermal stability of warm waters. Similarity of foot structure links hyraxes and elephants; and tooth replacement patterns of elephants and manatees are similar. Elephants, which have the largest brains of all terrestrial mammals and are highly intelligent, spread from the Old World to the New World in the Pliocene and survived until the end of the Ice Age, when they were hunted by early Indians and possibly extinguished by them as well. Subungulates originated either from very early true ungulates or from condylarths, but again the details remain to be clarified. Mammalian diversity in time is summarized in Figure 20–28.

AN OVERVIEW OF MAMMAL EVOLUTION

Most orders of mammals trace their origins to the early part of the Cenozoic era. The major groups from which most orders derive are the condylarths, insectivores, and creodonts, which in turn trace back to the Late Cretaceous protoinsectivores. Perissodactyls were one of the most successful orders early in the Cenozoic, but they dwindled as artiodactyls and rodents expanded in the latter half of the Cenozoic. The fossil record of mammals is generally excellent and tells us much about the effects of geographic isolation on the one hand and patterns of dispersal and faunal exchange on the other. A recurring theme among mammals is that of evolutionary convergence: occupation of similar niches on different continents by unrelated animals. Let us consider, for instance, the niche of a terrestrial anteater. In Australia this niche is filled by a marsupial numbat (*Myrmecobius*); in New Guinea, by the monotreme echidna (*Tachyglossus*); in Southeast Asia, by pholidotan pangolins (*Manis*); in

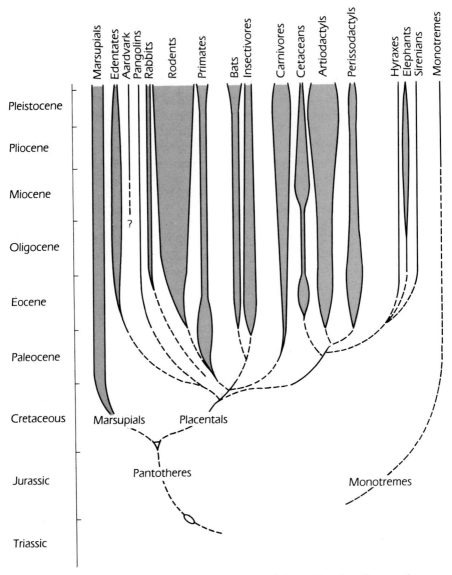

Figure 20–28.   *The origin and relationships of the major orders of mammals.*

Africa, by aardvarks (*Orycteropus*); and in South America by edentate anteaters (*Myrmecophaga*). North America and Europe cannot support anteating specialists because ants exist in insufficient numbers outside the tropics. Analysis of the anatomy, systematics, zoogeography, biochemistry, and paleontology of mammals provides some of the best evidence imaginable for the verification of the fact of evolution.

SUMMARY

Chordates are animals that possess a notochord, a dorsal nerve cord, and pharyngeal gill slits. Embryological and biochemical characters suggest that chordates are related to echinoderms. Two subphyla of chordates are marine invertebrates. The *Cephalochordata,* including amphioxus, are tadpolelike animals that show the chordate characters at all stages of development; the *Urochordata,* or tunicates, show these only as larvae.

The *Vertebrata* are chordates characterized by a brain with ten pairs of cranial nerves, ectodermal sensory placodes, neural crest tissue, pronephric kidneys, and hard parts consisting of bone, enamel, or dentine. The most primitive vertebrates belong to the class *Agnatha,* the jawless fishes. Living lampreys and hagfishes lack jaws, paired fins, or hard parts of any kind, and have but a single nostril. Fossil agnathans, the *ostracoderms,* had armor over the head. Fragmentary ostracoderms appear in the Late Cambrian and were extinct by the Early Devonian. The earliest jawed fishes, found in Upper Silurian and Devonian rocks, belonged to the extinct class *Placodermi,* the plate-skinned fishes. The primitive structure of the jaws and the lack of mobile paired fins limited their effectiveness, and they became extinct as progressive types appeared.

The class *Chondrichthyes,* or cartilage fishes, comprises the sharks, skates, and rays. These fishes appeared in the Middle Devonian and have a long history of progressive evolution. Modern sharks, appearing in the Late Cretaceous, employ enamel and dentine in teeth, scales, and fin spines, but they do not use bone in their skeletons, which remain cartilaginous. Paired fins are rigid in comparison with those of bony fishes. There are about 600 species of living chondrichthyans.

The *Osteichthyes,* or bony fishes, appeared in the Early Devonian. Progressive trends in the *Actinopterygii,* or ray-finned bony fishes, culminated with the appearance and explosive diversification in the late Mesozoic and early Cenozoic of the teleosts, of which there are at least 20,000 species today. *Teleosts* are characterized by a swim bladder, a covering of thin, flexible scales, a superficially symmetrical tail, highly flexible paired and median fins, and several types of protrusible jaws. The *Sarcopterygii,* or fleshy-finned bony fishes, arose in common with ray-finned fish in the Early Devonian, flourished in the late Paleozoic, then declined. Today only three living lungfishes and one coelacanth survive. Fleshy-finned fishes have a bony fin skeleton that resembles the bones of the tetrapod limb in varying degrees. Bones of the skull of the rhipidistians closely match those of early amphibians, and these fishes include the ancestors of all tetrapods.

Amphibians, the most primitive tetrapods, have good limbs and breathe air but are reproductively dependent on water. Their eggs are laid in water and fertilized externally, with the larvae passing an aquatic phase. There are 3,000 species of living frogs and salamanders, which are small and covered with thin, moist skin. Frogs date from the Triassic, and salamanders are of more recent origin. Fossil labyrinthodont amphibians, some of which were large and armored, date from the Late Devonian.

Reptiles are thoroughly terrestrial animals that developed a penis for internal fertilization and an amniotic egg that eliminates the need for free-swimming larvae. Reptiles appeared in the Pennsylvanian and have a rich and diverse history. They dominated the Mesozoic era and included many giant forms, both dinosaurian and nondinosaurian. Turtles, dating from the Triassic, are the most primitive living reptiles. Lizards, also from the Triassic, and snakes from the Cretaceous, form the bulk of the living reptilian fauna, about 6,000 species all told. Crocodilians are the living reptiles that are closest to dinosaurs.

Birds, which are small, warm-blooded, large-brained vertebrates that are best characterized by their feathers, are Late Jurassic derivatives of small meat-eating dinosaurs. Living birds, numbering 9,000 species, are all toothless, but a few Cretaceous types had teeth. In many aspects of their anatomy, birds are thoroughly reptilian.

Mammal-like reptiles were among the first reptiles to appear, and mammals themselves date from the Late Triassic. *Mammals* are warm-blooded, hairy animals that nurse their young with mammary glands. Parental care, as for birds, is a hallmark, and their brain is relatively large compared to that of reptiles. Mammals have only two generations of teeth. Both marsupial and placental mammals appeared in the Late Cretaceous. *Marsupials,* the pouched mammals, have been geographically and ecologically somewhat restricted; whereas *placentals* have radiated into a wide variety of carnivorous, herbivorous, and omnivorous types, including both flying and swimming types. There are some 300 species of living marsupials and 4,000 species of living placentals.

REFERENCES

Bond, C.E. 1979. *Biology of Fishes.* Saunders, Philadelphia.
   A valuable survey of the fishes.
Colbert, E.H. 1980. *Evolution of the Vertebrates,* 3rd ed. Wiley, New York.
   A highly readable, well-illustrated account of vertebrate history.
Denison, R.H. 1971. The origin of vertebrates: A critical examination of current theories. *Proceedings of the North American Paleontological Convention II,* vol. H:1132–1146.
   A good critical review of this important subject.
Eisenberg, J.F. 1981. *The Mammalian Radiations: An Analysis of Trends in Evolution, Adaptation, and Behavior.* University of Chicago Press, Chicago.
   As the subtitle suggests, the author takes a synthetic overview of mammals—stimulating.
Feduccia, A. 1980. *The Age of Birds.* Harvard University Press, Cambridge, Mass.
   Beautiful illustrations complement an informative text on the fossil history of birds.
Jollie, M. 1982. What are the "Calcichordata"? and the larger question of the origin of the chordates. *Zoological Journal of the Linnaean Society* 75:167–188.
   A critique of R.P.S. Jeffries's controversial ideas on the origin of the chordates.

Kurtén, B. 1972. *The Age of Mammals.* Columbia University Press, New York.
   A readable history of the Cenozoic era.
Kurtén, B., and E. Anderson. 1980. *Pleistocene Mammals of North America.* Columbia
   University Press, New York.
   *A fine account of the not very far removed period of time when the richness of the North
   American mammalian fauna resembled that of Africa today.*
Lilligraven, J.A., Z. Kielan-Jaworowska, and W.A. Clemens. 1980. *Mesozoic Mammals:
   The First Two-Thirds of Mammalian History.* University of California Press, Berkeley.
   A superb review of a subject that has expanded greatly in the past twenty years.
Løvtrop, S. 1977. *The Phylogeny of the Vertebrata.* Wiley, New York.
   *This volume challenges almost every assumption held by more orthodox workers. It is
   stimulating reading for those with a solid foundation on the subject.*
Northcutt, R.G., and C. Gans. 1983. The genesis of neural crest and epidermal
   placodes: A reinterpretation of vertebrate origins. *Quarterly Review of Biology* 58:
   1–28.
   *A very important reexamination of vertebrate origins that emphasizes the significance of
   neural crest and epidermal sensory placodes of the head in the evolution of active predation,
   the primitive feeding mode of vertebrates. It applies cladistic methods.*
Ostrom, J.H. 1976. *Archaeopteryx* and the origin of birds. *Biological Journal of the
   Linnaean Society of London* 8:91–182.
   A major statement by the leading modern student of the earliest bird.
Orr, R.T. 1976. *Vertebrate Biology,* 4th ed. Saunders, Philadelphia.
   A useful elementary survey of the vertebrates.
Panchen, A.L., ed. 1981. *The Terrestrial Environment and the Origin of Land Vertebrates.*
   Systematic Association Special Volume 15. Academic Press, New York.
   Contains some of the most recent findings on the early diversification of tetrapods.
Porter, K.R. 1972. *Herpetology.* Saunders, Philadelphia.
   Probably the best survey of amphibians and reptiles currently available.
Romer, A.S. 1966. *Vertebrate Paleontology,* 3rd ed. University of Chicago Press, Chicago.
   Still the first book many vertebrate paleontologists turn to.
Simpson, G.G. 1945. The principles of classification and the classification of the
   mammals. *Bulletin of the American Museum of Natural History,* vol. 85. New York.
   *Although old and in need of revision, it is still the most comprehensive review of the
   mammals available. Also included is a valuable treatment of the principles of classification.*
Stahl, B.J. 1974. *Vertebrate History: Problems in Evolution.* McGraw-Hill, New York.
   A good, biologically oriented text.
Thomas, R.D.K., and E.C. Olson, eds. 1980. *A Cold Look at the Warm-Blooded
   Dinosaurs.* Westview Press, Boulder, Colo.
   *The most comprehensive review of dinosaur biology generally and of the concept of endothermy
   in particular that dinosaurs are likely to get.*
Vaughan, T.A. 1978. *Mammalogy,* 2nd ed. Saunders, Philadelphia.
   A fine survey of mammals that covers fossils, anatomy, and ecology.
Welty, J.C. 1982. *The Life of Birds,* 3rd ed. Saunders, Philadelphia.
   A very good ornithology text.
Young, J.Z. 1981. *The Life of Vertebrates.* Oxford Clarendon Press, Cambridge, England.
   A remarkable original synthesis by one of Britain's leading vertebrate zoologists.

21

History of the Primates

We come now to the climax of this phylogenetic history—the order Primates, including lemurs, tarsiers, New World and Old World monkeys, apes, and our own kind, *Homo sapiens.* Before discussing the history of this group, let us review the classification and the major characteristics of the living members of this order.

CLASSIFICATION OF THE PRIMATES

The order Primates is singularly difficult to define. Its distinguishing characters are not so salient as the chisel-like incisors of rodents or the hooves with an odd number of toes of the Perissodactyla. More than a century ago, the British zoologist St. George Mivart defined the Primates as placental mammals with nails (or claws, in some), clavicles, orbits encircled by bone, three kinds of teeth, a brain whose posterior lobe has a fold called the calcarine fissure, a thumb and great toe having a flat nail or no nail, a large intestine with a blind pouch (the cecum), penis pendulous and the testes descended into a scrotum, and two pectoral mammary glands. This definition is impossible to apply to fossil primates (or near-primates); and among living members of the order, there are exceptions—for example, galagos have *three* pairs of mammary glands. Moreover, most of these traits are primitive mammalian characters; except for a tendency toward expansion of the brain, the primates are relatively unspecialized mammals. Primitively, they had two incisors, one canine, four premolars and three molars in each half jaw, for a total of forty teeth. (We express the dental formula as 2-1-4-3.) The teeth are adapted to a generalized diet, but this is itself a lack of specialization. The thumb and great toe are usually opposable to the other digits, which gives efficiency both in climbing and in grasping objects. The eyes of most mammals are on the sides of the head, so that each eye sees a different field. Those of primates are placed toward the front, permitting binocular, stereoscopic vision. Stereoscopic vision correlates with life in the trees (*arboreal* habits), a mode of life in which three-dimensional vision is obviously important. Vision, including color vision, is generally more highly developed than in other mammals; the sense of smell, less so. Finally, although enlargement of the brain is a general mammalian characteristic, it is most marked in the primates, particularly in the cerebral cortex.

A recent approach to defining the primates focuses on specialized characters of the ear region shared by all representatives of the order: peculiarities of the auditory bulla and of the pattern of circulation of arterial blood to the brain. Such characters can be identified on fossil skulls. When only advanced characters are used and primitive characters are eliminated, such extremely primitive animals as the tree shrews (*Tupaia*) drop out of the primate array, either to be subsumed as a family of insectivores or to stand in their own order, the Tupaioidea.

Lemurs and Lorises

There are two groups of prosimians: lemuriforms (Figure 21–1), now confined to Madagascar, and the lorisiforms of both Africa and Asia. They vary from the size of a mouse to that of a cat, and they exhibit the basic primate character of well-developed hands and feet, with opposable thumbs and great toes. Lemurs present a rather squirrel-like appearance. The snout is usually long and projecting. The ears are large and mobile, but there is little mobility of facial expression. They are generally nocturnal, hence large eyes have been favored by natural selection. Lorisiforms (lorises of Asia, pottos and galagos of Africa) are very small, have short faces, and lack tails.

Tarsiers

The living tarsier (Figure 21–2) of Indonesia and the Philippines is the last survivor of an old and important group of primates that was probably derived from lemuroid ancestors. The tarsier (*Tarsius spectrum*) is about the

Figure 21–1. *The gregarious ring-tailed lemur,* Lemur catta, *one of the endemic prosimians of Madagascar.*

size of a young kitten. An exclusively nocturnal animal, its eyes are immense relative to the size of its head; and they look forward, permitting binocular vision. The snout is correspondingly reduced, giving a monkeylike appearance. The hind legs are modified for jumping, and it can leap from branch to branch with considerable accuracy. The ears are large; the tail is long and naked, except for a terminal hairy segment. Although tarsiers resemble lemurs in many details, the structure of the brain and of the reproductive organs is essentially simian. Hence some students group them with the monkeys, apes, and humans as anthropoids, rather than with lemurs as prosimians.

New World and Old World Monkeys

Monkeys belong to two distinct evolutionary radiations in the New World and Old World, respectively. The New World monkeys (Ceboidea) include the tiny callothricids (tamarins, marmosets) and the cebids, the largest of which, the howler (*Alouatta*), weighs up to 16 kg. Familiar cebids include the capuchin (*Cebus*), the spider monkey (*Ateles;* Figure 21–3), and the woolly monkey (*Lagothrix*). A prehensile tail forms a spectacular fifth appendage in some cebids. All ceboids (members of the superfamily Ceboidea; cebids are members of the family Cebidae) are highly arboreal. Biologists have generally

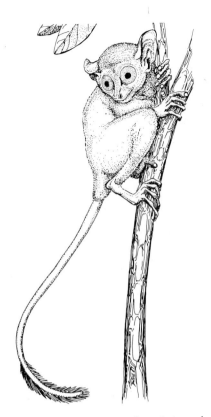

Figure 21–2. *A tiny, nocturnal prosimian, the tarsier,* Tarsius spectrum, *from Indonesia and the Philippi es.*

treated the New World monkeys as more primitive than their Old World counterparts. They possess three premolars as do prosimians, rather than two, as do Old World monkeys; in addition, marmosets have claws rather than nails. However, the idea is now seriously entertained, though by no means proven, that ceboids may have originated from a close relative of Old World monkeys by rafting across the then-narrow South Atlantic in the Oligocene, 35 million years ago.

Old World monkeys (Cercopithecoidea) are widely distributed from Africa across Asia to Japan and the Philippines. These include the leaf-eating colobus and langurs and the cercopithecines—macaques (Figure 21–4), vervets, baboons, and others. Old World monkeys lack prehensile tails; some lack tails altogether but have ischial callosities, hairless areas of insensitive skin on the buttocks that facilitate sitting. All cercopithecoids are diurnal, some are strongly terrestrial, and one, the patas (*Erythrocebus*), is a swift runner on the African savannah.

Figure 21–3. A spider monkey, Ateles, *representative of the highly arboreal New World cebid monkeys, some of which use the tail as a prehensile fifth appendage.*

Figure 21–4. A lion-tailed macaque, Macaca silenus, *representative of the semiterrestrial Old World monkeys.*

Large male baboons (*Papio*) attain weights of up to 60 kg. All monkeys run on the tops of branches; they do not hang by their arms beneath branches. Their chests are laterally compressed as is the case in dogs, and fore and hind limbs are about equal in length.

Apes

Of all the extant primates, the apes resemble us most closely. Only five genera are living: the gibbon of Asia, the siamang of Sumatra and Malaysia, the orangutan of Borneo and Sumatra, and the gorilla and chimpanzee of equatorial Africa. All these animals resemble humans in structure of skull and skeleton, dentition, physiology, blood groups, parasitic susceptibilities, and other characteristics. Like humans, all of them lack tails. They are adapted for two highly specialized modes of locomotion: arboreal arm-swinging (*brachiation*) and terrestrial *knuckle-walking*. In brachiators, the chest is broad and shallow, the shoulder blade is located on the back rather than on the side, and the arms are long compared to the legs. No ape provides a good model for our immediate hominid ancestors.

The gibbon (*Hylobates;* Figure 21–5) is the smallest of the apes, and in

Figure 21–5.   *A gibbon,* Hylobates, *of Southeast Asia, the smallest and most graceful of the apes. Note the very long arms.*

many respects it is the most primitive. It is completely arboreal, and it is more adroit on foot than most of the apes, for it can run along the branches skillfully. It is also capable of remarkably swift and accurate brachiation: using its greatly elongated arms, it swings from branch to branch, body and legs playing only an indirect role. Its thumbs are reduced, allowing the hand to function as a hook for grasping branches. The legs are much shorter than the arms, in contrast to the Old World monkeys and humans. The siamang (*Symphalangus*) is larger than the gibbon, but otherwise differs from it only in details. The orangutan (*Pongo*) is a much larger ape, often weighing over 45 kg; but it is still primarily arboreal, moving successfully by brachiation and rarely descending to the ground. The chimpanzee (*Pan*) is somewhat larger and less arboreal; it spends much time on the ground. Finally, the gorilla (*Gorilla;* Figure 21–6) is by far the largest of the apes. Adult males on the average weigh 135 to 180 kg, and some weigh considerably more. Although they retain the morphology of a brachiator, the huge size of the adults confines

Figure 21–6. *The largest of the great apes,* Gorilla, *a gentle herbivore of equatorial Africa.*

them to the ground, where, in common with chimpanzees, they use a peculiar quadrupedal gait called knuckle-walking. These gentle giants are mainly herbivorous.

These animals, then, are the living members of the order Primates, apart from *Homo sapiens,* the dominant member. Table 21–1 provides a summary of their taxonomic relationships.

PROSIMIANS IN THE FOSSIL RECORD

Primate remains have been found from rocks of Early Paleocene age, and one find may be of Cretaceous age. From this remote time, near the beginning of the Age of Mammals, teeth, jaws, and some skulls of small mammals suggesting affinities with lemuroids have been discovered (Figure 21–7). The structure of the molar teeth of these fossils is primate in character, and the skulls show a tendency toward expansion of the brain. However, these animals, called plesiadapoids, also had somewhat rodentlike incisors, and claws instead of nails on their toes. These rather squirrel-like primates were successful in arboreal niches, and their remains are fairly abundant in Upper Paleocene and

Table 21–1. Summary of primate classification

Order Primates

Suborder Prosimii
　　Superfamily Lemuriformes—lemurs, indriids, aye-ayes
　　Superfamily Lorisiformes—lorises, pottos, bush babies
　　Superfamily Tarsiiformes—tarsiers
Suborder Anthropoidea
　　Superfamily Ceboidea—New World monkeys
　　Superfamily Cercopithecoidea—Old World monkeys
　　Superfamily Hominoidea
　　　　Family Hylobatidae—lesser apes: gibbons, siamangs
　　　　Family Pongidae*—great apes: orangutan, gorilla, chimpanzee
　　　　Family Hominidae—humans

* In Chapter 8, the research of Yunis and Prakash on the karyotypes of hominoids was discussed. If specialists on primates accept their conclusions, the Pongidae will have to be included in the Hominidae. This will require a subfamily Ponginae for the orangutan and a subfamily Homininae for humans, gorillas, and chimpanzees.

Note: Like all taxonomic names, the above are Latin nouns. In less formal discussion, they may be used without capitalization as nouns or adjectives, the endings of which indicate taxonomic rank. Thus, *hominoid* refers to members of the superfamily Hominoidea; *hominid* refers to members of the family Hominidae. Later in the chapter it may be necessary to speak of members of subfamilies; the formal names end in -inae, hence less formally they are dryopithecines, pongines, and so on.

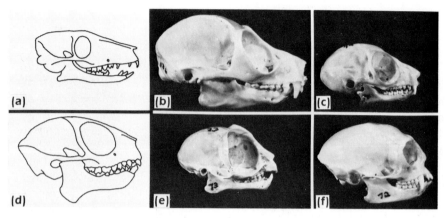

Figure 21–7. *A series of skulls showing morphological (but not phylogenetic) transition from a tree shrew (a) through two lemuriforms (b and c) and two tarsiers (d and e) to a primitive monkey (f). Notice the trends to shorten the face, rotate the eyes toward the front, and increase the relative size of the cranium.* (a) Tupaia; (b) Nycticebus; (c) Galago; (d) Necrolemur, *a fossil tarsier in spite of its name;* (e) Tarsius; (f) Callithrix.
Reprinted from E. Dodson, *The Phenomenon of Man Revisited,* by permission of Columbia University Press (New York, 1984).

Eocene deposits of Europe and North America. They were evidently a sterile side branch of primate evolution.

Lemurs first appear in the record in the Eocene, both in Europe (*Adapis,* probably the ancestral lemuriform) and in North America (*Notharctus*). These were comparable in size to modern lemurs, but the brain was smaller, and they had not yet formed certain specializations of the teeth that characterize modern lemurs.

Tarsiers were present in great abundance in the Eocene. Of many genera described from Europe and North America, only four are still considered valid. Some of these were similar to the modern tarsier; others were decidedly more primitive in skull pattern, brain, and limbs. Some retained the primitive insectivore dentition of forty-four teeth; others had the dentition reduced to thirty-two, the number characteristic of the anthropoids. Other changes in the teeth of these Eocene tarsiers also tended in the direction of the anthropoids. For example, they were the first primates to develop bicuspid premolars. The molars of the tree shrews and insectivores have three cusps, but some of the Eocene tarsiers, in common with all of the anthropoids, have molars with four cusps. Thus many of these primitive tarsiers had decidedly monkeylike features. Tarsiers probably derived from the Eocene family Anaptomorphidae, but their relationship to other prosimians is not clear. Primates flourished in North America and Europe until the end of the Eocene, then strangely disappeared almost completely, perhaps due to a cooling climate and increased competition from true rodents.

HIGHER PRIMATES IN THE FOSSIL RECORD

Fayum Primates

After the abundance of primates in the Eocene, the succeeding Oligocene is somewhat of a disappointment—only a single site of Early Oligocene age has so far been productive. The Fayum deposits of Egypt represent a 30-million-year-old, lush tropical forest, with associated river deposits. In the 1960s, E.L. Simons led expeditions to this currently desert terrain and recovered a considerable diversity of primate remains. Work has recently resumed here, following a hiatus caused by political problems. Six genera of primates have been described from the Fayum. Some are known from many specimens (*Aegyptopithecus, Apidium*) and others are very rare (*Aeolopithecus, Oligopithecus*). All are fairly small and anthropoid rather than lemuroid, but none show unequivocally the group of characters of either cercopithecoids or hominoids, although it is probable that ancestors of both are found here.

Aegyptopithecus (Figure 21–8) is known from a nearly complete skull and from numerous jaws and other skeletal parts. Its snout was relatively long, suggesting more reliance on smell than monkeys and apes show today; and relatively small, bony eye sockets suggest diurnal habits. Sexual dimorphism of the canine teeth has been described, and this suggests life in social groups with male aggressiveness. *Aegyptopithecus* had the skeleton of a tree dweller, possessed a tail, and was not a brachiator. Nonetheless, its teeth suggest relationship to the great apes. *Aeolopithecus* has been proposed as an ancestor of the gibbon on the basis of a striking resemblance of its lower jaw to that of gibbons. Similarly, *Parapithecus* has, on the basis of the tooth pattern of the lower jaw, been referred to the ancestry of the cercopithecoids. *Propliopithecus*

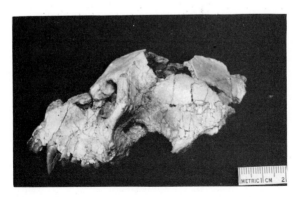

Figure 21–8. *A beautifully preserved skull of* Aegyptopithecus, *a thirty-million-year-old possible ancestor of the great apes, found in Oligocene deposits of Egypt.*
Photo courtesy of Dr. Ian Tattersall, American Museum of Natural History.

and *Oligopithecus* occur in the oldest Fayum deposits and may be ancestral to *Aegyptopithecus, Aeolopithecus,* and *Parapithecus. Apidium,* though abundant, is not readily referred to any particular group of living primates.

Miocene Differentiation

During the Miocene, undoubted cercopithecoids and hominoids appear. Initially, cercopithecoids were rare and hominoids were widespread and relatively abundant. By the end of the Miocene, this situation was more or less reversed: Old World monkeys were abundant and diverse, and they extended from ancestral Africa to India and Pakistan. Both leaf-eating colobines and fruit-eating cercopithecines were present. Late Miocene *Oreopithecus* is an interesting animal. This primate is known from numerous specimens found in a coal mine in Tuscany, northern Italy. Included is a complete skeleton that reveals an animal with a short face; a shallow, broad chest; long, slender arms; and relatively short legs—characters with a striking resemblance to those of gibbons. The pattern of the tooth cusps, however, suggests that *Oreopithecus* was an aberrant monkey and not really a hominoid, as had been claimed.

Pongid hominoids were important and widespread in the Miocene and are known from Europe, Africa, and Asia. The oldest of these was *Proconsul* (Figure 21–9) from the Early Miocene of Kenya, first discovered by Louis and Mary Leakey in the 1930s. Sites from Kenya and Uganda yielding *Proconsul* have been dated at 17 to 21 million years before the present (b.p.). Several species ranged in size from as small as 20 kg (*P. africanus*) to as large as 100 kg (*P. major*). They had the cheek teeth of hominoids but lacked the tusklike canines that characterize living pongids.

Dryopithecus was first found in France in 1856. Its remains extend from Spain to China, and it ranges in time from mid to late Miocene, 16 million to 8 million years b.p. Though *Dryopithecus* seems to be a descendant of *Proconsul,* it has not been found in Africa. The apparent geographic separation

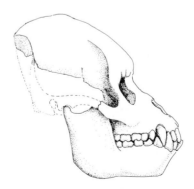

Figure 21–9. Proconsul, *a Lower Miocene pongid from East Africa.*

of *Dryopithecus* from *Proconsul* may exemplify allopatric speciation. (For a time, *Proconsul* was treated as a synonym of *Dryopithecus*, but it now appears to be generically distinct.) The teeth are very like those of living apes, and some limb bones suggest the beginnings of brachiation. The ancestors of the living apes are almost certainly among these Miocene apes, which comprise a subfamily, Dryopithecinae, of the family Pongidae, to which chimps, gorillas, and orangutans belong.

The Siwalik Hills of northern India and Pakistan yield rich mammalian fossils of Miocene and Pliocene age. Hominoids are rare but welcome finds in these beds. Several genera of mid and late Miocene dryopithecines have been described, based upon several maxillae and mandibles. *Sivapithecus* and *Ramapithecus* are the most famous of these; indeed, in 1961 Simons claimed that *Ramapithecus* was the earliest recognizable hominid. Also in 1961 Leakey discovered *Kenyapithecus* in Kenyan deposits dated at 14 million years b.p. In 1975 *Rudapithecus* was described on the basis of abundant fossils found at Rudabanya, Hungary. The last two "genera" are well within the range of variation of *Ramapithecus* and should be treated as African and European representatives of that genus. Some have even suggested that *Ramapithecus* should be included within *Sivapithecus*. Another genus, *Gigantopithecus*, has been found not only in the Siwalik Hills but also in Pleistocene deposits of South China. This huge animal exceeded the size of living gorillas. In all of the Miocene hominoids, there is a mixture of hominid and pongid traits. Some of their hominid traits are shortening of the face, thickening of the molar enamel, and modest development of the canines. Simons regarded *Ramapithecus* as the earliest recognizable hominid especially because of these dental characters. Their pongid characters include a dental arcade with parallel rows of cheek teeth, rather than the parabolic arcade of hominids, a gap (diastema) between the incisors and canines, and a sharpening surface on the first lower premolar, against which the upper canine was honed. Telling evidence against their hominid status came with the report in 1982 by D. Pilbeam of a beautifully preserved partial skull of *Sivapithecus*. This skull appears strikingly orangutanlike. It had already been concluded by P. Andrews that ramapithecines were pongids, ancestral to the orangutan, and that true hominids had not yet appeared in the late Miocene, up to 8 million years b.p. The study of late Miocene hominoids is an area of intense current interest and controversy. On the one hand, there is a tendency to proliferate names of species and genera; on the other hand, every comprehensive review reduces many of these to synonyms of some of the earlier described groups. No consensus exists, and to reach a consensus we will require more and better preserved fossils.

The Australopithecines

Most of the later hominoid fossils can be assigned with certainty either to the genus *Homo* or to the anthropoid apes, but one important group of fossils raises the question of what a human is. In a broad sense any member of the family Hominidae is human; in a narrow sense only members of the genus *Homo* are human. But what criteria do we use to assign an organism to its family and genus? For other mammals criteria are anatomical, with a certain range of variability marking a given genus, and a wider range, typically including several to many genera, marking a family. The particular characters that are useful for delimiting specific groups can only be determined by thorough knowledge of those groups. For humans, another criterion is added: culture in the broadest sense. By culture we mean the ability of individuals in a society to formulate abstract and arbitrary concepts and to communicate them to other members of society. Language of some sort is a fundamental medium of culture, though certainly not a fossilizable one. Culture seems to be a specifically human trait. For some anthropologists, the use of fire is the essential proof of humanity; and using fire appears to have begun with *H. erectus*. For others, the use of tools is critical, even though chimpanzees use simple tools. Some anthropologists define a human as a tool maker. At least some of the gracile australopithecines of two million years b.p. made tools; and, accordingly, L.S.B. Leakey treated them as the earliest humans, under the name *H. habilis*. Many others have rejected this interpretation.

R. Dart made the first of these ambiguous finds in 1924 at Taungs, South Africa. It was the skull of a child (Figure 21–10) of about six years, showing a curious mixture of human and simian features. He described it under the name *Australopithecus* (southern ape). The difficulties of study in this case were increased by the fact that most comparisons are based on adult specimens. But a considerable number of additional skeletons, more mature and some nearly complete, have since been found by R. Broom, R. Dart, J.T. Robinson,

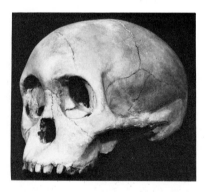

Figure 21–10. The Taungs skull, the first example of Australopithecus *to be described, was discovered by R. Dart in 1924. This is the skull of a six-year-old; later finds were all from adults.*
Courtesy of the National Museum of Natural Sciences, National Museums of Canada.

and others at a series of South African sites. Dating of these fossils has been difficult because they were all found in cave deposits, to which the standard stratigraphic techniques are not applicable.

Fossils of this type found in South Africa once were assigned to five species in three genera, but they are all now usually assigned to two species of the genus *Australopithecus* in the family Hominidae. The fossils fall into two groups: a lighter, more progressive group, described as *A. gracilis, A. africanus,* or *A. habilis,* and a heavier, less progressive group, described as *A. robustus.* A variety of theories of their relationships included the view that *A. robustus* was ancestral to the gracile types or even that it was a simple sex difference, the gracile ones being the females. When at last reliable radiometric dates were obtained, the picture was clarified. Neither of these theories is tenable, for the gracile species, now called *A. africanus,* flourished much earlier than did *A. robustus. A. africanus* is now dated from 2.7 to 2.1 million years b.p. (late Pliocene); and *A. robustus* ranged from 2.0 to 1.5 million years b.p. (early Pleistocene).

Starting in 1959, the Leakeys, working at Olduvai Gorge in Tanzania, uncovered an important record of australopithecine evolution in East Africa. The Olduvai beds range from about one million to two million years in age, are well-layered, and have volcanic ashes that provide radiometric dates. As in South Africa, two hominid types are represented, as are crude stone tools made by them. Their remains were described under the names *Zinjanthropus boisei* and *Homo habilis,* but most workers regard the former as *A. boisei.* Many additional finds have been made in East Africa. At Laetoli, Tanzania, M. Leakey found trails of footprints that clearly show a bipedal stride in hominids at this early date. They have been dated at 3 million years b.p. or older. In East Turkana in Kenya, R. Leakey has discovered a number of superb hominid skulls since 1968.

In 1974 D.C. Johanson made the most-electrifying find in recent years at Hadar, in the remote Afar region of Ethiopia. It consisted of about 40 percent of the skeleton of a diminutive (.9 m tall), gracile australopithecine, promptly nicknamed "Lucy" (and more formally named *Australopithecus afarensis* by Johanson and T.D. White in 1979). Two things made Lucy (Figure 21–11) so important. The first was that this was the first time that so extensive a part of a hominid skeleton had been found in Africa; and the second was the great antiquity of the Hadar beds, older by far than any previously known hominid sites (unless *Ramapithecus* be accepted as hominid). The following year, the remains of thirteen individuals, perhaps constituting a single family group, were found at a single site. As we would expect from its great age, *A. afarensis* seems to be primitive in many of its characters. But it also shows, by the structure of the pelvis and femur, that posture was quite upright, as it is in modern humans.

The skulls of the australopithecines resemble those of modern chimpanzees; but the differences are significant. First, the braincase is larger in the fossil.

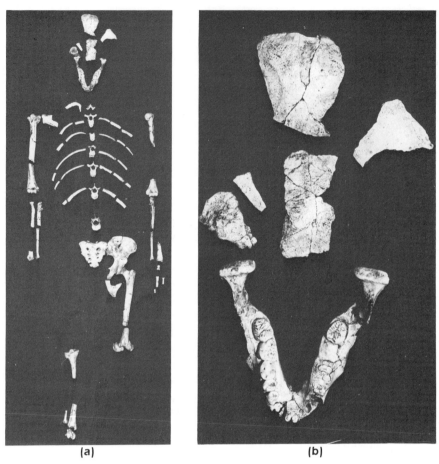

(a)   **(b)**

Figure 21–11.   Lucy, formally known as Australopithecus afarensis. *This is the oldest and most complete skeleton of* Australopithecus *yet found. (a) The entire specimen. (b) The remains of the skull.*
Courtesy of the National Museum of Natural Sciences, National Museums of Canada.

Its capacity of 400 to 600 cubic centimeters (cc) is somewhat larger than that of gorillas and it is much larger in proportion to body size (about 1.2 m, gracile forms weighing around 23 kg). The forehead is more rounded out than in chimpanzees, possibly indicating greater development of the highest centers of the brain. The eyebrow ridges are very prominent, but less so than in chimpanzees. The jaws protrude prominently, but again less so than in modern apes. The dentition is quite human in character. The canines are larger than in modern humans but much smaller than in any modern ape. Further, the shape of the tooth rows is altogether different. In apes, the canines, premolars, and molars form parallel rows, with the incisors set at right angles to them at the front of the jaws. In humans and in the australopithecines, the entire tooth row is parabolic (Figure 21–12). Finally, the occipital condyles, by which the skull articulates with the spinal column, are set much farther forward on the under surface of the skull in the australopithecines than in any living ape. This position correlates with erect posture.

The rest of the skeleton gives evidence that corroborates that of the skull. The limb bones also indicate that this southern ape was erect in posture. There is little indication of the overdevelopment of the arms, which goes with brachiation. The hipbone is characteristically long and narrow in apes; but in humans and in the australopithecines, it is broad and flat, an anatomical feature that is associated with erect posture and bipedal walking. The great toe lies parallel to the other toes, as in humans and unlike apes, and both longitudinal and transverse arches are developed. Anatomical details of the hand and sacrum suggest that *Australopithecus* did not pass through a knuckle-walking stage but that its ancestors descended directly from the trees, so to speak.

These, then, are some of the major facts about the australopithecines. Regarding the factual findings there is no disagreement, but there has been much disagreement regarding their interpretation. Initially the conclusions of the South African workers, Dart and Broom, that *Australopithecus* was on the direct line of human descent, received little or no support from the rest of the scientific world. In the late 1940s, the distinguished British scientist Sir

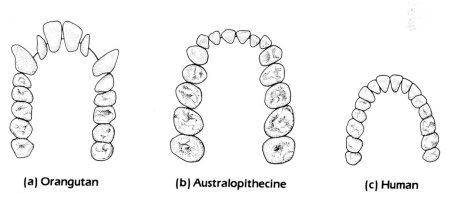

(a) Orangutan   **(b) Australopithecine**   **(c) Human**

Figure 21–12. *Tooth rows of (a) orangutan, (b) australopithecine, and (c) human upper jaw. Notice the parallel rows of cheek teeth in the ape and the gently curved arch of the human tooth rows. The autralopithecine is intermediate, but much closer to the human. In most instances, the australopithecine dentition would be much smaller than the human, but this drawing was made from A. boisei, a species with massive jaws and teeth.*
Specimen of A. *boisei* loaned by Dr. J.S. Cybulski and the National Museums of Canada. Reprinted from E. Dodson, *The Phenomenon of Man Revisited,* by permission of Columbia University Press (New York, 1984).

Wilfred LeGros Clark visited South Africa to look at the material. He became convinced of the importance of the fossils and lent the weight of his prestige to the concept that *Australopithecus* lay on the direct line of human descent.

The point of view has long since prevailed that Africa is indeed the cradle of humanity. Darwin predicted this in 1871, even though there were then no fossils to support this position. Many lines of evidence, both anatomical and biochemical, point to the special relationship between *Homo* and the African apes. Karyotype analysis (see Chapter 8) and DNA hybridization indicate that humans and the African apes share as much as 98 percent of their genetic material. As more and more fossils accumulate, the question is no longer *whether* African fossil hominids are relevant to human origins, but rather *which* ones are. The robust forms, *A. robustus* and *A. boisei,* are too late in time and too specialized in morphology. They are believed to be terminal members of a line beginning with *A. afarensis* and passing through *A. africanus.* A major question at present is whether *Homo* also derived from *A. africanus* or whether the line leading to *Homo* had already separated from a common hominid ancestor before the origin of the australopithecines. Perhaps the fossils necessary to resolve this issue will be found during the 1980s.

HUMANS IN THE FOSSIL RECORD

The outstanding climatic events of the Pleistocene were the advances and retreats of the great continental ice sheets that covered the Northern Hemisphere. Human fossils in Europe, and to varying extents elsewhere, are dated with respect to these profound physical events. There appear to have been four principal ice advances, separated by warm interglacial intervals. Dating of the glacial events is still remarkably uncertain, but we may provisionally accept the following estimates. The first glaciation began some 1.3 million years ago and lasted perhaps some 600,000 years; the second glaciation began about 500,000 years ago, lasting about 250,000 years; the third glacial advance commenced 225,000 years ago and lasted 100,000 years. The ice last advanced 100,000 years ago, retreating a scant 11,000 years ago. We are now in an interglacial period, and may expect the ice to begin spreading southward again in a few thousand years. Against this background we will trace the fossil record of our own kind.

Homo habilis

In 1962 Louis Leakey found four poorly preserved partial skulls and dentitions at Olduvai in beds, dated at about 1.75 million years b.p., from which *A. boisei* had come. Also found in the same beds were crude stone tools. Careful study of these fragmentary skulls suggested a brain size of 640 cc, about 50 percent larger than typical australopithecines. The teeth were also less massive than those of typical australopithecines. Leakey was delighted to name the animal *Homo habilis,* "the handy man"—in other words, the maker of tools. The designation of these animals as the oldest humans was controversial for years, and the resolution was brought about by new discoveries made by Richard Leakey, beginning in 1972. At Koobi Fora, near Lake Turkana, he found magnificent skulls, among the finest of all fossil hominid skulls; and these proved to have cranial capacities of up to 810 cc, so much larger than any ape or australopithecine that few would deny their designation as *Homo.* Tools were also found here. These finds were initially thought to be about 3 million years old, but they are now known to fit in the time range of 2.0 to 1.5 million years b.p. Thus they are roughly contemporaneous with the finds at Olduvai. The assignment of these large-brained hominids to our own genus no longer elicits controversy.

Homo erectus

In contrast to *Homo habilis, H. erectus* has been known to science for nearly a century. It was the first human fossil that was found not by chance but as the result of a deliberate search. In 1891 E. Dubois, a Dutch physician assigned to the Dutch East Indies, made an important find at Trinil, on the Solo River of central Java. The fossil consisted of a skull cap, a jaw fragment, and a femur. These remains sufficed to show that their possessor had heavy, apelike eyebrow ridges, a low, sloping forehead, and a low cranial vault but a much larger braincase (about 775 cc) than any known ape and occipital condyles apparently set far enough forward to permit erect posture. The femur was indistinguishable from that of modern humans. Dubois regarded this creature as the "missing link," which was then so much under discussion, and therefore he named it *Pithecanthropus erectus* (erect ape-man).

For nearly fifty years, the true nature of the Java fossils was controversial, but in 1938, G.H.R. von Koenigswald found a second, more complete skull. Additional finds have raised the total to eight skulls, several jaw fragments, and a femur. As more was learned about australopithecines, it became evident that *Pithecanthropus* was in fact truly human and that it should properly be named *Homo erectus.* The age of the fossils is estimated to be 900,000 to 600,000 years.

Meanwhile, an important group of fossils was discovered in China. In 1927, while excavating a cave near Choukoutien, the Canadian anatomist D. Black and his collaborators found three teeth that they identified as probably human. Two years later they found a nearly complete skull, including parts of the lower jaw and teeth. Subsequent finds by Black, F. Weidenreich, and their collaborators have raised the total to fifteen skulls and other bones representing some forty-five individuals. They were originally described under the name *Sinanthropus pekinensis,* but detailed study by Weidenreich and von Koenigswald has shown that the Chinese fossils do not differ significantly from the Javanese fossils, and so they also are included in *Homo erectus*. The Chinese finds are dated at between 600,000 and 400,000 years b.p. (Figure 21–13).

Collectively, the Chinese and Javanese fossils give a fair picture of these primitive humans. They were of moderate stature, and the straight limb bones, broad hip bone, and position of the occipital condyles all show that they stood erect or nearly so. The relative proportions of arms and legs were modern, suggesting that this may be a primitive character, whereas the elongated arms of the apes are specialized. The forehead was retreating and the jaws projecting, but much less so than those of any ape. The jaws and teeth were rather large, and there was no chin, a structure usually regarded as specifically human. The teeth, although larger than usual in humans, agreed with those of humans rather than with those of apes in all specific differences. The size of the braincase was variable. In the Javanese skulls it varies from 775 to 900 cc, with an average at 860; in the Chinese skulls it varies from 850 to 1,225 cc, with an average of 1,075 cc. These measurements fall between the averages of 500 cc in the gorilla and 1,350 cc in modern humans. Intelligence is only loosely correlated with brain size at best, and measurements on badly damaged and incomplete skulls are very crude estimates, yet *Homo erectus* was probably very clever by ape standards and very dull by human standards. No cultural remains have been found with the Javanese fossils, but the Chinese fossils are associated with crude tools of chipped stone and bone. Peking humans used fire, and charred deer bones indicate that they had learned to cook. Such cultural attainments clearly indicate some human intelligence, and,

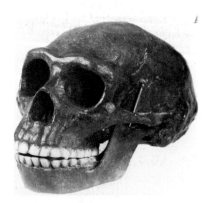

Figure 21–13. Skull of Homo erectus *from China, photo of a cast of a reconstruction. The protruding jaws, strong eyebrow ridges, and flat cranial vault are characteristic of the species.*
Reprinted from E. Dodson, *The Phenomenon of Man Revisited,* by permission of Columbia University Press (New York, 1984).

coupled with the fact that the brain size overlaps the lower end of the normal range for modern humans, they confirm the opinion that *H. erectus* was truly human.

As *H. erectus* from Asia became well characterized, finds from Africa and Europe were referred to it. *H. erectus* has been identified in Africa at sites ranging from South Africa, to the Turkana region in the east, and to the Atlas Mountains of Morocco and Ternifine in Algeria. In 1907, a fine jaw was found near Heidelberg, Germany, and this is now also referred to *H. erectus*. The back of a skull was found in Vertesszollos, Hungary in 1965. The finest European find is a skull from a cave in Petralona, Greece, discovered in 1960. This skull, dated at 300,000 years b.p., has a higher cranial vault than does typical *H. erectus;* and the cranial capacity is 1,200 cc, well within the range of modern humans. Some workers classify the Petralona skull as advanced *H. erectus,* others as early *H. sapiens.* In either case, it is indicative of the progressive evolution that resulted in the appearance of our own species.

These fossils also illustrate the two major differences between us and the great apes: adaptations to erect posture and the development of the brain and associated changes in the skull. Of these, enlargement of the brain has lagged behind in human evolution, as shown by the study of australopithecines and *H. erectus.*

Adaptations to erect posture were quite far advanced in *Australopithecus,* and were complete in *H. erectus.* Many lines of evidence bear upon even such a seemingly simple character. The ankle bones bear the main weight of the erect body. Ours are large, those of the apes are small. Correspondingly, the toes of the apes are long and freely movable, the great toe being opposable to the others. Human toes are short, and the great toe is held in line with the others, making a more rigid support for the weight of the body. As already pointed out, the human leg bones are longer than the arm bones; the reverse is true in the apes. The human trunk is short relative to the legs; the opposite is true in the apes. The human situation is obviously mechanically more stable for erect posture. The hipbone of the apes is long and narrow; that of humans is broad, giving optimum support to the viscera in the erect position. The spinal column of the apes, like that of any four-footed mammal, follows a single, sweeping, outward curve. This tends to throw the animal off balance when standing erect. But the curvatures of the human spinal column alternate in direction, averaging out to a straight line. In the upright position, the human knee and hip joints are held straight; those of the apes are flexed slightly. Finally, human occipital condyles are near the center of the base of the skull and are directed downward; those of apes are more posterior and are not vertical. Thus the human adaptations to erect posture affect every part of the skeleton, as well as the viscera in many ways that have not been discussed. These adaptations appear to be largely completed in the most primitive humans known.

The most important change in the skull has been the increase in size of the braincase. The largest braincase known in apes has a capacity of 685 cc.

The average capacity of the Javanese fossils is about 900 cc; that of modern humans is about 1,350 cc. This greater capacity reflects an increase in the height of the vault of the skull and an increase in the diameter of the skull above the ear line. In the apes, the australopithecines, and *H. erectus,* the greatest width of the skull is at the level of the ears. In later human races, the skull has become wider in the parietal region. As a result the shape of the skull has become more nearly globular. This change started later than the adaptation to erect posture, and it has proceeded more slowly. As the braincase has enlarged, the jaws have decreased in size, with the result that the face has gradually receded to a position *under* the braincase, rather than in front of it as in all other mammals. As already pointed out, the premolar and molar teeth form parallel rows in the apes, in humans all of the teeth form a gently curved arch (Figure 21–12). Several other differences in the teeth may be mentioned, apart from size. Human canines are no larger than the adjacent teeth; those of apes are powerfully developed tusks. The first lower premolar of the apes—morphologically the third—is modified as a shearing tooth to work against the upper canine; whereas in humans the same tooth is a typical bicuspid grinding tooth. Also, the surface of the molars of the apes is rather elaborately etched, in contrast to that of humans. Finally, humans have recently developed the chin, a protuberance on the lower jaw that is unknown in any other mammal. The chin evidently resulted from an unequal regression of different parts of the lower jaw.

An important feature of *H. erectus* was the association with stone tools, which defines the Lower Paleolithic age. A number of traditions have been recognized. Tools of the Olduwan industry (named for Olduvai Gorge) are very simple, unspecialized, and geographically restricted. The later Olduwan industry represents considerably more skill and sophistication and is recognized in South Africa as well as in East Africa. The Acheulian industry is characterized by large (15 cm), tear-shaped hand axes that show fine workmanship. Tools of the Acheulian industry (named after the village of St. Acheul in France) are found in abundance from France to India and through much of Africa. At Ternifine, they are associated with *H. erectus* fossils. The Acheulian industry persisted for a million years, and tools of the Acheulian industry were probably used by early *H. sapiens* as well.

Piltdown Fraud

In 1911 and 1912, Charles Dawson, an amateur collector, recovered parts of a skull and lower jaw from a gravel pit near Piltdown, England. The skull was thick but remarkably human, the jaw quite anthropoid, and the bones soon became the center of controversy. Proponents regarded it as the earliest known fossil human (which flattered the national pride), claiming that such a combination of human and simian characters was to be expected in the most primitive human. Their opponents considered it to be a spurious association of human and ape bones. Stone and bone implements found in the same deposit were claimed (or disclaimed) as evidence of a simple culture. The existence of the Piltdown fossil was a serious impediment to the acceptance of the hominid status of *Australopithecus,* described by Dart in 1925.

In 1955, results of a reinvestigation of this fossil were published by the British Museum, to which Dawson had presented the material. Tests of fluorine content (fluorine accumulates in buried bone) showed that whereas the skull was less than 50,000 years old, hardly a relic of the early Pleistocene, the jaw was modern! Chemical tests showed that it had been stained to simulate a fossil of great age. The teeth showed atypical wear, and microscopic examination revealed file marks. X rays showed that the roots of the teeth were too long for the crowns, being actually the size of those of chimpanzees. In short, it was proven that a chimpanzee or orangutan jaw had been deliberately modified to simulate a transitional stage between apes and humans. Thus the Piltdown discovery was proven to be a clever fraud. The story of its exposure, principally by K.P. Oakley, J.S. Weiner, and W.E. LeGros Clark, is one of fascinating scientific detective work.

Early *Homo sapiens*—275,000 to 125,000 Years before Present

There is difficulty in determining precisely the time of appearance of true *Homo sapiens.* The Petralona skull appears to be that of a *Homo erectus* skull close to the level of *H. sapiens.* A series of European and African skulls appear to be early *H. sapiens.* Collectively they date from about 275,000 years to 125,000 years b.p. (although, as always, there is no consensus on these dates). Perhaps the oldest of these is an incomplete skull from Swanscombe, England, part of which was found in 1935 and more of it in 1955. The thickness of the bone suggests *H. erectus,* the brain size of 1,100 cc is low for *H. sapiens;* but the high, rounded vault of the skullcap is characteristic only of *H. sapiens.* A crushed skull found at Steinheim, Germany in 1933 is roughly the same age as the skull from Swanscombe. It shows brow ridges, a vaulted cranium, and a more-reduced facial region than that of *H. erectus.* The cave of Arago in the French Pyrenees has yielded skull material of this general morphological level. In Africa, representatives of this morphology and time period come from Broken Hill in Zambia, Saldanha in South Africa, and Bodo in Ethiopia. From the Solo River in Java, a series of specimens was found that included eleven partial skulls that ranged in cranial capacity from 1,150 to 1,300 cc. There is no sharp break in morphology of these specimens from advanced *H. erectus,* and the available evidence points to a gradual transition.

Neanderthals

As early as 1848, a large portion of a skull was found at Gibraltar. The bones were the first hominid fossil known, and they seemed incredibly primitive to nineteenth-century minds. The bones of this skull were very thick. The eyebrow ridges were very prominent, the nose was broad, and the jaws were massive. This skull (Figure 21–14) did not attract much attention, but eight years later, in 1856, a similar skullcap together with a few ribs and limb bones were recovered from a cave in the Neander Valley (Neander Tal), near Düsseldorf, Germany. The remains became well known under the name of *Homo neanderthalensis,* or the Neanderthal man, which was popularly regarded as "the prehistoric man." The relative dating of Neanderthal was first established in 1886, when two skeletons were found in Namur, Belgium, in association with bones of the mammoth and the woolly rhinoceros, characteristic European animals of the last Ice Age. Since then, a large number of Neanderthal fossils, some of them quite complete, have been found throughout the Palearctic region. Neanderthals flourished from 75,000 to 35,000 years ago, but Neanderthal morphology is evident in specimens as old as 135,000 years b.p.

A rather complete picture of the appearance of Neanderthals can be constructed on the basis of available skeletons and from cultural evidence (Figure 21–15). The skull was large and thick-boned. The eyebrow ridges were prominent but less so than in earlier humans. The forehead was receding. Although the cranial capacity was greater than that of modern humans (the average was about 1,450 cc), the roof of the skull was rather flat. The teeth and jaws were very large and heavy by modern standards, and the chin was receding. It was often said that the posture of Neanderthal was stooped, but this was a misinterpretation based upon the study of an arthritic skeleton. The spinous processes of the cervical vertebrae were exceptionally large, indicating that the neck musculature was powerfully developed. The limb bones were robust, their ends somewhat exaggerated in comparison with those of moderns. The skeleton was adapted to higher levels of activity and stress than is our own—hardly surprising given the rigors of Ice Age Europe! Anatomy of the hand suggests a powerful grip. The standing height of a Neanderthal could not have been much more than 1.5 m. Biologically very successful, they were

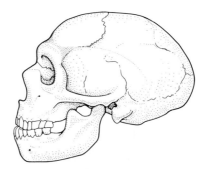

Figure 21–14. Neanderthal skull. Note the prominent eyebrow ridges, the low forehead, and the lack of a chin. Courtesy of Dr. Ian Tattersall, American Museum of Natural History.

Figure 21–15. *Neanderthal family group. Note erect posture.*
Courtesy of the Field Museum of Natural History, Chicago.

a homogeneous, widely distributed, morphologically distinctive, and fairly abundant people.

We have insights into certain aspects of the culture of Neanderthals. They inhabited caves, where their remains are sometimes abundant; they also inhabited open areas, particularly in summer, but here their remains were rarely preserved. They buried their dead with reverence—evidence of flowers was found at a burial in Shanidar Cave in Iraq. Other ritualistic burials are known from Russia, Italy, and France. Also found in Shanidar was an individual with a crippled shoulder and an amputated elbow, evidence of a social structure that was sufficiently benign to care for such a person. Neanderthals made stone tools of greater diversity and skill than were known before. Stone tools of the Mousterian industry (after Le Moustier, France) characterize the Middle Paleolithic age, beginning around 125,000 years b.p. This industry was fully developed by 75,000 years b.p. Although many types of tools are known, points are most characteristic of the Mousterian industry. Art was not an attribute of Neanderthal life.

Modern Humans

Anatomically modern humans, *Homo sapiens sapiens,* appeared in Western Europe about 33,000 years ago. They are called Cro-Magnons from the most famous site, the Cro-Magnon shelter in Les Eyzies, France, where they were first found. Additional sites are known in France, Italy, and the Middle East. Skulls (Figure 21–16) exhibit much reduced brow ridges, steep forehead, high, rounded cranial vault, short face, and pronounced chin. Their stature was more gracile than that of Neanderthals; they were simply less bulky. Structurally, Cro-Magnons were well within the range of variability of modern Europeans. Associated with these essentially modern people are stone artifacts of high technological perfection, which herald the Late Paleolithic age. Long, thin blades of diverse types are its hallmark. Art was very much a part of Cro-Magnon life. Cro-Magnons made beads, carved statues, and engraved pictures on reindeer antlers. They left a record of cave paintings of great beauty, the finest of which is in Lascaux Cave (Figure 21–17) in the Dordogne Valley of France. Their ceremonial burials (Figure 21–18) are a major source of Cro-Magnon skeletal material. Cro-Magnon culture was the source of the earliest historical cultures of Europe.

Since the appearance of modern humans, morphological evolution has been less important, and cultural evolution has been the hallmark of human progress. Beginning about 10,000 years ago, a profound shift in the pattern of human activity occurred. This shift was from hunting and gathering to agriculture— the planting and harvesting of crops, the domestication of plants and animals, and the storing of food. Cultural evolution was marked by the development of new technologies, first of bronze, then of iron; the development of specialized occupations; the rise of cities some 5,000 years ago; the development of writing, of history, of wealth, of leisure, of science, and of the arts. These

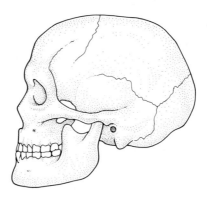

Figure 21–16. *A Cro-Magnon skull. Note that with prominent chin, weak eyebrow ridges, and high forehead, it is essentially modern.*

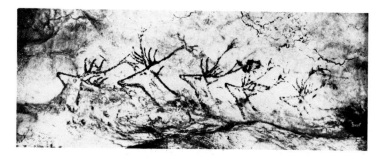

Figure 21–17. *Swimming deer. Photo of a cave drawing by Cro-Magnon artists in Lascaux Cave, near Montignac, France. The age of this drawing is estimated at 15 to 20 thousand years.*

are the flowers of *Homo sapiens sapiens.* This is how modern humans arrived; where we are going is up to us.

PROBLEMS IN HOMINID PHYLOGENY

We have reviewed the highlights of the fossil record of the Primates and of *Homo sapiens.* Can we now synthesize the path of human ancestry? The family Hominidae probably arose from dryopithecine ancestors, and *Proconsul africanus* and *Ramapithecus* may have been near the point of origin. Indeed,

Figure 21–18.   A Cro-Magnon burial.

the latter was once named as the earliest recognizable hominid, but it now appears to be closer to the orangutan. There is a dearth of primate fossils in late Miocene and early Pliocene. Then in late Pliocene *Australopithecus afarensis*, the earliest indisputable hominid, appeared. There are two principal views of the relationship of *A. afarensis* to later humans. One is that the entire australopithecine line was a side branch of hominid evolution and left no descendants in the modern world. According to this view, human ancestry derives from a contemporary of *A. afarensis*, a species of which no fossils have been found, but which gave rise to *Homo habilis*, and through the latter to *H. erectus*, and finally to *H. sapiens*. The alternative is that *A. afarensis* gave rise to *A. africanus*, which then divided into two lineages. One of these terminated with *A. robustus* and *A. boisei*. The other, more progressive, branch gave rise to *H. habilis*, and through the latter to *H. erectus*, and finally to *H. sapiens*.

The later history of *Homo* has also been interpreted in many ways. At one extreme, each of the present human races has been treated as a distinct species with separate ancestry going back to early Pleistocene time. This viewpoint has no support today. At the other extreme, all members of the genus *Homo* have been assigned to one species, *H. sapiens*. This viewpoint, too, is difficult to defend, for *H. habilis* and *H. erectus* are as distinct as are the species of other mammalian genera. Unlike other mammalian genera, however, there does not seem to have been more than one species of *Homo* at any one time, although subspecies have been contemporaneous. *H. sapiens* probably arose from *H. erectus* in the mid-Pleistocene.

The origin and fate of the Neanderthals are also interesting problems. The classical Neanderthal remains were all found in Western Europe and were of rather late date. However, many skeletons of earlier date and of less-marked Neanderthaloid type have been found in Eastern Europe and Asia. As early as the second interglacial period, a modern but highly variable race occupied much of the Old World. As the fourth glaciation set in, those with the more extreme Neanderthaloid traits were isolated from the rest in southwestern Europe. Developing in isolation and probably aided by considerable inbreeding, they developed the classical Neanderthaloid habitus. Meanwhile the main population to the east (including moderate Neanderthaloids) developed along different lines and produced *Homo sapiens sapiens*. Subsequently, this more progressive race, which we know as the Cro-Magnons, invaded Western Europe and replaced its Neanderthal cousins.

In France, the latest Mousterian site (La Quina) is 32,500 years old, whereas Upper Paleolithic sites are as much as 33,000 years old. The inference is that an abrupt replacement occurred. By contrast there is evidence of a more gradual transition in the Middle East. Important caves, Tabūn and Skhūl, on Mount Carmel and a third at Qafzeh, have been found in Israel. Two skeletons from Tabūn, dated at 45,000 b.p. (at least) are Neanderthals, associated with Mousterian tools. At Skhūl, ten skeletons show a mixture of Neanderthaloid and modern features. Skhūl is tentatively dated at 40,000 years or less. At Qafzeh, probably the same age as Skhūl, there are skeletons that resemble the most advanced at Skhūl, and they are associated with Mousterian tools. E. Trinkaus and W.W. Howells suggest that modern humans first appeared in the Middle East, spread out from there, and replaced Neanderthals rapidly. Whatever the dynamics, it is probable that intermating assured the transmission of Neanderthal genes to the Cro-Magnons.

THE STUDY OF PRIMATES AND HUMANS

An important component of human nature is curiosity, including the desire to know our past, for ours is the only species that knows that it has a history. Accordingly, the study of primates (primatology) and of humans (anthropology) and their paleontological equivalents are subjects of enduring popularity. Great strides have been made in the past twenty years to elucidate the path of descent from 60-million-year-old animals, barely recognizable as primates, to ourselves. In some cases discoveries of new fossils have dramatically affected our view of evolution. Such is the case of Lucy, the oldest indisputable hominid. In other cases reinterpretation of long-known fossils has been important, as in the case of the new views on Neanderthals. A third way in which our knowledge has grown is by the incorporation of findings from other fields. Thus in Chapter 4 the molecular clock of Sarich and Wilson was introduced, with the suggestion that the separation of hominids from pongids might have been as recent as 4 to 7 million years ago. In Chapter 7 we saw that studies of primate genotypes by the method of electrophoresis had shown the human and pongid genotypes to be more than 98 percent identical, for the proteins tested. Again, this suggests recent separation with minimal differentiation. Finally, in Chapter 8 we reviewed the research of Yunis and Prakash on the karyotypes of the higher primates, which showed that human and pongid karyotypes differ by only a few rearrangements.

Against this array of evidence for recent separation, the fossil record had been interpreted, until well into the 1970s, in favor of separation of hominid and pongid lines of descent as early as mid-Miocene. The discovery of Lucy in 1974 focused paleontological interest upon the possibility of late Pliocene separation, and recent studies have tended to ally *Ramapithecus* more with the orangutans than with hominids. Thus the controversy between molecular and paleontological viewpoints is in the process of resolution. Such controversy may be indicative of intellectual health and vitality. The progress of the last decade provides much reason to hope for great strides in the next decade.

SUMMARY

The order *Primates* comprises lemurs, tarsiers, New World and Old World monkeys, apes, and humans. Primates possess a suite of characters that relate to aboreal habits, including a shortening of the face, binocular vision, increase in cerebral cortex, opposable thumb and great toe, and nails instead of claws on the digits. The most primitive living primates are in the suborder *Prosimii*. The prosimians include the lemurs of Madagascar; the lorises, pottos, and bush babies of Africa and Asia; and the tarsiers of the Philippines. New World monkeys (*Ceboidea*) represent an endemic radiation of arboreal monkeys, possibly derived from the Old World in the Oligocene. Prehensile-tailed forms are characteristic. Old World monkeys (*Cercopithecoidea*), which lack prehensile tails and have ischial callosities, are widely distributed on three continents; all are diurnal, many are terrestrial. The *Hominoidea* include the lesser apes (gibbons and siamangs), the great apes (orangutans, chimpanzees, and gorillas), and humans. Gibbons and orangutans are arboreal armswingers (*brachiators*), and adult gorillas and chimpanzees are terrestrial *knuckle-walkers*. Hominids have perfected bipedal terrestrial walking.

Primates first appeared in the fossil record at the end of the Cretaceous or the beginning of the Paleocene. Lemurs, tarsiers, and archaic, rodentlike primates were fairly abundant in the fossil record of Europe and North America during the Eocene, but then they disappeared. Oligocene primates are found in abundance only in the Fayum deposits of Egypt (30 million years b.p.), where ancestors of cercopithecoid monkeys and hominoids have been identified. In the Miocene, both monkeys and hominoids were in evidence. *Proconsul* (17 to 21 million years b.p.) was an African early Miocene hominoid. Its mid and late Miocene successor, *Dryopithecus* (16 to 8 million years b.p.), was widely distributed from Spain to China. *Ramapithecus* and *Sivapithecus* were contemporaries. Collectively, these genera show a mixture of pongid and hominid traits, and a recently found partial skull of *Sivapithecus* is strikingly like that of an orangutan.

The oldest undoubted hominids are referred to the genus *Australopithecus*, specimens of which range from Ethiopia to South Africa and are of Pliocene and Pleistocene age (1 to 4 million years b.p.). Both gracile and robust species are known, and anatomy of the pelvis and leg plus footprints show that australopithecines walked bipedally. Brain sizes ranged from 400 to 600 cc, greater than those of living apes.

The earliest species of *Homo, H. habilis*, comes from East African beds dated at 1.5 to 2 million years b.p. Brain sizes of these specimens range from 640 to 810 cc. *H. habilis*, whose name means "handy man," is associated with crude stone tools. The famous Java ape-man and Peking man are referred to *H. erectus*. *H. erectus* is known from sites as far removed from each other as South Africa, Europe, China, and Java, in deposits ranging in age from 1 million years b.p. to 300,000 years b.p. Brain sizes range from 775 cc to 1,225 cc with an average of 1,000 cc. *H. erectus* used fire as well as stone tools.

H. sapiens, our own species, appeared around 275,000 years b.p., as documented by numerous finds in Europe, Africa, and Asia. Distinguishing features are the rounded cranial vault and a cranial capacity often exceeding 1,200 cc. Neanderthals were a distinctive race, *H. sapiens neanderthalensis*, that flourished in Europe and the Near East from 75,000 to 35,000 years b.p. They were robust, large-brained humans, with weak chins and prominent brow ridges. They manufactured fine stone points, cared for their crippled, and buried their dead with reverence. Neanderthals were suddenly replaced by anatomically modern humans, *H. sapiens sapiens,* about 35,000 years ago. *Cro-Magnons* appeared during the height of the last glaciation in Europe and are renowned for their cave drawings and other works of art. Twentieth-century humans do not differ morphologically in any significant way from Cro-Magnons.

REFERENCES

Andrews, P., and J.F. Cronin. 1982. The relationships of *Sivapithecus* and *Ramapithecus* and the evolution of the orang-utan. *Nature* 297:541–546.
  Based on new fossils and molecular data, the authors argue that "ramapiths" are not hominids after all.
Campbell, B.G. 1982. *Humankind Emerging.* 3d ed. Little, Brown, Boston.
  A useful survey.
Cartmill, M. 1974. Re-thinking primate origins. *Science* 184:436–443.
  An important critique of some long-held ideas.
Clark, W.E. LeGros. 1970. *History of the Primates.* 10th ed. British Museum (Natural History), London.
  Still a valuable, well-illustrated account of much of the material of this chapter.
Graziosi, P. 1960. *Paleolithic Art.* Faber and Faber, London; and McGraw-Hill, New York.
  An authoritative and richly illustrated work.
Hay, R.L., and M.D. Leakey. 1982. The fossil footprints of Laetoli. *Scientific American* 246 (2):50–57.
  An account of a new kind of evidence on the australopithecines.
Johanson, D.C., and M. Edey. 1981. *Lucy—The Beginnings of Humankind.* Simon and Schuster, New York.
  A firsthand account of the discovery and analysis of the oldest known hominid. The authors skillfully situate Lucy in relation to the whole context of hominid evolution in Africa—good reading!
Leakey, R.E., and R. Lewin. 1977. *Origins.* Dutton, New York.
  An account of African hominid evolution by another of its most active contributors.
Lovejoy, C.O. 1981. The origin of man. *Science* 211:341–350.
  The reasoning is based on reproductive behavior and demographics.
Napier, J.P., and P.H. Napier. 1967. *A Handbook of Living Primates.* Academic Press, New York.
  A valuable and well-illustrated survey of all living species of primates.
Robinson, J.T. 1972. *Early Hominid Posture and Locomotion.* University of Chicago Press, Chicago.
  The viewpoint of a longtime student of the australopithecines.
Szalay, F.S., and E. Delson. 1979. *Evolutionary History of the Primates.* Academic Press, New York.
  Although some of the results of the cladistic analyses are controversial, this ambitious, comprehensive survey of fossil primates has been widely praised.

Trinkaus, E., and W.W. Howells. 1979. The Neanderthals. *Scientific American* 241 (6):118–133.
   Renewed analysis of these first-discovered and most famous fossil humans.
Weiss, M.L., and A.E. Mann. 1985. *Human Biology and Behavior—An Anthropological Perspective*, 4th Little, Brown, Boston.
   Combines detailed, well-illustrated analysis of the primate fossil record with an account of human biology.
Wolpoff, M.H. 1980. *Paleoanthropology*. Knopf, New York.
   A leading paleoanthropologist surveys his field.
Young, J.Z., E.M. Jope, and K.P. Oakley, eds. 1981. The emergence of man. *Philosophical Transactions of the Royal Society* (London), Series B, 292:1–216.
   A broad survey of such topics as the fossil record, teeth and diet, locomotion, behavior, genetics, and the brain, by some of the most-distinguished students of these subjects.

22

Some Evolutionary Generalizations

We have now reviewed the broad outlines of the history of life, beginning with the separation of the procaryote and eucaryote series, each of which is subdivided into kingdoms that differ profoundly. The procaryotes were, until recently, all referred to one kingdom, the Monera, but currently there is much interest in the recognition of an even more ancient kingdom, the Archaebacteria. Among the eucaryotes, biologists first recognized the kingdoms Animalia and Plantae but it has long been apparent that a vast array of microorganisms, as well as some of their larger relatives, cannot be assigned to one or the other unequivocally. Some biologists prefer to recognize only the two eucaryote kingdoms and treat the difficult groups as intermediates that are the logical consequence of origin by evolution. Others prefer to define four eucaryote kingdoms on the basis of nutrition: kingdoms Plantae (autotrophic), Animalia (heterotrophic), Fungi (saprotrophic), and Protista, or better, Protoctista, (microorganisms of varied nutritive habit, including a wide variety of protozoans and algae). Each of the kingdoms in turn includes many phyla, classes, orders, and lower categories reviewed in the preceding chapters. Does this survey offer a basis for generalizations on evolution? In this chapter we attempt to derive such evolutionary generalizations.

RETROGRESSIVE EVOLUTION

Evolution is not always "upward" or "progressive." Many examples are known of the evolution of simpler or more degenerate types from originally complex types. Thus the fungi may have evolved from algae by the loss of chlorophyll. Grasses have evolved from lilylike ancestors by simplification of parts, especially of flowers. Mistletoe, an angiosperm parasitic on trees, has undoubtedly evolved from free-living ancestors.

Similarly, in the animal kingdom, many examples of retrogressive or degenerative evolution are known. The development of sexual reproduction and its great evolutionary importance was emphasized in Chapter 17. The Rotifera were undoubtedly evolved from bisexual ancestors, and some species are still bisexual. Nonetheless there are species in which males are unknown. These species still reproduce sexually, for the ova develop parthenogenetically, but the major advantages of sexual reproduction are lost. The same is true of some lizards. The development of the parasitic habit almost always involves degenerative evolution. The tapeworms are an extreme example. Although derived from free-living flatworms, with well-developed digestive, nervous, reproductive, muscular, and other systems, the tapeworm is reduced substantially to an absorptive integument containing gonads and a few rudimentary organs. *Sacculina* and *Enteroxenos* are even more extreme examples (Figure 22–1). While such degenerative evolution is characteristic of parasites, it is by no means confined to them. As already pointed out, the class Archiannelida most probably derived from polychaete ancestors by a process of simplification and loss of parts (and a good case can be made for its inclusion in the class Polychaeta as a subclass or even as an order). Such evidence shows that evolution can be retrogressive as well as progressive. In every case, however, retrogressive evolution is a specialization for an ecological niche or adaptive zone.

ORIGIN OF NEW GROUPS FROM PRIMITIVE ANCESTORS

A second very important generalization relates to the form of the tree of life. It is more properly a shrub than a tree, for new groups do not arise from the most-advanced and specialized members of their parent groups but from the primitive, unspecialized ones. Thus the primitive flagellates have given rise to many additional plant and animal groups, whereas the more specialized protozoan and algal groups are generally terminal. Again, if indeed the Hemichordata and Chordata were derived from echinoderms, the more advanced phyla must certainly have arisen from the ancestral Dipleurula in a very primitive stage before radial symmetry developed. It was not the highly advanced teleost fishes that ventured onto the land to become amphibians, but the more primitive rhipidistians. Similarly, the first reptiles took origin among the most primitive amphibians.

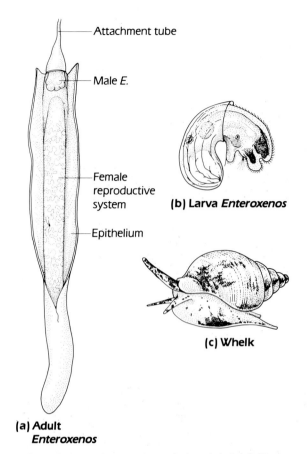

Attachment tube

Male *E.*

Female
reproductive
system

(b) Larva *Enteroxenos*

Epithelium

(c) Whelk

**(a) Adult
*Enteroxenos***

Figure 22–1.   Retrogressive evolution. (a) Adult Enteroxenos, *a parasite of sea cucumbers that is so degenerate that
it was recognizable as a snail only after the discovery of
(b) its larva, which is a typical snail larva with a lightly
coiled shell. (c) A whelk, which is a typical marine snail.
Contrast this with* Enteroxenos. *Note the following labels:
attachment tube, by which the female* E., *attaches to the host,
usually to the host's esophagus; the male* E., *a parasite within
the female; the epithelium, laid back to expose the internal
organs; and the reproductive system of the female, comprising
ovarian tubules, oviducts, and uterus.*

One more qualification relative to the shape of the phylogenetic tree is
necessary for plants. Because of the phenomenon of allopolyploidy, hybridization
of related species may result in new species. Hence branches may form, then
fuse again, so that a network results. This phenomenon is of great importance
for plants, less so for animals.

RATES OF EVOLUTION

Rates of evolution have varied greatly among different kinds of organisms and at different times in the same line of descent, so it is difficult to generalize about them. For instance, Simpson estimated the average duration of genera of clams as 78 million years, versus only 8 million years for mammalian carnivores; Westoll demonstrated that 90 percent of the morphological change in lungfishes occurred during the first third of their 350 million year history (Figure 22–2). Environmentally stable, offshore, shallow marine habitats have

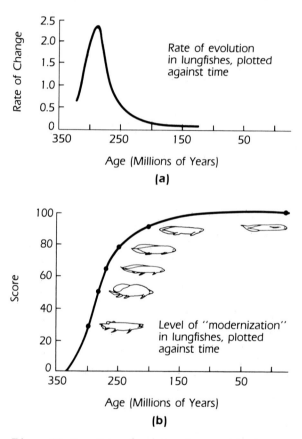

Figure 22–2. *Rates of evolution documented in the more than 300-million-year history of the lungfishes (order Dipnoi). (a) Early in their history, lungfishes evolved rapidly, then slowed drastically. (b) The morphology of lungfishes changed progressively until about 175 million years ago, by which time lungfishes had a very modern appearance. Subsequent changes were very minor.*

been the site of more rapid evolution than the more stressful intertidal and near-shore habitats. Two residents of the latter, the brachiopod *Lingula* and the horseshoe crab *Limulus*, are among the oldest living fossils. Although the intertidal zone undergoes severe stress, it has existed in much its present form for long ages, so that its short-term instability is superimposed upon a long-term stability.

In general, periods of major geological change, with attendant climatic disruption and rapid change of selective forces, have been times of rapid evolution. Random mutation proceeds as usual and presents a great array of types to the test of natural selection. In the presence of new and varied environments, selection favors those types capable of exploiting the new environments, and rapid adaptive radiation results. Thus the vertebrates invaded the land at the end of the Devonian, when uplift of the earth's surface caused the drying of many low-lying ponds; and dinosaurs and mammals appeared during the Triassic, following times of glaciation and aridity. A situation favoring particularly rapid evolution occurs when an entirely new mode of life is adopted for the first time, thus evoking a major adaptive shift, such as when plants and animals invaded the land for the first time. Rapid diversification follows, as new niches are occupied without competition. Following the initial adaptive radiation, many of the experimental types are eliminated by selection, which favors a moderate number of efficient adaptive types. Thus, in the Triassic, numerous and diverse types of archosaurs flourished briefly; then a few of them evolved into the highly successful reptiles that ruled the world for the rest of the Mesozoic Era.

At the beginning of such a major adaptive transition, the organisms may be very imperfectly adapted; hence strong selection pressure favors rapid change, and only those that can respond will leave descendants. Survival will be greatly facilitated if, in the earlier environment, characteristics developed that are *preadaptive* to the later habitat. Thus many Devonian fishes developed lungs, which were adaptive to life in stagnant, drying ponds. This preadaptation later speeded the adaptation to life on land. Simpson has called this process "quantum evolution," as it involves sudden change from one "adaptive orbit" to another. He concedes that the analogy to physics is incomplete, because quantum events in physics are very small, while quantum evolution is on a large scale. Perhaps *macroadaptive* evolution would be a better term.

PUNCTUATED EQUILIBRIA

The preceding discussion suggests great variability of rates of evolution, yet Darwin thought in terms of slow, steady change; and the idea that such *phyletic gradualism* is the norm, with examples such as the above being deviants, has been very persistent. Nonetheless, the fossil record often shows abrupt intrusion of new groups, with no clear connections to their predecessors. The incompleteness of the fossil record has usually been used to explain such intrusions. N. Eldredge and S.J. Gould, however, maintain that the gaps in the fossil record are real and that they result from the normal mode of speciation, a pattern of evolution they call *punctuated equilibrium*. Their proposal is that species (and higher groups) remain stable over long periods of time (equilibrium, or *stasis*), then they change rapidly (punctuation) in small, peripherally isolated populations (as required by Mayr's theory of allopatric speciation). The new species may then spread back into the range of the parent species. Another long period of stasis follows.

The Snails of Lake Turkana

Perhaps the best example of this proposal is P.G. Williamson's study of snails at Lake Turkana, Kenya (Figure 22–3). Williamson followed the snail fossils through a column of sediment some 400 meters deep, deposited over many millions of years. In general a narrow range of variability was maintained, but twice during this time the water level of the lake fell severely. Populations of snails were isolated in small ponds and were subjected to severe environmental stress. The snails responded with a great increase in variability, and new species emerged over periods of 5,000 to 50,000 years, a mere moment of geological time and very short periods compared to the long periods of stasis. In this example, when the water level again rose and connections with other ponds were reestablished, the original species reinvaded Lake Turkana, and the new species soon became extinct. In other cases, newly originated species may be more successful.

Evaluation of Punctuated Equilibria

The more enthusiastic advocates of punctuated equilibria believe that the concept is so radical a departure from earlier evolutionary thinking that it constitutes a totally new evolutionary synthesis. Others, however, point out that the evolutionary synthesis of the middle third of this century brought together aspects of all branches of biology, including ecology, genetics, and paleontology; that it included Simpson's profound analysis of rates of evolution, which demonstrated both stasis and rapid origin of new groups as extremes of a continuous spectrum of evolutionary rates; and that the concept of punctuated equilibria can and should be integrated into that great evolutionary synthesis. Further, although gradualism is usually considered to be central to Darwin's thought, he described (but did not name) punctuated equilibrium as a typical process of evolution in the fourth edition of the *Origin of Species*. Most evolutionary biologists are prepared to acknowledge that punctuated equilibrium is an important phenomenon, even if somewhat less so than its more enthusiastic advocates claim. And population geneticists, who have labored mainly to clarify the genetic basis of evolutionary change, may now have to give greater attention to the problem of evolutionary stasis.

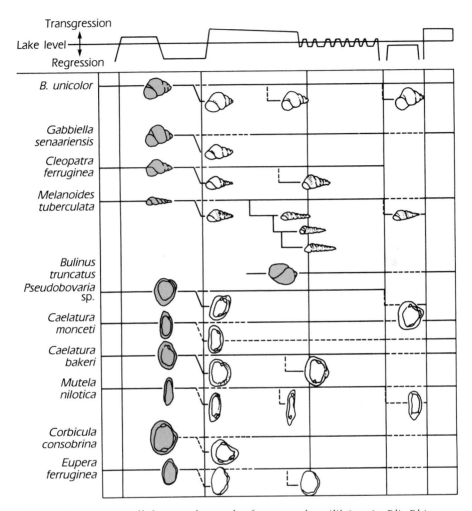

Figure 22–3. *A well-documented example of punctuated equilibrium in Plio-Pleistocene snails of Lake Turkana, studied by P.G. Williamson. Twice, a drop in the lake level caused isolation of populations and rapid speciation.*

TRENDS IN SIZE

A very common trend in evolution, sometimes called Cope's law, is toward the increasing size of individuals. The original studies of the phenomenon were made on mammals, but comparable studies have shown the same tendency in many other vertebrates as well as in invertebrates and plants. A review of the paleontology of almost any group shows that its largest representatives are not its earliest ones, though not necessarily its latest ones either. S.M. Stanley regards the trend as being *away* from small size, at which evolutionary transitions usually occur, rather than *toward* large size. N.D. Newell has pointed out that species now living are the largest known representatives of the vertebrates, crustaceans, echinoderms, pelecypods, gastropods, cephalopods, and annelids. Yet the tendency toward size increase has by no means been universal. As already mentioned, herbs and shrubs have only recently arisen from trees and other large plants. D.A. Hooijer has pointed out that progressive size decrease has been characteristic of many vertebrate groups during the Quaternary period, which is now in progress.

ANAGENESIS, CLADOGENESIS, AND TAXONOMIC METHODS

When Devonian fishes first crawled out of drying ponds to seek more water, they had either to exploit the terrestrial environment through mutations adaptive to air breathing and to locomotion by crawling, or they had to revert to the water. In fact they adapted and became amphibians. Such progressive evolution, which, if favored by strong selection pressure, may transform whole populations, is called *anagenesis*. Commonly, however, a species is broken up into a series of subspecies, and peripheral populations may be isolated. An isolated population may form a new species, whereas the main population is changed much less. Such splitting to form sister species is called *cladogenesis*. These concepts are central to the development of several competing methodologies of taxonomy. Linnaeus designed his taxonomic system purely as a systematic catalog of natural entities. The categories of classification were conceived as closed sets, an idea that is not readily adaptable to evolutionary thinking. Darwin introduced the concept that the taxonomic categories represent different levels of (genetic) relationship. Members of the same species are so closely related that they are potentially interbreeding. As we ascend the taxonomic scale—genus, family, order, class—the relationship becomes progressively more distant, until finally distant members of the same phylum (e.g., fish and mammal) share only the most basic characters.

Evolutionary Systematics

For nearly a century after Darwin, taxonomists worked without clearly defined methodology. They tried to evaluate similarities and differences among organisms, with homology as the most important guide to relationships and with the goal of a classification in maximal agreement with phylogeny. They placed more or less equal importance on anagenesis and cladogenesis. In time, this fine old science became stultified (Chapter 6). It was revitalized in the thirties and forties by participation in the then-new evolutionary synthesis. Thus the traditional data from comparative morphology and embryology were coordinated with data from genetics, ecology, and the many other facets of biology to form the new evolutionary synthesis. The resulting taxonomy came to be known as *evolutionary systematics*. E. Mayr has been one of the most effective advocates of evolutionary systematics in North America.

Phylogenetic Systematics

Even in the new systematics, however, subjective judgment of the investigator remained an important factor, and other systems have attempted to eliminate it. In 1950 the late W. Hennig published a book in which he advocated *cladism,* or *phylogenetic systematics.* He maintained that two-by-two branching (each branch being a *clade*) is the most important evolutionary process and that the points of branching can be identified by careful comparison of the organisms in a series. The distinction between primitive and derived characters is important for cladism. Shared primitive characters (*synplesiomorphies*) do not indicate close relationship because their genetic basis may be inherited by any or all of a descendant group, hence they indicate fairly remote relationship. Shared derived characters (*synapomorphies*), however, indicate close relationships of the groups compared because they shared a recent ancestor that had the derived character. Because new groups always result from splitting of an antecedent group, cladists propose that any group always has a *sister group* of the same taxonomic rank, a direct consequence of cladogenesis. The more enthusiastic cladists claim that their method is *the* evolutionary taxonomy.

Cladists, who rely on derived morphological characters, often distrust the fossil record. C. Patterson has gone so far as to state that fossils should be fitted into phylogenetic schemes only after the relationships of living forms have been determined! The cladists' results are presented on branching diagrams called *cladograms* (Figures 22–4 and 22–5), in which closeness of relationship between pairs of taxa is expressed by points of branching, without specifying ancestors. The American Museum of Natural History in New York and the British Museum (Natural History) in London are two important institutions in which cladism has been championed.

Numerical Taxonomy

Yet another taxonomic method developed during the 1960s. *Numerical,* or *phenetic,* taxonomy depends on comparison of a large number of characters (typically 50 to 100) among a series of taxa to be classified. In the phenetic jargon, the series under study is called a series of operational taxonomic units, or OTUs. A numerical value is given to each character, and each is usually given equal weight. The data are then analyzed by a suitable cluster analysis, usually by computer. Because the method tends to eliminate the subjective biases of the investigator, pheneticists believe that the analysis will yield natural phyletic associations. The results of the analysis are often displayed

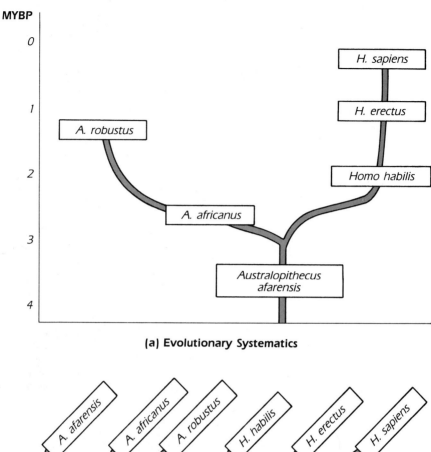

(a) Evolutionary Systematics

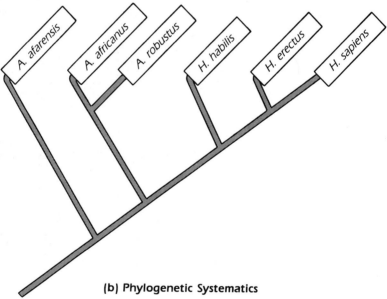

(b) Phylogenetic Systematics

Figure 22–4. The relationships of australopithecines and humans as expressed by two different methods of phylogenetic analysis. (a) A classic phylogeny of evolutionary systematics, with both ancestor-descendant relationships and time of separation specified. (b) The same relationships shown as a cladogram without the implication that any species known is actually ancestral to any other and without time scale.

524

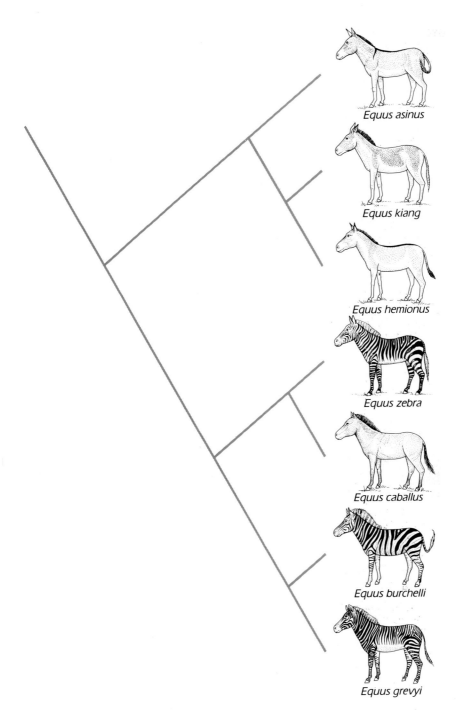

Figure 22–5. *Cladogram showing relationships among living members of the genus* Equus. *One conclusion of this interesting but controversial cladistic analysis is that horses,* Equus caballus, *are more closely related to zebras than to asses.*
Data from a study by D.K. Bennett.

on a hierarchical diagram called a *dendrogram* or a *phenogram* (Figure 22–6). P.H.A. Sneath and R.R. Sokal were important in the rise of numerical taxonomy.

All three taxonomic schools agree that a major function of taxonomy is to express evolutionary relationships as clearly as the taxonomic system will permit. In part their disagreements on taxonomic methods are based upon disagreement on evolutionary processes (for example, the relative roles of anagenesis and cladogenesis). All three schools have been severely criticized by the others; nonetheless, they agree on the major features of the classification of the living world. All have benefited from exchange of ideas. Our perceptions of evolution and especially of macroevolution are so closely allied to taxonomic concepts that we cannot avoid these problems here. Mayr points out that classification is a complex process and that the methods of the three schools may find their proper applications at different stages of the process. The final resolution of these controversies is still in the future.

COMPLEXITY AND EFFICIENCY

The general progress of evolution has involved the development of new organ systems and increasing complexity. Yet development of increased efficiency often involves reduction in number and complexity of parts. This consideration may account for vestigial organs; it applies equally to actively functional structures. For example, in fishes the teeth are numerous and usually indefinitely replaceable. They are less numerous in amphibians and reptiles, still less so in mammals, where they reach their maximum degree of specialization and efficiency. Much the same is true of the vertebrae and of the bones of the skull. This tendency can also be exemplified in plants, for example by the reduction of numbers of stamens in specialized plants.

Increased efficiency is also often obtained by fusion of originally separate parts. Thus the sacrum of mammals is formed by the fusion of three to five originally separate vertebrae, thereby making a much stronger attachment of the hind limb to the vertebral column. Another good example is afforded by the pectoralis muscles, which in tetrapods arise near the midline of the chest region and attach to the humerus. Slips of muscles from many adjacent body segments join to form these muscles. Among plants, the corollas of flowers such as the cucurbits and petunias are formed by the fusion of originally separate petals.

DOLLO'S LAW

Many times during the long history of life, advanced organisms have returned to ancestral habitats and modes of life. This return gives selective

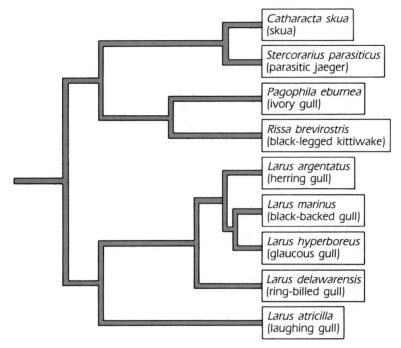

Figure 22–6. *Phenogram of relationships among some gulls and their relatives.*
Data from a study by G.D. Schnell.

value to adaptations similar to those of the ancestral species and raises the question whether evolution might be reversible. Study of such cases always shows that a gross similarity between ancestral and descended structures is achieved without any genuine reversal at all. Thus, many reptiles (turtles, mosasaurs, and plesiosaurs) and mammals (seals, otters, and whales) have reverted to an aquatic mode of life. They have assumed a generally streamlined, fishlike form, and the limbs have become shortened, webbed, and finlike. Yet the skeleton of such flippers is always distinctly that of the class to which the animal belongs rather than that of a fish fin. Similarly, many angiosperms have returned to the water and assumed algalike appearances, but their morphology is still that of flowering, vascular plants. The evidence indicates that major evolutionary steps, once taken, are never reversed. This generalization is known as Dollo's law (Figure 22–7). It might even be expected a priori, for major evolutionary steps are compounded of many smaller steps, each preserved by natural selection. That such a sequence, occurring by chance once, should by chance be exactly reversed would be most extraordinary. If not impossible, it is at least most improbable for whole organisms. Attempts to apply Dollo's

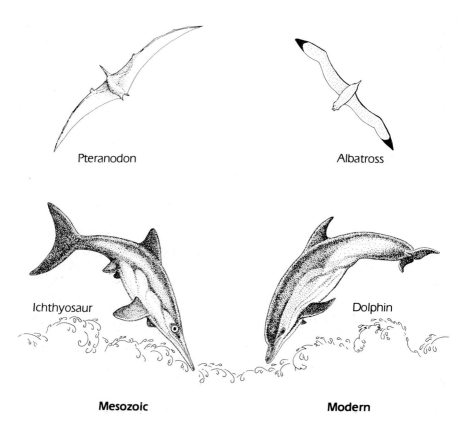

Pteranodon

Albatross

Ichthyosaur

Dolphin

Mesozoic

Modern

Figure 22–7. Dollo's law, the irreversibility of evolution. Even though body form may be similar, anatomical details always differ. The animals on the left are Mesozoic reptiles: an ichthyosaur swimming and a pterosaur (Pteranodon) flying. The animals on the right are mammal (dolphin) swimming and a bird (albatross) flying.

law to individual characters have failed, for these are, indeed, reversible by mutation and selection.

LINKS—MISSING AND MANIFEST

During early post-Darwinian times, much controversy revolved around "missing links." Those who favored evolution searched for these connections between species and higher groups, confident that their discovery would provide decisive proof of evolution. Opponents contended that such links were missing because they had never existed and hence that all of the indirect evidences of evolution had led to an erroneous conclusion. Speciation is the normal mode of evolution; and, if the theory of punctuated equilibria is correct, then we would not expect these evanescent moments of transition to be caught in the fragmentary samples of the great span of geological time. Further, each fossil is generally incomplete, so that identification down to species is often not possible. Hence, even if generic (or higher level) data strongly suggest links between groups, the actual sequence of transitional species may be impossible to discover. Just as light microscopy cannot reveal the ultrastructure of protoplasm, which remained unknown until the advent of electron microscopy, so taxonomy at the levels of families and genera cannot reveal links at the species level.

Links between Phyla

At the highest level, a point of definition of any phylum is that its structural plan differs fundamentally from those of all other phyla and that it is difficult to derive from any other. Thus the very definition of *phylum* militates against links among them. So far as the fossil record goes, no evidence of links between phyla exists, unless the Calcichordata be accepted as a link between echinoderms and chordates, a proposition that has not gained much support (Figure 22–8). Nonetheless, data based on living organisms do suggest links between several pairs of phyla. Some of these are supported by strong evidence, and it is quite probable that the links are valid. Thus, considerable assurance exists that the vascular plants were derived from green algae (Chapters 17 and 18); that flagellate protozoans gave rise to the Porifera (Chapter 19); that coelenterates and ctenophores arose from a common ancestor, probably a planuloid organism comparable to a free-living mesozoan (Chapter 19); that flatworms gave rise to nemerteans or that both arose from a common ancestor (Chapter 19), that annelids, onychophorans, and arthropods constitute one related series (which a few zoologists prefer to unite in a single phylum, the Articulata; Chapter 19); and, finally, that the echinoderms, hemichordates, and chordates constitute a cluster of three phyla derived from a common ancestor, usually visualized as a dipleurula (Chapter 20).

Rather less secure lines of evidence link several more phyla, all of which we discussed in Chapter 19. The first is the aschelminth group. L.H. Hyman defined a phylum Aschelminthes, comprising an array of animals, mostly wormlike and bilaterally symmetrical, with an unsegmented body clothed in a cuticle. They have a straight digestive tract that has no definite muscular wall and terminates in an anus, usually located at the posterior end. She defined six classes in the phylum: Nematoda, Nematomorpha, Rotifera, Gastrotricha, Echinodera, and Priapulida. She conceded that these classes are more distantly related than is typical for classes of a phylum and that zoologists who were unable to accept so loose a relationship were justified in treating them as separate phyla. Most zoologists have taken this liberty, but the fact

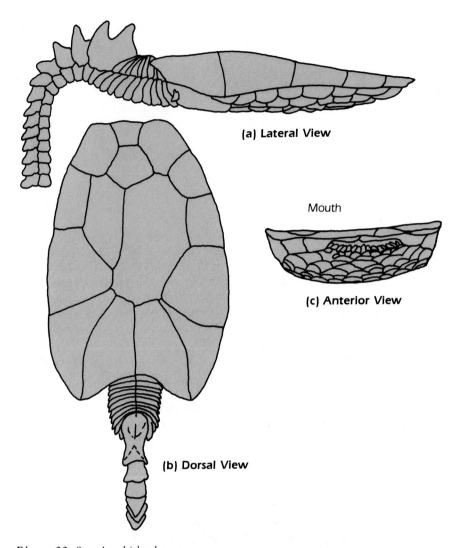

(a) Lateral View

Mouth

(c) Anterior View

(b) Dorsal View

Figure 22–8.   A calcichordate.
Redrawn from Jefferies, by courtesy of the Trustees of the British Museum (Natural History).

that so astute and profound a zoologist as Hyman saw them as the classes of one phylum provides an inference of links among them.

Embryological characters of their trochophore larvae suggest links between molluscs and annelids. This link was formerly weakened by the contrast between the segmental organization of annelids and the striking lack of segmentation among molluscs. The discovery of living monoplacophorans (*Neopilina,*

Figure 22–9), however, which are clearly segmented molluscs, has suggested that the primitive molluscs may have been segmented. Hence, the evidence of rather remote links between molluscs and annelids has been strengthened. Finally, the lophophore bearers (bryozoans, phoronids, and brachiopods) share enough characteristics to suggest relationship, even though their differences are great. To this group, C. Nielsen would add the entoprocts, although one of the present authors (E.O.D.) has adduced evidence that the relationship is, at best, a very remote one.

In summary, then, although paleontology has almost nothing to offer on links between phyla, neontology affirms, with varying degrees of assurance, links among more than two-thirds of the phyla of animals, not a bad record considering that the definition of *phylum* and the concept of taxonomic groups as closed sets militate against even so modest a degree of success.

Links between Classes

At a somewhat lower level, a few good links are known among major groups. Bactritid molluscs appear to lead into the ammonites, and the edrioasteroids (Figure 22–10) may be intermediate between globular echinoderms and asteroids. The best examples, however, are among the vertebrates. There is a strong case for the origin of the earliest amphibians from rhipidistian fishes, based upon common structural features of limbs, skulls, scales, vertebrae, and labyrinthodont teeth (Chapter 20). The exact genera involved in the transition cannot be specified, but *Eusthenopteron* and *Sauripterus* cannot be far off the line of transition. Similarly, the transition from primitive amphibians to captorhinomorph reptiles is indicated as strongly as possible in the absence of identification of the transitional species and genera. In fact, the same

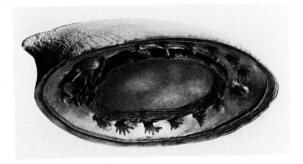

Figure 22–9. Neopilina, *a living fossil among molluscs. Note the paired, segmentally arranged gills. Internal structures are similarly arranged, in contrast to all modern molluscs.* Courtesy of the National Museum of Natural Sciences, National Museums of Canada.

Figure 22–10.   Edrioaster bigsbyi.
Courtesy of the National Museum of Natural Sciences,
National Museums of Canada.

organism has been classed as an amphibian by some paleontologists and as a reptile by others. The same authority has even classified one organism differently at different times, a fact that strongly suggests transitional status. *Archaeopteryx* appears to be a good connecting link between birds and reptiles. The skeleton of *Archaeopteryx* is easily confused with that of *Compsognathus,* a small (turkey-sized) coelurosaurian dinosaur (Figure 22–11). In fact, a few specimens that had been assigned to the latter had to be transferred to the former when it was discovered that impressions of feathers had been overlooked on the fossil-bearing stones. Again, this speaks strongly for transitional status.

Perhaps the best-documented transition between classes is that of the origin of the mammals from cynodont reptiles. During a first great reptilian radiation in Late Pennsylvanian and Early Permian time, the ancestral capto-rhinomorphs gave rise to three lines of descent: more-evolved captorhinomorphs, sauropsids, and the mammal-like synapsids. In a later radiation, the sauropsids produced dinosaurs, typical reptiles, and birds. The diversification of the synapsids during the Late Permian and the Triassic included an array of mammal-like reptiles, including the cynodonts, the group of reptiles within which the mammals actually arose. Mammals are commonly defined on the basis of such characters as endothermy, nursing of the young, and possession of hair, none of which is fossilized (with rare exceptions in the case of hair). Paleontologists must, therefore, rely on skeletal characteristics to diagnose the classes. Typical reptiles have a lower jaw made up of as many as seven separate bones, with the joint to the skull through the quadrate and articular bones.

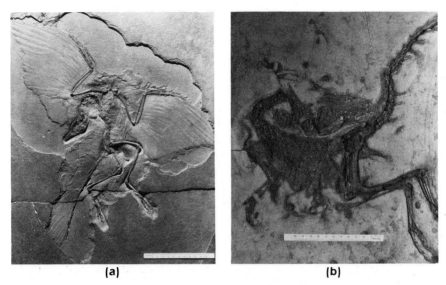

(a) **(b)**

Figure 22–11. *(a)* Archaeopteryx *and (b)* Compsognathus. *The former has been mistaken for the latter when feather impressions are overlooked.*
Courtesy of Dr. John H. Ostrom, Yale University.

In typical mammals, one of these bones, the dentary, makes up the entire mandible, and the quadrate-articular pair has been reduced to the incus and malleus of the auditory ossicles. A variety of stages in this transition is known in many fossil genera, and it is controversial among specialists which of these genera should be considered reptiles and which early mammals. A genus often named as the earliest mammal is *Morganucodon* (Figure 22–12).

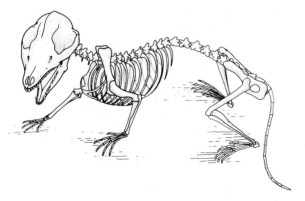

Figure 22–12. *A reconstruction of a skeleton of a mor-ganucodont, a very early mammal.*

Links between Orders and Families

At the ordinal and familial levels, one of the best examples is provided by the Perissodactyla, hoofed mammals with an odd number of toes. *Hyracotherium* is an Early Eocene genus, which we presented in Chapter 5 as the earliest member of the family Equidae (horses). In fact, *Hyracotherium* is a nearly ideal ancestor for *all* of the families of perissodactyls. Eocene perissodactyls are usually partitioned into families according to their descendants, but if the descendants were not known, they would probably all be assigned to one family. Further, that family could not be easily separated from the condylarths, unspecialized mammals that were intermediate between herbivores and carnivores. An order Condylarthra is recognized for them. It is probable that early condylarths, perhaps *Desmatoclaenus* (Figure 22–13), may have given rise to *Hyracotherium* and through the latter to the entire order Perissodactyla.

Links between Genera and Species

Finally, are there good links at generic and specific levels? As discussed here, the relatively poor resolution of the fossil record makes it difficult to identify such links. The actual transitions take place at the level of species or even of population. The best opportunity to document the transition from one species to another is in those cases in which populations of fossils rather than single specimens are available from successive stratigraphic horizons that encompass significant intervals of time. Because of the enormous effort involved, few such studies have been made. Nonetheless, there are some persuasive cases.

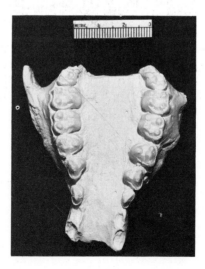

Figure 22–13. Tetraclaenodon, *a condylarth. Photo of a cast of the upper jaw and teeth.*

Among the ammonites, generic transitions are clear, and specific links are highly probable. Excellent examples also occur among foraminiferans and some other microorganisms. P.D. Gingerich has documented numerous specific transitions among Early Eocene primates and condylarths from Wyoming. Williamson's snails present a whole series of fully documented transitions from species to species.

The understanding of the paleontology of the family Hominidae has advanced greatly within the past few years, and we can now make a strong case for links at both generic and specific levels. *Australopithecus afarensis* appears to have arisen in mid- to late Pliocene time from some species of dryopithecine. This is the weakest link in the series at the current level of knowledge. The adaptive radiation of *A. afarensis* included *A. africanus* and another species that is sufficiently intermediate between genera that it is classed as *A. habilis* by some paleoanthropologists and as *Homo habilis* by others. This species blended into *H. erectus,* which in turn blended into *H. sapiens,* the only organism that studies all of the others, the only organism that knows that it has a history. As a morphological series, this example is undisputed among paleoanthropologists; whether it is also a phylogenetic series is very much disputed.

Although evidence of interspecific links is rare in the fossil record, there is much evidence from living organisms that cannot be readily interpreted in any other frame of reference. One example concerns transitions in the genus *Drosophila* (fruit flies). Observers can identify very fine details of structure in the giant chromosomes of these flies (see Chapters 8 and 12). By comparing overlapping inversions (Chapter 8), Dobzhansky and his collaborators were able to specify the exact sequence in which three species and a large number of subspecies were formed.

A second example concerns allopolyploidy, the multiplication of whole sets of chromosomes so that organisms are formed with multiples of the basic set of chromosomes (Chapter 13). Allopolyploidy is a very common phenomenon among plants. Wheat, for example, includes species with 14, 28, or 42 chromosomes. In many instances, as is the case with wheat, polyploidy has resulted in the formation of new species within a genus. In others, new genera are formed. As breeders can duplicate the hand of nature by synthesizing such species and genera, there can be no doubt how the natural groups were formed.

The assertion that links between groups never existed is a form of the universal negative. As such, it is inherently unprovable. The demonstration of even one such link, however, demolishes that argument against evolution. We might be disappointed that well over a century of post-Darwinian paleontology has not produced a greater number and variety of clearly demonstrated links between groups. Nonetheless, the examples above were drawn from many different groups at several taxonomic levels. They also represent several levels of assurance, from tentative suggestion to near certainty or even to absolute certainty in the last two examples. Collectively, they provide very strong assurance of the validity of evolution.

SIGNIFICANCE OF EXTINCTION

Some closing remarks should be made upon the subject of extinction, the fate of most species since the origin of life. Extinction does not imply failure any more than does death of an individual. Dinosaurs were highly successful for 140 million years, yet they are now extinct. Extinction is an important, though poorly understood biological phenomenon that occurs at the level of populations, species, or even orders and higher levels at times of crisis in the earth's history. Yet the major adaptive types—phyla and classes—rarely become extinct. Extinction of species may result from competition between two species for the same resources, but this cannot explain wholesale extinctions of unrelated groups. The best that can be said at present is that profound changes in the surface of the earth have at times caused climatic changes that have wreaked havoc with complex food chains on land and in the seas. For instance, at the same time that the dinosaurs were eliminated, there were also extinctions of microscopic foraminiferans, bivalves, gastropods, cephalopods, various reptiles in the seas, and even marsupials on land. During that same period, plants, turtles, crocodiles, fishes, birds, and placental mammals were comparatively unaffected, a fact that is not understood.

There is important evidence that, at the Cretaceous-Cenozoic boundary some 65 million years ago, an exceptionally large meteorite, perhaps as large as ten kilometers in diameter, struck the earth. Of course, all life would be destroyed in the area of impact of such a strike, but extinction may have been much more extensive because the strike raised a dust cloud that obscured the sun worldwide for several months, thus seriously inhibiting photosynthesis and disrupting the food chain. Because it came too late, however, this catastrophe cannot account for the extinction of the dinosaurs.

There are two quite distinct kinds of extinction, one resulting when a new group arises from an old one, as mammals from mammal-like reptiles; the other resulting from the termination of a lineage, as when trilobites or dinosaurs disappeared without leaving descendants. The two should be distinguished, for their significance is quite different.

L. Van Valen, J.P. Grime, and G.J. Vermeij have noted that organisms differ widely in competitive ability and susceptibility to extinction. Organisms that are tolerant of stressful, low quality habitats, where temperature, moisture, and salinity fluctuate widely, tend to grow slowly and compete poorly; such *stress-tolerant* organisms tend to be relatively resistant to extinction. Lungfishes and horseshoe crabs are two good examples. In contrast, *biotically competent* species are highly successful competitors for favorable habitats to which they are finely attuned; such organisms, exemplified by the dinosaurs, are relatively more susceptible to extinction. Mayr believes that biotically competent species are characterized by inflexibility of the genome that leaves them vulnerable to extinction when conditions change. In view of the near universality of genetic polymorphism demonstrated by electrophoresis, to accept this explanation for the vulnerability of biotically competent species is difficult. D.M. Raup has questioned whether extinction results from "bad genes" or "bad luck"— random unfavorable events for which no particular cause should be sought.

SUMMARY

We have derived some evolutionary generalizations from our review of the history of life. First, *evolution is not always progressive.* It has obviously been retrogressive in the case of parasitic groups, and some free-living groups (grasses, for example) are simplified by comparison to their probable ancestors.

The form of the evolutionary "tree" is more that of a *shrub,* for new groups arise not from the advanced members of their predecessors but from primitive members. Thus reptiles arose from amphibians that had themselves only just emerged from crossopterygian fishes, whereas advanced amphibians have given rise to no further groups.

Rates of evolution have been highly *variable.* Some groups (e.g., brachiopods) have remained stable over long reaches of time; others have differentiated rapidly (e.g., mammals in the Tertiary). Periods of geological stability favor evolutionary stability, and periods of geological upheaval provide the selective pressure for rapid change. When plants or animals move from one adaptive orbit to another—as when mainland plants colonized Hawaii in the Pliocene or when crossopterygian fishes first invaded the land in the Devonian—the transitions may be very rapid in a geological sense. This has been called *quantum,* or *macroadaptive, evolution.*

Whether rates of evolution are generally rather uniform (*gradualism*) or whether *punctuated equilibria* are the rule is currently much disputed. According to the latter theory, species and higher groups remain stable over long periods of time (*equilibrium*), then they diversify rapidly during geologically short periods of 5,000 to 50,000 years (*punctuation*). The best example is that of the snails of Lake Turkana, but many other examples are known. Thus, the question is not whether punctuated equilibria occur, but how general they are and whether they can be absorbed into the modern evolutionary synthesis.

A supposed evolutionary tendency toward ever-increasing size has been called *Cope's law.* It is true that the largest species known for many groups are now living, but some groups have specialized for the exploitation of small size, and many groups of vertebrates have decreased in size through the Quaternary period. Thus Cope's "law" is really only a rule of variable applicability.

In a general way, evolution has been marked by increasing complexity, yet efficiency has often been gained by simplification. Examples are the derivation of the mammalian skull from that of the reptiles or the reduction of the number of stamens in advanced as compared to primitive flowers.

Often, organisms have reverted to an ancestral habitat, as the return of the plesiosaurs to the seas, and this has raised the question of whether evolution is reversible. Plesiosaurs simulated the streamlined form of fishes while retaining distinctively reptilian morphology. Comparable facts apply to all other cases studied, and so it was long ago concluded that evolution, at least in its major steps, is not reversible—a conclusion that is called *Dollo's law*.

Anagenesis, progressive evolution favored by strong selection pressure, may transform whole populations. In *cladogenesis*, a species or higher group may split to form two sister groups of equal taxonomic rank. These concepts are basic to different taxonomic methodologies. In the early years of the modern synthesis, taxonomists who gave equal importance to anagenesis and cladogenesis tried to synthesize data from comparative morphology and embryology, genetics, ecology, and other biological sciences to form an *evolutionary systematics*.

Cladists emphasize cladogenesis, and they maintain that for every group there is a sister group of equal rank. They look for shared derived characters (synapomorphies) as evidence of such two-by-two splitting. The *numerical*, or *phenetic*, school of taxonomy depends on computer analysis of a large number of variable characters. In this way, numerical taxonomists expect to eliminate the subjective biases of the investigator. Proponents of the three schools have been sharply critical of each other, yet they agree on the major features of classification of the world of life, and all have contributed to the understanding of evolution.

Much evolutionary controversy has revolved around "missing links," transitional forms between species and higher groups that would prove their evolutionary origin. Opponents of evolution maintained that they were missing because they had never existed; proponents confidently worked to find them. Links between phyla are almost unknown in the fossil record, but neontology gives evidence of extensive links among the phyla. There is good fossil evidence of links among various classes and orders of invertebrates; but the best examples at this level are among the vertebrates—for instance, the transitions from crossopterygian to amphibian, amphibian to reptile, and reptile to bird and mammal.

At ordinal and family levels, the best examples are in the perissodactyls, including the horse family. Generic and specific transitions are well demonstrated in various Eocene mammals and in the snails of Lake Turkana as well as by the numerous examples of *allopolyploidy* in plants and of overlapping inversions in *Drosophila*.

Demonstrated links among groups of plants and animals are less numerous than might be desired, yet they are found at all taxonomic levels and in diverse taxonomic groups. Collectively, they give strong testimony to the fact of evolution.

Although *extinction* has been the fate of the overwhelming majority of species that have ever lived, the concept is not well understood. Extinction may represent the final failure of a group to adapt to changing conditions, as in the case of the passenger pigeon. It may also be the seal of success, as in the case of the cynodont reptiles, which were succeeded by their own descendants, the adaptively superior mammals. Finally, extinction may be simply a matter of bad luck—the occurrence of a disaster that is so severe that no conceivable adaptation could prepare the organism to cope with it.

REFERENCES

Bennett, D.K. 1980. Stripes do not a zebra make. Part I: A cladistic analysis of *Equus*. *Systematic Zoology* 29:272–287.
A clearly reasoned exposition of the cladistic method as applied to an interesting problem.

Ehrlich, P., and Ehrlich, A. 1981. *Extinction: The Causes and Consequences of the Disappearance of Species*. Random House, New York.
A highly readable and well researched study of this important problem.

Eldredge, N., and J. Cracraft. 1980. *Phylogenetic Patterns and the Evolutionary Process*. Columbia University Press, New York.
An exposition of cladistics as seen by two paleontologists.

Gould, S.J., and N. Eldredge. 1977. Punctuated equilibria: The tempo and mode of evolution reconsidered. *Paleobiology* 3:115–151.
A thorough exposition of this subject.

Hennig, W. 1966. *Phylogenetic Systematics*. University of Illinois Press, Urbana. Translated from German and edited by D.D. Davis and R. Zangerl.
The source of cladistic taxonomy.

Mayr, E. 1981. Biological classification: Toward a synthesis of opposing methodologies. *Science* 214:510–516.
A comparison and analysis of phenetic, cladistic, and evolutionary taxonomic methods. Mayr favors the latter, but he advocates the use of some of the methods of the first two.

Olson, E.C. 1981. The problem of missing links: Today and yesterday. *Quarterly Review of Biology* 56:405–442.
A current review of an old problem.

Simpson, G.G. 1953. *The Major Features of Evolution*. Columbia University Press, New York.
After thirty years, this book is still of great importance. The work of a genetically oriented paleontologist, it has, together with an earlier book by the same author, influenced two generations of paleontologists. It includes much material applicable to the present chapter. (Cope, Dollo).

Sneath, P.H.A., and R.R. Sokal. 1973. *Numerical Taxonomy: The Principles and Practice of Numerical Classification*. Freeman, San Francisco.
The keystone work of this branch of taxonomy.

Stanley, S.M. 1973. An explanation for Cope's law. *Evolution* 27:1–26.
Explained in terms of ecology and morphology.

Stanley, S.M. 1981. *The New Evolutionary Timetable: Fossils, Genes, and the Origin of Species*. Basic Books, New York.
An enthusiastic presentation of punctuated equilibria.

Wiley, E.O. 1981. *Phylogenetics, the Theory and Practice of Phylogenetic Systematics*. Wiley (Interscience), New York.
A good exposition from the viewpoint of a vertebrate zoologist.

Evolutionary Perspectives

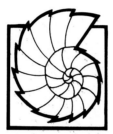

In the preceding chapters, we have defined evolution, explored the forces and processes by which it works, and reviewed its historical course. Now, in the final chapter, we shall try to put it in perspective and to consider its relationship to some contemporary problems.

23

Retrospect and Prospect

We have reviewed the main outlines of the concept of organic evolution as modern scientists have developed it, as well as the evidence and the reasoning that led them to this viewpoint. We have reviewed the fossil record and seen that it is most impressive evidence of evolution, in spite of its grave deficiencies. Finally, we have reviewed the modern attempts to analyze the genetic, cytological, ecological, geographical, and other causative factors in evolution, and we have seen that these complement one another to form a synthetic whole.

DOMINANT THEORIES

The dominant theme of recent work in evolution has been the neo-Darwinian, or modern synthetic, theory, according to which the basic phenomenon of evolution is the slow accumulation of small mutations, the screening out of combinations of these by ecological factors that comprise natural selection to form subspecies, and finally the formation of good species by the same processes, aided by isolating mechanisms that prevent the subspecies (incipient species) from merging with the general population from which they came.

A second theme has been the search for a mechanism for rapid evolution, typified currently by the punctuated equilibria of N. Eldredge and S.J. Gould. They direct attention to the discontinuities of the fossil record, which they explain not by incompleteness but by long periods of stasis punctuated by occasional geologically brief periods of rapid change. They relate the changes to mutations of regulatory genes, reminiscent of the systemic mutations of R.B. Goldschmidt in the early years of the modern synthesis. His viewpoint was determined partly by his conception of the gene as a physiological unit in an integrated chromosome, at a time when most geneticists envisioned the gene in morphological and atomistic terms. The gene is now known to be a specific sequence of nucleotides in the DNA, functioning according to a molecularly determined sequence of processes. Thus, elements of both competing theories have been combined in the modern understanding of the gene. Studies in molecular phylogeny show that slow accumulation of mutations does in fact continue across specific barriers and much further. Speciation by systemic mutation is still undemonstrated, unless polyploidy be included here; but it remains a possibility, even though a less cogent one than before.

It is too early to predict the future of the concept of punctuated equilibria, whether it will replace the modern synthesis as a new paradigm or whether it will be integrated into it. Many geneticists have pointed out that the geologically brief periods of origin of new groups by punctuated equilibrium are nonetheless long on a genetic time scale, long enough for origin by neo-Darwinian processes. In view of this and of the durability that the modern synthesis has shown in the face of decades of exposure to criticism and experimental test, it seems probable that punctuated equilibria may be integrated into the modern synthesis.

Many other accelerating mechanisms are also available. For example, character displacement, acting simultaneously on an array of characters and under strong selective pressure, might simulate systemic mutation, as W.L. Brown has said. Again, selection forces may be very strong, and this should produce rapid changes. For example, both physical and biotic factors in many parts of the world have been profoundly changed during the past 400 years. Many species have become extinct as a result of changed selection pressures, and many others must have undergone great changes adaptive to the new conditions.

Such strong selective forces are probably always acting during major transitions, as from fish to amphibian or from reptile to bird, for the intermediates are perhaps ill-adapted to both modes of life and under strong selection pressure to complete the transition. This may be why such transitions often seem abrupt in the fossil record. For predominant direction of mutation and selection to coincide in any particular case is improbable; but if this should happen occasionally, evolutionary change might be very rapid indeed.

Each of these accelerating mechanisms has probably played a role in evolution, and it is perhaps premature to estimate their relative importance. Collectively, they provide an answer to one of the apparent contradictions of a few years ago, and they help to provide middle ground between evolutionary hypotheses that once seemed to be irreconcilable.

The most significant aspect of current research in evolution is the effort to synthesize data from all aspects of biology and from many of the physical sciences into a meaningful whole. Great advances have been made in all branches of biology under the stimulus of the modern synthesis, and the latter's productivity may be expected to continue well into the future. This continued productivity may result in profound modification of the major evolutionary theories of today, but these theories will have served science well by providing the basis for such fruitful investigations.

HUMANITY AND THE FUTURE

In prospect, what does the future hold? What influence will we have on the future of evolution, and what will be the character of our own future evolution? Obviously, to answer these questions is impossible; but, as speculation on them is always fascinating, let us briefly examine some possibilities.

Plant and Animal Breeding

One of the major ways in which we have influenced evolution is through plant and animal breeding for agricultural and other purposes. The achievements in this field are great. Probably not a single plant now growing in the farms and gardens of the world is the same as when people first cultivated it. Indeed, the very survival of domesticated plants may be due to the protection we have given them. For the plants that thrive best in the wild are often the most difficult to grow under cultivation, and those that thrive best under cultivation, such as corn (Figure 23–1), may quickly die out in competition with wild species. Because of these facts, P.C. Mangelsdorf has suggested that people have domesticated plants with character complexes already unsuited for competition in the wild, so that it may be said that we have rescued them from impending extinction, to the mutual advantage of plant and rescuer.

By artificial selection, agriculturally desirable characters, such as volume of seed production in grain plants, have been accentuated, while other characters, perhaps more important in a wild state, have dwindled. The bearing season of many plants has been much extended. Even the biochemical characters of plants have been altered by artificial selection, for protein and vitamin content of many plants has been significantly increased by selection of favorable breeding stocks. By the same method, the range over which particular kinds of plants can be successfully grown has been greatly extended. The development of new, resistant strains of plants is one of the chief weapons in combating plant diseases, such as wheat rust. Although not generally so described, many of our agricultural productions may properly be described as good subspecies,

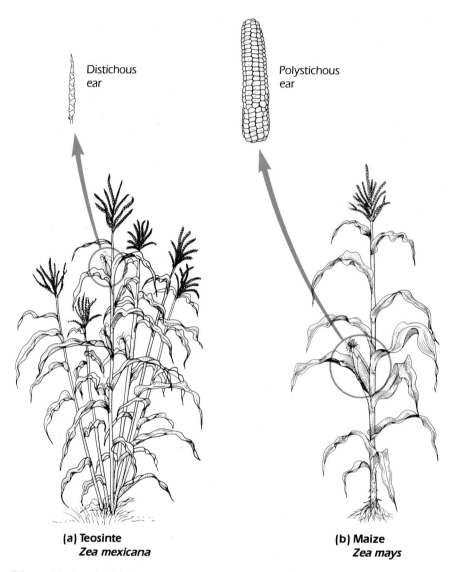

Distichous ear

Polystichous ear

(a) Teosinte
Zea mexicana

(b) Maize
Zea mays

Figure 23–1. (*a*) *Teosinte,* Zea mexicana *is the closest relative of* (*b*) *maize,* Zea mays. *Teosinte produces small, distichous (two-rowed) ears, but maize produces large, polystichous ears.*

not qualitatively different from naturally produced subspecies. But the new wheats and other plants that have been produced by the induction of allopolyploidy are best regarded as artifically produced species, or even genera, again not qualitatively different from those that occur in nature. Indeed, in some cases, such as *Galeopsis tetrahit,* the natural species itself has been resynthesized artificially.

Some of the successes of plant breeding also give rise to dangers. In the late 1960s, varieties of wheat and rice were developed that are especially responsive to favorable conditions of fertilizer and water. These plants form the basis of the much publicized "Green Revolution." They were developed especially for the Third World. Farmers who formerly planted many different varieties of grain now tend to concentrate on these alleged miracle grains. They can produce heavy crops only if the fields are well fertilized and if there is abundant rainfall or irrigation, either of which may fail to happen. There is also a further danger: plant diseases spread easily when a single variety of a plant is widely used (monoculture). If wheat rust or another disease should mutate to a form capable of attacking these "miracle" grains, the miracle will turn into a disaster, for famine is an almost certain result.

The achievements in animal breeding have been more modest but still significant. From wild horses typified by the Przewalski's horse such divergent breeds as the Thoroughbred racehorse, the Clydesdale draft horse, and the Shetland pony have been formed by selection of breeding stock for the purposes desired (Figure 23–2). Different breeds of cattle have been perfected for high

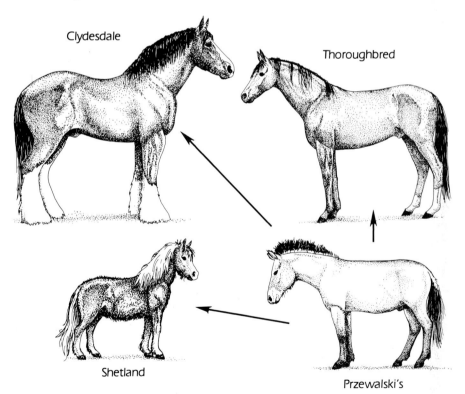

Figure 23–2. *Domestication of the horse,* Equus caballus, *from a Pleistocene progenitor resembling Przewalski's horse has resulted in a wide range of body sizes and conformations, from diminutive ponies to huge workhorses such as the Clydesdale.*

milk production or for high beef production. Each has been adapted to a wide variety of climates, from tropical to subarctic. Sheep have been specialized for production of wool or of mutton. Chickens with much increased egg production have been developed. The immense variety of dogs, from Pekinese to St. Bernard, have been produced by artificial selection. The source of all of these is the wolf, *Canis lupus* (Figure 23–3). If these extremes were found in nature, we would not hesitate to call them distinct species, but because we know their history, we refer them all to a single species, *Canis familiaris*.

Compared to parental species in the wild, domesticated animals show enhanced fertility and great morphological variability. Dogs, for instance, have two estrus periods per year compared to one for wolves. Human aboriginals at a low nutritional plane have a markedly lower level of fertility than do well-nourished Westerners. Humans exemplify well the morphological variability of domesticated species! All of the changes described have been brought about in the last 6,000 to 10,000 years.

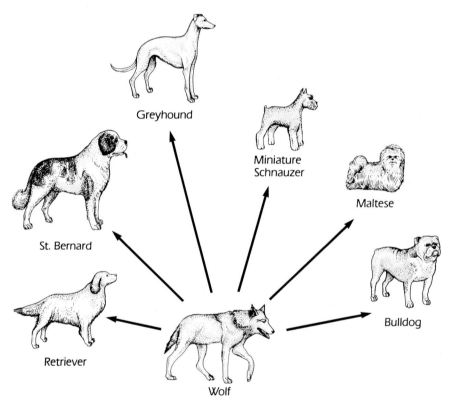

Figure 23–3. *Domestication of the dog,* Canis familiaris, *from the wolf has released enormous latent variability in size, shape, color, and even behavior. Illustrated are representatives of the six classes of dogs recognized by the American Kennel Club.*

Although the past achievements of plant and animal breeding are great, there is every reason to believe that far greater achievements lie ahead, for researchers accomplished the greatest past achievements, such as the development of hybrid corn, during the present century by using the tools of genetics, which were previously not available. The continued application of these principles will probably lead to even better agricultural products, both plant and animal. Thus we are truly achieving dominion over the world of life.

Distribution and Evolution of Wild Species

Humans have also had a great influence on the distribution and evolution of wild species, and that influence may be expected to continue and to increase. So far as effects on other organisms were concerned, primitive humans probably were not much more important than were many of the larger wild animals. As proficiency in the use of tools and the tilling of the soil increased, so also did their influence on other organisms. Soon the presence of humans became a major selective force to which wild species, both plant and animal, had to adapt in order to survive. With the development of the industrial age and the tilling of a very large portion of the arable soil in all civilized countries, our own selective influence on other organisms has reached a peak. In order to survive now, all living things must either adapt to the presence and activities of humans or else they must be restricted to those dwindling refuges in which our influence is least prominent.

We have caused the extinction of many species and the extreme reduction of others within recent times. The case of the passenger pigeon, *Ectopistes migratorius* (Figure 23–4), is a well-known example. Within recent memory the passenger pigeons were so numerous that flocks of them literally blackened the sky for hours at a time. The supply of pigeons was regarded as inexhaustible,

Figure 23–4. *Passenger pigeon,* Ectopistes migratorius. Photo courtesy of Dr. Leonard E. Munstermann, University of Notre Dame.

and they were intensively hunted and trapped for the market. By 1880 passenger pigeons were noticeably less numerous. By 1890 their numbers were seriously depleted. By 1900 they were a rare bird, and movements were afoot to save the pigeons. But all efforts to save them failed, and the last one died in the Cincinnati Zoo in 1914. Extinction of the passenger pigeons is commonly attributed to intensive hunting, but the equally intensive destruction of the forests in which they found cover may have been quite as important a factor.

Bison were also prodigiously numerous when Europeans invaded North America. They are now reduced to a few isolated and protected herds and to exhibition specimens. Again, intensive hunting may have been an important factor in the decimation of the bison but the fencing of the range was equally important. Comparable stories could be recited for many other species that inhabited North America before the Europeans came. They all simply emphasize the statement with which this discussion began, that if organisms are to survive, they must either adapt to the presence and activities of humans or they must be restricted to those dwindling refuges in which our influence is not great.

Disheartening as these facts are, the worst may still be in the future. An article in *Science* in 1981 spoke of the probable extinction of no less than a million species before the year 2000! It is problematical whether *H. sapiens* can survive in so impoverished a world.

But we are not totally insensitive. We have established wildlife reserves, and we are nurturing many endangered species in zoological and botanical gardens with a view to reestablishing them in the wild. Thus, some endangered species may be saved from extinction. Moreover, some organisms have profited greatly by our activities. This is obvious in the case of domesticated plants and animals, but it is also true of many wild species. Rats and mice, originally Palearctic animals, have been inadvertently carried to all parts of the world on cargo ships. They live in direct competition with us, invading our buildings for shelter and robbing our larder for food. On the whole they have been so successful that they must be much more numerous than they were before the rise of civilization. Crows similarly have profited from farmlands; and small carnivores, such as foxes, coyotes, and weasels, have not only learned to live in close proximity to us, but they have learned to attack our small domestic animals, such as chickens and rabbits, and still escape our wrath sufficiently well to maintain themselves in competition. Gulls and pigeons have prospered. The success with which many weeds have invaded civilized habitats is too familiar to require discussion.

The Ecological Crisis

Indeed, our predominance throughout the ecosystems of the world and our disruptive influence in many of them are at the heart of what many people now describe in apocalyptic terms as an ecological disaster. They point to many facts, of which those in the preceding section are a small sample, that suggest that we, through our many activities, but especially through pollution of the environment by industry, by fuel fumes, by pesticides, by acidifying the rain, by building up CO_2 in the atmosphere, and by the wastes of individual consumption, are rapidly making our world uninhabitable for ever-greater numbers of species. Because we ourselves are a part of the web of life, we may well be working toward our own extinction. Miners used to carry canaries, which are very sensitive to carbon monoxide poisoning, into the mines. Death of a canary would then warn the miners of a dangerous level of carbon monoxide in time for the miners to escape. Each species that becomes extinct because of deterioration of the environment may now provide a similar warning.

Nowhere has human influence been ecologically more pernicious than in the tropics. Slash-and-burn agriculture is typical in the tropics. First, large areas of forest are cut down and burned, then the land is planted in crops. As nutrients fall upon the laterite soil of the rain forest, organic materials are quickly decomposed and resorbed by the trees along with any inorganic nutrients. Accordingly, the soil is always poor and is soon exhausted. The farm is then abandoned and replaced by more slash-and-burn destruction. The resulting desolation is so great that the tropical rain forests of the world may disappear before the end of the century, with consequent loss of many thousands of species.

In a sense, the exploitation of our environment is normal and desirable: every species exploits its environment to whatever extent it can, without regard to the consequences. Typically, however, each species is checked by others, so that an equilibrium of sorts is maintained in the complex web of nature. Human knowledge and technology, however, have made this a one-way street: our influence on other species greatly exceeds theirs on us. Although continued exploitation may be biologically normal and even necessary, it may be carried out wisely or unwisely. Wise use of our resources need not make the world less habitable and less desirable for us, but because we are part of and dependent on a complex web of life, it must include protecting and promoting the good health of our ecosystem as a whole. Thanks to the labors of many ecologists, we have been advised of the nature and dimensions of the problem. The technology for its solution is at hand, and the peoples and governments of the world are awakening to the necessity for appropriate action. A cautious optimism for the future seems justifiable.

Radiation and Evolution

Another human activity with great potential importance for evolution is the induction of mutations by radiation, including atomic radiation. Since the discovery by Muller that mutations could be induced by X-radiation, the possibilities have been intensively explored. Any high-energy radiation will cause mutations in numbers directly proportional to the total dosage of radiation.

Although the possibility of using radiation to induce useful mutations has been extensively investigated, especially in crop plants, almost all of the experimentally produced mutations have been deleterious. A brilliant exception is provided by the work of M. Demerec on *Penicillium notatum,* the mold which produces penicillin. Previous to Demerec's experiments, the best available cultures of *Penicillium* produced about 70,000 units of the drug per pint of culture. He X-rayed large numbers of the organisms and carefully tested penicillin production of pure cultures derived from X-rayed mold. Most of the X-rayed stock was of no special value, but one highly productive strain, yielding 280,000 units of penicillin per pint of culture, was obtained. Most commercial penicillin is now produced from this stock. While the possibility that other valuable mutations may be produced by radiation cannot be entirely ruled out, still radiation is, on the whole, a genetic and evolutionary hazard. In a society in which radiation is playing an increasing role in medicine, experimental science, and even industry, it is impossible to estimate what the effects of inadvertent radiation of the gonads may be, both for humans and for associated plants and animals.

Sociobiology

A currently controversial aspect of human evolution is a new discipline called *sociobiology.* E.O. Wilson introduced the term in a book of that title published in 1975. The book, a study of animal behavior, presents the thesis that the main aspects of animal behavior are genetically determined and that behavior, therefore, is also a product of the evolutionary history of the species. The study is concerned almost exclusively with lower animals, but in the last chapter Wilson tentatively suggested that a similar approach to human behavior might be profitable. For this, Wilson was severely attacked. His opponents have commonly claimed that he is supporting racist and other prejudicial policies and that his motivation is not so much scientific as reactionary-political. His defenders claim that this is nonsense, that his opponents have overreacted, and that his ideas should be judged on their merits without such emotional overtones. Wilson had the opportunity to clarify and calm this debate when he published a second book, *On Human Nature.* Instead, however, he seems to have given substance to some of the charges against him. The place of sociobiology in evolutionary studies is thus shrouded in controversy and intemperate verbiage. It will be interesting to watch the resolution of this problem in the future.

Human Society and Evolution

We are the product of a long evolutionary history, of natural selection operating over generations beyond reckoning. But to what extent are we still constrained by the evolutionary process? Are we best understood in terms of gene frequencies and selective pressures? Is it mere vanity to think that we differ from other animals? One of the most-striking aspects of our species is the ability to cooperate, to form societies of increasing size and complexity. Human societies evolve, and the process may be viewed as decidedly Lamarckian, for acquired knowledge is most definitely transmitted; because of this, societies can evolve with astonishing speed. Genetic engineering may give us partial control of our future evolution.

Individual actions may starkly contradict the evolutionary "good." "Increase and multiply, and fill the earth" is an evolutionary mandate; birth control, the intentional reduction of natural fertility, is inconceivable in an evolutionary sense. Altruism is quintessentially human. It is not merely kin selection, which is favored by natural selection, but genuine, heroic, self-sacrifice, whether the soldier's for his comrades on the battlefield or the lifetime commitment of Albert Schweitzer or Mother Teresa. Such altruism makes a mockery of the evolutionary process, yet it is the glory of humanity. We are motivated by concern for human misery half a globe away; whereas a wildebeest female will watch an orphaned calf starve to death before her eyes, for in wild nature there is no such thing as compassion.

Warfare was perhaps once an agent of natural selection in human affairs, but now it is an indiscriminate horror. Natural selection still maintains sickle-cell anemia in Africa (Chapters 9 and 10), it is often at work when miscarriages occur, and it continues to work in many other ways in modern society. We have by no means "escaped" entirely from genetic evolution, but we clearly are not in bondage to it in the way that other organisms are. Spirituality and free choice characterize us as much as does primate anatomy. As Lewis Thomas said,"the genetic message is like distant music, and some of us are hard of hearing."

FUNDAMENTALISM AND SCIENTIFIC CREATIONISM

The concept of evolution has had great impact on other aspects of culture. The relationship of Darwinism to religion has been a *cause célèbre*. Most of the concepts of religion were formulated in the days of fixity, and naturally a certain harmony had developed in relation to static biology. The new Darwinian biology seemed to be a direct challenge to the familiar interpretation of Genesis. After the initial shock, however, many theologians and scientists showed that, if both avoided unjustified incursions into the areas of competence of the other, Genesis and evolution were not necessarily incompatible. Thus, a century after Darwin, C.C. Gillispie*, a leading historian of science, wrote:

> It is the theologians who have learned to live with evolution. Whereas the ones who have not so learned, and cannot, are those atheists who want to substitute nature for God as the source of morality and ethics . . . . It

* C.C. Gillispie, *The Edge of Objectivity* (Princeton, N.J.: Princeton University Press, 1960), p. 350.

is not the conflict between science and religion which has proved fundamental. It is the conflict between science and naturalistic social or moral philosophy. We must be clear about the nature of science as a description of the world, declarative but never normative.

Fundamentalism

One group of Christians, however, did not participate in this rapprochement. The fundamentalists are religious conservatives who interpret the creation narrative in Genesis literally. During the first quarter of the present century, they mustered sufficient strength in the less well educated parts of the American South to pass laws in four states forbidding the teaching of evolution in the public schools. A test case was sought for the Tennessee law in 1925; and John Scopes, a high school teacher in the small town of Dayton, volunteered to assign a book that taught evolution. By mutual agreement, he was indicted and brought to trial. He was expected to be quickly convicted so that the case could be appealed to the supreme court of the state for decision on the constitutionality of the law. This expectation was frustrated by sensational news coverage. Also, two very distinguished lawyers volunteered their services: William Jennings Bryan, a leading fundamentalist and three times a Democratic candidate for the presidency of the United States, for the prosecution; and Clarence Darrow, a highly successful criminal lawyer, for the defense. The case was a failure for both sides. Scopes was found guilty and fined $100 by the judge. Because the offense was a misdemeanor, for which the judge could not levy a fine of more than $50, the Supreme Court of Tennessee dismissed the case *without judgment on the validity of the law.* No further charges were brought under this law, and it was repealed many years later.

Recently, there has been a resurgence of fundamentalism, and its strength is no longer among the less well educated. On the contrary, many are university graduates, including lawyers, engineers, and physical scientists. There are few biologists among them, but those few may be important. Their organizations include the Creation Research Society, whose president has a doctorate in botany from a leading university, and the Institute for Creation Research, whose director holds a doctorate in hydraulic engineering and whose associate director has a doctorate in biochemistry, both from leading universities.

One result is that the creationists of today are far more sophisticated than were their predecessors of the Scopes era. They no longer ask that the teaching of evolution be outlawed, but rather that "scientific creationism" be given equal time whenever evolution is taught. Campaigns for laws requiring equal time have been mounted, and such laws were enacted in 1981 in Arkansas and Louisiana. Similar legislation is pending in nearly twenty other states. A suit was promptly filed to contest the Arkansas law on the grounds that it violates the constitutionally required separation of church and state. It is noteworthy that among the complainants who filed this suit are the Catholic and Episcopal bishops of the state. The case went to trial in December 1981, and in January 1982 the verdict was rendered: the court declared the law unconstitutional on the grounds that "scientific creationism" was actually sectarian religion (and on two related grounds). The Louisiana law was declared unconstitutional under the state constitution in November 1982, but a revised law was passed and was also declared invalid in federal court (fall, 1984).

Scientific Creationism

Advocates of equal time laws claim that scientific creationism is no more a religion than is evolution because they can present their arguments without reference to the Bible; and they describe evolution as the religion of secular humanism. They use much the same array of examples that were presented in the first five chapters of this book, but they offer them as evidence of creation on a common plan rather than as evidence of common descent. This view is logically defensible to a point, but in most of the lines of evidence, there are critical cases that do not fit well under the common planning explanation. Examples are vestigial organs in comparative anatomy, recapitulations in embryology, as well as the whole body of paleontology.

In practice, however, scientific creationism is primarily an attempt to exploit the flaws—real or imagined—in evolutionary science and to exploit disagreements among evolutionary biologists. Creationists declare that, if the case for evolution is so defective and its advocates so divided, special creation as described in Genesis must be right. We believe that they are wrong on all of these points. It is, of course, possible to criticize evolutionary biology or anything else without reference to the Bible. As soon, however, as the creationists try to make a positive proposal, they have to attribute the origin of biological groups to the direct action of God, and that is unavoidably religious, if not necessarily biblical.

Further, it cannot be merely coincidence that the advocates of this view are all fundamentalists and that their conclusion is favorable to their religious outlook, an outlook from which other religious authorities, such as the bishops cited earlier, dissociate themselves. Nor is it self-evident that evolution is the religion of secular humanism. Professional students of evolution cover the whole spectrum with regard to religious belief and disbelief. T. Dobzhansky, the dean of modern students of evolution, was a very devout man. The present authors are practicing Christians, but our respect for the Bible does not accord it scientific authority.

A REPLY TO CREATIONIST CRITICISMS

Law of Entropy

A detailed reply to the fundamentalist criticisms of evolutionary science is beyond the scope of this book, although the preceding chapters are applicable. Let us take a few examples. Because evolution has proceeded from simple to complex, they claim that it violates the law of entropy (that matter and energy proceed to ever-greater disorder). In fact there is no contradiction whatever here, for the law of entropy is strictly applicable only to closed systems (which neither gain nor lose energy). Presumably, the universe as a whole is such a closed system, but the earth is not. The earth is bathed in a prodigal outflow of energy from the sun, which provides the power for evolution as for almost all of the processes of life.

Micro- and Macroevolution

Some fundamentalists are much impressed by the distinction between micro- and macroevolution. Microevolution is concerned with changing gene frequencies to form subspecies adapted to different niches in the total range of a species. The more knowledgable ones among them concede that this is valid science, and some will even extend this concession to the origin of the species of a genus. But, they maintain, the evolution of higher groups—macroevolution—is beyond the possibility of a scientifically valid, experimental approach because no observer was there when it happened. They discount or disregard all of the evidence cited in the preceding chapters. Alternatively, they explain this as evidence of creation upon a common plan.

It is, of course, true that the evidence of macroevolution is very much less complete than that of microevolution. It is in the nature of science, however, to be always incomplete, always subject to revision in the light of new data. Nonetheless, there are enough data considered valid by scientists to convince the majority of unbiased students that macroevolution is genuine, even if inadequately understood. Future data may require replacement of current hypotheses of macroevolution, but macroevolution itself will stand as a fact of nature.

What Is a Theory?

Fundamentalists sometimes belittle evolution with the statement that "it is only a theory." One important dictionary gives no fewer than seven definitions for *theory,* among them the following three: (1) a plan in the mind, based upon principles verifiable by experiment or observation; (2) a set of fundamental principles underlying a science; (3) mere speculation, an individual idea or guess.

When scientists speak of the theory of evolution, they are using one or both of the first two definitions; and when fundamentalists denigrate evolution as "only a theory," they are using the third definition. Although the taunt that evolution is only a theory is still common (it figured in the United States presidential election of 1980), modern fundamentalists often bypass the term by referring to evolutionary and creationist *models of origins.* In this context, *model* has no clear scientific meaning, and evolutionary scientists do not use the term.

Fair Play

Finally, fundamentalists often appeal for equal time for scientific creationism on the basis of fair play. Such an appeal is difficult to refuse, but it is reasonable to question whether that is what they are really asking. No law requires the teaching of evolution. Schools are generally authorized to teach a science curriculum that includes chemistry, physics, and often biology. Professionals in these fields then decide what specific topics should be included in each course. Because most modern biologists regard evolution as a central—perhaps *the* central—aspect of biology, it is likely to be included, but it is never mandated by law. Yet the proposal of the fundmentalists is that teachers who teach evolution should also teach scientific creationism even if it violates their professional judgment. It is not self-evident that this is fair play.

EVOLUTION: A YOUNG SCIENCE

Most sciences have only gradually emerged from their predecessors or from natural philosophy, but the science of evolution is one of a few for which a fairly definite time of origin is known. Although the concept of origin of species by modification is an ancient one, only since the publication of the *Origin of Species* in 1859 has it had a firm scientific basis, capable of commanding the respect of competent scientists. The date of origin might be set at the earliest in the fall of 1834, when Darwin visited the Galápagos Islands, or at the latest on November 24, 1859, when the *Origin of Species* was published.

The new science developed rapidly indeed. Scientists of the latter decades of the nineteenth century labored mightily to establish the fact of evolution, especially through studies in comparative anatomy and comparative embryology. However, the other basic evidences of evolution also began to be developed at this time, as we saw in Part 1.

Yet it was not only because of the great labors of many people that the new science prospered. Represented by such men as Spallanzani, Redi, Ray, Wolff, and Harvey, biology had participated in the great upsurge of science in post-Renaissance times. Yet the vast store of data these men established was largely chaotic, for there was no unified, theoretical framework within which their diverse contributions could be marshalled. Linnaeus tried to fill this need with his taxonomic system, and for a time he seemed to succeed in giving biology an appearance of order. In the end, however, he stimulated a great deal more exploration and fact-finding without satisfying the need for a basic theory, because the taxonomic system itself was inexplicable. Thus Darwin found biology a burgeoning chaos of more or less unrelated data without any comprehensive theory to make it a cohesive whole. The theory of the origin of species by means of natural selection filled that need. It gave meaning to the taxonomic system of Linnaeus and to the many other biological sciences that Linnaeus had sought to clarify by his system. Today, almost all biological work is based, directly or indirectly, on recognition of the fact that the plant or animal of today is the most recent product of an historical process. *Time* has become an essential dimension of biology.

Thus the role of evolutionary theory in biology is comparable to that of the laws of thermodynamics in physics: it propounds the basic laws from which a major branch of science, biology, derives its comprehensibility. It is for this reason especially that the new science of evolution has developed into so great a body of experimental and theoretical knowledge in little more than a century. We may confidently hope for even greater progress during the second century of evolutionary studies!

REFERENCES

Creation Research Society Quarterly. Ann Arbor, Mich.
   The principal journal for papers supporting the creation model of origins.
Day, D. 1981. *The Doomsday Book of Animals: A Natural History of Vanished Animals.* Ebury Press, London, and Viking Press, New York.
   The poignant story of the extinction of over 300 species in the past 300 years.
Dixon, D. 1981. *After Man—A Zoo of the Future.* St. Martin's Press, New York.
   A carefully reasoned, imaginative, and beautifully illustrated speculation on what birds and mammals might look like 50 million years from now. Entertaining and stimulating.
Dodson, E.O. 1984. *The Phenomenon of Man Revisited: A Biological Viewpoint on Teilhard de Chardin.* Columbia University Press, New York.
   Teilhard was at once a priest and a paleontologist. In The Phenomenon of Man, *he synthesized the two aspects of his career to make a meaningful whole, but it is often very difficult reading. This reference provides a more easily understood introduction to Teilhard's thought.*
Ehrlich, P., and A. Ehrlich. 1981. *Extinction: The Causes and Consequences of the Disappearance of Species.* Random House, New York.
   Good reading; a well-researched and well-written study of an important, but neglected, subject.
Fisher, J., N. Simon, and J. Vincent. 1969. *Wildlife in Danger.* Viking Press, New York.
   A study of endangered species, published for the International Union for the Conservation of Nature and Natural Resources.
Gish, D. 1978. *Evolution—The Fossils Say No!* Creation–Life Publishers, San Diego, Calif.
   A vigorous, if not always well-informed, attack on evolutionary biology by a leading scientific creationist.
Kitcher, P. 1982. *Abusing Science: The Case against Creationism.* MIT Press, Cambridge, Mass.
   A philosopher's well-reasoned assessment of this problem.
Marsden, G.M. 1981. *Fundamentalism and American Culture: The Shaping of Twentieth Century Evangelism.* Oxford University Press, London and New York.
   A thorough probing of the historical roots of this aspect of contemporary America.

Moore, J.R. 1979. *The Post-Darwinian Controversies: A Study of the Protestant Struggle to Come to Terms with Darwin in Great Britain and America, 1870–1900.* Cambridge University Press, London.
An excellent study, which, unfortunately, stops far short of the present time.

Morris, H.M. 1974. *Scientific Creationism.* Creation–Life Publishers, San Diego, Calif.
Scientific creationism presented by one of its leading advocates.

Newell, N. 1982. *Creation and Evolution.* Columbia University Press, New York.
A popularly written discussion of the current controversy as seen by a leading paleontologist.

Overton, Judge W.R. 1982. Creationism in schools: The decision in McLean versus the Arkansas Board of Education. *Science* 215:934–943.
The complete text of the decision in this landmark case.

Simon, J.L. 1980. Resources, population, environment: An oversupply of false bad news. *Science* 208:1431–1437.
A refreshing article!

Soule, M., and B. Wilcox, eds. 1980. *Conservation Biology: An Evolutionary-Ecological Perspective.* Sinauer Associates, Sunderland, Mass.
An excellent presentation of the subject indicated by the title.

Teilhard de Chardin, P. 1955. *Le Phénomène Humain.* Editions du Seuil, Paris. Translated by Bernard Wall and published in 1959 as *The Phenomenon of Man.* Collins & Co., London, and Harper & Row, New York.
This remarkable book explores the future of man from a different viewpoint than those cited. While it is beyond the scope of this book, it may be of great interest to many of our readers.

Williams, E.L. 1981. *Thermodynamics and the Development of Order.* Creation Research Society Books, Norcross, Ga.
A creationist viewpoint on the law of entropy.

Wilson, E.O. 1975. *Sociobiology: The New Synthesis.* Harvard University Press, Cambridge, Mass.
The starting point for a major controversy. An abridged edition was published in 1980.

Wilson, E.O. 1978. *On Human Nature.* Harvard University Press, Cambridge, Mass.
Continuing the same controversy.

Zitterberg, J.P. 1983. *Evolution versus Creationism: The Public Education Controversy.* Oryx Press, Phoenix, Ariz.
A collection of essays by many authors on both sides of this question. Like most collections, the quality of the essays is variable, but some are truly excellent.

Glossary

Acquired character A character developed during the lifetime of the individual because of the direct effect of some environmental factor rather than because of the genotype (e.g., tanning of the skin upon exposure to ultraviolet radiation).

Adaptation Mutual fitting of structure, function, and environment so that the organism may better compete in its normal habitat.

Adaptive radiation The tendency of successful species (or higher groups) to fan out into all available ecological niches.

Adenine One of the four bases of DNA; always paired with thymine.

Aerobic Having a metabolic system that requires oxygen.

Allele An alternative form of a gene.

Allen's rule The rule stating that in a given species or genus, the extremities (limbs and tail) tend to be smaller in cooler than in warmer climates.

Allometry Differential growth of one part with respect to another. (adj., *allometric*)

Allopatry Occupying quite separate territories.

Allopatric speciation Formation of new species by populations of the parent species that occupy allopatric territories.

Alloploid A polyploid based on doubling of the chromosomes of an organism that contains chromosome sets from two or more unlike sources (different species).

Allotetraploid A tetraploid in which two different genomes are each present twice; sometimes called an amphidiploid.

Allozyme An alternative form of an enzyme.

Alternation of generations In plants, a haploid, sexually reproducing generation, the gametophyte, alternates with a diploid, asexually reproducing generation, the sporophyte. In coelenterates, a free-swimming, sexually reproducing medusa alternates with a sessile, asexually reproducing polyp, but both are diploid.

Amino acid An organic compound of the general formula $NH_2 \cdot R \cdot COOH$; the 20 naturally occurring amino acids are the building blocks of proteins.

Anaerobic Having a metabolic system that does not require oxygen.

Anagenesis Transformation of whole populations in evolution. (ant., *cladogenesis*)

Analogy Similarity of structures because of similar use, even though the organisms may be unrelated (e.g., wings of insects and birds). (adj., *analogous;* ant., *homology*)

Anticodon A sequence of three nucleotides in transfer RNA, complementary to the codon for a specific amino acid; the latter is part of the messenger RNA molecule.

Apomorphy A shared derived character in cladistic taxonomy; an advanced character shared by two or more taxa that may be used to analyze relatedness.

Aposematic (or warning) coloration Conspicuous colors or patterns on an otherwise well-protected organism, so that potential predators are forewarned.

Archetypal theory An earlier alternative to evolution, according to which actual species were more or less imperfect copies of supernal models that were essentially ideas in the mind of God.

Autocatalysis   Promotion of a chemical reaction the end product of which is more of the catalyst itself. The classical biological example is the replication of the chromosomes.

Autopolyploidy   Polyploidy in which all of the genomes are substantially identical, that is, from the same species. (or, *autoploidy;* ant., *allopolyploidy*)

Autotetraploid   Presence of a single genome in quadruplicate.

Autotrophic   Capable of synthesizing all food requirements from the elements; true of green plants. (ant., *heterotrophic*)

Balance theory of population structure   The genetic structure of a population is the result of a balance between origin of new variability by mutation and elimination of alleles by selection and by chance factors.

Balanced polymorphism   Maintenance of equilibrium in the proportions of several forms in a population because heterozygotes are favored by selection over both homozygous types.

Batesian mimicry   Mimicry of a well-protected model by an unprotected mimic species, which then shares the protection of the model.

Bergmann's rule   The rule stating that in a given species or genus, populations in cooler climates have larger mean body size than do those in warmer climates.

Biogenetic law   The proposition that embryos of more highly evolved species repeat in their development, in abbreviated form, their ancestral history; "ontogeny recapitulates phylogeny."

Biogeography   The study of the distribution of organisms over the earth and of the principles that govern this distribution.

Bipolar mirrorism   The occurrence of similar plants, and to a lesser extent animals, in widely separated places on either side of the equator. Sweepstakes dispersal has been invoked to account for such examples.

Bivalent   Paired chromosomes at meiosis.

Bottleneck phenomenon   The reduction of populations to very small numbers in a bad year, with a rebound to large numbers later. The small "bottleneck" population can carry only a small part of the genetic variability of its species, yet that small sample must have a limiting influence on the larger population to which it will give rise.

Brachiation   Locomotion by arm-swinging through trees, with power provided by the arms and shoulders; best exemplified by the gibbon.

Catastrophism   The theory that all life has been destroyed periodically, with a new creation following each catastrophe.

Character displacement   The tendency to stronger differentiation in sympatry; when two related species are largely allopatric but are sympatric in a limited part of their range, they may be strongly differentiated in sympatry, yet quite similar in allopatry.

Chlorophyll   The green pigment by which plants use solar energy for the synthesis of carbohydrates, thus making solar energy available to organisms. Chemically, it is based on the porphyrin ring in combination with magnesium.

Chromosome   A stainable, threadlike body in the cell nucleus, consisting of DNA complexed with large amounts of protein. The chromosomes are made up of genes, and their number is characteristic for each species.

Ciliate-flatworm theory   The theory that ciliates gave rise to very simple flatworms, from which all other groups of metazoans derive.

Clade A homogeneous (monophyletic) line of descent.

Cladistic evolution Splitting of a line of descent into two. (ant., *phyletic evolution*)

Cladistics The practice of phylogenetic systematics.

Cladogenesis Splitting of a taxon to form two sister taxa.

Cladogram A branching diagram showing sister groups, without necessarily specifying the parent groups from which they came.

Class A taxonomic unit (taxon) comprising related orders; the major subdivision of a phylum.

Cline Progressive variation of a character in parallel with some aspect of the environment, as altitude or latitude.

Coacervate A complex system formed when colloids of opposite charge precipitate each other. Mayonnaise is a familiar and highly complex example. Coacervates play a role in the structure of protoplasm.

Codon A sequence of three nucleotides in DNA or RNA, determining a specific amino acid in the synthesis of a protein.

Coevolution Coordinated evolution of two or more ecologically related species.

Colchicine An alkaloid derived from the autumn crocus, *Colchicum autumnale,* which interferes with formation of the mitotic spindle and causes polyploidy.

Competition Contention of two or more species for the same limited resource.

Continental drift Movements of the continents because of plate tectonics.

Convergence The evolving of superficial similarity in spite of basic differences among different organisms occupying similar habitats; convergence of ichthyosaurs and porpoises is a classic example.

Cope's law The theory that all organisms tend to increase in size during their evolutionary history.

Corridor A broad, varied, and long-lasting connection between two regions, so that whole floras and faunas may be exchanged.

Cosmozoic theory The theory that the spores of life reached the earth from some other part of the universe.

Crossing-over Exchange of genes between members of an homologous pair of chromosomes.

Cryptic coloration Matching of the background so that the organism is rendered inconspicuous.

Cytochrome *c* A respiratory pigment found in all cells, and the subject of some illuminating studies in molecular evolution.

Cytosine One of the four bases of DNA, always paired with guanine.

D Genetic distance.

Deletion Loss of a segment from a chromosome. (syn., *deficiency*)

Dendogram A branching diagram used to express the results of numerical taxonomy. (syn., *phenogram*)

Derived Pertaining to characters arising during the history of a group and not included in its primitive ancestors. (ant., *primitive*)

Deuterostomia A group of phyla culminating in the Echinodermata and the Chordata. They are united by an array of embryological traits.

Dichotomous Characterized by division into two, generally equal or subequal, parts, as dichotomous veins or appendages.

Diploid The occurrence of chromosomes of somatic tissues in pairs, so that each cell nucleus contains two complete sets of equivalent chromosomes.

Directional selection   Selection favoring adaptation to new conditions so that change is produced.

Disruptive selection   Selection for two or more character states so that diversity is produced. (also, *diversifying selection* or *fractionating selection*)

Division   In plant taxonomy, the most inclusive taxon below kingdom, equivalent to the phylum of zoology.

DNA   *D*eoxyribo*n*ucleic *a*cid; nucleic acid based on deoxyribose; the active substance of the gene.

DNA polymerase   An enzyme that catalyzes the polymerization of DNA; it also "proofreads" the complementarity of the bases and makes corrections as necessary.

Dollo's law   The law of irreversibility, that the major steps in evolution are not reversed.

Dominant   Of genes, the allele that is expressed in a heterozygote to the exclusion of its homologue.

Duplication   A repeated segment in a chromosome.

Ecotype   A specific character complex that is associated with a specific habitat.

Electrophoresis   The separation of proteins (or other electrically charged compounds) by migration in an electrical field.

Endemic   Narrowly restricted in geographic distribution.

Entropy, law of   The second law of thermodynamics, which states that, in a closed system, matter and energy proceed to ever-greater disorder; in an open system, energy may be expended to oppose that tendency.

Epoch   In geological time, the principal subdivisions of the periods.

Equilibrium theory of island biogeography   The theory that a given island has a capacity to support a limited number of species, roughly proportional to the size of the island so that establishment of a new species results in the extinction of another.

Era   One of the great divisions of geological time.

Ethology   The science of animal behavior.

Eucaryote   Organisms with a nuclear membrane separating nucleus and cytoplasm, in which cytoplasmic organelles are present, and in which mitosis and often meiosis occur; all nuclear organisms. (also, *eukaryote*)

Evolution   "Descent with modification"; process in which closely related species resemble one another because of their common inheritance and differ from one another because of hereditary differences accumulated since the separation of their ancestors; processes by which related populations diverge, giving rise to new species and higher groups.

Evolutionary systematics   The systematics developed in the early years of the modern synthesis, based on the traditional data of comparative anatomy and embryology coordinated with genetics, ecology, and such other sciences as seem to be applicable to a given problem.

Exon   The actively coding part of the mRNA molecule of eucaryotes.

Family   A cluster of related genera.

Filter bridge   A connection between two regions that is ecologically fairly uniform and geologically of fairly short duration so that it will serve for exchange only of species that can exploit its ecological conditions and that can migrate fast enough to complete the crossing while the bridge lasts.

Fitness   A measure of the state of adaptation of an organism, expressed by the contribution of its genotype, relative to that of other genotypes, to succeeding generations.

Fossil   Any remains of organisms of the past.

Founder principle   The principle that populations on oceanic islands and other isolated places may be established by a very small sample from a continent or another island. Such a sample can include only limited variability (i.e., it is a special case of the bottleneck phenomenon).

Fundamentalism   A conservative religious position that precludes acceptance of evolution because of a literal interpretation of the first three chapters of Genesis.

Gastrea theory   Haeckel's theory of the origin of the Metazoa, based on embryology interpreted according to the biogenetic law. Because he thought that all embryos of higher animals passed through a gastrula stage, like a very simple coelenterate, he erected a hypothetical primitive genus *Gastrea* from which all metazoans were derived.

Gene   The unit of heredity; a segment of the DNA molecule having a specific sequence of nucleotides and so determining a specific polypeptide or protein.

Gene flow   Exchange of genetic information within and between local populations of a species.

Gene pool   All of the genes of a given population.

Genetic code   The pattern of nucleotide triplets on the strands of DNA by which genetic information is encoded and transmitted.

Genetic distance   The extent to which two populations differ consistently in their alleles.

Genetic drift   Change in gene frequency because of random processes, such as meiosis and random fertilization.

Genetic identity   The extent to which two populations share the same alleles.

Genome   One complete set of chromosomes, comprising one member from each pair.

Genome analysis   A series of crosses made to establish the source species of the several genomes of suspected allopolyploids.

Genotype   The genetic constitution of an organism, irrespective of its appearance.

Genus   A group of closely related species that share more characteristics with each other than with members of other related genera. (pl., *genera*)

Gloger's rule   The rule stating that animals in cool, dry regions tend to be lighter in color than those in warm, humid regions.

Group selection   The theory according to which it is not the individual that is selected but the entire population of which it is a part.

Guanine   One of the four bases of DNA; always paired with cytosine.

Haploid   The occurrence in gametes (and gametophytes) of only one chromosome of each pair.

Hardy-Weinberg law   The law stating that if alleles of a gene are in given proportions in a population and if random mating and equal viability of all genotypes apply, then the original proportions will be maintained in all subsequent generations unless disturbed by mutation, selection, or some other factor.

Heredity   Genetic transmission of traits from parents to offspring.

Heterotrophic   Requiring organic compounds as food because the organism is incapable of synthesizing these compounds from the elements; true of all animals.

Heterozygote   Having the two genes of a pair unlike, as *Aa*. (adj., *heterozygous*)

Holophytic Characterized by autotrophic nutrition; true of most plants.

Holozoic Characterized by heterotrophic nutrition; true of all animals.

Homology In a series of related organisms, similarity of structures because of descent from common ancestors, without regard to function.

Homozygote Having both genes of a pair alike, as *AA*. (adj., *homozygous*)

Hybridization The combining of different genotypes within or between species.

I Genetic identity.

Industrial melanism Many species of moths that were formerly light colored developed melanic variants during the industrial revolution, and these variants have largely replaced the original forms in industrial regions.

Introgressive hybridization Transfer of genes from one species to another by hybridization, followed by repeated backcrossing to one of the parental species.

Intron A noncoding part of the mRNA molecule of eucaryotes, the prospective fate of which is elimination.

Inversion Reversal of order of a segment of the chromosome.

Isolating mechanism Mechanisms that prevent the flow of genes between populations, classified as premating or postmating.

K Carrying capacity of the habitat. (ant., *r*)

K selection Selection favoring fine tuning of adaptations rather than high reproductive rate as a strategy for survival.

K strategy A type of life cycle relying on finely tuned adaptation rather than on high reproductive rate.

Karyotype The total pattern of the chromosomes of a species.

Kin selection Selection that favors altruistic behavior of an individual toward its own kin (i.e., those that have some of its own genes); forms the genetic basis for sociobiology.

Knuckle-walking A quadrupedal gait characteristic of chimpanzees and gorillas, in which the fingers are curled so that the second phalanx of each digit contacts the ground.

Lamarckism The evolutionary theory of J.B. Lamarck, based on inheritance of acquired characters and direct action of the environment in producing adaptive characters.

Linkage Association of transmission of different pairs of genes because they are parts of the same chromosome.

Macroevolution Evolution above the specific level, especially as seen in the fossil record.

Microevolution Evolution at the level of genetic analysis, usually at the levels of populations and subspecies and over short spans of time.

Mimicry The sharing of an aposematic pattern for protection. In Batesian mimicry, an edible species assumes the cloak of a noxious, aposematic species and thus shares its protection. In Müllerian mimicry, two noxious species share one aposematic pattern, thus gaining double insurance.

Missing link An unknown intermediate in an evolutionary sequence; basis for much controversy.

Molecular clock The claim that, in a given group, neutral changes in the proteins are accumulated at a constant rate and hence that the amount of molecular differentiation between two species indicates the time since their separation.

Molecular drive   Cohesion of species, especially homogeneity of repeated sequences of DNA, because of unequal crossing-over, transposition, and gene conversion.

Monophyly   Designation of a taxon all members of which are derived from a common ancestor; a clade.

Mosaic evolution   The tendency for various parts of complex organisms to evolve at different rates.

Müllerian mimicry   Mutual mimicry of two or more well-protected species so that they share their protection, thus gaining double insurance.

Multiple alleles   The formation of a whole series of alleles, often subserving an extensive range of variant phenotypes, resulting from gene mutation. A population may include many alleles of a given gene; an individual can carry no more than two of the series.

Mutation   An inheritable change in a gene.

Mutational distance   The number of mutations necessary to account for the transition from one state to another, as from the cytochrome *c* composition of a horse to that of a cow.

Mutator gene   A gene that causes a high rate of mutation in other genes because of production of defective DNA polymerase.

n   The haploid, or gametic, number of chromosomes.

Natural selection   Changing genetic makeup of populations because of differential reproductive success; the various factors that cause this.

Neo-Darwinism   Currently the dominant theory of evolution, combining classical Darwinism with modern genetics as well as with other biological (and even nonbiological) sciences. (also, *synthetic theory* and *biological theory*)

Neontology   The study of organisms now living. (ant., *paleontology*)

Neoteny   Achievement of sexual maturity at an embryonic or juvenile stage.

Neutral evolution   The accumulation of mutants arising from the fact that natural selection would not distinguish among about a quarter of all base substitutions in DNA resulting in synonymous codons and might not distinguish among still others resulting in different but functionally similar amino acids.

Niche   The sum of the ecological requirements of a species.

Nucleic acid   A macromolecule formed by the polymerization of nucleotides.

Nucleotide   A complex molecule formed by the union of a pentose (5-carbon sugar) with an organic base (purine or pyrimidine) and a phosphate radical. These polymerize to form macromolecules, the nucleic acids.

Numerical taxonomy   Taxonomy based on computer analysis of large numbers of characters. (syn., *phenetic taxonomy*)

Ontogeny   The path of development of an individual through its life cycle from embryo through maturity to death.

Order   A taxonomic unit made up of related families.

Orthogenesis   The theory that evolution follows a predetermined course to an inevitable end.

p   The proportion of one allele, usually the dominant one, in a Mendelian population.

Paleontology   The study of life of the past through its fossil remains. (ant., *neontology*)

Period   In geological time, the longest subdivisions of an era, ranging up to many millions of years in extent.

Peripatric speciation   Formation of new species by small, peripherally located isolates from the main population of a species.

Peripatry   Living in a territory peripheral to that of the main population of a species.

Phenotype The trait manifested by an individual, without regard to the genotype. (adj., *phenotypic*)

Pheromone Chemical substance released to influence the behavior of other members of the same species. For example, females of many species release pheromones that stimulate sexual activity of males of the same species.

Photosynthesis The process by which green plants synthesize carbohydrates with the release of oxygen, using chlorophyll as a catalyst. Because it makes available the energy of sunlight, it is this that powers most of the processes of life, including evolution.

Phyletic evolution Sequential changes in a single line of descent, as opposed to cladistic evolution.

Phyletic gradualism The idea that evolutionary change is slow and steady. (ant., *punctuated equilibrium*)

Phylogenetic systematics Cladism, a methodology of systematics based on two-by-two branching so that every taxon has a sister group of equal rank.

Phylogeny The evolutionary history of a species or group; the sequence leading from an ancestral to a derived group.

Phylum The broadest taxon below the kingdom, characterized by a structural plan that differs fundamentally from all other such plans and is difficult to derive from any other.

Picogram The unit of measure for quantities of a substance present in extremely small traces, such as the amount of DNA in a single nucleus; 10^{-12} gram.

Plasmid Small, accessory chromosomes, with a few nonessential genes, transmitted infectively among bacteria; capable of exchanging genes with the bacterial chromosome by crossing-over.

Plate tectonics The mechanical basis for continental drift. The crust of the earth comprises a number of great plates. Along certain midoceanic ridges, new crustal material wells up and pushes the plates apart; at midoceanic trenches old crustal material plunges to the interior of the earth. The movements of the plates are responsible for important geological features, such as earthquakes and mountain building.

Pleiotropy Multiple phenotypic effects of a single gene.

Plesiomorphy A shared primitive character in cladistic taxonomy; regarded as unsuitable for analysis of relatedness.

Polymorphism The presence of several contrasting phenotypes within the same species.

Polyphyly Designation of an unnatural taxon, that is, one whose members do not share common ancestry.

Polyploidy Duplication of whole sets of chromosomes so that the nucleus contains three or more basic genomes.

Population A geographically circumscribed group of conspecific organisms whose members breed with each other more frequently than they do with members of other such groups of the species.

Position effect Mutation due to chromosomal rearrangement, that is, due to changed spatial relationships within the chromosome.

Postmating Of isolating mechanisms, taking effect after mating.

Preadaptation Said of characters that, developed in one habitat, may prove to be adaptive to another habitat (e.g., lungs developed in stagnant ponds proved to be adaptive to life on land).

Premating   Of isolating mechanisms, taking effect before mating.

Primitive   Pertaining to the earliest members of a given group; sharing characters of the ancestral group. (ant., *derived*)

Prion   Very small protein particles that have been implicated in such "slow" diseases as kuru and scrapie. Nucleic acids have not been demonstrated in association with prions, and the possibility is open that they may be organisms of protein only.

Procaryote   Unicellular organisms without nucleocytoplasmic differentiation, in which mitosis and meiosis are unknown and in which no cytoplasmic organelles other than ribosomes are present; bacteria, including blue-green bacteria, and viruses. (also, *prokaryote*)

Prodigality of nature   The tendency of all species to reproduce more offspring than can possibly survive.

Protein   A macromolecule formed by linking many amino acids by peptide bonds.

Protostomia   An array of phyla united by a series of embryological traits and culminating in the Mollusca, Annelida, and Arthropoda.

Punctuated equilibrium   The concept that species (and higher groups) remain stable for long periods of time (equilibrium), then change rapidly (punctuation).

q   The proportion of a recessive allele in a Mendelian population.

Quantitative inheritance   Inheritance of characters that must be defined by measurement, based on segregation of numerous Mendelian genes.

Quantum evolution   Sudden change from one "adaptive orbit" to another (e.g., the transition from aquatic to terrestrial life). (also, *rapid adaptive evolution*)

r   Reproductive capacity.

r **selection**   Selection favoring reproductive rates high enough to offset high mortality rates.

r **strategy**   A type of life cycle exploiting high reproductive rate to achieve survival.

Radiometric dating   Dating of geological deposits and fossils by the rate of decay of radioactive elements, resulting in "absolute" ages.

Rassenkreis   The geographically replacing array of subspecies of a species (pl., *Rassenkreise*)

Recapitulation   Repetition in abbreviated form of ancestral stages by the embryo.

Regulatory gene   A gene that controls the expression of one or more structural genes. Those acting early in development may alter whole patterns of development and simulate systemic mutations.

Relict distribution   The pattern of geographic distribution of a species left after extinction over much of its former range.

Retrogression   Evolution toward a less-developed state; characteristic of parasitic groups. (adj., *retrogressive*)

RNA   *Ribonucleic acid*; nucleic acid based on ribose; occurs in three forms: messenger RNA (mRNA), ribosomal RNA (rRNA), and transfer RNA (tRNA).

s   Selection pressure.

Saprotrophic   Absorbing nutrients from decaying material; characteristic of fungi.

Scientific creationism   The study of origins from a creationist viewpoint, according to which the data of evolution is explained on the basis of creation on a common plan rather than by common descent.

Segregation   The separation of sister alleles one from the other into sister gametes when gametes are formed; the formation of separate genotypic and phenotypic classes in the F_2 of Mendelian crosses.

Serial homology In metameric animals, the construction of successive segments on a single plan.

Sex linkage The association of different genes in the X chromosome; a modified pattern of inheritance that results because there are two X chromosomes in females and only one in males.

Sexual selection Selection of secondary sex characters through competition for mates.

Sibling species Species that resemble each other so closely that their diagnosis is very difficult, sometimes requiring statistical analysis of populations or other special techniques, thus suggesting recent divergence from a common ancestor.

Sociobiology The study of animal behavior from a genetic viewpoint.

Speciation Formation of new species; splitting of a lineage to form new species; cladogenesis.

Species The basic unit of taxonomy; the stage in the evolutionary process at which potentially interbreeding populations become reproductively isolated.

Species synthesis Formation of species in the laboratory (or experimental garden) by crossing species, then inducing polyploidy.

Spontaneous generation The theory that living organisms, even complex ones, may come directly from nonliving matter.

Stabilizing selection Natural selection that maintains a well-adapted condition by eliminating any marked deviations from it. (also, *normalizing selection*)

Stratigraphy The study of the layers of rocks, particularly with respect to the relative ages of the fossils they contain.

Structural gene A gene that controls the synthesis of a specific protein.

Subspecies Regional populations that have distinctive patterns of the variable characters of their species, are interfertile with members of other such subspecies, and produce intergrades wherever their ranges meet.

Sweepstakes Very low probability crossings between regions in the absence of an actual connection; crossing may be made by rafting, wind dispersal, and other chance methods of low probability.

Symbiosis Unlike organisms living together to their mutual advantage; a symbiont is one of the symbiotic partners. (adj., *symbiotic*)

Sympatric speciation The formation of new species that occupy the same territory as that of the parental species.

Sympatry Occupying the same territory.

Synapomorphy In cladism, a shared derived character, considered to indicate close relationship of the species (or higher groups) that share the synapomorphies.

Synapsis The point-by-point union of homologous chromosomes preparatory to the meiotic divisions.

Synplesiomorphy In cladism, a shared primitive character, not considered to indicate close relationship.

Systemic mutation Mutations with primary effects on early embryonic processes, with secondary effects cascading throughout development. Goldschmidt believed that one or a few such mutations might establish new species or higher groups rapidly.

Taphonomy The study of fossilization and of the processes that bias the record.

Taxon Any named taxonomic unit, as species, genus, and so on.

Tetraploidy The most common type of polyploidy, in which there are four genomes present in somatic tissues and two in gametes.

Thymine   One of the four bases of DNA; always paired with adenine.

Translocation   Exchange of segments between nonhomologous chromosomes; "illegitimate" crossing-over.

Trisomic   The presence of a chromosome in triplicate rather than the usual duplicate.

Univalent   An unpaired chromosome at meiosis.

Vestigial organ   A structure that is much reduced, often apparently functionless, but may be adapted to a secondary function (e.g., wings of penguins will not support flight, yet they are effective flippers for swimming).

Vicariance   A pattern of distribution in which related populations replace each other (i.e., are vicars for each other) in widely separated areas.

Vicariance biogeography   Biogeography based on the belief that the most important process in geographic distribuion is the cleaving of widely dispersed biotas by geological processes.

Virus   An organism comprising a loop of DNA or RNA within a sheath of protein. They differ from small bacteria principally by the lack of ribosomes, hence they are all dependent on a host organism for intermediary metabolism.

Wild type   The "normal" genotype and phenotype for a species, from which mutations were once thought to be more or less exceptional deviants.

Index

DATE DUE
DATE DE RETOUR

FEB 1 3 1990		
MAR 2 7 1991		
APR 1 7 1991		
APR 24/91		
APR 2 6 1991		
APR 1 2 1994		
NOV - 2 1999		
NOV 0 2 1999		
APR 1 0 2000		
APR 0 3 2000		
FEB 0 4 2004		

LOWE-MARTIN No. 1137